Algebraic and Geometric Combinatorics on Lattice Polytopes

Algebraic and Geometric Combinatorics on Lattice Polytopes

Proceedings of the Summer Workshop on Lattice Polytopes

Osaka, Japan, 23 July – 10 August 2018

Editors

Takayuki Hibi
Osaka University, Japan

Akiyoshi Tsuchiya
Osaka University, Japan

World Scientific

NEW JERSEY • LONDON • SINGAPORE • BEIJING • SHANGHAI • HONG KONG • TAIPEI • CHENNAI • TOKYO

Published by

World Scientific Publishing Co. Pte. Ltd.
5 Toh Tuck Link, Singapore 596224
USA office: 27 Warren Street, Suite 401-402, Hackensack, NJ 07601
UK office: 57 Shelton Street, Covent Garden, London WC2H 9HE

British Library Cataloguing-in-Publication Data
A catalogue record for this book is available from the British Library.

ALGEBRAIC AND GEOMETRIC COMBINATORICS ON LATTICE POLYTOPES
Proceedings of the Summer Workshop on Lattice Polytopes

ISBN 978-981-120-047-2

For any available supplementary material, please visit
https://www.worldscientific.com/worldscibooks/10.1142/11284#t=suppl

Printed in Singapore

Preface

Historically, the study of convex polytopes originated with Euler's polyhedral formula $v - e + f = 2$, which was discovered by Leonhard Euler in 1752, and with Pick's formula $A = b/2 + c - 1$ to calculate the area of a lattice polygon, which was discovered by Georg Alexander Pick in 1899. The former gave birth to enumerative combinatorics on the numbers of faces of convex polytopes and the latter caused a variety of attractive theory on counting the number of lattice points of lattice polytopes.

In the modern theory of enumerating the numbers of faces of convex polytopes, the upper bound theorem (Peter McMullen, 1970) and the lower bound theorem (David Barnette, 1973) had played a major role. However, a new trend broke out in 1975, when Richard Stanley proved the upper bound conjecture for spheres by means of commutative algebra, especially the theory of Cohen–Macaulay rings. Stanley's work is a monumental achievement and changed the course of history of combinatorics on convex polytopes. Soon after, the exciting area called "combinatorics and commutative algebra" emerged all of a sudden.

On the other hand, in 1971, Peter McMullen proposed the "g-conjecture," which characterizes the possible face numbers of simplicial convex polytopes. In 1981, Richard Stanley succeeded in proving the necessity of the g-conjecture. Stanley's proof relied on the hard Lefschetz theorem of projective toric varieties. Its sufficiency was proved by Louis J. Billera and Carl W. Lee in 1981.

In 1960s, in order to generalize Pick's formula in higher dimensions, Eugène Ehrhart started the study of counting the numbers of lattice points of dilations of lattice polytopes and succeeded in establishing the beautiful theory of Ehrhart polynomials. Again, to modern development of Ehrhart polynomials, Cohen–Macaulay rings and Gröbner bases have contributed greatly. Nowadays, it is fashionable among researchers of convex polytopes to study lattice polytopes and their Ehrhart polynomials in various ways coming from various areas of mathematics.

Taking current trends on lattice polytopes into consideration, in order to promote international exchange on the study of lattice polytopes, we acted as host at the summer workshop named "Summer Workshop on Lattice Polytopes," which was held for three weeks from July 23 to August 10, 2018, at the Department of Pure and Applied Mathematics of Osaka University, Suita, Osaka, Japan. Especially it is remarkable that 32 young researchers from overseas were invited.

The overall schedule of the workshop is worth keeping a record. The first week was devoted to a school for graduate students, where the following two series of expository lectures were presented: "Toric geometry with a view towards lattice polytopes" by Johannes Hofscheier and "Valuations on lattice polytopes" by Katharina Jochemko. Each of the series of expository lectures consisted of five 90-minute lectures together with five 60-minute exercise classes.

In the second week, a conference entitled "Algebraic and Geometric Combinatorics on Lattice Polytopes" was organized. Each of the 32 invited speakers gave a 30-minute talk. In particular, among the invited speakers, 30 speakers came from overseas. The audience could understand new trends on lattice polytopes.

During the third week, there were no official event, except for an open problem session that was organized. Imitating the Oberwolfach style, a heated and serious discussion took place among the participants.

The present volume plays a role as the proceedings of the workshop. All papers in this volume have been refereed and are in their final form. No version of any of them will be submitted for publication anywhere else.

The source of funding for the workshop is the JSPS (Japan Society for the Promotion of Science) Grant-in-Aid for Scientific Research (S) entitled "The Birth of Modern Trends on Commutative Algebra and Convex Polytopes with Statistical and Computational Strategies" (JP 26220701).

Osaka, 20 September 2018
Takayuki Hibi
Akiyoshi Tsuchiya

Contents

Introduction to toric geometry with a view towards lattice polytopes

Johannes Hofscheier

Department of Mathematics and Statistics
McMaster University
1280 Main Street West
Hamilton, Ontario L8S4K1
Canada
E-mail: johannes.hofscheier@math.mcmaster.ca

These are notes of a five-talk lecture during the "Summer Workshop on Lattice Polytopes", at Osaka University on July 23rd – August 10th, 2018.

Toric varieties are algebraic varieties defined by combinatorial data, and there is a rich interplay between algebra, combinatorics and geometry. The goal of the lecture was to give an introduction to this wonderful theory to an audience with a strong background in combinatorics and convex geometry. We covered affine toric varieties, projective toric varieties and then looked into the exciting problems concerned with projective normality and Oda's conjecture.

Keywords: Toric variety, affine variety, projective variety, projective normality, Frobenius splitting, semigroup, lattice polytope.

Introduction

We expect some elementary experience with algebraic geometry from the reader. Good texts on algebraic geometry are, e.g., [Har77, Sha13a, Sha13b, GH78] or [Vak]. We refer to [CLS11], [Ful93] or [Mus], [Oda08], [Dan78] for further details and references on toric varieties. Our notes closely follow [CLS11] and [Mus].

The power of toric geometry has made it possible to achieve combinatorial results that are challenging and difficult to prove by convex-geometric arguments only, such as the characterization of face numbers of simplicial polytopes [Sta80]. The goal of these notes is to introduce the reader with a background in combinatorics and convex geometry to this powerful machinery.

In Section 1, we recall the definition of the algebraic torus and its fundamental properties. In Section 2, we define toric varieties and explain how to associate an affine toric variety to a semigroup, i.e., we discuss the transition from combinatorics to algebraic geometry. In Section 3, we explain the reverse direction, i.e., we discuss how to associate a semigroup to an affine variety. The two Sections 2 and 3, thus establish the fundamental relationship between combinatorics and algebraic geometry for the affine case. In Section 4, we then study projective toric varieties and how lattice polytopes are related to them. Further, we study smooth projective toric varieties and how this property translates to convex geometry. In Section 5, we conclude these notes with a discussion of Oda's conjecture, one of the big open questions in the field, and its relation to projective normality.

Note that Table 1 contains a list of notations.

Acknowledgments

The author would like to thank Takayuki Hibi and Akiyoshi Tsuchiya for their hospitality and providing great working conditions at Osaka University. We are grateful to Katharina Jochemko and Sebastian Manecke for their support during the summer school. Finally, we would also like to thank all the participants of the "Summer Workshop on Lattice Polytopes" for their ample feedback and interest which resulted in such a good atmosphere.

Table 1. Notations

Notation	Description
$\mathbb{N}$	The natural numbers $\mathbb{Z}_{\geq 0} = \{0, 1, 2, 3, \ldots\}$.
$\lfloor x \rfloor$	The floor function, i.e., the biggest integer less than or equal to the real number x.
M, N	Throughout $M = \mathbb{Z}^n$ denotes a lattice and $N = \mathrm{Hom}(M, \mathbb{Z})$ denotes its dual lattice. We have a dual pairing $\langle \cdot, \cdot \rangle \colon M \times N \to \mathbb{Z}$.
$\mathbb{Z}A$	If $A \subseteq M$ is a subset in a lattice M, then $\mathbb{Z}A$ denotes the linear span $\mathrm{span}_{\mathbb{Z}}\{A\}$.
$\mathbb{N}A$	If $A \subseteq M$ is a subset in a lattice M, then $\mathbb{N}A$ denotes the set $\{\sum_{i=1}^{r} k_i a_i \colon r > 0, k_i \in \mathbb{N}, a_i \in A\}$
$\mathrm{cone}(A)$	The positive hull of a finite subset A in a lattice M, i.e., $\mathrm{cone}(A) = \{\sum_{a \in A} \lambda_a a \colon \lambda_a \geq 0\}$.
$\check{\sigma}$	The cone dual to the rational polyhedral cone σ.
$H_{u,b}$	The affine hyperplane with normal vector $u \in N_{\mathbb{R}}$ on level $b \in \mathbb{R}$, i.e., $H_{u,b} = \{m \in M_{\mathbb{R}} \colon \langle m, u \rangle = b\}$.
H_u	The affine hyperplane $H_{u,0}$.
$H^+_{u,b}$	The affine half-space $H^+_{u,b} = \{m \in M_{\mathbb{R}} \colon \langle m, u \rangle \geq b\}$ where $u \in N_{\mathbb{R}}$ and $b \in \mathbb{R}$.
H^+_u	The half-space $H^+_{u,0}$.
$\mathrm{conv}(A)$	The convex hull of a finite subset A in a lattice M, i.e., $\mathrm{conv}(A) = \{\sum_{a \in A} \lambda_a a \colon \lambda_a \geq 0, \sum_{a \in A} \lambda_a = 1\}$.
$\mathbb{C}[X]$	Algebra of regular functions on a variety X.
$\mathrm{Spec}(\mathbb{C}[X])$	The affine variety with coordinate ring $\mathbb{C}[X]$.
$\mathbb{C}[S]$	Semigroup algebra of a semigroup S.
χ^m	Character associated to $m \in M$ where $T = (\mathbb{C}^*)^n$ is an algebraic torus with character lattice M, i.e., a morphism $\chi^m \colon T \to \mathbb{C}^*$.
λ^u	1-parameter subgroup associated to $u \in N$ where $T = (\mathbb{C}^*)^n$ is an algebraic torus with lattice of 1-parameter subgroups N, i.e., a morphism $\lambda^u \colon \mathbb{C}^* \to T$.

1. The Algebraic Torus

Let $\mathbb{C}^*$ be the multiplicative group of nonzero complex numbers. For $n \in \mathbb{N}$, the multiplicative group $T = (\mathbb{C}^*)^n$ is called *n-dimensional (algebraic) torus*. It contains the topological compact torus $(S^1)^n$ as a subgroup where S^1 denotes the unit circle in $\mathbb{C}^*$ (see Figure 1). Using polar coordinates and the logarithm map, we see that $(\mathbb{C}^*)^n = (S^1 \times \mathbb{R})^n$. It is an affine variety with coordinate ring the algebra of Laurent polynomials $\mathbb{C}[x_1^{\pm}, \ldots, x_n^{\pm}]$.

Fig. 1. The compact torus $S^1 \times S^1$. We have $(\mathbb{C}^*)^2 \cong (S^1 \times \mathbb{R})^2$.

One reason for working with the multiplicative group $((\mathbb{C}^*)^n, \cdot)$ instead of the additive group $(\mathbb{C}^n, +)$ is that any linear representation of $(\mathbb{C}^*)^n$ is completely reducible while the same is not true for $\mathbb{C}^n$ (cf. Problem 1.1).

Problem 1.1. *Consider the linear action of* $\mathbb{C}$ *on* $\mathbb{C}^2$ *given by* $t \mapsto \left(\begin{smallmatrix} 1 & t \\ 0 & 1 \end{smallmatrix}\right)$. *Show that the x-axis is an invariant subspace which does not have a complementary invariant subspace.*

Definition 1.1. A *toric variety* is a normal irreducible variety X equipped with a T-action for some algebraic torus T such that there is a Zariski open orbit isomorphic to T.

Example 1.1. The projective space $\mathbb{P}^n$ is a toric variety with respect to the following T-action where $T = (\mathbb{C}^*)^n$:

$$(t_1, \ldots, t_n) \cdot [x_0 : \ldots : x_n] = [x_0 : t_1 x_1 : \ldots : t_n x_n].$$

Indeed, $\mathbb{P}^n$ is a normal irreducible variety and $T \cdot [1 : \ldots : 1] \cong T$ is an open dense subset in $\mathbb{P}^n$ obtained by removing all coordinate hyperplanes $\{[x_0 : \ldots : x_n] \in \mathbb{P}^n : x_i = 0\}$ for $i = 1, \ldots, n$.

Problem 1.2. *Consider the subvariety* $Y = \{[x : y : z : w] \in \mathbb{P}^3 : xy - zw = 0\}$ *of* $\mathbb{P}^3$.

(1) Find a linear action of the torus $T = (\mathbb{C}^*)^2$ *on* $\mathbb{P}^3$ *such that Y is a T-invariant subvariety and a toric variety with respect to this T-action.*

(2) *Show that* Y *is* T*-equivariantly isomorphic to* $\mathbb{P}^1 \times \mathbb{P}^1$ *where* T *acts on* $\mathbb{P}^1 \times \mathbb{P}^1$ *by* $(t_1, t_2) \cdot ([x_1 : y_1], [x_2 : y_2]) = ([t_1 x_1 : y_1], [t_2 x_2 : y_2])$. *(Recall that a morphism* $\varphi\colon V \to W$ *of algebraic* T*-varieties is called* equivariant *if* $\varphi(t \cdot v) = t \cdot \varphi(v)$ *for any* $v \in V$ *and* $t \in T$.*)*

(3) *Find all the* T*-orbits in* Y.

For an algebraic torus $T = (\mathbb{C}^*)^n$, we denote by M the *character group* of T, i.e., the collection of all algebraic homomorphisms $\chi\colon T \to \mathbb{C}^*$.

Problem 1.3. *For* $m = (m_1, \ldots, m_n) \in \mathbb{Z}^n$, *we have a character* $\chi^m(x) = x^m$ *where* $x^m = x_1^{m_1} \cdots x_n^{m_n}$. *Show that all characters are of the form* χ^m *for some* $m \in \mathbb{Z}^n$. *Hence, we can identify* M *with* $\mathbb{Z}^n$ *via* $\chi^m \mapsto u$.

Similarly, we denote by N the group of 1*-parameter subgroups* of T, i.e., algebraic homomorphisms $\gamma\colon \mathbb{C}^* \to T$.

Problem 1.4. *Let* $T = (\mathbb{C}^*)^n$. *For each* $u = (u_1, \ldots, u_n) \in \mathbb{Z}^n$, *we have a* 1*-parameter subgroup* $\gamma^u(t) = (t^{u_1}, \ldots, t^{u_n})$. *Show that all* 1*-parameter subgroups of* T *are of this form, and thus we can identify* N *with the additive group* $\mathbb{Z}^n$ *via* $u \mapsto \gamma^u$.

There is a natural nondegenerate pairing $\langle \cdot, \cdot \rangle\colon N \times M \to \mathbb{Z}$ defined as follows: For $\gamma \in N$ and $\chi \in M$, we have $(\chi \circ \gamma)(t) = t^{\langle \gamma, \chi \rangle}$. Under the identifications $N \cong \mathbb{Z}^n$, $M \cong \mathbb{Z}^n$ from above, this pairing coincides with the usual dot product in $\mathbb{R}^n$.

2. Semigroups $\longrightarrow$ Affine Toric Varieties

A *semigroup* is a set S endowed with a binary operation "$+$" which is associative. All our semigroups will satisfy additional properties: We will always assume that "$+$" is commutative and S has a unit element denoted by "0".

Example 2.1. $(\mathbb{C}, \cdot)$ and $(\mathbb{C}, +)$ are two semigroups (in the above sense).

We also assume that S is *integral*, i.e., it can be embedded as a sub-semigroup in some lattice M (recall that a *lattice* is a finitely generated free abelian group, i.e., $M = \mathbb{Z}^n$). A semigroup S is *finitely generated* if there are $m_1, \ldots, m_r \in S$ such that any $m \in S$ can be written as $m = a_1 m_1 + \ldots + a_r m_r$ for some $a_i \in \mathbb{N}$.

To a semigroup S, we associate its *semigroup algebra* $\mathbb{C}[S]$ which has a $\mathbb{C}$-vector space basis indexed by the elements of S, denoted by χ^m for $m \in S$. The multiplication is defined by $\chi^{m_1} \cdot \chi^{m_2} = \chi^{m_1 + m_2}$. If we put $1 = \chi^0$, it straightforwardly follows that $\mathbb{C}[S]$ is a $\mathbb{C}$-algebra. The following problem should make it clear why we use the same symbol for characters χ^m of $T = (\mathbb{C}^*)^n$ associated to $m \in \mathbb{Z}^n$ and these basis vectors.

Problem 2.1. *Show that the coordinate ring* $\mathbb{C}[T]$ *of the algebraic torus* $T = (\mathbb{C}^*)^n$ *is isomorphic to the semigroup algebra* $\mathbb{C}[\mathbb{Z}^n] = \mathbb{C}[M]$ *where* $M = \mathbb{Z}^n$ *denotes the character lattice of* T.

Problem 2.2. *Let* A *be a* $\mathbb{Z}^n$*-graded* $\mathbb{C}$*-algebra and domain, i.e.,* $A = \bigoplus_{\alpha\in\mathbb{Z}^n} A_\alpha$. *Moreover, suppose for every* $\alpha \in \mathbb{Z}^n$ *we have* $\dim_{\mathbb{C}}(A_\alpha) \leq 1$. *Show that* A *is isomorphic to the semigroup algebra* $\mathbb{C}[S]$ *where* $S = \{\alpha \mid A_\alpha \neq \{0\}\}$. *Note that we are not assuming that* A *is finitely generated.*

If S is a finitely generated, integral semigroup, then $\mathbb{C}[S]$ is a finitely generated $\mathbb{C}$-algebra with no zero-divisors, so it defines a variety over $\mathbb{C}$, $X = \mathrm{Spec}(\mathbb{C}[S])$. Recall that closed points of X correspond to maximal ideals in $\mathbb{C}[S]$ which in turn correspond to $\mathbb{C}$-algebra homomorphisms $f\colon \mathbb{C}[S] \to \mathbb{C}$.

Problem 2.3. *Show that* $\mathbb{C}$*-algebra homomorphisms* $f\colon \mathbb{C}[S] \to \mathbb{C}$ *correspond to semigroup morphisms* $\varphi\colon S \to (\mathbb{C}, \cdot)$ *with* $\varphi(m) = f(\chi^m)$ *for* $m \in S$. *(Recall that a* semigroup morphism *between two semigroups is a map* $\varphi\colon S \to S'$ *such that* $\varphi(m_1 + m_2) = \varphi(m_1) + \varphi(m_2)$ *for any* $m_1, m_2 \in S$ *and* $\varphi(0) = 0$.*)*

Note that, by the previous problem, we also get a bijection between closed points in $X = \mathrm{Spec}(\mathbb{C}[S])$ and semigroup morphisms $\varphi\colon S \to (\mathbb{C}, \cdot)$.

Problem 2.4. *Assume that* $X = \mathrm{Spec}(\mathbb{C}[S])$ *for a finitely generated, integral semigroup* S. *Let* $p \in X$ *be a closed point corresponding to the semigroup morphism* $\varphi\colon S \to \mathbb{C}$. *Show that for* $m \in S$, *the value of* $\chi^m \in \mathbb{C}[X] = \mathbb{C}[S]$ *at* p *is* $\varphi(m)$.

Proposition 2.1. *Let* T *be an algebraic torus with character lattice* M. *If* $S \subseteq M$ *is a finitely generated semigroup with* $\mathbb{Z}S = M$, *the morphism* $T = \mathrm{Spec}(\mathbb{C}[M]) \to X := \mathrm{Spec}(\mathbb{C}[S])$ *embeds* T *as a principal affine open subset of* X. *In particular,* $\dim(X) = \mathrm{rk}(M)$.

Proof. Let $m_1, \ldots, m_r$ be generators for the semigroup S. Then $\mathbb{Z}S$ is generated (as a group) by S and $-(m_1 + \ldots + m_r)$. This implies that $\mathbb{C}[M]$ is the localization of $\mathbb{C}[S]$ at the element $\chi^{m_1+\ldots+m_r}$. □

Let M, S, T and X like in the previous proposition. We define an operation $T \times X \to X; (t, x) \mapsto t \cdot x$ which at the level of $\mathbb{C}$-algebras is given by the morphism $\alpha\colon \mathbb{C}[S] \to \mathbb{C}[M] \otimes \mathbb{C}[S]$ such that

$$\alpha(\chi^m) = \chi^m \otimes \chi^m, \text{ for } m \in S.$$

At the level of points, it associates to a pair (φ, ψ) of semigroup morphisms $\varphi\colon M \to (\mathbb{C}, \cdot)$ and $\psi\colon S \to (\mathbb{C}, \cdot)$, the morphism $\varphi \cdot \psi\colon S \to (\mathbb{C}, \cdot)$ defined by $(\varphi \cdot \psi)(m) = \varphi(m) \cdot \psi(m)$ for $m \in S$.

Proposition 2.2. *The morphism* $T \times X \to X$ *defines an action of* T *on* X, *i.e.,*

(1) $1 \cdot x = x$ *for any* $x \in X$;

(2) $g_1 \cdot (g_2 \cdot x) = (g_1 g_2) \cdot x$ *for any* $g_1, g_2 \in G$ *and* $x \in X$.

Proof. Note that the semigroup morphism $\mathbb{1}: M \to (\mathbb{C}, \cdot)$ given by $m \mapsto 1$ corresponds to the unit element in T. It follows that for $x \in X$ corresponding to $\psi: S \to (\mathbb{C}, \cdot)$, we have $1 \cdot x$ corresponds to $\mathbb{1} \cdot \psi = \psi$, i.e., $1 \cdot x = x$.

Similarly, if $\varphi_i: M \to (\mathbb{C}, \cdot)$ corresponds to $g_i \in T$ and $\psi: S \to (\mathbb{C}, \cdot)$ corresponds to $x \in X$, then $\varphi_1 \cdot (\varphi_2 \cdot \psi)$ corresponds to $g_1 \cdot (g_2 \cdot x)$ while $(\varphi_1 \cdot \varphi_2) \cdot \psi$ corresponds to $(g_1 g_2) \cdot x$ and clearly these two products coincide (just use the associativity of the multiplication in the field of complex numbers). □

Problem 2.5. *Let T be an algebraic torus with character lattice M. Show that the characters M, i.e., the set of all homomorphisms $T \to \mathbb{C}^*$, form a linearly independent subset of the space of all $\mathbb{C}$-valued functions on T.*

Suppose that S is a sub-semigroup of a lattice M. We say that S is *saturated in* M if for every $m \in M$, such that there is a positive integer k with $km \in S$, we have $m \in S$. If S is saturated in $\mathbb{Z}S$, we will simply say that S is *saturated.* In Figures 2 and 3, we illustrate two examples of semigroups and consider the question whether they are saturated.

Fig. 2. The sub-semigroup S in $\mathbb{Z}$ generated by $\{2, 3\}$ is not saturated. Its *saturation*, i.e., the smallest saturated semigroup containing S, is $\mathbb{Z}_{\geq 0}$.

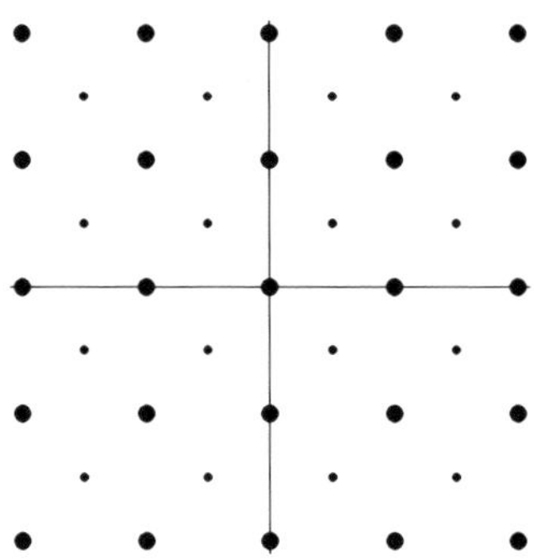

Fig. 3. If M is a lattice, then the semigroup $S = 2M$ is saturated (as $\mathbb{Z}S = S$), but S is not saturated in M.

A rational polyhedral cone in $M_{\mathbb{R}} := M \otimes \mathbb{R}$ is a set of the form

$$\sigma = \operatorname{cone}(A) = \left\{ \sum_{m \in A} \lambda_m m : \lambda_m \geq 0 \right\} \subseteq M_{\mathbb{R}},$$

where $A \subseteq M$ is a finite subset. We say that σ is generated by A and set cone$(\emptyset) = \{0\}$.

Proposition 2.3 (Gordan's Lemma). *If $\sigma \subseteq M_{\mathbb{R}}$ is a rational polyhedral cone, then $S = \sigma \cap M$ is a finitely generated semigroup.*

Proof. Let $\sigma = \text{cone}(A)$ for some finite set $A \subseteq M$. We set

$$\Pi_A := \left\{ \sum_{m \in A} \delta_m m \colon 0 \leq \delta_m < 1 \right\} \subseteq M_{\mathbb{R}}.$$

The closure $\overline{\Pi_A} \subseteq M_{\mathbb{R}}$ is a bounded region, and thus compact, so that $\overline{\Pi_A} \cap M$ is finite. Clearly $A \subseteq (\overline{\Pi_A} \cap M)$ is contained in S and we claim that this set generates the semigroup S.

Let $w \in S$ and write $w = \sum_{m \in A} \lambda_m m$ for $\lambda_m \geq 0$. For any real number x, we denote the biggest integer which is less than or equal to x by $\lfloor x \rfloor$. Then $\lambda_m = \lfloor \lambda_m \rfloor + \delta_m$ with $\lfloor \lambda_m \rfloor \in \mathbb{N}$ and $0 \leq \delta_m < 1$, so that

$$w = \sum_{m \in A} \lfloor \lambda_m \rfloor m + \sum_{m \in A} \delta_m m \qquad \text{(see Figure 4).}$$

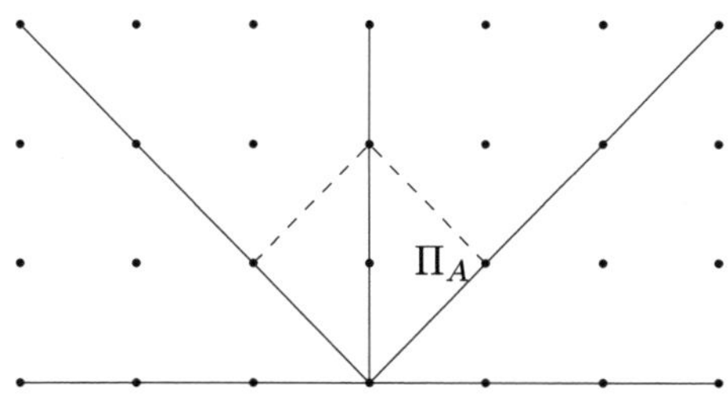

Fig. 4. The cone σ can be tiled by translates of Π_A.

□

Problem 2.6. *Show that a finitely generated sub-semigroup S of a lattice M is saturated in M if and only if it is given by $S = \sigma \cap M$ for some rational polyhedral cone $\sigma \subseteq M \otimes \mathbb{R}$.*

Let $N = \text{Hom}_{\mathbb{Z}}(M, \mathbb{Z})$ be the lattice dual to M and let $N_{\mathbb{R}} = \text{Hom}(M, \mathbb{R})$ the vector space dual to $M_{\mathbb{R}}$. Note that we have a natural dual pairing $\langle \cdot, \cdot, \rangle \colon M \times N \to \mathbb{Z}$. Given a polyhedral cone $\sigma \subseteq M_{\mathbb{R}}$, its *dual cone* is defined by

$$\check{\sigma} = \{u \in N_{\mathbb{R}} \colon \langle m, u \rangle \geq 0 \text{ for all } m \in \sigma\}.$$

Problem 2.7. *Let $\sigma \subseteq M_{\mathbb{R}}$ be a rational polyhedral cone. Show that $\check{\sigma}$ is a rational polyhedral cone in $N_{\mathbb{R}}$ and $(\check{\sigma})^{\vee} = \sigma$.*

For a given $u \neq 0$ in $N_{\mathbb{R}}$, we define the following hyperplane and closed half-space in $M_{\mathbb{R}}$:

$$H_u = \{m \in M_{\mathbb{R}} \colon \langle m, u\rangle = 0\} \subseteq M_{\mathbb{R}}, \quad H_u^+ = \{m \in M_{\mathbb{R}} \colon \langle m, u\rangle \geq 0\} \subseteq M_{\mathbb{R}}.$$

We say H_u is a *supporting hyperplane* of a rational polyhedral cone $\sigma \subseteq M_{\mathbb{R}}$ if $\sigma \subseteq H_u^+$, in which case H_u^+ is called a *supporting half-space.* Note that $\sigma \subseteq H_u^+$ is equivalent to $u \in \check{\sigma}$. We extend the definition of H_u and H_u^+ to include the case $u = 0$, i.e., $H_0 = H_0^+ = M_{\mathbb{R}}$, although in this case H_u is not a hyperplane and H_u^+ is not a half-space.

The sets $\tau = H_u \cap \sigma$ for $u \in \check{\sigma}$ are the *faces* of the polyhedral cone σ, written $\tau \preceq \sigma$. Using $u = 0$ shows that σ is a face of itself. A face of codimension 1 is called a *facet* and a face of dimension 1 is called a *ray.*

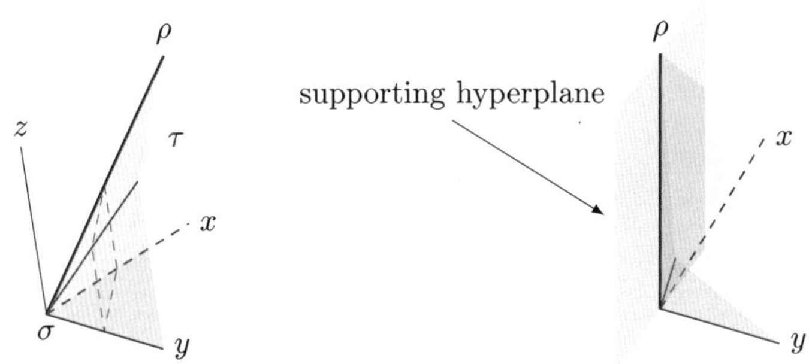

Fig. 5. A rational polyhedral cone σ together with a ray ρ and a facet τ. On the right side a supporting hyperplane (a plane in this case) cuts out a ray ρ of the cone.

Problem 2.8. *Let M be a lattice and $\sigma =$ cone(A) be a polyhedral cone in $M_{\mathbb{R}}$ where $A \subseteq M$ is a finite subset. Then*

(1) Every face of σ is a polyhedral cone.
(2) An intersection of two faces of σ is again a face of σ.
(3) A face of a face of σ is again a face of σ.

Problem 2.9. *Let M be a lattice and $\sigma \subseteq M \otimes \mathbb{R}$ a rational polyhedral cone. For any face $\tau \preceq \sigma$, we define*

$$\tau^{\perp} := \{u \in N_{\mathbb{R}} \colon \langle m, u\rangle = 0 \textit{ for all } m \in \tau\} \quad \textit{and} \quad \tau^* := \check{\sigma} \cap \tau^{\perp}.$$

Prove the following statements (see also Figure 6):

(1) τ^* is a face of $\check{\sigma}$.
(2) The assignment $\tau \mapsto \tau^*$ induces an inclusion-reversing correspondence between the faces of σ and the faces of $\check{\sigma}$.
(3) $\dim \tau + \dim \tau^* = \dim \sigma$.

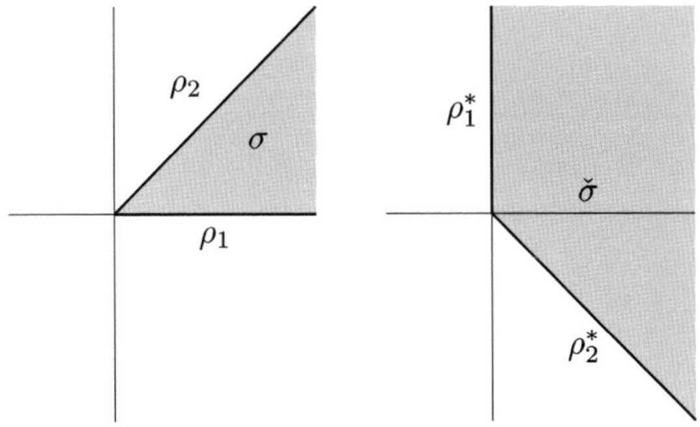

Fig. 6. The correspondence between the faces of σ and the faces of its dual cone $\check{\sigma}$.

A rational polyhedral cone is called *strongly convex* if it contains no lines of $M_{\mathbb{R}}$ (see Figure 7).

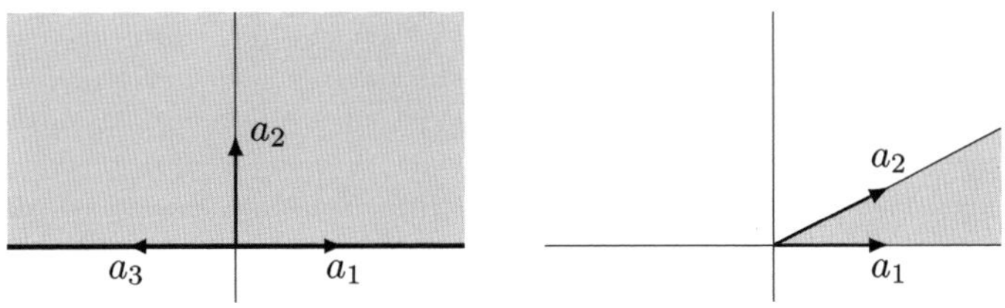

Fig. 7. The cone on the left contains a line, and thus is not strongly convex. The cone on the right is strongly convex. The arrows A indicate a minimal set of generators such that $\sigma = \operatorname{cone}(A)$.

Problem 2.10. *Let M be a lattice and let $\sigma \subseteq N_{\mathbb{R}}$ be a rational polyhedral cone in the dual space $N_{\mathbb{R}} = \operatorname{Hom}(M, \mathbb{R})$. Show that the following statements are equivalent:*

(1) σ is strongly convex.
(2) $\sigma \cap (-\sigma) = \{0\}$.
(3) The origin is a face of σ.
(4) $\dim \check{\sigma} = \dim M_{\mathbb{R}}$.

Let M be a lattice and $S \subseteq M$ a semigroup which is saturated in M. By Problem 2.6, $S = \sigma \cap M$ for some rational polyhedral cone $\sigma \subseteq M \otimes \mathbb{R}$. The reason why one usually works with the dual cone $\check{\sigma}$ instead of $\sigma \subseteq M_{\mathbb{R}}$ stems from the last equivalence in the previous exercise. Usually one assumes that the lattice M is generated by the semigroup $S = \sigma \cap M$, i.e., $M = \mathbb{Z}S$. Hence $\sigma \subseteq M_{\mathbb{R}}$ is full-dimensional, but in general not strongly convex. The exercise above shows that the dual cone $\check{\sigma}$ is strongly convex, a usually desired property.

Here is a reminder on normal affine varieties:

(1) Let R be a ring and S an R-algebra. An element $s \in S$ is called *integral over R* if it satisfies a monic polynomial equation with coefficients in R, i.e., there

is a monic polynomial $p \in R[t]$ such that $p(s) = 0$. The set $\widetilde{R}$ consisting of all elements in S which are integral over R is called the *integral closure* of R in S (or the *normalization* of R in S). If $\widetilde{R} = R$ then R is *integrally closed* in S.

(2) Let us consider the special case where R is an integral domain and S is its quotient field. If the subalgebra of elements in S integral over R is equal to R, we will simply say that R is a *normal domain* (or *integrally closed*).

(3) An irreducible affine variety $X \subseteq \mathbb{C}^n$ is *normal* if its coordinate ring $\mathbb{C}[X]$ is integrally closed.

Problem 2.11. *Give examples of R-algebras S and elements $s \in S$ which are not integral over R.*

Problem 2.12. *Show that any* unique factorization domain *(*UFD*) is integrally closed.*

Proposition 2.4. *Let R_i for $i \in I$ be a family of normal domains with the same field of fractions K. Then the intersection*[a] *$S := \bigcap_{i \in I} R_i$ is normal.*

Proof. Note that the field of fractions of S is contained in K. Indeed, if $f/g \in Q(S)$ for $f, g \in S$, then $f/g \in Q(R_i) = K$.

Let $f/g \in Q(S) \subseteq K$ be integral over S. Then $f/g \in K$ is integral over R_i for any $i \in I$. By assumption, it follows that $f/g \in R_i$, and thus $f/g \in S$. □

Problem 2.13. *Let R be a normal domain with field of fractions K and let $S \subseteq R$ be a multiplicative subset. Prove that the localization R_S is normal. In particular, $\mathbb{C}[x_1^{\pm}, \ldots, x_n^{\pm}] = \mathbb{C}[x_1, \ldots, x_n]_{x_1 \cdots x_n}$ is normal (recall that the polynomial ring $\mathbb{C}[x_1, \ldots, x_n]$ is normal as it is a UFD).*

Problem 2.14. *Let N be a lattice.*

(1) Let $N_1 \subseteq N$ be a sublattice such that N/N_1 is torsion-free. Prove that there is a "complementary" sublattice $N_2 \subseteq N$, i.e., $N = N_1 \oplus N_2$.

(2) Let $u \in N$ be primitive, i.e., $\frac{1}{k}u \notin N$ for all $k > 1$. Prove that N has a basis $e_1, \ldots, e_n$ with $e_1 = u$.

Proposition 2.5. *Let S be a finitely generated sub-semigroup of a lattice M. The variety $X = \operatorname{Spec}(\mathbb{C}[S])$ is normal if and only if S is saturated*[b].

Proof. We may assume that $M = \mathbb{Z}S$ (otherwise restrict the ambient lattice). It follows that the rings $\mathbb{C}[S] \subseteq \mathbb{C}[M]$ have the same fraction field. As $\mathbb{C}[M] \cong \mathbb{C}[t_1^{\pm}, \ldots, t_n^{\pm}]$ for $n = \operatorname{rk}(M)$, $\mathbb{C}[M]$ is normal (cf. Problem 2.13) and therefore $\mathbb{C}[S]$ is a normal domain if and only if it is integrally closed in the $\mathbb{C}[S]$-algebra $\mathbb{C}[M]$.

[a] As $R_i \subseteq K$, the intersection $\bigcap_{i \in I} R_i$ can be taken inside K.

[b] Recall this means that S is saturated in $\mathbb{Z}S$.

Suppose first that $\mathbb{C}[S]$ is a normal domain. If $m \in M$ and $km \in S$ for some $k \in \mathbb{Z}_{>0}$, then $(\chi^m)^k \in \mathbb{C}[S]$ and $\chi^m \in \mathbb{C}[M]$. As $\mathbb{C}[S]$ is integrally closed in $\mathbb{C}[M]$, it follows that $\chi^m \in \mathbb{C}[S]$, so $m \in S$.

Next assume that S is saturated. By Problem 2.6, there is a rational polyhedral cone $\sigma \subseteq M \otimes \mathbb{R}$ such that $S = \sigma \cap M$. Let $u_1, \ldots, u_k \in N$ be the primitive inner facet normals of σ, i.e., we can write $\sigma = H^+_{u_1} \cap \ldots \cap H^+_{u_k}$ and the supporting hyperplanes $H^+_{u_i}$ cut out the facets of σ. As $S = \sigma \cap M = (H^+_{u_1} \cap \ldots \cap H^+_{u_k}) \cap M = \bigcap_{i=1}^k (H^+_{u_i} \cap M)$, it follows that $\mathbb{C}[S] = \bigcap_{i=1}^k \mathbb{C}[H^+_{u_i} \cap M]$. By Proposition 2.4, it suffices to show that $\mathbb{C}[H^+_{u_i} \cap M]$ is normal for all $i = 1, \ldots, k$. Since u_i is primitive, we can find a basis $e_1, \ldots, e_n$ of N with $e_1 = u_i$, so that $\mathbb{C}[H^+_{u_i} \cap M] \cong \mathbb{C}[x_1, x_2^{\pm}, \ldots, x_n^{\pm}] = \mathbb{C}[x_1, \ldots, x_n]_{x_2 \cdots x_n}$ which by Problem 2.13 is a normal domain. □

Corollary 2.1. *For any integral finitely generated saturated semigroup S, the variety $X = \mathrm{Spec}(\mathbb{C}[S])$ defines an affine toric variety with acting torus $T = \mathrm{Spec}(\mathbb{C}[\mathbb{Z}S])$.*

Proof. Follows from Propositions 2.1, 2.2 and 2.5. □

Problem 2.15. *Let $S = \mathbb{N}2 + \mathbb{N}3 \subseteq \mathbb{Z}$.*

(1) Embed $X = \mathrm{Spec}(\mathbb{C}[S])$ in some affine space $\mathbb{C}^r$.
(2) Understand the correspondence

$$\{\textit{closed points in } X\} \leftrightarrow \{\mathfrak{m} \subseteq \mathbb{C}[S] \textit{ maximal ideal}\}.$$

(3) Understand the correspondence

$$\{\textit{closed points in } X\} \leftrightarrow \{\phi\colon S \to \mathbb{C} \textit{ semigroup morphism}\}.$$

(4) Understand the action of $T = \mathbb{C}^*$ on X on the level of semigroup morphisms.
(5) Is X normal?
(6) What is $R = \{f \in \mathbb{C}[\mathbb{Z}]\colon f$ integral over $\mathbb{C}[S]\}$? The algebra R is also a semigroup algebra for some semigroup $S' \subseteq \mathbb{Z}$ (check this!). What do you guess is the corresponding semigroup S' in general?

Problem 2.16. *For an integral finitely generated semigroup S, let*

$$S^{\mathrm{sat}} := \{m \in \mathbb{Z}S : km \in S \textit{ for some } k \geq 1\}.$$

Prove that S^{sat} is a saturated semigroup. In fact, it is the smallest saturated subsemigroup of $\mathbb{Z}S$ which contains S.

Show that the morphism $\mathrm{Spec}(\mathbb{C}[S^{\mathrm{sat}}]) \to \mathrm{Spec}(\mathbb{C}[S])$ induced by $S \hookrightarrow S^{\mathrm{sat}}$ is the normalization of $\mathrm{Spec}(\mathbb{C}[S])$.

3. Affine Toric Varieties ⟶ Semigroups

Let X be an affine toric T-variety where $T = (\mathbb{C}^*)^n$ is an algebraic torus with character lattice M. The action of T on X turns its coordinate ring $\mathbb{C}[X]$ in a

T-algebra[c] via

$$t \cdot f(x) = f(t^{-1} \cdot x) \qquad \text{for } t \in T, x \in X, f \in \mathbb{C}[G].$$

Theorem 3.1. *There is a finitely generated, integral semigroup $S \subseteq M$ and an isomorphism $X \cong \operatorname{Spec}(\mathbb{C}[S])$. Moreover $\mathbb{Z}S = M$.*

Proof. The open, dominant embedding $T \hookrightarrow X$ induces an inclusion $\mathbb{C}[X] \hookrightarrow \mathbb{C}[T]$. Consider the following commutative diagram with its corresponding diagram of comorphisms:

$$\begin{array}{ccc} X & \longleftarrow & T \times X \\ \uparrow & & \uparrow \\ T & \longleftarrow & T \times T \end{array} \qquad \begin{array}{ccc} \mathbb{C}[X] & \xrightarrow{\psi} & \mathbb{C}[T] \otimes \mathbb{C}[X] \\ \downarrow & & \downarrow \\ \mathbb{C}[T] & \xrightarrow{\varphi} & \mathbb{C}[T] \otimes \mathbb{C}[T] \end{array}$$

where all comorphisms are injective. Note that $\{\chi^m \otimes \chi^{m'} : m, m' \in M\}$ is a $\mathbb{C}$-vector space basis of $\mathbb{C}[T] \otimes \mathbb{C}[T]$. Recall the explicit definition of the morphism φ from the paragraph after Proposition 2.1 where we have to take $M = S$.

Let $f \in \mathbb{C}[X]$. We can write[d] $\psi(f) = \chi^{m_1} \otimes g_1 + \ldots + \chi^{m_r} \otimes g_r$ for $m_i \in M$ (pairwise distinct) and $g_i \in \mathbb{C}[X]$. Further, we can write $f|_T = \mu_1 \chi^{m'_1} + \ldots + \mu_s \chi^{m'_s}$ for $\mu_j \in \mathbb{C}$ and $m'_j \in M$ (pairwise distinct). As $\varphi(f|_T) = \sum_{j=1}^{s} \mu_j \chi^{m'_j} \otimes \chi^{m'_j}$, it follows that $r = s$, $\chi^{m_i} = \chi^{m'_i}$ (up to reordering) and $g_i|_T = \mu_i \chi^{m'_i}$. In particular, the characters $\chi^{m_i} = \chi^{m'_i} = g_i|_T / \mu_i$ can uniquely be extended to regular functions on X.

We set $S := \{m \in M : \chi^m \text{ can be extended to a regular function on } X\} \subseteq M$. From the above, we obtain $\mathbb{C}[X] = \mathbb{C}[S]$ (via the inclusion $\mathbb{C}[X] \hookrightarrow \mathbb{C}[T]$). As $\mathbb{C}[X]$ is a finitely generated $\mathbb{C}$-algebra, S is a finitely generated semigroup.

Note that we have $\operatorname{rk}(\mathbb{Z}S) = \operatorname{rk}(M) = \dim(X)$. By the Elementary Divisor Theorem, there is a basis $e_1, \ldots, e_n$ of M such that $a_1 e_1, \ldots, a_n e_n$ is a basis of $\mathbb{Z}S$ where $a_1, \ldots, a_n$ are positive integers. The inclusions $S \subseteq \mathbb{Z}S \subseteq M$ induce dominant morphisms $T \to \operatorname{Spec}(\mathbb{C}[\mathbb{Z}S]) \to X$ whose composition coincide with the injective morphism $T \to X$.

Note that the first morphism $T \to \operatorname{Spec}(\mathbb{C}[\mathbb{Z}S])$ coincides with $(\mathbb{C}^*)^n \to (\mathbb{C}^*)^n; (t_1, \ldots, t_n) \mapsto (t_1^{a_1}, \ldots, t_n^{a_n})$ while the second morphism is the embedding of the algebraic torus $\operatorname{Spec}(\mathbb{C}[\mathbb{Z}S])$ on a principal affine open subset in X (see Proposition 2.1). It follows that $a_1 = \ldots = a_n = 1$. □

A *face* of a semigroup S is a sub-semigroup F such that whenever $m_1, m_2 \in S$ with $m_1 + m_2 \in F$, then we have $m_1, m_2 \in F$. In particular, $(S \setminus F) \cup \{0\}$ is a sub-semigroup of S and if S is generated by $m_1, \ldots, m_r$, then F is generated

---

[c] i.e., T acts linearly on $\mathbb{C}[X]$ by automorphisms of a $\mathbb{C}$-algebra.

[d] Clearly, $\psi(f) = \alpha_1 \otimes \beta_1 + \ldots + \alpha_k \otimes \beta_k$ for $\alpha_i \in \mathbb{C}[T]$ and $\beta_i \in \mathbb{C}[X]$. Expressing the α_i as linear combinations of the basis vectors $\{\chi^m : m \in M\}$ of $\mathbb{C}[T]$ and then collecting terms, yields the above presentation of $\psi(f)$.

as a semigroup by a subset of these generators. In particular, if S is an integral finitely generated semigroup, every face is an integral, finitely generated semigroup. Moreover, it follows that S has only finitely many faces if it is finitely generated. See Figure 8 for an illustration.

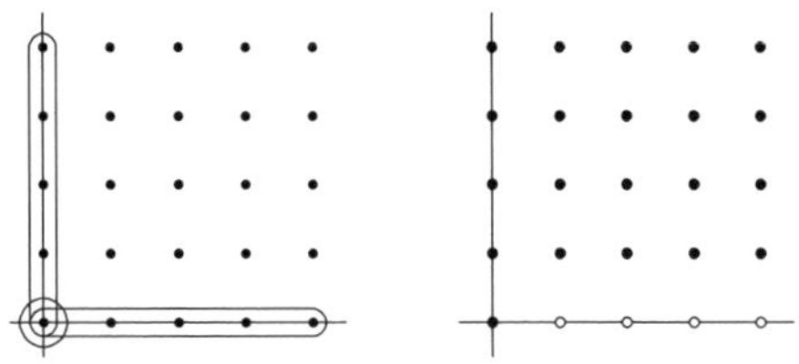

Fig. 8. The semigroup $S = \mathbb{N}e_1 + \mathbb{N}e_2$, where e_1, e_2 denotes the standard basis of $\mathbb{Z}^2$, has 3 proper faces, namely $\mathbb{N}e_1$, $\mathbb{N}e_2$ and $\{0\}$. Note that $(S \setminus \mathbb{N}e_1) \cup \{0\}$ is a sub-semigroup of S which is not a face and not finitely generated.

A subvariety $Y \subseteq X$ is called *invariant* if for any $y \in Y$ and any $t \in T$, we have $t \cdot y \in Y$.

Problem 3.1. *Let S be a finitely generated sub-semigroup of a lattice M which is saturated in M, i.e., $S = \sigma \cap M$ for some rational polyhedral cone $\sigma \subseteq M_{\mathbb{R}}$. Show that the faces of the semigroup S are in 1-to-1 correspondence with the faces of the cone σ.*

Let X be an affine toric T-variety with respect to an algebraic torus T with character lattice M. In Theorem 3.1, we have seen that $X = \operatorname{Spec}(\mathbb{C}[S])$ for some finitely generated sub-semigroup $S \subseteq M$ wit $\mathbb{Z}S = M$.

Theorem 3.2. *The irreducible invariant subvarieties of the toric variety $X = \operatorname{Spec}(\mathbb{C}[S])$ are in an inclusion-preserving bijection with the faces of S, such that the ideal of the variety Y_F corresponding to the face F is given by $\bigoplus_{m \in S \setminus F} \mathbb{C}\chi^m$. In particular, the invariant subvariety Y_F has coordinate ring $\mathbb{C}[F]$, i.e., it is a toric variety as well.*

Proof. Let $Y \subseteq X$ be an irreducible invariant subvariety with vanishing ideal $I = I(Y) \subseteq \mathbb{C}[X]$, i.e., $\mathbb{C}[Y] = \mathbb{C}[X]/I$. Note that $T \times Y \to Y$ is a well-defined morphism, so that we obtain the following commutative diagrams:

$$\begin{array}{ccc} Y & \longleftarrow & T \times Y \\ {\scriptstyle \iota}\downarrow & & \downarrow{\scriptstyle \mathrm{id}_T \times \iota} \\ X & \longleftarrow & T \times X \end{array} \qquad \begin{array}{ccc} \mathbb{C}[Y] & \longrightarrow & \mathbb{C}[T] \otimes \mathbb{C}[Y] \\ {\scriptstyle \iota^*}\uparrow & & \uparrow{\scriptstyle \mathrm{id} \otimes \iota^*} \\ \mathbb{C}[X] & \xrightarrow{\psi} & \mathbb{C}[T] \otimes \mathbb{C}[X] \end{array}$$

As $\ker(\iota^*) = I$ and $\ker(\mathrm{id} \otimes \iota^*) = \mathbb{C}[T] \otimes I$, it follows from the commutativity of the above diagram that $\psi(I) \subseteq \mathbb{C}[T] \otimes I$. Now we proceed as in the proof of

Theorem 3.1: Consider the following commutative diagram where all morphisms are injective:

$$\begin{array}{ccc} I & \xrightarrow{\psi|_I} & \mathbb{C}[T] \otimes I \\ \downarrow & & \downarrow \\ \mathbb{C}[T] & \xrightarrow{\varphi} & \mathbb{C}[T] \otimes \mathbb{C}[T] \end{array}$$

Note that $\{\chi^m \otimes \chi^{m'} : m, m' \in M\}$ is a $\mathbb{C}$-vector space basis of $\mathbb{C}[T] \otimes \mathbb{C}[T]$. Let $f \in I$ and write $\psi(f) = \chi^{m_1} \otimes g_1 + \ldots + \chi^{m_r} \otimes g_r$ for $m_i \in M$ (pairwise distinct) and $g_i \in I$. On the other hand, we can write $f|_T = \mu_1 \chi^{m'_1} + \ldots + \mu_s \chi^{m'_s}$ for $\mu_j \in \mathbb{C}$ and $m'_j \in M$. As $\varphi(f|_T) = \mu_1 \chi^{m'_1} \otimes \chi^{m'_1} + \ldots + \mu_s \chi^{m'_s} \otimes \chi^{m'_s}$, it follows that $s = r$, $\chi^{m_i} = \chi^{m'_i}$ (up to reordering) and $g_i|_T = \mu_i \chi^{m'_i}$.

We set $S' := \{m \in M : \chi^m$ can be extended to a regular function on $X, \chi^m|_Y = 0\} \subseteq M$. From the above it follows that $I = \bigoplus_{m \in S'} \mathbb{C}\chi^{m'}$, i.e., I is M-graded.

Let $S \subseteq M$ be the semigroup such that $\mathbb{C}[X] = \mathbb{C}[S]$ (see Theorem 3.1). As Y is irreducible, the complement $F := S \setminus S'$ is a semigroup[e]. Note that if $m_1, m_2 \in S$ such that $m_1 + m_2 \in F$ then[f] neither $m_1 \in S'$ nor $m_2 \in S'$, i.e., $m_1, m_2 \in F$. Hence F is a face of S.

It is straightforward to verify that if F is a face of S, then $J := \bigoplus_{m \in S \setminus F} \mathbb{C}\chi^m$ is a prime ideal of $\mathbb{C}[X] = \mathbb{C}[S]$ whose vanishing locus is an irreducible and invariant subvariety of X. □

Problem 3.2. *Complete the proof of Theorem 3.2.*

See Figure 9 for an illustration of Theorem 3.2.

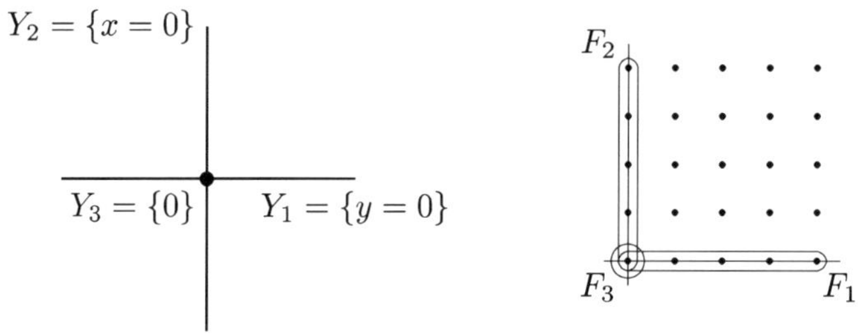

Fig. 9. The affine plane $\mathbb{A}^2$ is a toric variety with respect to the action $T \times \mathbb{A}^2 \to \mathbb{A}^2; ((t_1, t_2), (x_1, x_2)) \mapsto (t_1 x_1, t_2 x_2)$ where T is the algebraic torus $(\mathbb{C}^*)^2$. Indeed, we have $\mathbb{A}^2 = \mathrm{Spec}(\mathbb{C}[S])$ where $S = \mathbb{Z}^2_{\geq 0}$. It is straightforward to check that there are exactly 3 proper irreducible invariant subvarieties, namely the coordinate axis and the origin. On the right side, we illustrate the corresponding faces (here F_i corresponds to Y_i).

[e]Note that $0 \notin S'$, and thus $0 \in F$.

[f]Use that I is an ideal.

The following statement is usually referred to as the "orbit cone correspondence"[g].

Corollary 3.1. *The bijection in Theorem 3.2 yields a bijection between the faces of S and the T-orbits in X. To be more precise, the correspondence is given by*

$$\{T\text{-orbits}\} \to \{\text{faces of } S\}; T \cdot \varphi \mapsto \varphi^{-1}(\mathbb{C}^*),$$

where $\varphi\colon S \to (\mathbb{C}, \cdot)$ corresponds[h] *to a closed point of X.*

Proof. It is straightforward to show that the map from T-orbits to faces is well-defined.

We construct the inverse map. By Theorem 3.2, the faces F of S are in bijection with irreducible invariant subvarieties $Y = \operatorname{Spec}(\mathbb{C}[F])$ of X which are also toric varieties with acting torus $T_Y := \operatorname{Spec}(\mathbb{C}[\mathbb{Z}F])$. To a face F of S, we associate the unique open T_Y-orbit in the closed invariant subvariety Y of X. Note that the inclusion $\mathbb{Z}F \subseteq \mathbb{Z}S$ induces a surjective morphism $T \to T_Y$ of algebraic tori, so that T_Y is an orbit for the action of T on X.

Using Lemma 3.1, it is straightforward to check that the two maps above are inverse to each other. □

Lemma 3.1. *We continue to use the notation and assumptions from above. Let Y be an irreducible invariant subvariety of X which corresponds, by Theorem 3.2, to a face F of S. A semigroup morphism $\varphi\colon S \to (\mathbb{C}, \cdot)$ corresponds to a closed point in the unique open T-orbit T_Y of Y if and only if $\varphi^{-1}(\mathbb{C}^*) = F$.*

Proof. The point p corresponding to φ is in Y if and only if $I(Y) = \bigoplus_{m \in S \setminus F} \mathbb{C}\chi^m \subseteq \mathfrak{m}_p = f^{-1}(0)$ where $f\colon \mathbb{C}[S] \to \mathbb{C}$ is the algebra morphism associated to φ, i.e., $S \setminus F \subseteq \varphi^{-1}(0)$.

If $p \in Y$, then $p \in T_Y$ if and only if $\varphi(m) \neq 0$ for all $m \in F$. □

Remark 3.1. In the situation of Corollary 3.1, let F be a face of S. The inclusion $F \subseteq S$ induces a morphism of algebras $\mathbb{C}[F] \to \mathbb{C}[S]$ which in turn induces a morphism of algebraic varieties $f\colon X \to Y$. Let $\mathbb{C}[S] \to \mathbb{C}[F] \cong \mathbb{C}[S]/(\bigoplus_{m \in S \setminus F} \mathbb{C}\chi^m)$, the comorphism of the inclusion $Y \subseteq X$. Then $\mathbb{C}[F] \hookrightarrow \mathbb{C}[S] \to \mathbb{C}[S]/(\bigoplus_{m \in S \setminus F} \mathbb{C}\chi^m) = \mathbb{C}[F]$ is the identity. In other words: $f\colon X \to Y$ is a retract[i].

Let M be a lattice and $\sigma \subseteq N \otimes \mathbb{R}$ a strongly convex rational polyhedral cone in the dual space, so that $S := \check{\sigma} \cap M \subseteq M$ is a finitely generated integral semigroup which is saturated in M. Corollary 3.1 motivates, why instead of considering the cone σ itself, we will study the collection Σ of all faces[j] of σ (see Figure 10).

---

[g] Recall that a face of a polyhedral cone is a polyhedral cone.

[h] See Problem 2.3.

[i] Recall that a *retract* of a topological space X onto a subspace Y is a continuous map $r\colon X \to Y$ such that the restriction of r to Y is the identity map on Y.

[j] Recall that the full cone σ is a face of itself.

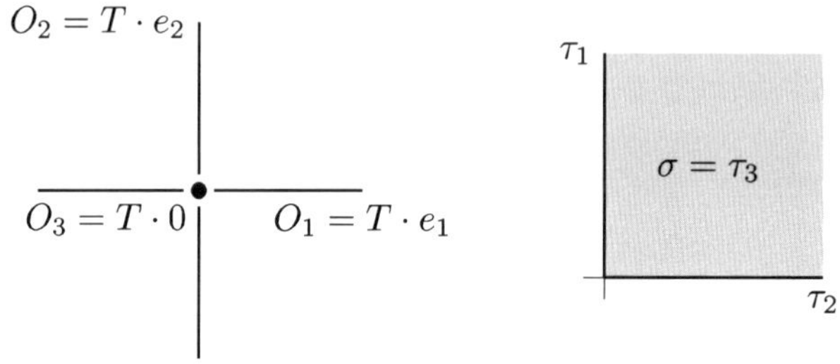

Fig. 10. Recall from above that $\mathbb{A}^2$ is a toric variety with respect to the algebraic torus $T = (\mathbb{C}^*)^2$ with character lattice $M = \mathbb{Z}^2$ and that $\mathbb{A}^2 = \text{Spec}(\mathbb{C}[S])$ for $S = \mathbb{Z}^2_{\geq 0} \subseteq M$. Let e_1, e_2 be the standard basis of $N = \mathbb{Z}^2$. Then $S = \check{\sigma} \cap M$ for $\sigma = \text{cone}(e_1, e_2) \subseteq N \otimes \mathbb{R}$. We illustrate the correspondence between the faces of σ and the T-orbits of $\mathbb{A}^2$ (here O_i corresponds to τ_i). Recall from Problem 2.9 that the faces of σ are in bijection with the faces of $\check{\sigma}$ which are in correspondence with the faces of S by Problem 3.1.

3.1. *Smooth Affine Toric Varieties*

Let M be a lattice, $X = \text{Spec}(\mathbb{C}[S])$ be an affine toric variety with $S = \check{\sigma} \cap M$ for some strongly convex rational polyhedral cone $\sigma \subseteq N \otimes \mathbb{R}$ in the dual space.

Problem 3.3. *Let $S = \check{\sigma} \cap M$ be a finitely generated semigroup saturated in some lattice M where $\sigma \subseteq N_{\mathbb{R}}$ is a strongly convex rational polyhedral cone in the dual vector space. Show that $M = \mathbb{Z}S$.*

Recall that X is *smooth* at a closed point $x \in X$ if the dimension of the vector space $\mathfrak{m}_x/\mathfrak{m}_x^2$ coincides with the dimension[k] of the variety X where $\mathfrak{m}_x$ is the maximal ideal corresponding to the closed point x. The $\mathbb{C}$-vector space $T_xX := (\mathfrak{m}_x/\mathfrak{m}_x^2)^*$ is called the *Zariski tangent space* to the affine variety X at x.

Problem 3.4. *Let $X = \text{Spec}(\mathbb{C}[S])$ where $S = \mathbb{Z}_{\geq 0}2 + \mathbb{Z}_{\geq 0}3 \subseteq \mathbb{Z}$. Determine the smooth and singular points.*

We know from algebraic geometry that the singular locus[l] X_{sing} of X is a closed subvariety of X. Note that the action of $t \in T = \text{Spec}(\mathbb{C}[M])$ induces an automorphism of X, namely $\lambda_t \colon X \to X; x \mapsto t \cdot x$. Since λ_t maps singular points to singular points (the automorphism λ_t induces an isomorphism of the respective Zariski tangent spaces), X_{sing} is T-invariant.

Problem 3.5. *Let $\sigma \subseteq N_{\mathbb{R}}$ be a strongly convex rational polyhedral cone of maximal dimension. Set $S = \check{\sigma} \cap M$. An element $m \neq 0$ of S is* irreducible *if $m = m' + m''$ for $m', m'' \in S$ implies $m' = 0$ or $m'' = 0$. Then $\mathcal{H} := \{m \in S \colon m$ is irreducible$\}$ has the following properties:*

---

[k]Recall that the dimension of X coincides with the Krull dimension of $\mathbb{C}[S]$ which coincides with the rank of $\mathbb{Z}S$.

[l]Recall that X_{sing} consists of all singular points of X.

(1) $\mathcal{H}$ is finite and generates S.
(2) $\mathcal{H}$ contains the primitive ray generators of $\check{\sigma}$.
(3) $\mathcal{H}$ is the minimal generating set of S with respect to inclusion.

The set $\mathcal{H} \subseteq S$ is called the Hilbert basis *of S.*

Problem 3.6. *Show that the affine toric variety $X = \mathrm{Spec}(\mathbb{C}[S])$ has a fixed point x_0 if and only if $\{0\}$ is a face of S, i.e., S is* pointed, *in which case the unique fixed point is given by the semigroup homomorphism*

$$S \to (\mathbb{C}, \cdot); m \mapsto \begin{cases} 1 & \text{if } m = 0 \\ 0 & \text{if } m \neq 0. \end{cases}$$

In particular, the maximal ideal is given by $\mathfrak{m}_{x_0} = \bigoplus_{m \in S \setminus \{0\}} \mathbb{C}\chi^m$.

Let us first discuss a special case: Assume $S = \check{\sigma} \cap M$ for a strongly convex rational polyhedral cone $\sigma \subseteq N_{\mathbb{R}}$ of maximal dimension. By Problem 2.10, $\check{\sigma}$ is also strongly convex, and thus the origin is a face of $\check{\sigma}$. By Problem 3.6, X has a fixed point $x_0 \in X$ and, by Problem 3.5, the semigroup S has a Hilbert basis which we denote by $\mathcal{H}$.

Lemma 3.2. *Let $X = \mathrm{Spec}(\mathbb{C}[S])$ be an affine toric variety with $S = \check{\sigma} \cap M$ for a strongly convex rational polyhedral cone $\sigma \subseteq N \otimes \mathbb{R}$ of maximal dimension. Let $x_0 \in X$ be the unique closed point in X fixed under the torus action. Then* $\dim_{\mathbb{C}}(T_{x_0}X) = |\mathcal{H}|$.

Proof. By Problem 3.6, the maximal ideal of $\mathbb{C}[S]$ corresponding to x_0 is $\mathfrak{m}_{x_0} = \bigoplus_{0 \neq m} \mathbb{C}\chi^m$. Since $\{\chi^m\}_{m \in S}$ is a basis of $\mathbb{C}[S]$, we obtain

$$\mathfrak{m}_{x_0} = \bigoplus_{m \neq 0} \mathbb{C}\chi^m = \bigoplus_{m \text{ irreducible}} \mathbb{C}\chi^m \oplus \bigoplus_{0 \neq m \text{ reducible}} \mathbb{C}\chi^m = \left(\bigoplus_{m \in \mathcal{H}} \mathbb{C}\chi^m \right) \oplus \mathfrak{m}_{x_0}^2.$$

It follows that $\dim_{\mathbb{C}}(\mathfrak{m}_{x_0}/\mathfrak{m}_{x_0}^2) = |\mathcal{H}|$. Since $T_{x_0}X$ is the dual space of $\mathfrak{m}_{x_0}/\mathfrak{m}_{x_0}^2$ the result follows. □

Recall that a strongly convex rational polyhedral cone $\sigma \subseteq N_{\mathbb{R}}$ is *smooth* if the primitive ray generators of σ form part of a $\mathbb{Z}$-basis of N (see Figure 11).

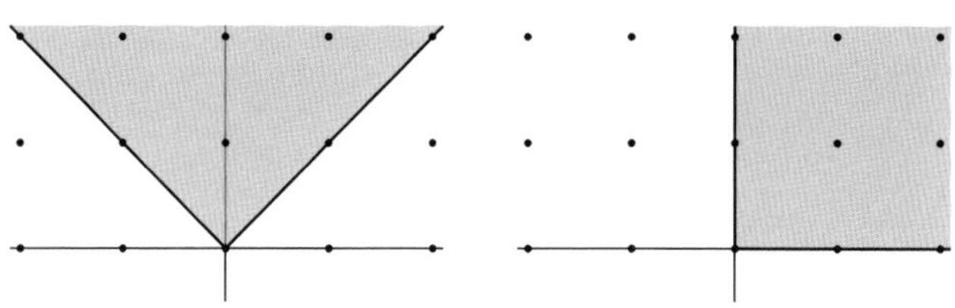

Fig. 11. The cone on the left side is not smooth. Every 2-dimensional rational polyhedral cone is unimodularly equivalent[m] to the cone on the right side.

[m]Two cones $\sigma, \tau \subseteq N \otimes \mathbb{R}$ are called *unimodularly equivalent* if there is a linear endomorphism $\varphi\colon N \otimes \mathbb{R} \to N \otimes \mathbb{R}$ which induces an isomorphism of lattice $\varphi|_N\colon N \to N$ such that $\varphi(\sigma) = \tau$.

Theorem 3.3. *Let $X = \operatorname{Spec}(\mathbb{C}[S])$ be an affine toric variety with $S = \check{\sigma} \cap M$ for a strongly convex rational polyhedral cone $\sigma \subseteq N_{\mathbb{R}}$ (not necessarily maximal dimensional). Then X is smooth if and only if σ is smooth.*

Proof. If σ is smooth, then there is a basis $e_1, \ldots, e_n$ of N such that $\sigma = \operatorname{cone}(e_1, \ldots, e_r)$ for some $0 \leq r \leq n$. Let $f_1, \ldots, f_n$ be the dual basis of M. Then $\check{\sigma} = \operatorname{cone}(f_1, \ldots, f_r, \pm f_{r+1}, \ldots, \pm f_n)$ and the corresponding affine toric variety is $X = \operatorname{Spec}(\mathbb{C}[x_1, \ldots, x_r, x_{r+1}^{\pm}, \ldots, x_n^{\pm}]) = \mathbb{C}^r \times (\mathbb{C}^*)^{n-r}$. In particular, X is smooth.

Now assume that X is smooth. First consider the case that σ has maximal dimension $\operatorname{rk}(N) = n$ and let $x_0 \in X$ be the unique fixed point under the torus action. Since X is smooth at x_0, the Zariski tangent space $T_{x_0}X$ has dimension n. By Lemma 3.2, $n = \dim_{\mathbb{C}}(T_{x_0}X) = |\mathcal{H}|$ where $\mathcal{H}$ is the Hilbert basis of $S = \check{\sigma} \cap M$. Thus

$$n = |\mathcal{H}| \geq |\{\text{rays of } \check{\sigma}\}| \geq n,$$

where the first inequality holds by Problem 3.5 and the second holds since $\check{\sigma}$ is strongly convex with $\dim \check{\sigma} = n$. It follows that $\check{\sigma}$ has n rays and $\mathcal{H}$ consists of the primitive ray generators of these rays. Since $M = \mathbb{Z}S$ (see Problem 3.3), the ray generators of $\check{\sigma}$ span the lattice M and hence form a basis of M. Thus $\check{\sigma}$ is smooth, and then $\sigma = (\check{\sigma})^{\vee}$ is smooth as well.

Next suppose $\dim \sigma = r < n$. We reduce to the previous case: Let $N_1 = \operatorname{span}_{\mathbb{Z}}\{\sigma \cap N\} \subseteq N$. By the Elementary Divisor Theorem[n], there is a sublattice $N_2 \subseteq N$ with $N = N_1 \oplus N_2$. Note that $\operatorname{rk}(N_1) = r$ and $\operatorname{rk}(N_2) = n - r$. The cone σ is contained in both $N_1 \otimes \mathbb{R}$ and $N_{\mathbb{R}}$. The decomposition $N = N_1 \oplus N_2$ induces a decomposition $M = M_1 \oplus M_2$, so that we obtain affine toric varieties $X = \operatorname{Spec}(\mathbb{C}[S])$ and $X' = \operatorname{Spec}(\mathbb{C}[S'])$ where $S = \check{\sigma} \cap M$ and $S' = \check{\sigma} \cap M_1$ (σ considered as cone in $N_1 \otimes \mathbb{R}$) respectively. It is straightforward to show that $S = S' \oplus M_2$ which in terms of semigroup algebras can be written as

$$\mathbb{C}[S] \cong \mathbb{C}[S'] \otimes_{\mathbb{C}} \mathbb{C}[M_2].$$

The right-hand side is the coordinate ring of $X' \times T'$ where $T' = \operatorname{Spec}(\mathbb{C}[M_2])$ is an algebraic torus. Thus $X \cong X' \times T' \cong X' \times (\mathbb{C}^*)^{n-r} \subseteq X' \times \mathbb{C}^{n-r}$. From algebraic geometry, we know that $T_{(x,y)}X' \times (\mathbb{C}^*)^{n-r} = T_x X' \oplus T_y(\mathbb{C}^*)^{n-r}$ for any $x \in X'$ and $y \in (\mathbb{C}^*)^{n-r}$. It follows that X' is smooth, so that the previous case implies that σ is smooth in N_1 (since $\dim \sigma = \dim(N_1 \otimes \mathbb{R})$). Hence σ is smooth in $N = N_1 \oplus N_2$. □

Problem 3.7. *Let $X = \{(x_0, x_1, x_2) \in \mathbb{C}^3 \colon x_0 x_2 - x_1^2 = 0\} \subseteq \mathbb{C}^3$.*

(1) Find an action of $T = (\mathbb{C}^*)^2$ on X such that X becomes a toric variety with respect to this action.

[n] Note that $\operatorname{span}_{\mathbb{R}}\{\sigma\} \cap N = \operatorname{span}_{\mathbb{Z}}\{\sigma \cap N\}$.

(2) *Let M be the character lattice of T. Find the corresponding semigroup $S \subseteq M$ of X, i.e., such that $X = \operatorname{Spec}(\mathbb{C}[S])$.*
(3) *Is X smooth? If not, what are the singular points?*
(4) *Determine the T-orbits in X by using the "orbit-cone-correspondence".*
(5) *Describe the T-orbit closures.*

4. Toric Subvarieties of the Projective Space

Let T be an algebraic torus with character lattice M. A choice of a finite subset $A = \{m_0, \dots, m_s\} \subseteq M$ induces an action of T on $\mathbb{P}^s$ by

$$t \cdot [x_0 : \dots : x_s] = [\chi^{m_0}(t)x_0 : \dots : \chi^{m_s}(t)x_s].$$

Consider the orbit $T \cdot [1 : \dots : 1]$ and its closure $X_A := \overline{T \cdot [1 : \dots : 1]}$ in $\mathbb{P}^s$.

Proposition 4.1. *X_A is a T-invariant algebraic subvariety.*

Proof. Let $t \in T$ and note that $t \cdot T = T$. Then $T \cdot x_0 \subseteq tX_A$ and taking closures yields $X_A \subseteq t \cdot X_A$. Similarly, we obtain $X_A \subseteq t^{-1}X_A$. Multiplying the last inclusion by t, yields $t \cdot X_A \subseteq X_A$. □

Let $(\mathbb{C}^*)^{s+1}$ act naturally on $\mathbb{C}^{s+1} \setminus \{0\}$. Note that this action induces a natural action of $T_{\mathbb{P}^s} := (\mathbb{C}^*)^{s+1}/(\mathbb{C}^*(1, \dots, 1)) \cong (\mathbb{C}^*)^s$ on $\mathbb{P}^s$. Furthermore, it is straightforward to check that the character lattice of $T_{\mathbb{P}^s}$ is given by the sublattice $\mathcal{M} := \{(a_0, \dots, a_s) \in \mathbb{Z}^{d+1} \colon \sum_{i=0}^s a_i = 0\}$.

Recall that the *affine hull* $\operatorname{aff}_{\mathbb{R}}\{A\}$ of a subset A in a vector space $M_{\mathbb{R}}$ is defined as

$$\operatorname{aff}_{\mathbb{R}}\{A\} = \left\{ \sum_{m \in A} \lambda_m m \colon \lambda_m \in \mathbb{R},\ \sum_{m \in A} \lambda_m = 1, \text{ only finitely many } \lambda_m \neq 0 \right\}.$$

Proposition 4.2. *If X_A is as above, then $T \cdot [1 : \dots : 1]$ is the open subset of X_A where all homogeneous coordinates are non-zero. Moreover, the dimension of X_A equals the dimension of $\operatorname{aff}_{\mathbb{R}}\{A\} \subseteq M_{\mathbb{R}}$. Moreover, the stabilizer $T_{[1:\dots:1]}$ of the point $[1 : \dots : 1]$ is the intersection of the kernels of the characters $\chi^{m-m'}$ for $m, m' \in A$.*

Proof. The map $T \to \mathbb{P}^s$ from above induces a map $\Phi \colon T \to T_{\mathbb{P}^s}$ of tori. The image is a torus T' which is closed in $T_{\mathbb{P}^s}$, and thus $X_A \cap T_{\mathbb{P}^s} = T'$. It follows that $T' = X_A \setminus \{x_0 \cdots x_s = 0\}$ is an open and dense subset of X_A. As T' is irreducible, the same is true for its Zariski closure X_A. In particular, $\dim X_A = \dim T'$, so that it remains to compute the character lattice M' of T'. Consider the following commutative diagram of tori:

$$\begin{array}{ccccc} T & \xrightarrow{\Phi} & T_{\mathbb{P}^s} & \hookrightarrow & \mathbb{P}^s \\ & \searrow & \uparrow & & \\ & & T' & & \end{array}$$

where $\twoheadrightarrow$ denotes a surjective map and $\hookrightarrow$ an injective map. This diagram of tori induces a commutative diagram of character lattices

$$\begin{array}{ccc} M & \xleftarrow{\Phi^*} & \mathcal{M} \\ & \nwarrow & \downarrow \\ & & M'. \end{array}$$

Let $e_0, \ldots, e_s$ be the standard basis of $\mathbb{Z}^{d+1}$. Note that $\mathcal{M}$ is generated by the differences[o] $e_i - e_j$ for $i, j = 0, \ldots, s$. Since $\Phi^*\colon \mathcal{M} \to M$ maps the differences $e_i - e_j$ to the differences $m_i - m_j$, the image of Φ^* is $\operatorname{aff}_{\mathbb{Z}}\{A\} - m_i$ (for $i = 0, \ldots, s$). By the diagram, we obtain $M' \cong \operatorname{aff}_{\mathbb{Z}}\{A\} - m_i$ and we are done since the dimension of a torus equals the rank of its character lattice which equals the dimension of $\operatorname{aff}_{\mathbb{R}}\{A\}$ (note that M is torsion-free).

Finally, if $t \in \bigcap_{m,m' \in A} \ker \chi^{m-m'}$, then

$$\begin{aligned} t \cdot [1 : \ldots : 1] &= [\chi^{m_0}(t) : \ldots : \chi^{m_s}(t)] \\ &= \left[1 : \chi^{m_1 - m_0}(t) : \ldots : \chi^{m_s - m_0}(t)\right] = [1 : \ldots : 1], \end{aligned}$$

i.e., $t \in T_{[1:\ldots:1]}$. On the other hand, if $t \in T_{[1:\ldots:1]}$, then

$$[1 : \ldots : 1] = [\chi^{m_0}(t) : \ldots : \chi^{m_s}(t)] = [\chi^{m_0 - m_i}(t) : \ldots : 1 : \ldots : \chi^{m_s - m_i}(t)].$$

It follows that $\chi^{m-m'}(t) = 1$ for all $m, m' \in M$, and thus the statement follows. □

Problem 4.1. *Show that X_A may well be non-normal (Hint: Take, e.g., $T = \mathbb{C}^*$ and $A = \{0, 2, 3\} \subseteq \mathbb{Z} = M$).*

We have a similar affine version of the construction above. Indeed, for $A = \{m_0, \ldots, m_s\} \subseteq M$, we can also consider the induced action of T on $\mathbb{C}^{s+1}$:

$$t \cdot (x_0, \ldots, x_s) = (\chi^{m_0}(t)x_0, \ldots, \chi^{m_s}(t)x_s).$$

Consider the orbit $T \cdot (1, \ldots, 1)$ and its Zariski closure $Y_A := \overline{T \cdot (1, \ldots, 1)} \subseteq \mathbb{C}^{s+1}$. Like above one can show that T acts on Y_A. Moreover, we have the following:

Proposition 4.3. *If Y_A is as above, then $T \cdot (1, \ldots, 1)$ is the open subset of Y_A where all coordinates are non-zero. Moreover, the dimension of Y_A equals the dimension of $\operatorname{span}_{\mathbb{R}}\{A\} \subseteq M_{\mathbb{R}}$. The quotient torus[p] $T/T_{(1,\ldots,1)}$ naturally acts on Y_A with an open dense orbit and this torus has character lattice $\mathbb{Z}A$.*

Proof. The proof of this statement works analogously to the proof of Proposition 4.2 with the difference that this time we do not have a relation $\sum_{i=0}^{s} a_i = 0$, so that we get the linear span instead of the affine hull. □

[o] Recall the relation $\sum_{i=0}^{s} a_i = 0$ in the definition of $\mathcal{M}$.
[p] Here $T_{(1,\ldots,1)}$ denotes the T-stabilizer of the point $(1, \ldots, 1)$.

Let $A = \{m_0, \ldots, m_s\} \subseteq M$ and $e_0, \ldots, e_s$ be the standard basis of $\mathbb{Z}^{s+1}$. The map $e_i \mapsto m_i$ induces an exact sequence (here L is the kernel of the map $e_i \mapsto m_i$)

$$0 \to L \to \mathbb{Z}^{s+1} \to M. \tag{1}$$

We set $I_L := \langle T^\alpha - T^\beta \colon \alpha, \beta \in \mathbb{N}^{s+1} \text{ and } \alpha - \beta \in L \rangle \subseteq \mathbb{C}[T_0, \ldots, T_s]$ where as usual, $T^{(a_0,\ldots,a_s)} = T_0^{a_0} \cdots T_s^{a_s}$ for $(a_0, \ldots, a_s) \in \mathbb{N}^{s+1}$. The following alternative description of I_L will be useful.

Problem 4.2. *Show that $I_L = \langle T^{\ell_+} - T^{\ell_-} : \ell \in L \rangle$ where for $\ell = (l_0, \ldots, l_s) \in L$, we set $\ell_+ = \sum_{l_i > 0} l_i e_i$ and $\ell_- = -\sum_{l_i < 0} l_i e_i$.*

Proposition 4.4. *The ideal of the affine variety $Y_A \subseteq \mathbb{C}^{s+1}$ is $I(Y_A) = I_L$.*

Proof. "$\supseteq$": Let $T^{\ell_+} - T^{\ell_-} \in I_L$. Then for $(\chi^{m_0}(t), \ldots, \chi^{m_s}(t)) \in T \cdot (1, \ldots, 1)$, we obtain

$$\begin{aligned}(T^{\ell_+} - T^{\ell_-})(t \cdot (1, \ldots, 1)) &= \left(\chi^{\sum_{l_i>0} l_i m_i}(t) - \chi^{\sum_{l_i<0} l_i m_i}(t)\right) \\ &= \chi^{\sum_{l_i<0} l_i m_i}(t) \underbrace{\left(\chi^{\sum_{i=0}^s l_i m_i}(t) - 1\right)}_{=0} = 0.\end{aligned}$$

"$\subseteq$": The action of T on $\mathbb{C}^{s+1}$ induces a morphism $\Phi \colon T \to (\mathbb{C}^*)^{s+1}$. By choosing a basis of M (which yields an isomorphism $T \cong (\mathbb{C}^*)^n$), we may regard $\Phi \colon (\mathbb{C}^*)^n \to (\mathbb{C}^*)^{s+1}$ as given by Laurent monomials t^{m_i} in variables $t_1, \ldots, t_n$ (now we can assume that $m_i \in \mathbb{Z}^n$). Pick a monomial order "$\geq$" on $\mathbb{C}[T_0, \ldots, T_s]$ (see [CLO15, Definition 2.2.1]) and assume by contradiction that $I_L \neq I(Y_A)$. Then we can find some $f \in I(Y_A) \setminus I_L$ with minimal leading monomial $T^\alpha = \prod_{i=0}^s T_i^{a_i}$ (see [CLO15, Definition 2.2.7]). By rescaling if necessary, T^α becomes the leading term of f.

Since $f(t^{m_0}, \ldots, t^{m_s})$ is identically zero as polynomial in $t_1, \ldots, t_n$, it follows that f must contain a monomial $T^\beta = \prod_{i=0}^s T_i^{b_i} < T^\alpha$ such that

$$\prod_{i=0}^s (t^{m_i})^{a_i} = \prod_{i=0}^s (t^{m_i})^{b_i}.$$

This implies that

$$\sum_{i=0}^s a_i m_i = \sum_{i=0}^s b_i m_i,$$

so that $\alpha - \beta = \sum_{i=0}^s (a_i - b_i) e_i \in L$ where $\alpha = (a_0, \ldots, a_s)$ and $\beta = (b_0, \ldots, b_s)$. Then $T^\alpha - T^\beta \in I_L$ and it follows that $f - T^\alpha + T^\beta$ also lies in $I(Y_A) \setminus I_L$ and has a strictly smaller leading term. A contradiction. □

Proposition 4.5. *The coordinate ring of the affine variety $Y_A \subseteq \mathbb{C}^{s+1}$ is $\mathbb{C}[S]$ where $S = \mathbb{N}A \subseteq M$.*

Proof. The comorphism of $\Phi\colon T \to \mathbb{C}^{s+1}; t \mapsto t \cdot (1, \dots, 1)$ is the $\mathbb{C}$-algebra homomorphism $\pi\colon \mathbb{C}[T_0, \dots, T_s] \to \mathbb{C}[M]; T_i \mapsto \chi^{m_i}$. It is straightforward to show that $\ker(\pi) = I(Y_A)$ (see also Problem 4.3). The image of π is $\mathbb{C}[\chi^{m_0}, \dots, \chi^{m_s}] = \mathbb{C}[S]$, and then the coordinate ring of Y_A is

$$\mathbb{C}[Y_A] = \mathbb{C}[T_0, \dots, T_s]/I(Y_A) = \mathbb{C}[T_0, \dots, T_s]/\ker \pi \cong \operatorname{im}(\pi) = \mathbb{C}[S].$$

□

Problem 4.3. *Fix an affine variety V. Then $f_1, \dots, f_s \in \mathbb{C}[V]$ give a polynomial map $\Phi\colon V \to \mathbb{C}^s$ with comorphism*

$$\Phi^*\colon \mathbb{C}[T_1, \dots, T_s] \to \mathbb{C}[V]; T_i \mapsto f_i.$$

Let $Y \subseteq \mathbb{C}^s$ be the Zariski closure of the image of Φ. Prove that $I(Y) = \ker(\Phi^*)$.

Recall the definition of the affine cone over a projective variety: Let $\pi\colon \mathbb{C}^{s+1} \setminus \{0\} \to \mathbb{P}^s$ be the usual quotient morphism. For any subvariety $X \subseteq \mathbb{P}^s$, one defines its *affine cone* $\widehat{X}$ as the closure of the preimage $\pi^{-1}(X)$.

Proposition 4.6. *Given Y_A, X_A and I_L as above, the following are equivalent:*

(1) $Y_A \subseteq \mathbb{C}^{s+1}$ is the affine cone $\widehat{X}_A$ of $X_A \subseteq \mathbb{P}^s$.
(2) $I_L = I(X_A)$.
(3) I_L is homogeneous.
(4) There is $u \in N$ and $k > 0$ in $\mathbb{N}$ such that $\langle m_i, u \rangle = k$ for $i = 0, 1, \dots, s$.

Proof. As $I(X_A) = I(\widehat{X}_A)$ and $I_L = I(Y_A)$, we obtain (1) ⇔ (2). Moreover, the implication (2) ⇒ (3) is clear.

"(3) ⇒ (4)": Let $T^{\ell_+} - T^{\ell_-} \in I_L$ for $\ell \in L$. If T^{ℓ_+} and T^{ℓ_-} had different degrees, then $T^{\ell_+}, T^{\ell_-} \in I_L = I(Y_A)$ would vanish on Y_A which is not possible as $(1, \dots, 1) \in Y_A$. Hence the degrees of T^{ℓ_+} and T^{ℓ_-} coincide, and thus (note that $\ell_+ \cdot (1, \dots, 1) = \deg T^{\ell_+}$ and $\ell_- \cdot (1, \dots, 1) = \deg T^{\ell_-}$) $\ell \cdot (1, \dots, 1) = 0$ for all $\ell \in L$, i.e., $(1, \dots, 1) \in \mathbb{Z}^{s+1}$ induces a linear functional on L which vanishes everywhere. Tensoring the sequence (1) with $\mathbb{Q}$ and taking duals yields the following exact sequence

$$N_{\mathbb{Q}} \to \mathbb{Q}^{s+1} \to \operatorname{Hom}_{\mathbb{Q}}(L_{\mathbb{Q}}, \mathbb{Q}) \to 0.$$

It follows that $(1, \dots, 1) \in \mathbb{Q}^{s+1}$ maps to zero in $\operatorname{Hom}_{\mathbb{Q}}(L_{\mathbb{Q}}, \mathbb{Q})$ and hence comes from an element $\tilde{u} \in N_{\mathbb{Q}}$, i.e., $\langle m_i, \tilde{u} \rangle = 1$ for all i. Clearing denominators gives the desired $u \in N$ and $k > 0$ in $\mathbb{N}$.

"(4) ⇒ (1)": Since $Y_A \subseteq \widehat{X}_A$ and $\widehat{X}_A$ is irreducible, it suffices to show that $\widehat{X}_A \cap (\mathbb{C}^*)^{s+1} \subseteq Y_A$. Let $x \in \widehat{X}_A \cap (\mathbb{C}^*)^{s+1}$. By Proposition 4.2,

$$x = \mu \cdot (\chi^{m_0}(t), \dots, \chi^{m_s}(t))$$

for some $\mu \in \mathbb{C}^*$ and $t \in T$. The element $u \in N$ yields a 1-parameter subgroup of T, namely $\tau \mapsto \lambda^u(\tau)$ for $\tau \in \mathbb{C}^*$. Then, under the morphism $T \to T \cdot [1 : \ldots : 1]$, the torus element $\lambda^u(\tau)t \in T$ maps to the following point $y \in Y_A$

$$y = (\chi^{m_0}(\lambda^u(\tau)t), \ldots, \chi^{m_s}(\lambda^u(\tau)t)) = \left(\tau^{\langle m_0, u\rangle}\chi^{m_0}(t), \ldots, \tau^{\langle m_s, u\rangle}\chi^{m_s}(t)\right).$$

The hypothesis of part (4) implies that $y = \tau^k \cdot (\chi^{m_0}(t), \ldots, \chi^{m_s}(t))$. As $k > 0$, we can choose τ so that $x = y \in Y_A$. □

Problem 4.4. *Let $T = (\mathbb{C}^*)^2$ be an algebraic torus with character lattice $M = \mathbb{Z}^2$ and let $A = \{\binom{1}{0}, \binom{1}{2}, \binom{2}{1}\} \subseteq M$.*

(1) Determine Y_A, X_A and the affine cone $\widehat{X}_A$.
(2) Determine the corresponding vanishing ideals.
(3) Compare Y_A with $\widehat{X}_A$.

Next, we want to understand the local structure of the projective varieties X_A. Recall that the projective space $\mathbb{P}^s$ has a covering by affine open standard charts $U_i = \{[x_0 : \ldots : x_s] \in \mathbb{P}^s : x_i \neq 0\}$ with $U_i \cong \mathbb{A}^s$ for $i = 0, \ldots, s$.

Proposition 4.7. *Let $X_A \subseteq \mathbb{P}^s$ for $A = \{m_0, \ldots, m_s\} \subseteq M$. Then the affine piece $X_A \cap U_i$ is the affine variety*

$$X_A \cap U_i = Y_{A_i} = \operatorname{Spec}(\mathbb{C}[S_i])$$

where $A_i = A - m_i$ and $S_i = \mathbb{N}A_i$.

Proof. Note that the affine open sets $U_i = \mathbb{P}^s \setminus V(x_i)$ contain the torus $T_{\mathbb{P}^s}$. Thus, by Proposition 4.2,

$$T \cdot [1 : \ldots : 1] = X_A \cap T_{\mathbb{P}^s} \subseteq X_A \cap U_i.$$

Let us write T_A for the torus $T \cdot [1 : \ldots : 1]$. Since X_A is the Zariski closure of T_A in $\mathbb{P}^s$, it follows that $X_A \cap U_i$ is the Zariski closure of T_A in $U_i \cong \mathbb{C}^s$. Recall that the isomorphism $U_i \cong \mathbb{C}^s$ is given by

$$[a_0 : \ldots : a_s] \mapsto (a_0/a_i, \ldots, a_{i-1}/a_i, a_{i+1}/a_i, \ldots, a_s/a_i).$$

Combining this and $\chi^{m_j}/\chi^{m_i} = \chi^{m_j - m_i}$ with the composition of the action morphism $T \to T \cdot (1, \ldots, 1) \subseteq \mathbb{C}^{s+1}$ and the projection $(\mathbb{C}^*)^{s+1} \to T_{\mathbb{P}^s}$, we see that $X_A \cap U_i$ is the Zariski closure of the image of the map $T \to T_{\mathbb{P}^s}$ given by

$$t \mapsto \left(\chi^{m_1 - m_i}(t), \ldots, \chi^{m_{i-1} - m_i}(t), \chi^{m_{i+1} - m_i}(t), \ldots, \chi^{m_s - m_i}(t)\right).$$

If we set $A_i = A - m_i = \{m_j - m_i\}$, it follows, by Proposition 4.3, that

$$X_A \cap U_i \cong Y_{A_i} = \operatorname{Spec}(\mathbb{C}[S_i]),$$

where $S_i = \mathbb{N}A_i$ (note that $0 \in A_i$ just yields an additional coordinate which is constant 1). □

Corollary 4.1. *Let $A = \{m_0, \ldots, m_s\} \subseteq M$ be a finite set which affinely spans the lattice M. Then the variety X_A is normal if and only if all the semigroups $S_i = \mathbb{N}(A - m_i)$ are saturated in M. If this is the case, the variety X_A is a projective toric variety.*

Problem 4.5. *Find a finite set $A \subset M$ which affinely spans the lattice M such that X_A is not normal. (Hint: Consider $M = \mathbb{Z}$ and $A = \{0, 2, 3\}$.)*

Next, let us discuss how the affine pieces $X_A \cap U_i$ of $X_A \subseteq \mathbb{P}^s$ patch together, i.e., we give a description of the inclusions

$$X_A \cap U_i \supseteq X_A \cap U_i \cap U_j \subseteq X_A \cap U_j$$

when $i \neq j$.

Proposition 4.8. *The inclusion $X_A \cap U_i \cap U_j \subseteq X_A \cap U_i$ is given by*

$$\operatorname{Spec}(\mathbb{C}[S_i])_{\chi^m} \subseteq \operatorname{Spec}(\mathbb{C}[S_i]),$$

where $m = m_j - m_i$.

Proof. Let us denote the homogeneous coordinates of $\mathbb{P}^s$ by $x_0, \ldots, x_s$. The points in $X_A \cap U_i \cap U_j$ are exactly the points of $X_A \cap U_i$ where $x_j/x_i \neq 0$. In terms of $X_A \cap U_i = \operatorname{Spec}(\mathbb{C}[S_i])$ (see Proposition 4.7), this means those points where $\chi^{m_j - m_i} \neq 0$. Thus

$$X_A \cap U_i \cap U_j = \operatorname{Spec}(\mathbb{C}[S_i])_{\chi^{m_j - m_i}} = \operatorname{Spec}(\mathbb{C}[S_i]_{\chi^{m_j - m_i}}).$$

□

4.1. *The Projective Toric Variety of a Lattice Polytope*

In the previous section, we have seen how to associated a projective variety X_A to a finite set $A = \{m_0, \ldots, m_s\} \subseteq M$ and we have determined under which combinatorial conditions the associated variety is toric (see Corollary 4.1). In this section, we make the connection to convex geometry.

Recall that the convex hull of a finite subset $A \subseteq M_{\mathbb{R}}$ is the *polytope* $P = \operatorname{conv}(A)$, i.e.

$$P = \operatorname{conv}(A) = \left\{ \sum_{a \in A} \lambda_a a \colon \lambda_a \in \mathbb{R}_{\geq 0}, \sum_{a \in A} \lambda_a = 1 \right\}.$$

We extend the notation of hyperplanes (namely H_u) and half-spaces (namely H_u^+) from above to include the analogous affine versions. Let $u \in N_{\mathbb{R}}$ and $b \in \mathbb{R}$. Then we define the affine hyperplane $H_{u,b}$ and the affine half-space $H_{u,b}^+$ in $M_{\mathbb{R}}$ as follows

$$H_{u,b} = \{m \in M_{\mathbb{R}} \colon \langle m, u \rangle = b\}, \qquad H_{u,b}^+ = \{m \in M_{\mathbb{R}} \colon \langle m, u \rangle \geq b\}.$$

Above we wrote $H_u = H_{u,0}$ and $H_u^+ = H_{u,0}^+$.

Let $P \subseteq M_{\mathbb{R}}$ be a polytope. An affine half-space $H^+_{u,b}$ is a *supporting half-space for P* if P is contained in it, i.e., $P \subseteq H^+_{u,b}$. A *face* of P is any set of the form $F = P \cap H_{u,b}$ where $H^+_{u,b}$ is a supporting half-space for P, and we will write $F \preceq P$. The *dimension* of a face is the dimension of its affine hull, i.e., $\dim(F) = \dim(\mathrm{aff}_{\mathbb{R}}(F))$. The faces of dimension 0 are called *vertices* and the faces of dimension $\dim(P) - 1$ are called *facets*.

We start with a discussion how the polytope $P = \mathrm{conv}(A) \subseteq M_{\mathbb{R}}$ relates to X_A. First note that $\dim X_A = \dim P$ by Proposition 4.2. Furthermore, the vertices of P give an especially efficient affine covering of X_A.

Proposition 4.9. *If $A = \{m_0, \ldots, m_s\} \subseteq M$, let $P = \mathrm{conv}(A) \subseteq M_{\mathbb{R}}$ and set $J = \{j \in \{0, \ldots, s\} \colon m_j$ is a vertex of $P\}$. Then*

$$X_A = \bigcup_{j \in J} (X_A \cap U_j).$$

Proof. We show that if $i \in \{0, \ldots, s\}$, then $X_A \cap U_i \subseteq X_A \cap U_j$ for some $j \in J$.

Let $i \in \{0, \ldots, s\}$, we have $m_i \in P \cap M$, so that m_i is a convex $\mathbb{Q}$-linear combination of the vertices m_j (for $j \in J$). Clearing denominators, we get integers $k > 0$ and $k_j \geq 0$ (for $j \in J$) such that

$$km_i = \sum_{j \in J} k_j m_j \qquad \text{and} \qquad \sum_{j \in J} k_j = k.$$

From the equation $\sum_{j \in J} k_j(m_j - m_i) = 0$, we get for $k_j > 0$ that $k_j(m_i - m_j) = \sum_{j' \in J \setminus \{j\}} k_{j'}(m_{j'} - m_i) \in S_i$. Fix such a j. Then $\chi^{m_j - m_i} \in \mathbb{C}[S_i]$ is invertible, so $\mathbb{C}[S_i]_{\chi^{m_j - m_i}} = \mathbb{C}[S_i]$. By Proposition 4.8, $X_A \cap U_i \cap U_j = \mathrm{Spec}(\mathbb{C}[S_i]) = X_A \cap U_i$, giving $X_A \cap U_i \subseteq X_A \cap U_j$. □

On the other hand, a lattice polytope P gives a finite set of lattice points $P \cap M$, which give a projective variety $X_{P \cap M}$. Let us collect some immediate consequences from our previous discussion.

Definition 4.1. A lattice polytope $P \subseteq M_{\mathbb{R}}$ whose lattice points affinely span the ambient lattice M is called *spanning*.

Note that this is a mild assumption for lattice polytopes, as we can always coarsen the ambient lattice. Furthermore, from a geometric point of view this is a natural assumption, compare, e.g., Proposition 2.5 and Theorem 3.1.

Proposition 4.10. *Let $P \subseteq M_{\mathbb{R}}$ be a spanning lattice polytope. Then the projective variety $X_{P \cap M}$ is normal if and only if for every vertex $m \in P$, the semigroup $S_m = \mathbb{N}(P \cap M - m)$ generated by the set $P \cap M - m = \{m' - m \colon m' \in P \cap M\}$ is saturated in M.*

Proof. This is an immediate consequence of Proposition 2.5 and Proposition 4.7. □

Definition 4.2. A lattice polytope $P \subseteq M_{\mathbb{R}}$ is called *very ample*, if for every vertex $m \in P$, the semigroup $S_m = \mathbb{N}(P \cap M - m)$ generated by $P \cap M - m = \{m' - m \colon m' \in P \cap M\}$ is saturated in M.

Problem 4.6. *Prove that a very ample polytope is spanning. Find a spanning lattice polytope which is not very ample.*

Theorem 4.1. *Let $X_{P\cap M}$ be the projective toric variety of the very ample lattice polytope $P \subseteq M_{\mathbb{R}}$, and assume that P is full dimensional with* $\dim P = n$.

(1) For each vertex $m_i \in P \cap M$, the affine piece $X_{P\cap M} \cap U_i$ is the affine toric variety

$$X_{P\cap M} \cap U_i = \operatorname{Spec}(\mathbb{C}[\check{\sigma}_i \cap M])$$

where $\sigma_i \subseteq N_{\mathbb{R}}$ is the strongly convex rational polyhedral cone dual to the cone $\operatorname{cone}(P \cap M - m_i) \subseteq M_{\mathbb{R}}$. *Furthermore,* $\dim \sigma_i = n$.

(2) The torus of $X_{P\cap M}$ has character lattice M and hence is the torus T.

Proof. By Proposition 4.10, $X_{P\cap M}$ is normal, and hence a projective toric variety. The statement follows from Proposition 4.7, Problem 2.10, Problem 2.6 and Proposition 2.1. □

Problem 4.7. *The cones $\sigma_i \subseteq N_{\mathbb{R}}$ appearing in the previous theorem fit together in a remarkably nice way: Let Σ_P be the collection of the σ_i together with all faces of σ_i. Let us write the facet presentation of P as (here $u_F \in N_{\mathbb{R}}$ denotes an inner facet normal of the facet $F \preceq P$ and $a_F \in \mathbb{R}$)*

$$P = \{m \in M_{\mathbb{R}} \colon \langle m, u_F \rangle \geq -a_F \text{ for all facets } F \preceq P\}.$$

Show that Σ_P consists of all the cones

$$\sigma_Q = \operatorname{cone}(u_F \colon F \text{ contains } Q)$$

where Q ranges over the faces $Q \preceq P$. Then deduce the following properties of Σ_P for a full dimensional lattice polytope $P \subseteq M_{\mathbb{R}}$:

(1) For all $\sigma_Q \in \Sigma_P$, each face of σ_Q is also in Σ_P.
(2) The intersection $\sigma_Q \cap \sigma_{Q'}$ of any two cones in Σ_P is a face of each.

A finite collection of strongly convex rational polyhedral cones that satisfies these properties is called a fan. *As the cones in the fan Σ_P are built from the normal vectors of P, the fan Σ_P is the* normal fan *of P (see Figure 12).*

Lemma 4.1. *Let $\tau \preceq \sigma$ be a face of a strongly convex rational polyhedral cone $\sigma \subseteq N_{\mathbb{R}}$ and write $\tau = H_m \cap \sigma$ for some $m \in \check{\sigma} \cap M$. Then*

$$\mathbb{C}[\check{\tau} \cap M] = \mathbb{C}[\check{\sigma} \cap M]_{\chi^m}.$$

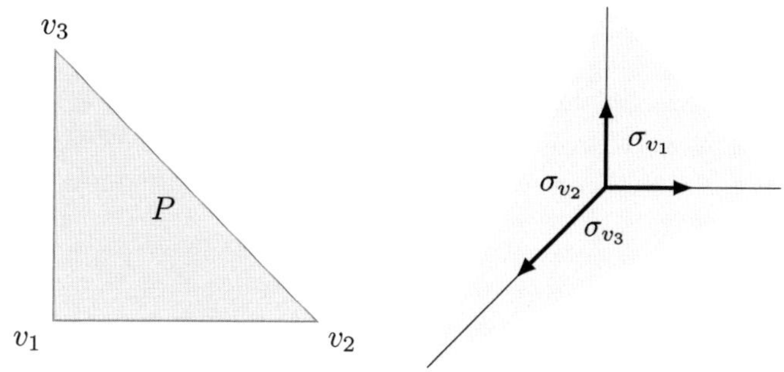

Fig. 12. A full-dimensional polytope $P \subseteq M_{\mathbb{R}}$ and its normal fan Σ_P in $N_{\mathbb{R}}$.

Proof. The inclusion $\tau \subseteq \sigma$ implies $\check{\sigma} \cap M \subseteq \check{\tau} \cap M$. As $\langle m, u\rangle = 0$ for all $u \in \tau$, we have $\pm m \in \check{\tau}$, and it follows that

$$(\check{\sigma} \cap M) + \mathbb{Z}m \subseteq \check{\tau} \cap M.$$

We claim that the previous inclusion is actually an equality. Let $S \subseteq N$ be a finite set such that $\sigma = \operatorname{cone}(S)$ and let $m' \in \check{\tau} \cap M$. Set $C := \max\{|\langle m', u\rangle| : u \in S\} \in \mathbb{N}$. We prove that $m' + Cm \in \check{\sigma} \cap M$, i.e., we have to show that $\langle m' + Cm, u\rangle \geq 0$ for any $u \in S$. If $\langle m, u\rangle = 0$, then $u \in \tau$, and thus $\langle m', u\rangle \geq 0$. It follows that $\langle m' + Cm, u\rangle \geq 0$. If $\langle m, u\rangle > 0$, then the choice of the constant C ensures that $\langle m' + Cm, u\rangle \geq 0$. This proves that $m' + Cm \in \check{\sigma} \cap M$, and thus

$$(\check{\sigma} \cap M) + \mathbb{Z}m = \check{\tau} \cap M,$$

from which $\mathbb{C}[\check{\tau} \cap M] = \mathbb{C}[\check{\sigma} \cap M]_{\chi^m}$ follows. □

Note that the inclusion $\tau \subseteq \sigma$ corresponds to the inclusion $\mathbb{C}[\check{\sigma} \cap M] \subseteq \mathbb{C}[\check{\tau} \cap M]$, and by the previous lemma

$$\operatorname{Spec}(\mathbb{C}[\check{\tau} \cap M]) = \operatorname{Spec}(\mathbb{C}[\check{\sigma} \cap M]_{\chi^m}) = \operatorname{Spec}(\mathbb{C}[\check{\sigma} \cap M])_{\chi^m} \subseteq \operatorname{Spec}(\mathbb{C}[\check{\sigma} \cap M]).$$

Thus $\operatorname{Spec}(\mathbb{C}[\check{\tau} \cap M])$ becomes an affine open subset of $\operatorname{Spec}(\mathbb{C}[\check{\sigma} \cap M])$ when $\tau \preceq \sigma$. In particular, if two cones σ and σ' intersect in a common face $\tau = \sigma \cap \sigma'$, then we have inclusions

$$\operatorname{Spec}(\mathbb{C}[\check{\sigma} \cap M]) \supseteq \operatorname{Spec}(\mathbb{C}[\check{\tau} \cap M]) \subseteq \operatorname{Spec}(\mathbb{C}[\check{\sigma'} \cap M]).$$

Problem 4.8. *Let $P \subseteq M_{\mathbb{R}}$ be a lattice polytope of maximal dimension and let $Q \neq \emptyset$ be a face of P. We continue to use the notation introduced in Problem 4.7. In particular, we write the facet presentation of P as*

$$P = \{m \in M_{\mathbb{R}} : \langle m, u_F\rangle \geq -a_F \text{ for all facets } F\}.$$

Let $H_{u,b}$ be a supporting affine hyperplane of P with $H_{u,b} \cap P \neq \emptyset$. Prove that $u \in \sigma_Q = \operatorname{cone}(u_F : F \text{ contains } Q)$ if and only if $Q \subseteq H_{u,b} \cap P$.

Let $P \subseteq M_{\mathbb{R}}$ be a polytope of maximal dimension. Recall that the finite set of lattice points $P \cap M$ yields a projective variety $X_{P\cap M} \subseteq \mathbb{P}^{|P\cap M|-1}$. In Proposition 4.9, we have seen that the vertices of P yield an affine cover of $X_{P\cap M}$. If we don't want to order the vertices in $P \cap M$, we will simple write U_v for the standard affine open chart of $\mathbb{P}^{|P\cap M|-1}$ corresponding to $v \in P \cap M$. For instance, Proposition 4.9, then becomes

$$X_{P\cap M} = \bigcup_{v\in \mathrm{Vert}(P)} (X_{P\cap M} \cap U_v),$$

where $\mathrm{Vert}(P)$ denotes the set of vertices of P.

Proposition 4.11. *Let $P \subseteq M_{\mathbb{R}}$ be a full dimensional and very ample lattice polytope. If v, w are two distinct vertices of P and Q is the smallest face of P containing v and w, then $X_{P\cap M} \cap U_v \cap U_w = \mathrm{Spec}(\mathbb{C}[\check{\sigma}_Q \cap M])$ and the inclusions*

$$X_{P\cap M} \cap U_v \supseteq X_{P\cap M} \cap U_v \cap U_w \subseteq X_{P\cap M} \cap U_w$$

are given by

$$\begin{aligned}\mathrm{Spec}(\mathbb{C}[\check{\sigma}_v \cap M]) \supseteq \mathrm{Spec}(\mathbb{C}[\check{\sigma}_v \cap M])_{\chi^{w-v}} &= \mathrm{Spec}(\mathbb{C}[\check{\sigma}_Q \cap M]) \\ &= \mathrm{Spec}(\mathbb{C}[\check{\sigma}_w \cap M])_{\chi^{v-w}} \subseteq \mathrm{Spec}(\mathbb{C}[\check{\sigma}_w \cap M]).\end{aligned}$$

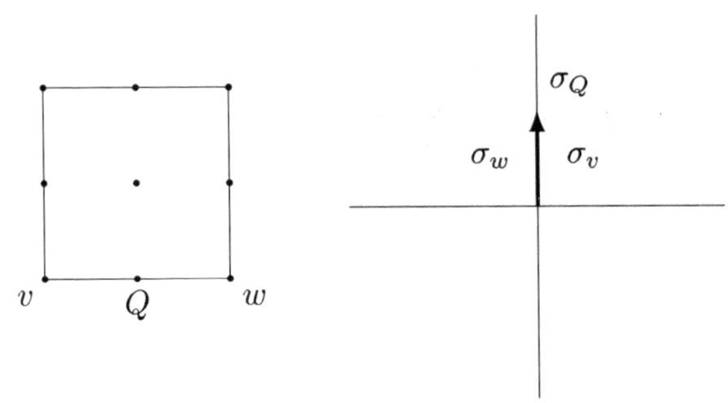

Fig. 13. Illustration of Proposition 4.11.

Proof. By Proposition 4.8, we have

$$X_{P\cap M} \cap U_v \cap U_w = \mathrm{Spec}(\mathbb{C}[\check{\sigma}_v \cap M])_{\chi^{w-v}} = \mathrm{Spec}(\mathbb{C}[\check{\sigma}_w \cap M])_{\chi^{v-w}}.$$

Thus it suffices to show that

$$\mathrm{Spec}(\mathbb{C}[\check{\sigma}_v \cap M])_{\chi^{w-v}} = \mathrm{Spec}(\mathbb{C}[\check{\sigma}_Q \cap M]).$$

Note that $w - v \in \check{\sigma}_v = \mathrm{cone}(P \cap M - v) \subseteq M_{\mathbb{R}}$, so that $\tau = H_{w-v} \cap \sigma_v$ is a face of σ_v. We claim $\sigma_Q = \tau$, so that, by Lemma 4.1, we obtain

$$\mathrm{Spec}(\mathbb{C}[\check{\sigma}_v \cap M])_{\chi^{w-v}} = \mathrm{Spec}(\mathbb{C}[\check{\tau} \cap M]) = \mathrm{Spec}(\mathbb{C}[\check{\sigma}_Q \cap M]).$$

We claim that $\sigma_Q = \sigma_v \cap \sigma_w$. Recall from Problem 4.7, that σ_Q, σ_v and σ_w can be written as the positive hull of certain inner facet normals of P. With this description the inclusion "$\subseteq$" straightforwardly follows. For the reverse inclusion recall that, by definition, σ_v is the cone dual to $\text{cone}(P \cap M - v)$ while σ_w is the cone dual to $\text{cone}(P \cap M - w)$. It follows that any $u \in \sigma_v \cap \sigma_w$ satisfies $\langle v - w, u \rangle = 0$. Again using the fact that $\sigma_v \cap \sigma_w$ is the cone spanned by certain inner facet normals of P, the reverse inclusion follows.

Since $\sigma_Q = \sigma_v \cap \sigma_w$, it suffices to prove that $H_{w-v} \cap \sigma_v = \sigma_v \cap \sigma_w$ (see Lemma 4.1). Let $u \in H_{w-v} \cap \sigma_v$. If $u \neq 0$, then $H_{u,b}$ is a supporting affine hyperplane of P where $b = \langle v, u \rangle \in \mathbb{R}$. As $\langle w - v, u \rangle = 0$, it follows that $w \in H_{u,b}$ as well. Applying Problem 4.8, we get $u \in \sigma_w$. For the reverse inclusion, let $u \in \sigma_v \cap \sigma_w$. As σ_v is the cone dual to $\text{cone}(P \cap M - v)$ and σ_w is the cone dual to $\text{cone}(P \cap M - w)$, it straightforwardly follows that $\langle w - v, u \rangle = 0$, i.e., $u \in H_{w-v}$. □

Theorem 4.2. *Let $P \subseteq M_{\mathbb{R}}$ be a very ample lattice polytope of maximal dimension n. Then the following are equivalent:*

(1) The projective toric variety $X_{P\cap M}$ is smooth.
(2) All cones in the normal fan of P are smooth.
(3) The primitive edge directions at every vertex of P form a basis of the lattice M. In particular, P is simple, *i.e., every vertex lies in exactly n edges.*

A polytope satisfying one of the equivalent conditions of the previous theorem is called a *smooth polytope* (see also Figure 14).

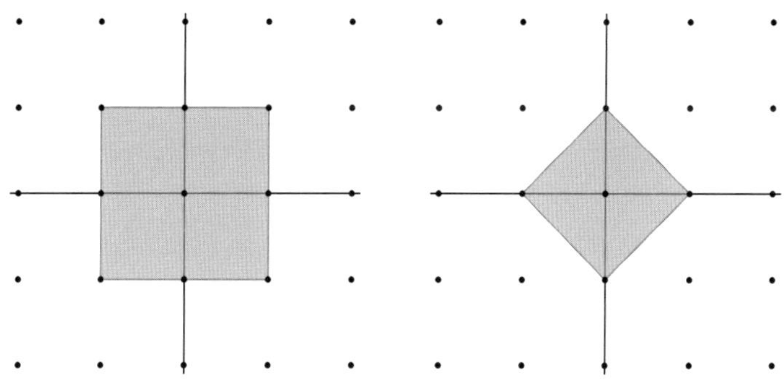

Fig. 14. The polytope on the left side is smooth while the one on the right side is not.

Proof. "(1) $\Leftrightarrow$ (2)": As smoothness is a local property, this equivalence straightforwardly follows from Theorems 3.3 and 4.1.

"(2) $\Leftrightarrow$ (3)": This equivalence follows from the observation that for a vertex v of P the cones σ_v are smooth if and only if their dual cones are smooth. It is straightforward to show that the cone dual to σ_v has primitive ray generators the primitive edge directions at the vertex v. □

Problem 4.9. *Let* $P = \operatorname{conv}(\binom{0}{0}, \binom{1}{0}, \binom{0}{1}) \subseteq M_{\mathbb{R}}$ *where* M *is the lattice* $\mathbb{Z}^2$.

(1) Check that P *is very ample.*
(2) What is $X_{P\cap M}$*? What are its local charts?*
(3) Is the variety smooth?

Problem 4.10. *Let* $P = \operatorname{conv}(\binom{1}{0}, \binom{0}{1}, \binom{-1}{0}, \binom{0}{-1}) \subseteq M_{\mathbb{R}}$.

(1) Check that P *is very ample.*
(2) Is the variety smooth?

Problem 4.11. *Prove that every smooth full dimensional lattice polytope is very ample.*

Recall that if $P \subseteq M_{\mathbb{R}}$ is a full dimensional polytope containing the origin, then its (*polar*) *dual* is given by (see also Figure 15)

$$P^* = \{u \in N_{\mathbb{R}} \colon \langle m, u\rangle \geq -1 \text{ for any } m \in P\}.$$

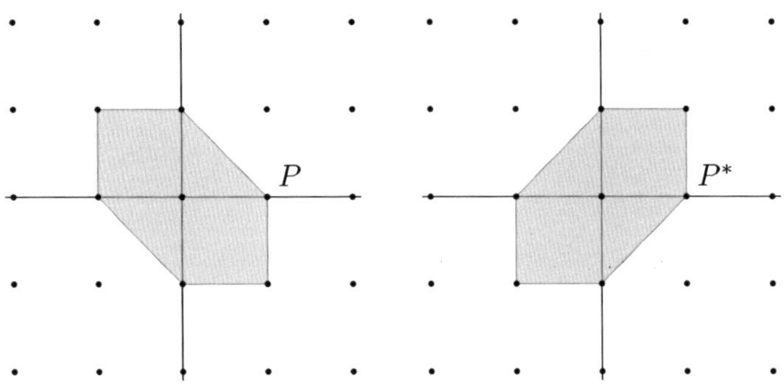

Fig. 15. A lattice polytope $P \subseteq M_{\mathbb{R}}$ and its dual polytope $P^* \subseteq N_{\mathbb{R}}$. Note that in general the dual polytope P^* need not to be a lattice polytope.

Problem 4.12. *Find a smooth reflexive polytope whose dual is not smooth. (Hint: Consider the polytope* $(n+1)\Delta_n - (1,\ldots,1) \subseteq \mathbb{R}^n$*, where* Δ_n *is the standard* n*-simplex.)*

5. Projective Normality

A projective variety $X \subseteq \mathbb{P}^n$ is *projectively normal* (with respect to the given embedding) if its homogeneous coordinate ring is an integrally closed domain, i.e., if $\mathbb{C}[x_0,\ldots,x_n]/I$ is integrally closed where $I = I(X)$. From the definition it follows that this property can be rather useful, as many theorems in algebra are true for normal rings but fail in general. In many cases projective normality makes it possible to work with the entire homogeneous coordinate ring instead of working locally, i.e., working with the localizations of the homogeneous coordinate ring and considering how the localizations patch together.

Problem 5.1. *Show that if $X \subseteq \mathbb{P}^n$ is projectively normal, then X is normal.*

Problem 5.2. *Show that there are normal varieties in projective space which are not projectively normal. (Hint: For example, consider the twisted quartic curve Y in $\mathbb{P}^3$ given parametrically by $[x : y : z : w] = [t^4 : t^3u : tu^3 : u^4]$. Note that Y is isomorphic to $\mathbb{P}^1$, which is projectively normal with respect to, e.g., the Veronese embedding. Thus projective normality depends on the embedding.)*

Assume that $X = X_{P\cap M} \subseteq \mathbb{P}^{|P\cap M|-1}$ is a projective toric variety for a very ample lattice polytope $P \subseteq M_\mathbb{R}$ where M is some lattice. Then projective normality translates into a combinatorial property of the lattice polytope P: A lattice polytope $P \subseteq M_\mathbb{R}$ has the *integer decomposition property* (or is *IDP*) if for any lattice point m in $(kP)\cap M$ there are lattice points $m_1, \ldots, m_k \in P\cap M$ such that $m = m_1+\ldots+m_k$ for all integers $k \geq 1$.

Problem 5.3. *Let M be a lattice and let $P \subseteq M_\mathbb{R}$ be a lattice polytope. Prove that P is IDP if and only if $(P\cap M)\times\{1\}$ generates the semigroup* $\operatorname{cone}(P\times\{1\})\cap(M\oplus\mathbb{Z})$.

Theorem 5.1. *Let M be a lattice and let $P \subseteq M_\mathbb{R}$ be a full dimensional very ample lattice polytope. Then $X_{P\cap M} \subseteq \mathbb{P}^{|P\cap M|-1}$ is projectively normal if and only if P has the integer decomposition property.*

Proof. Given $A := P \cap M \subseteq M$, there is a natural way to modify A such that the conditions of Proposition 4.6 are met: use $A \times \{1\} \subseteq M \oplus \mathbb{Z}$. The lattice $M \oplus \mathbb{Z}$ corresponds to the torus $T \times \mathbb{C}^*$, and since

$$(t,\mu)\cdot(1,\ldots,1) = (\chi^{m_0}(t)\mu,\ldots,\chi^{m_s}(t)\mu) = \mu\cdot(\chi^{m_0}(t),\ldots,\chi^{m_s}(t)),$$

it follows that $X_{A\times\{1\}} = X_A \subseteq \mathbb{P}^s$. Since $A\times\{1\}$ lies in an affine hyperplane missing the origin, we get that X_A has affine cone $Y_{A\times\{1\}} = \widehat{X}_A$. By Propositions 2.5 and 4.5, the coordinate ring of $Y_{A\times\{1\}}$ is integrally closed if and only if the semigroup $\mathbb{N}(A\times\{1\})$ is saturated in $M\oplus\mathbb{Z}$ (note that $\mathbb{Z}(A\times\{1\}) = M\oplus\mathbb{Z}$). This is equivalent to the assertion that the semigroup $\operatorname{cone}(A\times\{1\})\cap(M\oplus\mathbb{Z})$ is generated by $A\times\{1\}$ which in turn is equivalent to P having the integer composition property. $\square$

5.1. *Oda's Question*

On the Oberwolfach Conference, "Combinatorial Convexity and Algebraic Geometry" in October 1997, Oda asked a rather general and intriguing convex geometric question which is closely related to the projective normality of projective toric varieties $X_{P\cap M} \subseteq \mathbb{P}^{|P\cap M|-1}$ (see [OWR97]). The be able to state Oda's question, we need to recall the following standard constructions:

The *Minkowski sum* of two polytopes $P_1, P_2 \subseteq M_\mathbb{R}$ is (see also Figure 16)

$$P_1 + P_2 := \{m_1 + m_2 \colon m_1 \in P_1, m_2 \in P_2\}.$$

For non-negative real numbers r, s, we also have the distributive law

$$rP + sP = (r + s)P.$$

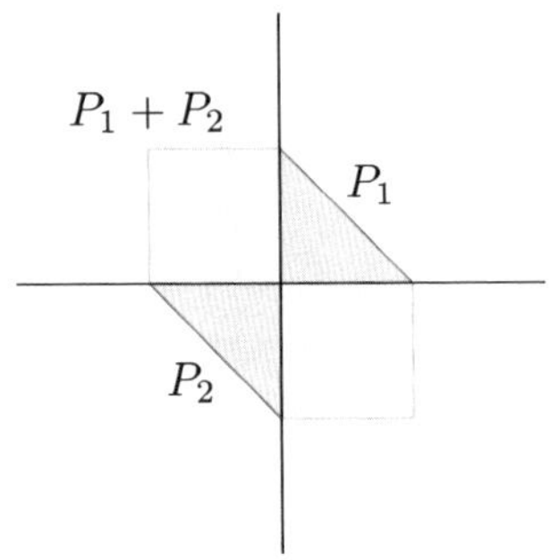

Fig. 16. The Minkowski sum $P_1 + P_2$ of two triangles P_1 and P_2.

On the above mentioned Oberwolfach Conference, Oda posed the following fundamental problem (see also [Oda08]):

Question 5.1 (Oda's question). *Given two lattice polytopes $P, Q \subseteq \mathbb{R}^d$, when can every lattice point p in the Minkowski sum $P + Q := \{x + y : x \in P, y \in Q\}$ be written as the sum of two lattice points $p_1 \in P$ and $p_2 \in Q$, i.e., $p = p_1 + p_2$?*

Problem 5.4. *Prove that for arbitrary lattice polytopes, not every lattice point in $P + Q$ is the sum of a lattice point in P and a lattice point in Q, not even in the special case $P = Q$. (Hint: For example, consider the convex hull P of $(0,0,0), (1,0,0), (0,0,1), (1,2,1)$ and determine the lattice points in $2P$.)*

Due to its relation with projective normality of projective toric varieties, the following specialization of Problem 5.1 was also asked by Oda (see [Oda08]). It has since become known as *Oda's Conjecture.*

Question 5.2 (Oda's Conjecture). *Is every smooth lattice polytope IDP?*

Although the previous question seems to be a special case of Question 5.1, it turns out that it actually implies a variant of Question 5.1.

Proposition 5.1. *If Question 5.2 is true in dimension $d + 1$, then the following variant of Questions 5.1 is true in dimension d.*

Let P and Q be two smooth lattice polytopes with the same normal fan. Can every lattice point $p \in P + Q$ be written as $p = u + v$ where u is a lattice point of P and v is a lattice point of Q?

This observation seems to be originally due to Mustaţă. We have taken the following argument from Maclagan's report (see [OWR07]).

Proof. Consider the polytope $R = \operatorname{conv}(P \times \{0\}, Q \times \{1\}) \subseteq \mathbb{R}^{d+1}$. As P and Q are smooth with the same normal fan, it straightforwardly follows that R is smooth. The lattice points in the slice of $2R$ with the affine hyperplane $\{x_{d+1} = 1\} \subseteq \mathbb{R}^{d+1}$ are in bijection with the lattice points in $P + Q$. Moreover, note that the lattice points in R are in bijection with the lattice points in P respectively Q. The statement now straightforwardly follows. □

So far, every smooth polytope that has been found has the integer decomposition property. Moreover, several partial results of Oda's Conjecture are known.

Theorem 5.2 ([Koe93, Stu97, HNPS08]). *All 2-dimensional smooth lattice polygons are IDP.*

Note that the 2-dimensional case follows from several more general statements (check the given references).

Theorem 5.3 ([BHH$^+$]). *Any centrally symmetric 3-dimensional smooth lattice polytope is IDP.*

Recall that a finite set of lattice points in a lattice N is said to be *unimodular* if every maximal independent subset generates the lattice N, i.e., the corresponding matrix whose columns are given by the vectors in the maximal independent subset has determinant ± 1.

Theorem 5.4 ([CHP$^+$18, Theorem 1.1]). *A lattice polytope $P \subseteq M_{\mathbb{R}}$ such that the primitive inner facet normals in N form a* unimodular vector configuration, *has the integer decomposition property.*

Problem 5.5. *Find a lattice polytope such that the inner facet normals of P form a unimodular vector configuration while the vertices of P don't.*

In August 2007 the Oberwolfach Mini-Workshop: "Projective Normality of Smooth Toric Varieties" was devoted to the study of Oda's conjecture and its implications (see [OWR07]). On this mini-workshop, several approaches to prove or disprove this beautiful conjecture were presented. We believe that after more than 10 years, it is time to examine these proposed approaches and suggest further promising new ones.

We start this program in the next section where we show that the proposed approach through Frobenius splittings is not going to work.

5.2. *Frobenius-Splitting*

The concept of Frobenius splittings was developed by Mehta, Ramanathan and their collaborators (see [MR85]). Frobenius splittings yield elegant proofs that all ample line bundles on Schubert varieties (of all types) are very ample and give projectively normal embeddings (see [RR85, Ram87]). We refer to [BK05] for further details and references on Frobenius splittings.

Regarding the success story of Frobenius splittings, it seems very promising to try to use them to prove Oda's Conjecture. In [Pay09], Payne studies Frobenius splittings in the context of toric varieties. Let us recall the fundamental definitions and results from [Pay09].

As usual let $T = (\mathbb{C}^*)^n$ be an algebraic torus with character lattice M and let $P \subseteq M_{\mathbb{R}}$ be a smooth lattice polytope yielding a projective toric variety $X = X_{P\cap M} \subseteq \mathbb{P}^{|P\cap M|-1}$. Fix an integer $q > 1$ and note that multiplication by q induces a lattice homomorphism $N \to N$ which preserves the normal fan of P, and therefore gives a morphism $F\colon X \to X$.

Definition 5.1. A *splitting of* X is an $\mathcal{O}_X$-module map $\pi\colon F_*\mathcal{O}_X \to \mathcal{O}_X$ such that the composition $\pi \circ F^*$ is the identity on $\mathcal{O}_X$.

In the context of projective normality, the following splittings are of particular interest. We say that X is *diagonally split* if there is a splitting of $X \times X$, i.e., $\pi\colon F_*\mathcal{O}_{X\times X} \to \mathcal{O}_{X\times X}$ (here $F\colon X \times X \to X \times X$ denotes the morphism induced by multiplication by q on $N \oplus N$), such that $\pi(F_*\mathfrak{I}_X)$ is contained in $\mathfrak{I}_X$ where $\mathfrak{I}_X$ denotes the ideal sheaf of the diagonal (which is isomorphic to X). By classical arguments of Mehta, Ramanan and Ramanathan, if X is diagonally split then every ample line bundle on X is very ample and defines a projectively normal embedding. In particular, P is IDP.

In [Pay09], Payne finds the following polyhedral criterion for a toric variety to be diagonally split. Let $\{u_F \in N\colon F \preceq P \text{ facet}\}$ be the primitive inner facet normals of P, and let the *diagonal splitting polytope* $\mathbb{F}_X$ be defined by

$$\mathbb{F}_X = \{m \in M_{\mathbb{R}}\colon -1 \le \langle m, u_F\rangle \le 1 \text{ for all facets } F \preceq P\}.$$

We write $\frac{1}{q}M$ for the subgroup of $M_{\mathbb{R}}$ consisting of fractional points m such that qm is in M.

Theorem 5.5 ([Pay09, Theorem1.2]). *The toric variety X is diagonally split if and only if the interior of $\mathbb{F}_X$ contains representatives of every equivalence class in $\frac{1}{q}M/M$.*

Using this combinatorial characterization, Chou, Hering, Payne, Tramel and Whitney further studied diagonal split toric varieties (see [CHP+18]).

Theorem 5.6 ([CHP+18, Theorem 1.1]). *If the primitive inner facet normals u_F of P form a unimodular vector configuration, then X is diagonally split at q for all $q \ge 2$.*

For a finite subset $\mathcal{V}$ in N, we let A be the matrix whose columns are the vectors in this subset. Recall that $\mathcal{V}$ is unimodular if and only if every maximal non-singular square submatrix of A is invertible over $\mathbb{Z}$. In [CHP+18], the authors also consider the following generalization of this notion. We say that $\mathcal{V}$ is *k-regular* if every maximal nonsingular square submatrix of A is invertible over $\mathbb{Z}[\frac{1}{k}]$. We say that X is *not diagonally split* if there is no q such that X is diagonally split at q.

Theorem 5.7 ([CHP+18, Theorem 1.5]). *If the primitive inner facet normals of P form a vector configuration that is not 2-regular, then X is not diagonally split.*

With the previous characterization, it is straightforward to find a toric variety $X = X_{P\cap M}$ for a smooth polytope $P \subseteq M_{\mathbb{R}}$ that is not diagonally split (see Figure 17). Note that the polytope $P \subseteq \mathbb{R}^2$ is obtained by iterative chiseling of dilates of the cube (see [CLNP18]).

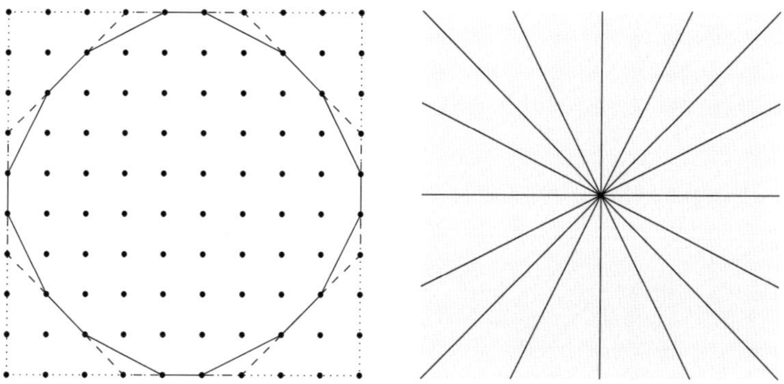

Fig. 17. A smooth polytope P whose primitive inner facet normals form a vector configuration that is not 2-regular, and hence the corresponding toric variety $X_{P\cap M}$ is not diagonally split.

Indeed, the primitive inner facet normals of P are given by

$$\binom{1}{0}, \binom{-1}{0}, \binom{0}{1}, \binom{0}{-1}, \binom{2}{1}, \binom{2}{-1}, \binom{-2}{1}, \binom{-2}{-1},$$
$$\binom{1}{2}, \binom{1}{-2}, \binom{-1}{2}, \binom{-1}{-2}, \binom{1}{1}, \binom{1}{-1}, \binom{-1}{1}, \binom{-1}{-1}.$$

The submatrix $\left(\begin{smallmatrix} 2 & 1 \\ 1 & 2 \end{smallmatrix}\right)$ is nonsingular, but has determinant 3 which is not invertible over $\mathbb{Z}[\frac{1}{2}]$. Thus the vector configuration given by the primitive inner facet normals of P is not 2-regular, and thus $X = X_{P\cap M}$ is not diagonally split. Note that P is smooth.

References

[BHH+] Matthias Beck, Christian Haase, Akihiro Higashitani, Johannes Hofscheier, Katharina Jochemko, Lukas Katthän, and Mateusz Michałek, *Smooth centrally symmetric polytopes in dimension 3 are IDP*, Annals of Combinatorics, to appear.

[BK05] Michel Brion and Shrawan Kumar, *Frobenius splitting methods in geometry and representation theory*, Progress in Mathematics, vol. 231, Birkhäuser Boston, Inc., Boston, MA, 2005.

[CHP+18] Jed Chou, Milena Hering, Sam Payne, Rebecca Tramel, and Ben Whitney, *Diagonal splittings of toric varieties and unimodularity*, Proc. Amer. Math. Soc. **146** (2018), no. 5, 1911–1920.

[CLNP18] Federico Castillo, Fu Liu, Benjamin Nill, and Andreas Paffenholz, *Smooth polytopes with negative Ehrhart coefficients*, J. Combin. Theory Ser. A **160** (2018), 316–331.

[CLO15] David A. Cox, John Little, and Donal O'Shea, *Ideals, Varieties, and Algorithms: An introduction to computational algebraic geometry and commutative algebra*, fourth ed., Undergraduate Texts in Mathematics, Springer, Cham, 2015.

[CLS11] David A. Cox, John B. Little, and Henry K. Schenck, *Toric varieties*, Graduate Studies in Mathematics, vol. 124, American Mathematical Society, Providence, RI, 2011.

[Dan78] Vladimir I. Danilov, *The geometry of toric varieties*, Russian Mathematical Surveys **33** (1978), no. 2, 97.

[Ful93] William Fulton, *Introduction to toric varieties*, Annals of Mathematics Studies, vol. 131, Princeton University Press, Princeton, NJ, 1993, The William H. Roever Lectures in Geometry.

[GH78] Phillip Griffiths and Joseph Harris, *Principles of algebraic geometry*, Wiley-Interscience [John Wiley & Sons], New York, 1978, Pure and Applied Mathematics.

[Har77] Robin Hartshorne, *Algebraic geometry*, Springer-Verlag, New York-Heidelberg, 1977, Graduate Texts in Mathematics, No. 52.

[HNPS08] Christian Haase, Benjamin Nill, Andreas Paffenholz, and Francisco Santos, *Lattice points in Minkowski sums*, Electron. J. Combin. **15** (2008), no. 1, Note 11, 5.

[Koe93] Robert Jan Koelman, *Generators for the ideal of a projectively embedded toric surface*, Tohoku Math. J. (2) **45** (1993), no. 3, 385–392.

[MR85] Vikram B. Mehta and Annamalai Ramanathan, *Frobenius splitting and cohomology vanishing for Schubert varieties*, Ann. of Math. (2) **122** (1985), no. 1, 27–40.

[Mus] Mircea Mustaţă, *Lecture notes on toric varieties*, `http://www-personal.umich.edu/~mmustata/toric_var.html`.

[Oda08] Tadao Oda, *Problems on Minkowski sums of convex lattice polytopes*, 2008, ArXiv preprint arXiv:0812.1418.

[OWR97] *Combinatorial Convexity and Algebraic Geometry*, Oberwolfach Rep. (1997), 20 pages, Abstracts from the Oberwolfach conference held October 26– November 1, 1997, Organized by Günter Ewald, Peter McMullen, Tadao Oda and Richard Stanley, Oberwolfach Reports; 1997, 41.

[OWR07] *Mini-Workshop: Projective Normality of Smooth Toric Varieties*, Oberwolfach Rep. **4** (2007), no. 3, 2283–2319, Abstracts from the mini-workshop held August 12–18, 2007, Organized by Christian Haase, Takayuki Hibi and Diane Maclagan, Oberwolfach Reports. Vol. 4, no. 3.

[Pay09] Sam Payne, *Frobenius splittings of toric varieties*, Algebra Number Theory **3** (2009), no. 1, 107–119.

[Ram87] Annamalai Ramanathan, *Equations defining Schubert varieties and Frobenius splitting of diagonals*, Inst. Hautes Études Sci. Publ. Math. (1987), no. 65, 61–90.

[RR85] Sundararaman Ramanan and Annamalai Ramanathan, *Projective normality of flag varieties and Schubert varieties*, Invent. Math. **79** (1985), no. 2, 217–224.

[Sha13a] Igor R. Shafarevich, *Basic algebraic geometry. 1*, third ed., Springer, Heidelberg, 2013, Varieties in projective space.

[Sha13b] ———, *Basic algebraic geometry. 2*, third ed., Springer, Heidelberg, 2013,

Schemes and complex manifolds, Translated from the 2007 third Russian edition by Miles Reid.

[Sta80] Richard P. Stanley, *The number of faces of a simplicial convex polytope*, Adv. in Math. **35** (1980), no. 3, 236–238.

[Stu97] Bernd Sturmfels, *Equations defining toric varieties*, Algebraic geometry—Santa Cruz 1995, Proc. Sympos. Pure Math., vol. 62, Amer. Math. Soc., Providence, RI, 1997, pp. 437–449.

[Vak] Ravi Vakil, *The rising sea: Foundations of algebraic geometry*, `http://math.stanford.edu/~vakil/216blog/`.

A brief introduction to valuations on lattice polytopes

Katharina Jochemko

Department of Mathematics, Royal Institute of Technology
Stockholm, SE – 100 44, Sweden
E-mail: jochemko@kth.se

These notes are based on a five-lecture summer school course given by the author at the "Summer Workshop on Lattice Polytopes" at Osaka University in 2018. We give a short introduction to the theory of valuations on lattice polytopes. Valuations are a classical topic in convex geometry. The volume plays an important role in many structural results, such as Hadwiger's famous characterization of continuous, rigid-motion invariant valuations on convex bodies. Valuations whose domain is restricted to lattice polytopes are less well-studied. The Betke-Kneser Theorem establishes a fascinating discrete analog of Hadwiger's Theorem for lattice-invariant valuations on lattice polytopes in which the number of lattice points — the discrete volume — plays a fundamental role. From there, we explore striking parallels, analogies and also differences between the world of valuations on convex bodies and those on lattice polytopes with a focus on positivity questions and links to Ehrhart theory.

Keywords: Lattice polytopes, valuations, Betke-Kneser Theorem, translation-invariance, combinatorial positivity, combinatorial mixed valuations, Ehrhart theory.

1. Lattice-invariant valuations

1.1. *Hadwiger's Characterization Theorem*

A **valuation** is a map φ from a family of convex bodies $\mathcal{P}$ in $\mathbb{R}^d$ containing the empty set into an abelian group G such that $\varphi(\emptyset) = 0$ and

$$\varphi(P \cup Q) = \varphi(P) + \varphi(Q) - \varphi(P \cap Q)$$

for all $P, Q \in \mathcal{P}$ for which $P \cup Q, P \cap Q \in \mathcal{P}$. The prototypical example of a valuation is the d-dimensional Euclidean volume $\mathrm{Vol}(P)$ which has many desirable properties. Besides being a valuation, it is rigid-motion invariant, positive, d-**homogeneous** (that is, $\mathrm{Vol}(tP) = t^d\mathrm{Vol}(P)$ for all convex bodies $P \subseteq \mathbb{R}^d$ and $t \geq 0$), and continuous with respect to the Hausdorff metric. A natural question is to determine all real-valued valuations with the same properties. Questions of that kind are a classical theme in valuation theory and an answer to this particular one was given by Hadwiger [12] who proved the following foundational result.

Theorem 1.1 (Hadwiger's Characterization Theorem [12]). *The family of continuous, real-valued, rigid-motion invariant valuations on convex bodies is a $(d+1)$-dimensional vector space spanned by the quermassintegrals.*

The quermassintegrals are the valuations $W_0, W_1, \ldots, W_d$ that are related to the volume via the **Steiner polynomial**

$$\mathrm{Vol}(tP + \mathcal{B}_d) \;=\; \sum_{i=0}^{d} \binom{d}{i} W_i(P)t^i \,.$$

Here, $\mathcal{B}_d$ denotes the unit ball, and "+" denotes the **Minkowski sum** which for two convex bodies P and Q is defined by $P + Q := \{p + q \colon p \in P, q \in Q\}$. It can furthermore be seen that all W_i are positive and, as coefficients of the Steiner polynomial, i-homogeneous.

We continue this section by considering another interesting valuation and recalling fundamentals from Ehrhart theory that serve us as base point and motivation in the following. We state the Betke-Kneser Theorem for lattice-invariant valuations and provide the main proof ideas given in [6]. Section 2 is devoted to translation-invariant valuations and their behavior under dilation. Our main objective is to recover a polynomiality and a reciprocity result due to McMullen [19]. In Section 3, we present a notion of positivity for translation-invariant valuations introduced in [14] that aligns with fundamental results in Ehrhart theory. In Section 4 we introduce combinatorial mixed valuations extending the notion of mixed volumes and address questions of positivity and monotonicity [13]. The purpose of these notes is to give an overview over the content of the summer school course. The focus is on results and proof ideas rather than giving full details (for which references are provided). No specific prerequisites are needed but familiarity with combinatorial concepts, in particular, with (lattice) polytopes is assumed. For further reading we recommend [1, 11, 21, 26].

1.2. *Ehrhart theory and the Betke-Kneser Theorem*

Of central interest in the following is the valuation $\mathrm{E}(P) := |P \cap \mathbb{Z}^d|$ counting the number of lattice points in a polytope $P \subset \mathbb{R}^d$. It is also called the **discrete volume** as it exhibits some strikingly parallel behavior to the volume, as we will see. For example, the discrete volume is certainly not homogeneous; we leave it to the reader to check small examples. However, if we view homogeneity as polynomiality in the dilation factor then, restricted to the class of lattice polytopes, this carries over to counting lattice points. The following result is due to Ehrhart [10] and constitutes the foundation of a field called *Ehrhart theory.*

Theorem 1.2 (Ehrhart [10]). *Let $P \subset \mathbb{R}^d$ be a lattice polytope. Then $|nP \cap \mathbb{Z}^d|$ is given by a polynomial $\mathrm{E}_P(n)$ of degree $\dim P$ for integers $n \geq 0$.*

The polynomial $\mathrm{E}_P(n)$ is called the **Ehrhart polynomial** of P. Since the discrete volume is a valuation on lattice polytopes, also the Ehrhart polynomial $\mathrm{E}_P(n) = \mathrm{E}_0(P) + \mathrm{E}_1(P)n + \cdots + \mathrm{E}_d(P)n^d$ itself as well as its coefficients $P \mapsto \mathrm{E}_i(P)$ define valuations. A fundamental question in Ehrhart theory is to characterize these coefficients. The coefficients E_i are homogeneous, however, in contrast to the quermassintegrals, they can be negative (see, e.g., [1]). Towards a characterization of Ehrhart polynomials, Stanley [23] showed that the coefficients of the Ehrhart polynomial of an r-dimensional lattice polytope with respect to the basis $\binom{n+r}{r}, \binom{n+r-1}{r}, \ldots, \binom{n}{r}$ are always nonnegative integers.

Theorem 1.3 (Stanley's Nonnegativity Theorem [23]). *Let P be a lattice polytope of dimension r. Then there are natural numbers $h_0^*(P), h_1^*(P), \ldots, h_r^*(P)$ such that*

$$\mathrm{E}_P(n) \;=\; h_0^*(P)\binom{n+r}{r} + h_1^*(P)\binom{n+r-1}{r} + \cdots + h_r^*(P)\binom{n}{r}.$$

The vector $h^*(P) = (h_0^*(P), \ldots, h_d^*(P))$ is called the h^*-vector, where $h_i^*(P) := 0$ for $i > r$ and $h_P^*(t) = h_0^*(P) + h_1^*(P)t + \cdots h_d^*(P)t^d$ is called the h^*-polynomial of P. Notice that, in contrast to the coefficients $\mathrm{E}_i(P)$ of the monomial basis, the coefficients $h_i^*(P)$ are not valuations in general. This is due to the fact that the chosen basis depends on the dimension of the polytope.

Unlike the volume, the discrete volume is not rigid-motion invariant. However, it is **lattice-invariant**, that is, invariant under transformations preserving the integer lattice $\mathbb{Z}^d$ (that is, **unimodular transformations**). This property carries over to the Ehrhart polynomial and its coefficients. Again, a natural question is to characterize all such valuations. The Betke-Kneser Theorem gives a characterization of lattice-invariant valuations and explains the particular role of the Ehrhart polynomial in valuation theory.

Theorem 1.4 (Betke-Kneser Theorem [6]). *The family of real-valued, lattice-invariant valuations on lattice polytopes is a $(d+1)$-dimensional vector space spanned by the coefficients of the Ehrhart polynomial.*

In the remainder of this section we outline the approach taken in [6] to prove this theorem.

1.3. *Valuations and groups*

A union $P = P_1 \cup \cdots \cup P_m$ of d-dimensional polytopes $P_1, \ldots, P_m$ is a **dissection** of a d-dimensional polytope P if $\dim(P_i \cap P_j) < d$ for all $i \neq j$. Let $(\tilde{F}^d, +)$ be the free abelian group generated by

$$\{[\![P]\!] \colon P \quad d-\text{dimensional lattice polytope in } \mathbb{R}^d\}$$

and let $\tilde{R}^d$ be the collection of the following two types of relations:

- $[\![P]\!] - [\![T(P)]\!]$ for any d-dimensional lattice polytope P and any unimodular transformation T.
- $[\![P]\!] - \sum_{i=1}^m [\![P_i]\!]$ for any dissection $P = P_1 \cup \cdots \cup P_m$ into lattice polytopes $P_1, \ldots, P_m$.

It turns out that $\tilde{\pi}^d := \tilde{F}^d/\tilde{R}^d$ has a very simple structure, namely that of a cyclic group. Let $\Delta_d = \operatorname{conv}(0, e_1, \ldots, e_d)$ be the d-dimensional standard simplex.

Theorem 1.5 ([6]). *The group $\tilde{\pi}^d$ is the infinite free cyclic group generated by $[\![\Delta_d]\!]$.*

For the proof the concept of visibility can be employed. A face F of a d-dimensional polytope P is **visible** from a point $q \in \mathbb{R}^d$ if $[q,p) \cap P = \emptyset$ for all points $p \in F$. Here, $[q,p)$ denotes the half-open segment $\{tq + (1-t)p \colon 0 < t \leq 1\}$. A face that is not visible is called **invisible**. A point $q \in \mathbb{R}^d$ is **general** with respect to P if q is not contained in any facet defining hyperplane of P. Visible and invisible facets can be used to dissect $\operatorname{conv}(P \cup \{q\})$ in two different ways. For that, let $F_1, \ldots, F_m$ be the facets of P and let $I_q(P) = \{i \in [m] \colon F_m \text{ visible}\}$. Then

$$\operatorname{conv}(P \cup \{q\}) = P \cup \bigcup_{i \in I_q(P)} \operatorname{conv}(F_i \cup \{q\}), \tag{1}$$

$$= \bigcup_{i \notin I_q(P)} \operatorname{conv}(F_i \cup \{q\}), \tag{2}$$

and the right hand side of (1) and (2) define dissections of $\operatorname{conv}(P \cup \{q\})$.

Proof idea of Theorem 1.5. Since every lattice polytope can be triangulated into empty lattice simplices (that is, simplices whose only lattice points are their vertices), it suffices to show that for every empty lattice simplex S, $[\![S]\!] = \mathrm{V}(S)[\![\Delta_d]\!]$ where $\mathrm{V}(S) = d!\mathrm{Vol}(S)$ denotes the normalized volume. The proof is by induction on $\mathrm{V}(S)$. If $\mathrm{V}(S) = 1$ then there is a unimodular transformation T with $T(\Delta_d) = S$, and we are done. If S is an empty lattice simplex with facets $F_1, \ldots, F_m$ and $\mathrm{Vol}S > 1$ then from (1) and (2) it follows that

$$[\![S]\!] = \sum_{i \notin I_q(S)} [\![\operatorname{conv}(F_i \cup \{q\})]\!] - \sum_{i \in I_q(S)} [\![\operatorname{conv}(F_i \cup \{q\})]\!] .$$

If now $q \in \mathbb{Z}^d$ is chosen in such a way that $\mathrm{Vol}(\operatorname{conv}(F_i \cup \{q\}q)) < \mathrm{Vol}(S)$ for all facets F_i then the claim follows by induction. In [6] such a q was explicitly constructed. □

Let $(F^d, +)$ be the free abelian group with generators

$$\{[\![P]\!] \colon P \text{ lattice polytope in } \mathbb{R}^d\}$$

and let R^d be the collection of relations:

- $[\![P]\!] - [\![T(P)]\!]$ for any lattice polytope P and any unimodular transformation T.
- $[\![P]\!] - \sum_{\emptyset \neq I \subseteq [m]} (-1)^{|I|-1} [\![\bigcap_{i \in I} P_i]\!]$ for any union $P = P_1 \cup \cdots \cup P_m$ such that $\bigcap_{i \in I} P_i$ is a lattice polytope for all $I \neq \emptyset$.

Let $\pi^d := F^d/R^d$. The second condition corresponds to the **inclusion-exclusion property** that is satisfied by any valuation φ on lattice polytopes [5, 20, 25]; for every union $P = P_1 \cup \cdots \cup P_m$ such that $\bigcap_{i \in I} P_i$ is a lattice polytope for all $I \neq \emptyset$

$$\varphi(P) = \sum_{\emptyset \neq I \subseteq [m]} (-1)^{|I|-1} \varphi\left(\bigcap_{i \in I} P_i\right).$$

It follows that every lattice-invariant valuation φ corresponds to a unique homomorphism of abelian groups $\bar{\varphi}\colon \pi^d \to G$ defined by $\bar{\varphi}([\![P]\!]) = \varphi(P)$. It turns out that also π^d has a very simple structure which can be seen by similar arguments as in Theorem 1.5.

Theorem 1.6 ([6]). *The group π^d is a free abelian group with generators $\{[\![\Delta_i]\!]\}_{i=0,1,\ldots,d}$.*

An immediate corollary is the following.

Corollary 1.1 ([6]). *Every lattice-invariant valuation is uniquely determined by its values on the standard simplices $\{[\![\Delta_i]\!]\}_{i=0,1,\ldots,d}$.*

Putting the pieces together we are now ready for the proof of the Betke-Kneser Theorem.

Proof of the Betke-Kneser Theorem. By Corollary 1.1, the space of real-valued rigid-motion invariant valuations is a vector space of dimension $d+1$. Thus, by observing that the coefficients of the Ehrhart polynomial are homogeneous of degrees $0, 1, \ldots, d$ and therefore linearly independent, the proof is complete. □

2. Translation-invariant valuations

2.1. *Polynomiality*

In the following let Λ denote $\mathbb{R}^d$ or $\mathbb{Z}^d$ and let $\mathcal{P}(\Lambda)$ be the family of polytopes with vertices in Λ called **Λ-polytopes**. A valuation $\varphi\colon \mathcal{P}(\Lambda) \to G$ is called **translation-invariant** (or a **Λ-valuation**) if $\varphi(P+t) = \varphi(P)$ for all $P \in \mathcal{P}(\Lambda)$ and all $t \in \Lambda$. Examples of $\mathbb{R}^d$-valuations include the volume and the **Euler characteristic** χ which evaluates to 1 on non-empty polytopes. An important example of a $\mathbb{Z}^d$-valuation is the discrete volume. McMullen [19] generalized Ehrhart's polynomiality theorem [10] to translation-invariant valuations.

Theorem 2.1 (McMullen [19]). *Let $\varphi\colon \mathcal{P}(\Lambda) \to G$ be a translation-invariant valuation and let $P \in \mathcal{P}(\Lambda)$ be a Λ-polytope. Then the function $\varphi(nP)$ agrees with a polynomial $\varphi_P(n)$ of degree at most* $\dim P$ *for all integers $n \geq 0$.*

Here, a polynomial is defined in the following combinatorial way. Let $G^{\mathbb{Z}}$ denote the collection of functions from $\mathbb{Z}$ to G. The shift operator $S\colon G^{\mathbb{Z}} \to G^{\mathbb{Z}}$ is defined by $(Sf)(n) = f(n+1)$ for all $f\colon \mathbb{Z} \to G$. The difference operator Δ is defined through

$(\Delta f)(n) = f(n+1) - f(n)$, that is, $\Delta = S - I$. A function $\mathbb{Z} \to G$ is a **polynomial** of degree at most d if and only if $\Delta^{d+1} f \equiv 0$. In terms of generating polynomials this can be characterized in the following way (see, e.g., Stanley [24]).

Theorem 2.2. *A function $f\colon \mathbb{Z} \to G$ is a polynomial of degree at most d if and only if*

$$\sum_{n\geq 0} f(n)t^n \;=\; \frac{h(t)}{(1-t)^{d+1}}$$

as rational functions where $h(t) = h_0 + h_1 t + \cdots + h_d t^d \in G[t]$ is a polynomial with $\deg h \leq d$. Equivalently,

$$f(n) \;=\; h_0\binom{n+d}{d} + h_1\binom{n+d-1}{d} + \cdots + h_d\binom{n}{d}$$

for all $n \geq 0$.

In particular, if f is the Ehrhart polynomial of a lattice polytope, then h is its h^*-polynomial. We will outline a proof of Theorem 4.3 given in [14] that uses Theorem 2.2 and provide an interpretation of the coefficients $h_0, h_1, \ldots, h_d$ in the case that $f(n) = \varphi_P(n)$ for arbitrary translation-invariant valuations φ and Λ-polytopes P.

Since every Λ-polytope can be triangulated into Λ-simplices, by the inclusion-exclusion property it is sufficient to prove polynomiality of $\varphi(nP)$ for arbitrary Λ-simplices P. However, to avoid inclusion-exclusion and thus considering lower dimensional polytopes, we consider half-open polytopes and half-open decompositions. This will come in handy in regards to positivity questions in later sections. Let P be a polytope with facets $F_1, \ldots, F_m$ and let $q \in \operatorname{aff} P$ be a point in general position to P. We obtain the **half-open polytope** $\mathrm{H}_q P$ by removing all visible faces of P:

$$\mathrm{H}_q P \;=\; P \setminus \bigcup_{i \in I_q(P)} F_i\,.$$

Proposition 2.1 ([15]). *Let $P = P_1 \cup \cdots \cup P_m$ be a dissection and let $q \in \mathbb{R}^d$ be general with respect to P_i for all $1 \leq i \leq m$. Then*

$$\mathrm{H}_q P \;=\; \mathrm{H}_q P_1 \sqcup \cdots \sqcup \mathrm{H}_q P_m$$

is a partition. In particular, if $q \in \operatorname{relint} P$, then P can be decomposed into half-open polytopes.

Any valuation on Λ-polytopes can be extended to half-open polytopes in a natural way by using the inclusion-exclusion property [20], namely

$$\varphi(\mathrm{H}_q P) := \varphi(P) - \sum_{\emptyset \neq J \subseteq I_q(P)} (-1)^{|J|-1}\varphi(F_J)\,,$$

where $F_J := \bigcap_{i\in J} F_i$. The following is an immediate consequence of Proposition 2.1.

Corollary 2.1. *Let $P = P_1 \cup \cdots \cup P_m$ be a triangulation into Λ-simplices and let $q \in \operatorname{relint} P$ be general with respect to P_i for all $1 \leq i \leq m$. Then*

$$\varphi(P) \;=\; \varphi(\mathrm{H}_q P_1) + \cdots + \varphi(\mathrm{H}_q P_k)\,.$$

Thus, since every Λ-polytope can be triangulated into Λ-simplices, Theorem 4.3 is a direct consequence of the following result.

Theorem 2.3. *Let $\tilde{S}$ be an half-open simplex in $\mathcal{P}(\Lambda)$ and φ be a translation-invariant valuation. Then $\varphi(n\tilde{S})$ agrees with a polynomial $\varphi_{\tilde{S}}$ of degree at most d for integers $n \geq 0$.*

Proof idea. We illustrate the argument given in [14] on a half-open triangle. We partition the dilated half-open triangle $\tilde{S}$ into congruent half-open triangles $\tilde{S}$ and $\hat{S}$ as shown in Figure 1.

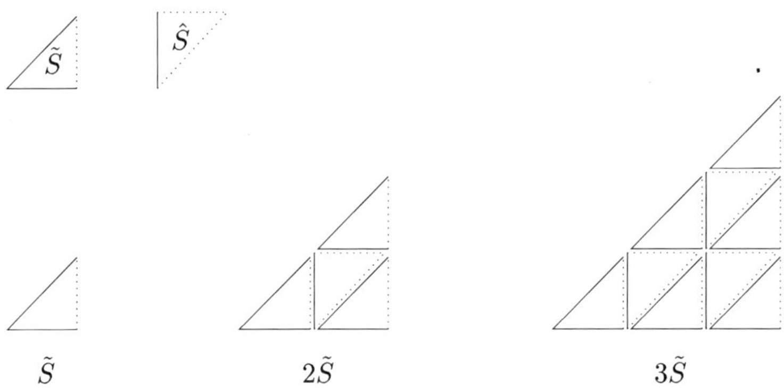

Fig. 1. Decomposition of integer dilates of $\tilde{S}$ into translates of $\tilde{S}$ and $\hat{S}$.

By translation-invariance and Corollary 2.1,

$$\varphi_{\tilde{S}}(n) \;=\; \varphi(\tilde{S})\binom{n+1}{2} + \varphi(\hat{S})\binom{n}{2}.$$

This method can be generalized to higher dimensions. More precisely, let $\tilde{S} = \mathrm{H}_q S$ for some Λ-simplex S and general point q with respect to S and let $I = I_q(S)$. Let $F_1, \ldots, F_{d+1}$ be its facets and $v_1, \ldots, v_{d+1}$ be the vertices labeled in such a way that $v_i \notin F_i$. Furthermore, let $\bar{v}_i := (v_i, 1)^T$ for all vertices v_i. Extending a standard method in Ehrhart theory we consider the half-open parallelepiped

$$\Pi \;=\; \{\mu_1\bar{v}_1 + \cdots + \mu_{d+1}\bar{v}_{d+1} \colon 0 \leq \mu_i < 1 \text{ if } i \notin I, 0 < \mu_i \leq 1 \text{ otherwise}\} \subset \mathbb{R}^{d+1}\,.$$

Then the coefficients can be expressed in terms of the values of φ on the half-open hypersimplices $\Pi_i \;=\; \Pi \cap \{x \in \mathbb{R}^{d+1} \colon x_{d+1} \;=\; i\}$; see [14] for further details. We collect the full statement in the next corollary for later reference. □

Corollary 2.2 ([14]). *Let $\tilde{S}$ be an half-open Λ-simplex and φ be a translation-invariant valuation. Then with the notation as in the proof of Theorem 2.3,*

$$\varphi(n\tilde{S}) = \sum_{r=0}^{d} \varphi(\Pi_r)\binom{n+d-r}{d}.$$

2.2. *Reciprocity*

By Theorem 4.3, $\varphi(nP)$ agrees with a polynomial $\varphi_P(n)$ for all integers $n \geq 0$. It is a natural question to ask for an interpretation for evaluating this polynomial at negative integers. A fundamental result in Ehrhart theory is the Ehrhart-Macdonald reciprocity theorem [10, 16] which relates the evaluation at negative integers to counting interior lattice points. The following theorem due to McMullen [19] generalizes the reciprocity to translation-invariant valuations.

Theorem 2.4 (McMullen [19]; Ehrhart-Macdonald reciprocity [10, 16]). *Let P be a lattice polytope and φ be a Λ-valuation. Then*

$$\varphi_P(-n) \;=\; (-1)^{\dim P}\varphi(\text{relint}(-nP)).$$

Here, $\varphi(\text{relint}P) := \sum_{F\subseteq P}(-1)^{\dim P-\dim F}\varphi(F)$.

Proof. We illustrate the proof idea given in [3, Chapter 5] considering the case that P is a lattice polygon. By considering the polygon $Q := mP$ we invite the reader to convince herself that it is sufficient to prove that $\varphi_P(-1) = (-1)^{\dim P}\varphi(\text{relint}(-P))$. Let again $\tilde{S}$ be a half-open lattice triangle and let $\hat{S}$ be the half-open triangle as in the proof of Theorem 2.3 (see Figure 1). We saw that

$$\varphi_{\tilde{S}}(n) \;=\; \varphi(\tilde{S})\binom{n+1}{2} + \varphi(\hat{S})\binom{n}{2}.$$

Evaluating the polynomial on the right hand side at -1 yields $\varphi(\hat{S})$. Assuming that $\tilde{S}$ was obtained from the triangle S by removing all faces visible from a point q we then observe that $\hat{S}$ is obtained from $-S$ by removing all facets that are invisible from $-q$. Now let $P = \tilde{S}_1 \cup \ldots \cup \tilde{S}_m$ be a decomposition into half-open Λ-triangles. Then

$$\varphi_P(-1) = \sum \varphi_{\tilde{S}_i}(-1) = \sum (-1)^d \varphi(\hat{S}_i) = \varphi(\text{relint}(-P)),$$

as illustrated in Figure 2. This argument can be generalized to higher dimensions. □

3. Combinatorial positive valuations

3.1. *Combinatorial positivity and combinatorial monotonicity*

In the following, let G always be an ordered abelian group. In the last section we have seen that for every translation-invariant valuation $\varphi\colon \mathcal{P}(\Lambda) \to G$ and every

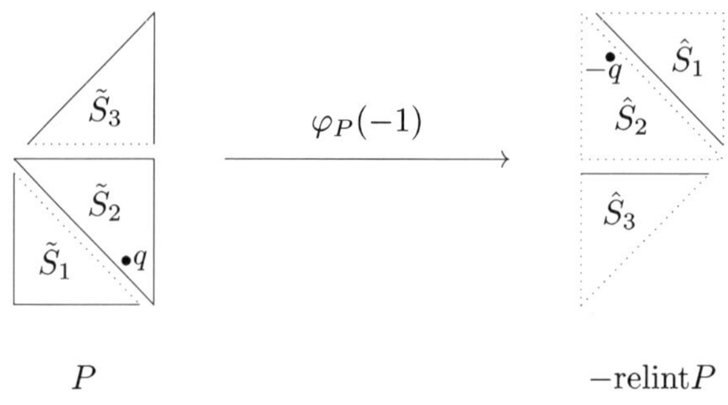

Fig. 2. Half-open partitions of P and relint$(-P)$.

Λ-polytope P of dimension r the function $\varphi(nP)$ is given by a polynomial $\varphi_P(n)$ of degree at most r for all integers $n \geq 0$. Thus, by Theorem 2.2 there exist coefficients $h_0^\varphi(P), h_1^\varphi(P), \ldots, h_r^\varphi(P) \in G$ such that

$$\varphi_P(n) \;=\; h_0^\varphi(P)\binom{n+r}{r} + h_1^\varphi(P)\binom{n+r-1}{r} + \cdots + h_r^\varphi(P)\binom{n}{r}.$$

The vector $h^\varphi(P) = (h_0^\varphi(P), h_1^\varphi(P), \ldots, h_d^\varphi(P))$ is called the **h^*-vector of P with respect to φ** where we set $h_i^\varphi(P) := 0$ for $i > r$. The polynomial $h_P^\varphi(t) = \sum_{i=0}^r h_i^\varphi(P)t^i$ is called the **h^*-polynomial of P with respect to φ**. By Stanley's Nonnegativity Theorem (Theorem 1.3), the h^*-polynomial has only nonnegative coefficients when considering the discrete volume of lattice polytopes. Moreover Stanley [22] showed the following.

Theorem 3.1 ([22]). *For all lattice polytopes $P, Q \in \mathcal{P}(\mathbb{Z}^d)$ satisfying $P \subseteq Q$*

$$h_i^*(P) \;\leq\; h_i^*(Q) \qquad \text{for all } i = 0, \ldots, d\,.$$

That is, the (Ehrhart) h^*-vector is componentwisely monotone with respect to inclusion.

In accordance with Stanleys results (Theorem 1.3 and Theorem 3.1) we define a translation-invariant valuation φ to be **combinatorially positive** if $h_i^\varphi(P) \geq 0$ for all $P \in \mathcal{P}(\mathbb{Z}^d)$ and all i, and **combinatorially monotone** if $h_i^\varphi(P) \leq h_i^\varphi(Q)$ for all i whenever $P \subseteq Q$. Examples of combinatorial positive and combinatorially monotone valuations are, of course, the discrete volume, the volume as it can be seen that the corresponding h^*-polynomial equals (up to a scalar) the Eulerian polynomial (see, e.g., [1]), and the so-called solid-angle polynomials [2]. It is left as an exercise to the reader that the Euler characteristic is not combinatorially positive.

In the spirit of the classification results in Section 3 we would like to characterize combinatorial positive and combinatorial monotone valuations. It turns out that both notions are equivalent and a simple characterization can be given [14].

Theorem 3.2 **([14]).** *Let φ be a translation-invariant valuation. Then the following are equivalent:*

(i) *φ is combinatorially positive.*
(ii) *φ is combinatorially monotone.*
(iii) *$\varphi(\text{relint}\Delta) \geq 0$ for all simplices $\Delta \in \mathcal{P}(\Lambda)$.*

Proof idea. We outline the proof given in [14]. The h^*-vector with respect to a valuation can be naturally extended to half-open Λ-polytopes. If $\tilde{S}$ is a half-open simplex of dimension r. Then, by Corollary 2.2, $h_r^\varphi(\tilde{S}) = \varphi(\Pi_r)$. In particular, if $S = \tilde{S}$ is a closed simplex, then $h_r^\varphi(S) = \varphi(-\text{relint}S)$ from which we see $(i) \Rightarrow (iii)$. The implication $(ii) \Rightarrow (i)$ is clear since the empty polytope is contained in any other polytope. If $P = \text{H}_q P_1 \sqcup \cdots \sqcup \text{H}_q P_m$ is a half-open decomposition, by the inclusion-exclusion principle it follows that $h^\varphi(P) = h^\varphi(\text{H}_q P_1) + \cdots + h^\varphi(\text{H}_q P_m)$. In particular, to show positivity of $h^\varphi(P)$ it is sufficient to assume that $P = \tilde{S}$ is an half-open simplex and $h_i^\varphi(\tilde{S}) = \varphi(\Pi_i)$ for all i. We observe that Π_i is a partially open polytope, (that is a polytope with certain faces removed). Using a triangulation of the corresponding closed polytope into Λ-simplices it can therefore be partitioned as $\Pi_i = \bigsqcup_l \text{relint}T_l$ where T_l are simplices in the triangulation contained in Π_i. Thus, $\varphi(\Pi_i) = \sum_l \varphi(\text{relint}T_l)$ and, assuming (iii), we obtain combinatorial positivity. To get combinatorial monotonicity a slightly more refined argument is needed and we refer the reader to [14]. □

3.2. *Combinatorially positive lattice-invariant valuations*

In case of lattice-invariant valuations on lattice polytopes combinatorial positivity has an even simpler characterization.

Theorem 3.3 **([14]).** *A lattice-invariant valuation $\varphi\colon \mathcal{P}(\mathbb{Z}^d) \to G$ is combinatorially positive if and only if $\varphi(\text{relint}\Delta_i) \geq 0$ for all $0 \leq i \leq d$.*

Proof. The proof will become apparent from the next two results; namely Theorem 3.4 and Lemma 3.1. □

That is, in case of lattice-invariant valuations condition (iii) in Theorem 3.2 needs only to be checked on the standard unimodular simplices. In particular, the cone of combinatorial positive valuations in polyhedral. We can describe this cone more concretely by considering the coefficients of the Ehrhart polynomial with respect to the basis $\binom{n-1}{0}, \binom{n-1}{1}, \ldots, \binom{n-1}{d}$.

$$\text{E}_P(n) \;=\; f_0^*(P)\binom{n-1}{0} + f_1^*(P)\binom{n-1}{1} + \cdots + f_d^*(P)\binom{n-1}{d}.$$

These coefficients $f_i^*(P)$ where first considered by Breuer [7] who coined the name **f^*-vectors** and showed that they are always nonnegative on relatively open complexes. The following theorem completely characterizes combinatorially positive lattice-invariant valuations.

Theorem 3.4 **([14]).** *Let φ be a lattice invariant valuation. Then φ is combinatorially positive if and only if*

$$\varphi \; = \; \alpha_0 f_0^* + \alpha_1 f_1^* + \cdots + \alpha_d f_d^*$$

for some $\alpha_0, \alpha_1, \ldots, \alpha_d \geq 0$.

This is parallel to the characterization of positive and monotone continuous rigid-motion invariant valuations which is a direct consequence of Hadwiger's Characterization [12].

Theorem 3.5. *Let φ be a real-valued continuous rigid-motion invariant valuation on convex bodies. Then φ is positive or monotone if and only if*

$$\varphi = \alpha_0 W_0 + \alpha_1 W_1 + \cdots + \alpha_d W_d$$

for some $\alpha_0, \alpha_1, \ldots, \alpha_d \geq 0$.

The following lemma clarifies the role of the f^*-vector and is left as an enjoyable exercise to the reader.

Lemma 3.1 **([14]).** *For all $0 \leq i, j \leq d$,*

$$f_j^*(\mathrm{relint}(\Delta_i)) \; = \; \delta_{i,j} \, .$$

In particular, $f_0^*, f_1^*, \ldots, f_d^*$ form a basis of the vector space of lattice-invariant valuations and for any lattice-invariant valuations φ,

$$\varphi \; = \; \varphi(\mathrm{relint}\Delta_0) f_0^* + \varphi(\mathrm{relint}\Delta_1) f_1^* + \cdots + \varphi(\mathrm{relint}\Delta_d) f_d^* \, .$$

From Lemma 3.1 and Theorem 3.2, it follows that Theorem 3.4 is equivalent to showing that $f_0^*, \ldots, f_d^*$ are combinatorially positive valuations. We will show even more: For all translation-invariant valuations φ consider $f_0^\varphi(P), \ldots, f_d^\varphi(P)$ such that

$$\varphi_P(n) \; = \; \sum_{i=0}^{d} f_i^\varphi(P) \binom{n-1}{i} ,$$

Theorem 3.6 **([14]).** *Let φ be a translation-invariant valuation. Then the following are equivalent:*

(i) φ is combinatorially positive.
(ii) f_i^φ is combinatorially positive for all $0 \leq i \leq d$.

Proof idea. One direction follows from the observation that $\varphi = f_0^\varphi$. For the other direction one can prove for all lattice polytopes P of dimension r that

$$f_{r-k}^\varphi(\mathrm{relint}P) \; = \; \sum_{i=k}^{r} h_i^\varphi(-P) \binom{i}{k}$$

for all $0 \leq k \leq r$ by applying Theorem 2.4. Since φ is combinatorially positive the right hand side of the equation is nonnegative and thus f_{r-k}^φ is combinatorially positive by Theorem 3.2. Further details may be found in [14]. $\square$

Proof of Theorem 3.2. The discrete volume is combinatorially positive by Stanley's Nonnegativity Theorem (Theorem 1.3). Thus $f_0^*, \ldots, f_d^*$ are combinatorially positive by Theorem 3.6. □

3.3. *Weak h^*-monotone valuations*

As will become apparent in the next section, it is also very natural to consider a weaker notion of combinatorial monotonicity that takes into account the dimension of Λ-polytopes. A translation-invariant valuation is called **weakly h^*-monotone** if

$$h_i^{\varphi}(P) \ \leq \ h_i^{\varphi}(Q)$$

for all $0 \leq i \leq d$ whenever $P \subseteq Q$ and $\dim P = \dim Q$. Using similar techniques as in the proof of Theorem 3.2, the following characterization and classification results were obtained in [14]:

Theorem 3.7 ([14]). *Let φ be a translation-invariant valuation. The following are equivalent:*

(i) φ is weakly h^*-monotone.
(ii) $\varphi(\mathrm{relint}\Delta) + \varphi(\mathrm{relint}F) \geq 0$ for every simplex $\Delta \in \mathcal{P}(\Lambda)$ and any facet F of Δ.
(iii) $\varphi(\tilde{S}) \geq 0$ for every half-open Λ-simplex $\tilde{S}$.

Restricted to the class of lattice-invariant valuations weak h^*-monotonicity can be characterized by considering the coefficients of the Ehrhart polynomial in (yet another) basis. Let

$$\mathrm{E}_P(n) \ = \ \tilde{f}_0(P)\binom{n}{0} + \tilde{f}_1(P)\binom{n}{1} + \cdots + \tilde{f}_d(P)\binom{n}{d}.$$

Theorem 3.8 ([14]). *Let φ be a lattice-invariant valuation. Then φ is weakly h^*-monotone if and only if*

$$\varphi \ = \ \alpha_0 \tilde{f}_0 + \alpha_1 \tilde{f}_1 + \cdots + \alpha_d \tilde{f}_d$$

for some $\alpha_0, \alpha_1, \ldots, \alpha_d \geq 0$.

We conclude this section by discussing the relationship of combinatorial positivity and weak h^*-monotonicity and monotonicity ($\varphi(P) \leq \varphi(Q)$ whenever $P \subseteq Q$) and positivity ($\varphi(P) \geq 0$). Clearly, combinatorial monotonicity/positivity implies weak h^*-monotonicity. It can be furthermore seen from Theorem 3.7 by using half-open decomposition that weak h^*-monotonicity implies **monotonicity**. Since every valuation is 0 on the empty polytope, monotonicity implies **positivity**. Figure 3 summarizes the chain of implications. We note that the reverse implications do not hold [14].

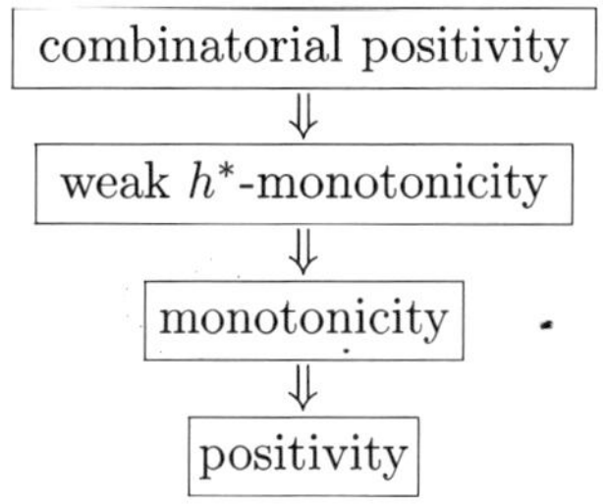

Fig. 3. Combinatorial positivity and its relatives.

4. Combinatorial mixed valuations

A fundamental notion in convex geometry is the mixed volume $\mathrm{MV}_d(P_1, \ldots, P_d)$ of convex bodies $P_1, \ldots, P_d$. It can be defined in different equivalent ways and each definition comes with certain advantages and disadvantages. A classical result is Minkowski's Identity (see, e.g., [11]) which states that $\mathrm{Vol}(t_1 P_1 + \cdots + t_m P_m)$ is given by a homogeneous polynomial of degree d for convex bodies $P_1, \ldots, P_m$ and $t_1, \ldots, t_m \geq 0$. More precisely, with the definition $\mathrm{MV}(P_1, \ldots, P_d) := \frac{1}{d!}[t_1 \cdots t_d]\mathrm{Vol}(t_1 P_1 + \cdots + t_d P_d)$, (that is, $\mathrm{MV}(P_1, \ldots, P_d)$ is defined as the coefficient of the monomial $t_1 \cdots t_d$ divided by $d!$) Minkowski's Identity states that

$$\mathrm{Vol}(t_1 P_1 + \cdots + t_m P_m) = \sum_{j_1, \ldots, j_d = 1}^{m} \mathrm{MV}(P_{j_1}, \ldots, P_{j_d}) t_{j_1} \cdots t_{j_d} .$$

In particular, MV is symmetric in its arguments and, by homogeneity, $\mathrm{MV}(P_1, \ldots, P_m) = 0$ if $m < d$. Furthermore, it can be seen that the mixed volume is multilinear in each argument.

An alternative definition is the following:

$$\mathrm{MV}(P_1, \ldots, P_m) := \frac{1}{m!} \sum_{I \subseteq [m]} (-1)^{d-|I|} \mathrm{Vol}(P_I)$$

where $P_\emptyset = \{0\}$ and $P_I = \sum_{i \in I} P_i$ for all $\emptyset \neq I \subseteq [m]$. This definition has the advantage that it is not necessarily trivial on less than d arguments and can be extended to valuations. We define the **combinatorial mixed valuation** [13] associated to a translation-invariant valuation φ of $P_1, \ldots, P_r \in \mathcal{P}(\Lambda)$ to be

$$\mathrm{CM}_r \varphi(P_1, \ldots, P_r) := \sum_{I \subseteq [r]} (-1)^{r-|I|} \varphi(P_I) ,$$

where in the following we suppress the index r in $\mathrm{CM}_r \varphi$ whenever the number of arguments is clear from the context. Similar to mixed volumes in Minkowski's Identity, combinatorial mixed volumes can be interpreted as coefficients of a polynomial. The Bernstein-McMullen Theorem [4, 17, 19] states that for any translation-invariant valuation φ and arbitrary Λ-polytopes $P_1, \ldots, P_r$, $\varphi(n_1 P_1 + \cdots n_r P_r)$ is given by a polynomial for integers $n_1, \ldots, n_r \geq 0$. The following theorem is a discrete version of Minkowski's Identity.

Theorem 4.1 ([13]). *Let $P_1, \ldots, P_r \in \mathcal{P}(\Lambda)$ and φ be a translation-invariant valuation. Then*

$$\varphi(n_1P_1 + \cdots + n_rP_r) = \sum_{\mathbf{k}\in\mathbb{Z}^r_{\geq 0}} \mathrm{CM}\varphi(P_1[k_1], \cdots, P_r[k_r])\binom{n_1}{k_1}\cdots\binom{n_r}{k_r}$$

where $P_i[k_i]$ denotes the k_i-fold appearance of P_i.

In particular, since the coefficients are uniquely determined,

$$P_i \mapsto \varphi(P_1[k_1], \cdots, P_r[k_r])$$

defines a translation-invariant valuation for all i and all $\mathbf{k}$.

A desirable, non-trivial property of the mixed volume is that it is always non-negative and, moreover, **monotone** with respect to inclusion, that is

$$0 \leq \mathrm{MV}(P_1, \ldots, P_r) \leq \mathrm{MV}(Q_1, \ldots, Q_r)$$

whenever $P_i \subseteq Q_i$ for all $1 \leq i \leq r$. It turns out, that this is true for combinatorial mixed valuations as well under the assumption of weak h^*-monotonicity.

Theorem 4.2 ([13]). *Let φ be a weakly h^*-monotone valuation. Then*

$$0 \leq \mathrm{CM}\varphi(P_1, \ldots, P_r) \leq \varphi(Q_1, \ldots, Q_r)$$

whenever $P_1 \subseteq Q_i$ for all $1 \leq i \leq r$.

The goal of this section is to proof Theorem 4.2. We will use the language of the polytope algebra introduced by McMullen [18].

Let $\mathbb{Z}\mathcal{P}(\Lambda)$ be the free abelian group with generators $[\![P]\!]$ for all $P \in \mathcal{P}(\Lambda)$. Let U be the subgroup generated by elements of the form

- $[\![P \cup Q]\!] + [\![P \cap Q]\!] - [\![P]\!] - [\![Q]\!]$ for all $P, Q \in \mathcal{P}(\Lambda)$ for which $P \cap Q, P \cup Q \in \mathcal{P}(\Lambda)$ and $[\![\emptyset]\!] = 0$, and
- $[\![P + t]\!] - [\![P]\!]$ for all $P \in \mathcal{P}(\Lambda)$ and $t \in \Lambda$.

Then the **polytope algebra** is defined as $\Pi(\Lambda) := \mathcal{P}(\Lambda)/U$. The polytope algebra has the universal property that for every translation-invariant valuation $\varphi\colon \mathcal{P}(\Lambda) \to G$ there is a unique homomorphism of abelian groups $\bar{\varphi}\colon \Pi(\Lambda) \to G$ such that $\bar{\varphi}([P]) = \varphi(P)$, and conversely. The product structure on $\Pi(\Lambda)$ is defined by the Minkowski sum of polytopes, that is, for $P, Q \in \mathcal{P}(\Lambda)$ we have $[\![P]\!] \cdot [\![Q]\!] := [\![P + Q]\!]$. Even though it is not relevant in the following, it allows us to express the combinatorial mixed valuation in a particularly nice form. The following corollary is obtained by applying Theorem 4.1 to the universal valuation $P \mapsto [\![P]\!]$.

Corollary 4.1 ([13]). *Let $P_1, \ldots, P_r$ be Λ-polytopes. Then*

$$[\![n_1P_1 + \cdots + n_rP_r]\!] = \sum_{\mathbf{k}\in\mathbb{Z}^r_{\geq 0}} \mathrm{CM}[\![P_1[k_1], \ldots, P_r[k_r]]\!]\binom{n_1}{k_1}\cdots\binom{n_r}{k_r},$$

where $\mathrm{CM}[\![P_1, \ldots, P_r]\!] := \sum_{I\subseteq[r]}(-1)^{r-|I|}[\![P_I]\!] = \Pi_{i=1}^r([\![P_i]\!] - 1)$.

In order to proof Theorem 4.2 we will need to interpret combinatorial mixed volumes geometrically. A Minkowski sum $P = P_1 + \cdots + P_r$ is called **exact** if $\dim P = \dim P_1 + \cdots + \dim P_r$. If $P_1, \ldots, P_r$ are simplices then their Minkowski sum is called a **cylinder** whenever it is exact. It is further called a k-cylinder if $\dim P_i > 0$ for exactly k-many summands P_i. For $k = 0, \ldots, d$ let

$$\tilde{Z}_k \; := \; \mathbb{Z}_{\geq 0}\{[\![S]\!] : S \text{ half-open } k\text{-cylinder}\}$$

be the cone generated by half-open k-cylinders, and let

$$W = \mathbb{Z}_{\geq 0}\{[\![\operatorname{relint} S]\!] + [\![\operatorname{relint} F]\!] : S \in \mathcal{P}(\Lambda) \text{ simplex}, F \subseteq S \text{ facet}\}.$$

Then, by Theorem 3.7, every weakly h^*-monotone φ evaluates nonnegatively on W. Thus, by the following lemma which is left as an exercise to the reader, every such valuation also evaluates nonnegatively on cylinders.

Lemma 4.1 ([13]).

$$\tilde{Z}_d \subseteq \tilde{Z}_{d-1} \subseteq \cdots \subseteq \tilde{Z}_1 = W.$$

It therefore suffices to prove that combinatorial mixed valuations are contained in $\tilde{Z}_k$ for some k.

Proposition 4.1 ([13]). *Let $\tilde{S}$ be a half-open simplex. Then*

$$[\![n\tilde{S}]\!] \; = \; \zeta_0 + \zeta_1 \binom{n}{1} + \cdots + \zeta_d \binom{n}{d}$$

where $\zeta_k \in \tilde{Z}_k$ for all $k = 0, \ldots, d$.

Proof. We illustrate the proof idea on a 2-dimensional half-open triangle. Similar to the proof of Theorem 2.3, the key idea is to partition the integer dilates in a suitable way with congruent pieces. The n-th dilate of the half-open triangle $\tilde{S}$ in Figure 4 can be dissected into $\binom{n}{1}$ translates of $\tilde{S}$ and $\binom{n}{2}$ translates of a half-open parallelepiped which is exact since it is a Minkowski sum of two segments. With the notation as in Figure 4, we then have

$$\varphi \; = \; \varphi(\tilde{S})\binom{n}{1} + \varphi(T)\binom{n}{2}.$$

In higher dimensions, one can assume that $\tilde{S}$ is obtained by removing facets from a simplex $S = \{x \in \mathbb{R}^d : 0 \leq x_1 \leq \cdots \leq x_d \leq 1\}$. For a generic point $p \in nS$, let $\bar{p} = (\lfloor x_1 \rfloor, \lfloor x_2 \rfloor, \ldots, \lfloor x_d \rfloor)$. Then $p - \bar{p}$ is contained in a cylinder of the form

$$\left\{ x \in \mathbb{R}^d : \begin{array}{c} 0 \leq x_1 \leq \cdots \leq x_{b_1} \leq 1 \\ 0 \leq x_{b_1+1} \leq \cdots \leq x_{b_2} \leq 1 \\ \vdots \\ 0 \leq x_{b_{n-1}+1} \leq \cdots \leq x_{b_n} \leq 1 \end{array} \right\},$$

for some $0 \leq b_1 \leq \cdots \leq b_{n-1} \leq b_n = d$. These cylinders dissect nS, and counting the number of occurrences up to translation of every such cylinder yields the result. For full details see [14]. □

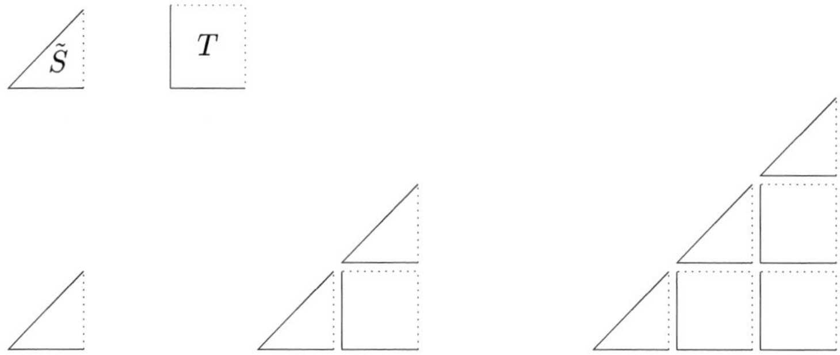

$\tilde{S}$ $2\tilde{S}$ $3\tilde{S}$

Fig. 4. Decomposition of integer dilates of $\tilde{S}$ into translates of $\tilde{S}$ and T.

More generally, by applying Proposition 4.1 to every Minkowski summand in a cylinder we obtain the following.

Corollary 4.2 ([13]). *Let $S_1 + \cdots + S_r$ be a half-open cylinder. Then there exist $\zeta_{\mathbf{k}} \in \tilde{Z}_{|k|}$ such that*

$$[\![n_1 S_1 + \cdots + n_r S_r]\!] = \sum_{\mathbf{k} \in \mathbb{Z}^k_{\geq 0}} \zeta_{\mathbf{k}} \binom{n_1}{k_1} \cdots \binom{n_k}{k_r}$$

where $\zeta_{\mathbf{k}} \in \tilde{Z}_{|\mathbf{k}|}$ for all $\mathbf{k}$.

By Corollary 4.2 it is therefore sufficient to see that every Minkowski sum $n_1 P_1 + \cdots + n_r P_r$ has a dissection into cylinders $R_i = R_{i_1} + \cdots + R_{i_r}$ with $R_{i_j} \subseteq P_j$. This can be achieved by the **Cayley-trick** (see [9]): the **Cayley polytope** of polytopes $P_1, \ldots, P_r \subset \mathbb{R}^d$ is defined as

$$\operatorname{Cay}(P_1, \ldots, P_r) := \operatorname{conv}\left(\bigcup_i P_i \times \{e_i\}\right) \subseteq \mathbb{R}^d \times \mathbb{R}^r .$$

One observes that $\frac{1}{r}(P_1 + \cdots + P_r) = \operatorname{Cay}(P_1, \ldots, P_r) \cap W$ where $W = \{(x, y) \in \mathbb{R}^d \times \mathbb{R}^r \colon y_i = \frac{1}{r} \text{ for all } i\}$. Now it can be shown that every triangulation of $\operatorname{Cay}(P_1, \ldots, P_r)$ restricts to a subdivision of $P_1 + \cdots + P_r$ into cylinders as we wanted.

Theorem 4.3 ([13]). *Let $P_1, \ldots, P_r, Q_1, \ldots, Q_r \subseteq \mathcal{P}(\Lambda)$ such that $P_i \subseteq Q_i$ for all $1 \leq i \leq r$. Then*

$$[\![n_1 Q_1 + \cdots + n_r Q_r]\!] - [\![n_1 P_1 + \cdots + n_r P_r]\!] = \sum_{\mathbf{k} \in \mathbb{Z}^r_{\geq 0}} \zeta_{\mathbf{k}} \binom{n_1}{k_1} \cdots \binom{n_r}{k_r}$$

with $\zeta_{\mathbf{k}} \in \tilde{Z}_{|\mathbf{k}|}$.

Proof idea. Since $P_i \subseteq Q_i$ for all $1 \leq i \leq r$ it follows that $\mathrm{Cay}(P_1, \ldots, P_r) \subseteq \mathrm{Cay}(Q_1, \ldots, Q_r)$. If $\dim(P_1 + \cdots + P_d) = \dim(Q_1 + \cdots + Q_r)$ then both Cayley polytopes have the same dimension. Triangulating $\mathrm{Cay}(P_1, \ldots, P_r)$ and extending it to a triangulation of $\mathrm{Cay}(Q_1, \ldots, Q_r)$ yields a subdivision into cylinders of $P_1 + \cdots + P_r$ that gets extended to a subdivision into cylinders of $Q_1 + \cdots + Q_r$. Choosing a general point in $P_1 + \cdots + P_r$ yields a partition of $(Q_1 + \cdots + Q_r) \setminus (P_1 + \cdots + P_r)$ into half-open cylinders and the result follows with Lemma 4.2. If $\dim(P_1 + \cdots + P_d) < \dim(Q_1 + \cdots + Q_r)$ a slightly more refined argument has to be applied. See [14] for the full details. □

We are now ready to proof the main theorem of this section (Theorem 4.2).

Proof of Theorem 4.2. By Theorem 4.1, $\zeta_{\mathbf{k}} = \mathrm{CM}[\![Q_1[k_1], \ldots, Q_r[k_r]]\!] - \mathrm{CM}[\![P_1[k_1], \ldots, P_r[k_r]]\!]$ in Theorem 4.3 which is evaluated to a nonnegative number for every weakly h^*-monotone valuation, by Lemma 4.1. □

5. Outlook

The route taken in this course only showed a glimpse of the past and current research on valuations on lattice polytopes and is strongly biased by the authors own research. To learn more about current research on valuations on lattice polytopes we recommend [8].

Acknowledgements: The author was supported by the Knut and Alice Wallenberg Foundation. She would like to thank Takayuki Hibi and Akiyoshi Tsuchiya for the organization and the invitation to speak at the "Summer Workshop on Lattice Polytopes" at Osaka University, and Sebastian Manecke, Raman Sanyal and Liam Solus for many helpful comments on this manuscript.

References

[1] M. Beck and S. Robins, *Computing the Continuous Discretely: Integer-point Enumeration in Polyhedra* Undergraduate Texts in Mathematics (Springer, New York, 2007).

[2] M. Beck, S. Robins and S. V. Sam, Positivity theorems for solid-angle polynomials, *Beiträge Algebra Geom.* **51**, 493–507 (2010).

[3] M. Beck and R. Sanyal, *Combinatorial reciprocity theorems: An invitation to enumerative geometric combinatorics* Graduate Studies in Mathematics, American Mathematical Society, to appear (2018).

[4] D. N. Bernstein, The number of lattice points in integer polyhedra, *Funkcional. Anal. i Priložen.* **10**, 72–73 (1976).

[5] U. Betke, Das Einschließungs-Ausschließungsprinzip für Gitterpolytope, *unpublished manuscript* (1984).

[6] U. Betke and M. Kneser, Zerlegungen und Bewertungen von Gitterpolytopen, *J. Reine Angew. Math.* **358** 202–208 (1985).

[7] F. Breuer, Ehrhart f^*-coefficients of polytopal complexes are non-negative integers, *Electron. J. Combin.* **19**, Paper 16, 22 (2012).

[8] K. J. Böröczky and M. Ludwig, Valuations on lattice polytopes, in *Tensor valuations and their applications in stochastic geometry and imaging*, Lecture Notes in Math. Vol. 2177 (Springer, Cham, 2017) pp. 213–234.

[9] J. A. De Loera, J. Rambau and F. Santos, *Triangulations*, Algorithms and Computation in Mathematics, Vol. 25 (Springer-Verlag, Berlin, 2010), Structures for algorithms and applications.

[10] E. Ehrhart, Sur les polyèdres rationnels homothétiques à n dimensions, *C. R. Acad. Sci. Paris* **254**, 616–618 (1962).

[11] P. M. Gruber, *Convex and Discrete Geometry* (Springer, Berlin, 2007).

[12] H. Hadwiger, *Vorlesungen über Inhalt, Oberfläche und Isoperimetrie* (Springer-Verlag, Berlin-Göttingen-Heidelberg, 1957).

[13] K. Jochemko and R. Sanyal, Combinatorial mixed valuations, *Adv. Math* **319**, 630–653 (2017).

[14] K. Jochemko and R. Sanyal, Combinatorial positivity of translation-invariant valuations and a discrete Hadwiger theorem, *J. Eur. Math. Soc. (JEMS)* **9** (2018).

[15] M. Köppe and S. Verdoolaege, Computing parametric rational generating functions with a primal Barvinok algorithm, *Electron. J. Combin.* **15**, Research Paper 16, 19 (2008).

[16] I. G. Macdonald, Polynomials associated with finite cell-complexes, *J. London Math. Soc. (2)* **4**, 181–192 (1971).

[17] P. McMullen, Metrical and combinatorial properties of convex polytopes, 491 (1975). *Proceedings of the International Congress of Mathematicians (Vancouver, B. C., 1974)*, Vol. 1, pp. 491–495. Canad. Math. Congress, Montreal, Que., 1975.

[18] P. McMullen, The polytope algebra, *Adv. Math.* **78**, 76–130 (1989).

[19] P. McMullen, Valuations and Euler-type relations on certain classes of convex polytopes, in *Proc. London Math. Soc.(3)*, (1) 113–135, 1977.

[20] P. McMullen, Valuations on lattice polytopes, *Advances in Mathematics* **220**, 303–323 (2009).

[21] R. Schneider, *Convex bodies: the Brunn-Minkowski theory*, Encyclopedia of Mathematics and its Applications, Vol. 151, expanded edn. (Cambridge University Press, Cambridge, 2014).

[22] R. P. Stanley, A monotonicity property of h-vectors and h^*-vectors, *European J. Combin.* **14**, 251–258 (1993).

[23] R. P. Stanley, Decompositions of rational convex polytopes, *Ann. Discrete Math.* **6**, 333–342 (1980), Combinatorial mathematics, optimal designs and their applications (Proc. Sympos. Combin. Math. and Optimal Design, Colorado State Univ., Fort Collins, Colo., 1978).

[24] R. P. Stanley, *Enumerative combinatorics. Volume 1*, Cambridge Studies in Advanced Mathematics, Vol. 49, second edn. (Cambridge University Press, Cambridge, 2012).

[25] R. Stein, *Additivität und Einschließungs-Ausschließungs-Prinzip für Funktionale von Gitterpolytopen*, PhD thesis, Dortmund, 1982).

[26] G. M. Ziegler, *Lectures on polytopes* (Springer Science & Business Media, 1995).

Ehrhart positivity and Demazure characters

Per Alexandersson

Department of Mathematics, KTH
SE-100 44 Stockholm, Sweden
E-mail: per.w.alexandersson@gmail.com

Elie Alhajjar

Department of Mathematics, USMA
West Point, New York, USA
E-mail: eliealhajjar@gmail.com

Demazure characters, also known as key polynomials, generalize the classical Schur polynomials. In particular, when all variables are set equal to 1, these polynomials count the number of integer points in a certain class of Gelfand–Tsetlin polytopes. This property highlights the interaction between the corresponding polyhedral and combinatorial structures via Ehrhart theory. In this paper, we give an overview of results concerning the interplay between the geometry of Gelfand–Tsetlin polytopes and their Ehrhart polynomials. Motivated by strong computer evidence, we propose several conjectures about the non-negativity of the coefficients of such polynomials.

Keywords: Demazure characters, key polynomials, Gelfand–Tsetlin polytopes, Ehrhart polynomial.

1. Introduction

The theory of Schur polynomials can be seen from two different sets of lenses. On one hand, the traditional approach begins with a definition involving quotients of matrix determinants. This method is mainly useful in representation theory, since it is derived as a special case of the Weyl character formula. On the other hand, the combinatorial approach uses the sum expansion over semi-standard Young tableaux of fixed shape. Note that it is not hard to show explicitly the equivalence of these two approaches.

Demazure characters [4, 5], also known as key polynomials, generalize the classical Schur polynomials. Key polynomials can be computed recursively via divided difference operators, and are closely related to Schubert polynomials. In particular, each Schubert polynomial can be expressed as a non-negative integer combination of key polynomials.

A combinatorial formula using semi-standard Young tableaux was discovered in [17], and this is where the notion of key polynomials come from. Key polynomials are specializations of non-symmetric Macdonald polynomials, [9] so the combinatorial formula of J. Haglund gives an alternative formulation, using *skyline fillings*. S. Mason [21] explores two variations of skyline fillings, both of which give key polynomials. It is possible to interpolate between the two skyline models, see [3, 16].

For each a partition λ, one can construct a Gelfand–Tsetlin polytope. These polytopes play a crucial role in representation theory, algebraic geometry and combinatorics. Their importance stems from the fact that their integer points are in bijection with semi-standard Young tableaux.

R. Stanley and A. Postnikov [24] study a certain subfamily of key polynomials, and prove that these are flagged Schur polynomials. This implies that key polynomials in this family can be computed as a certain sum over lattice points in a single face of a Gelfand–Tsetlin polytope.

The result by R. Stanley and A. Postnikov can be extended to the full family of key polynomials; V. Kirichenko, E. Smirnov and V. Timorin [12] show that key polynomials can be expressed as a sum over lattice pooints in a certain union of faces in a Gelfand–Tsetlin polytope. This way of thinking about key polynomials is not as well-known as the other interpretations, and the purpose of this survey is to emphasize the polyhedral aspect of key polynomials. A related result appears in [6], where Hall–Littlewood polynomials are expressed as a weighted sum over lattice points in Gelfand–Tsetlin polytopes.

Recent research has been focused on products involving key polynomials, see [25]. The main motivation for studying key polynomials is to gain insight about the expression of products of Schubert polynomials in terms of Schubert polynomials, a problem of main importance in representation theory. The close relationship between key and Schubert polynomials is emphasized in [27].

The purpose of the current paper is two-fold: on one hand, we aim to collect some of the main results related to the study of key polynomials. On the other hand, we propose several conjectures concerning the non-negativity of the coefficients of the 'stretched' version of such polynomials. In Section 2 below, we provide the basic material and fix the terminology for the remainder of the paper. Section 3 deals with the facial description of GT-polytopes and the formal definition of key polynomials, where we give several examples to illustrate the main ideas. In Section 4 we introduce the connection to Ehrhart theory through Kostka coefficients and in Section 5 we mention a sample of the computations that lead eventually to the main conjecture. Finally, we reconstruct a counterexample in Section 6 that shows the failure of the non-negativity argument in the case of arbitrary faces of GT-polytopes.

2. Preliminaries

Given an integer partition $\lambda_1 \geq \lambda_2 \geq \lambda_3 \geq \cdots$, we can associate a *Young diagram* of shape λ as a diagram in the plane with λ_i left-justified boxes in row i. For example,

$\lambda = (5, 3, 2, 2)$ gives rise to the following Young diagram:

$$\begin{array}{|c|c|c|c|c|}\hline & & & & \\ \hline & & \\ \cline{1-3} & \\ \cline{1-2} & \\ \cline{1-2} \end{array}$$

A *semi-standard Young tableau* of shape λ is an assignment of natural numbers to the boxes of the Young diagram, such that rows are weakly increasing from left to right, and columns are strictly increasing from top to bottom. Only the first of the following three assignments is a semi-standard Young tableau

$$\begin{array}{|c|c|c|c|c|}\hline 1 & 1 & 2 & 2 & 4 \\ \hline 2 & 3 & 3 \\ \cline{1-3} 4 & 5 \\ \cline{1-2} 5 & 7 \\ \cline{1-2} \end{array}\;, \quad \begin{array}{|c|c|c|c|}\hline 1 & 3 & 2 & 2 \\ \hline 2 & 4 & 4 \\ \cline{1-3} \end{array}\;, \quad \begin{array}{|c|c|c|c|}\hline 1 & 2 & 3 & 4 \\ \hline 2 & 3 & 3 \\ \cline{1-3} \end{array}\;.$$

2.1. *Gelfand–Tsetlin polytopes*

There are several families of polytopes which are referred to as Gelfand–Tsetlin polytopes, see for example [8] and [11, 20]. A *Gelfand–Tsetlin pattern* or GT-pattern for short is a triangular array (x_{ij}) visualized as

$$\begin{array}{ccccccc} x_{n1} & & x_{n2} & & \cdots & & x_{nn} \\ & \ddots & & \ddots & & \iddots & \\ & & x_{21} & & x_{22} & & \\ & & & x_{11} & & & \end{array} \tag{1}$$

satisfying the inequalities

$$x_{i+1,j} \geq x_{ij} \text{ and } x_{ij} \geq x_{i+1,j+1} \tag{2}$$

for all values of i, j where the indexing is defined. The inequalities simply state that down-right diagonals are weakly decreasing and down-left diagonals are weakly increasing.

Given an integer partition λ, the *Gelfand–Tsetlin polytope* $\mathcal{GT}(\lambda) \subset \mathbb{R}^{\frac{n(n+1)}{2}}$ is the convex polytope of Gelfand–Tsetlin patterns defined by the inequalities in Equation (2) together with the equalities $x_{ni} = \lambda_i$ for $i = 1, 2, \ldots, n$.

The polytope $\mathcal{GT}(\lambda)$ has integer vertices. In fact, it has a unimodular triangulation, see [2]. Also, note that $k \cdot \mathcal{GT}(\lambda) = \mathcal{GT}(k\lambda)$ for all $k \geq 0$.

2.2. *Bijection with semi-standard Young tableaux*

Note that (2) implies that any two adjacent rows in an integral GT-pattern form a skew Young diagram, see the standard textbook by R. Stanley [29] for terminology.

This property enables us to define a bijection with Young tableaux — the skew shape defined by row j and $j+1$ in an integral GT-pattern G describes which boxes in a Young tableau T have content j. In particular, tableaux of shape λ are in bijection with integral GT-patterns with topmost row equal to λ. See Figure 1 for an example of this correspondence.

$$\begin{array}{ccccccccccc} 5 & & 4 & & 2 & & 1 & & 1 & & 0 \\ & 5 & & 3 & & 2 & & 1 & & 0 & \\ & & 3 & & 3 & & 2 & & 1 & & \\ & & & 3 & & 3 & & 1 & & & \\ & & & & 3 & & 2 & & & & \\ & & & & & 3 & & & & & \end{array} \quad \longleftrightarrow \quad \begin{array}{|c|c|c|c|c|} \hline 1 & 1 & 1 & 5 & 5 \\ \hline 2 & 2 & 3 & 6 & \\ \cline{1-4} 3 & 4 & & & \\ \cline{1-2} 4 & & & & \\ \cline{1-1} 6 & & & & \\ \cline{1-1} \end{array} \tag{3}$$

Fig. 1. The GT-pattern corresponding to a Young tableau. For example, the third row tells us that the shape of the entries ≤ 3 in the tableau is $(3, 3, 1)$.

Note that in any integral GT-pattern, $x_{i+1,j} - x_{ij}$ counts the number of boxes with content i in row j in the corresponding tableau. Given a GT-pattern G, we define the *weight* $w(G)$ as the vector

$$w_i(G) := \sum_{j=1}^{i+1} x_{i+1,j} - \sum_{j=1}^{i} x_{ij}, \tag{4}$$

where $x_{0j} := 0$. Thus, an integral GT-pattern with weight w is in bijection with a semi-standard Young tableau with w_i entries equal to i. Hence, integer points in $\mathcal{GT}(\lambda)$ are in bijection with $\mathrm{SSYT}(\lambda, n)$ — the set of semi-standard Young tableaux of shape λ with maximal entry n. Given λ and w, let $\mathcal{GT}(\lambda, w) \subseteq \mathcal{GT}(\lambda)$ be the intersection of $\mathcal{GT}(\lambda)$ with the hyperplanes defined by (4). The lattice points in $\mathcal{GT}(\lambda, w)$ are enumerated by the Kostka coefficients, see Section 4.1 below.

One can then define the *Schur polynomials* $\mathrm{s}_\lambda(z_1, \dots, z_n)$ as

$$\mathrm{s}_\lambda(z_1, \dots, z_n) := \sum_{G \in \mathcal{GT}(\lambda) \cap \mathbb{Z}^{\frac{n(n+1)}{2}}} z_1^{w_1(G)} \cdots z_n^{w_n(G)}. \tag{5}$$

In particular, by using the Weyl dimension formula [29, Eq. 7.105] we have that

$$\mathrm{s}_\lambda(\underbrace{1, 1, \dots, 1}_{n}) = |\mathcal{GT}(\lambda) \cap \mathbb{Z}^{\frac{n(n+1)}{2}}| = |\mathrm{SSYT}(\lambda, n)| = \prod_{1 \leq i < j \leq n} \frac{\lambda_i - \lambda_j + j - i}{j - i}. \tag{6}$$

The fact that the right hand side is a polynomial in the λ_i is not a complete surprise. Gelfand–Tsetlin polytopes belong to a larger family of so called *marked order polytopes*, generalizing the notion of order polytopes introduced by R. Stanley in [28]. K. Jochemko and R. Sanyal proved in [10] that marked order polytopes give rise to Ehrhart functions which are piecewise polynomial in the markings, which in our case are the parts of λ.

2.3. *Skew GT-polytopes*

In analogy with the triangular case, one can define so called skew Gelfand–Tsetlin polytopes using parallelogram arrangements of non-negative numbers,

$$\begin{array}{ccccccccccc} x_{n1} & & x_{n2} & & \cdots & & \cdots & & x_{nm} & & \\ & \ddots & & \ddots & & \ddots & & \ddots & & \ddots & \\ & & x_{11} & & x_{12} & & \cdots & & \cdots & x_{1m} & \\ & & & x_{01} & & x_{02} & & \cdots & & \cdots & x_{0m} \end{array} \tag{7}$$

satisfying the inequalities

$$x_{i+1,j} \geq x_{ij} \text{ and } x_{ij} \geq x_{i+1,j+1} \tag{8}$$

for all i, j where the indexing is defined.

Consider an $(n+1) \times m$ GT-pattern with top row λ and bottom row μ — that is, $x_{ni} = \lambda_i$ and $x_{0i} = \mu_i$ for $i = 1, \ldots, m$. These equalities together with the above inequalities define a convex polytope, the *skew Gelfand–Tsetlin polytope*, $\mathcal{GT}(\lambda/\mu) \subset \mathbb{R}^{(n+1)m}$. Note that the vertices of such polytopes have integer coordinates. In fact, the skew Gelfand–Tsetlin polytopes admit a unimodular triangulation.

The weight of a GT-pattern in the parallelogram form is defined in the same way as in the triangular form (4). Similar to the bijection described in Section 2.2, the integer points in $\mathcal{GT}(\lambda/\mu)$ correspond to skew Young tableaux with shape λ/μ, where the entries belong to the set $\{1, 2, \ldots, n\}$. This allow us to define the *skew Schur polynomials* as

$$\mathrm{s}_{\lambda/\mu}(z_1, \ldots, z_n) := \sum_{G \in \mathcal{GT}(\lambda/\mu) \cap \mathbb{Z}^{(n+1)m}} z_1^{w_1(G)} \cdots z_n^{w_n(G)}. \tag{9}$$

We note that there is no simple formula for computing the specialization $\mathrm{s}_{\lambda/\mu}(1, 1, \ldots, 1)$ in general.

By specifying a weight vector w, we intersect $\mathcal{GT}(\lambda/\mu)$ with a set of hyperplanes, and denote the resulting polytope by $\mathcal{GT}(\lambda/\mu, w)$. As with $\mathcal{GT}(\lambda, w)$, the polytopes $\mathcal{GT}(\lambda/\mu, w)$ are also not integral in general.

3. Reduced Kogan faces and key polynomials

By imposing some additional equalities on the coordinates of a GT-polytope, one can obtain faces of the polytope. There is a particular interest with equalities of the form $x_{ij} = x_{i,j+1}$. By imposing a set of such equalities, we obtain a *Kogan face* of the GT-polytope. To each equality of the form $x_{ij} = x_{i,j+1}$, we associate the transposition $s_{n-i+j-1}$, as shown in Figure 2. We then construct a word from these transpositions by reading the entries from bottom to top, left to right. If this word is *reduced*, we say that the corresponding Kogan face is reduced. Note that the same word might be constructed from equalities in several different ways. The *type*

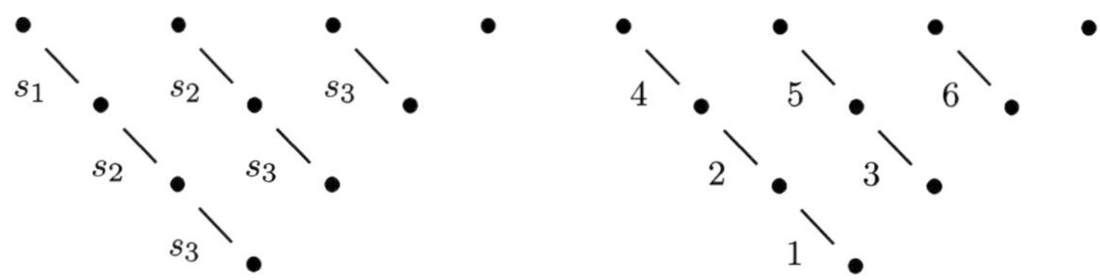

Fig. 2. The transposition corresponding to different equalities between entries in a Kogan face and the reading order of these.

of a Kogan face is the permutation obtained from the word. We should really view the equalities that define a Kogan face as some special set of equalities — a point in this face might satisfy some additional equalities present, if it is also a member of some sub-face. This implies that a point in the GT-polytope can be a member of *several* (reduced) Kogan faces. For example, the point where all equalities are present is the unique face of type w_0, the longest permutation in $\mathfrak{S}_n$. This point is a sub-face of all other Kogan faces. Furthermore, the full GT-polytope has the identity permutation as type.

Example 3.1. Consider the face in Figure 3. All marked equalities are in the south-east direction, so we obtain the word $s_3s_1s_2s_3$. It is straightforward to verify that this word is reduced, so this face is a reduced Kogan face.

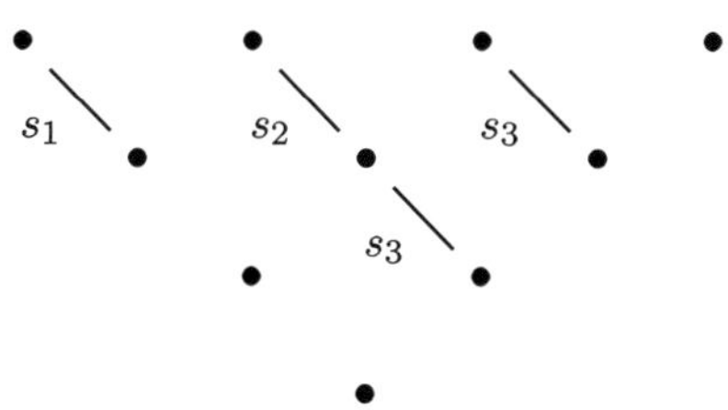

Fig. 3. A reduced Kogan face with the word $\omega = s_3s_1s_2s_3$.

Suppose we wish to examine the GT-polytope $\mathcal{GT}(\lambda)$ with $\lambda = (4, 3, 3, 2)$. The lattice points in the reduced Kogan face with the word $s_3s_1s_2s_3$ are the following GT-patterns:

$$\begin{array}{ccccccc} 4 & & 3 & & 3 & & 2 \\ & 4 & & 3 & & 3 & \\ & & 3 & & 3 & & \\ & & & 3 & & & \end{array} \qquad \begin{array}{ccccccc} 4 & & 3 & & 3 & & 2 \\ & 4 & & 3 & & 3 & \\ & & 4 & & 3 & & \\ & & & 3 & & & \end{array} \qquad \begin{array}{ccccccc} 4 & & 3 & & 3 & & 2 \\ & 4 & & 3 & & 3 & \\ & & 4 & & 3 & & \\ & & & 4 & & & \end{array}$$

It is evident from the bubble-sort algorithm [14] that there is at least one reduced Kogan face of every type $\sigma \in \mathfrak{S}_n$.

Proposition 3.1 (See [15]). *If $\sigma \in \mathfrak{S}_n$ avoids the permutation pattern* 132 *(such permutations are known as Kempf permutations), then there is a unique reduced Kogan face of type σ.*

See Section 5.1 for more background on this special case.

Example 3.2. There are 11 reduced Kogan faces for $n = 4$ that are covered by Proposition 3.1. These faces are illustrated in Figure 4.

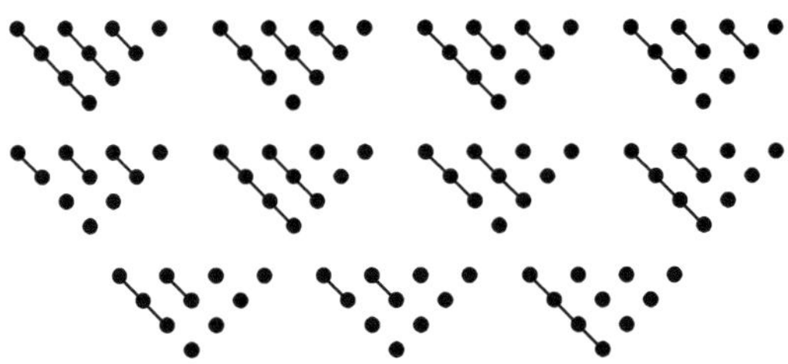

Fig. 4. The 11 Kogan faces for $n = 4$ whose type avoids the pattern 132.

3.1. *Key polynomials*

One possible generalization of Schur polynomials is the so called *Demazure characters*, also known as *key polynomials*. The latter name was introduced by V. Reiner and M. Shimozono in [27], and refers to the combinatorial model for Demazure characters introduced in [17], where they use the so called *key tableaux*. In order to define key polynomials, we need some preliminary terminology.

Whenever $s_i \in \mathfrak{S}_n$ is a simple transposition, with $i \in \{1, 2, \dots, n-1\}$, we let s_i act on $\mathbb{C}[z_1, \dots, z_n]$ by permuting the indices:

$$s_i \circ f(z_1, \dots, z_n) = f(z_1, \dots, z_{i-1}, z_{i+1}, z_i, z_{i+2}, \dots, z_n).$$

Define the *divided difference operator* ∂_i as

$$\partial_i(f) = \frac{f - s_i(f)}{z_i - z_{i+1}}.$$

Note that $\partial_i(f)$ is indeed a polynomial, and one can check that $\partial_i(f)$ is symmetric in the variables z_i and z_{i+1}. For example,

$$\begin{aligned}\partial_2(z_1^2 z_2^5 z_3^3 z_4) &= \frac{z_1^2 z_2^5 z_3^3 z_4 - z_1^2 z_2^3 z_3^5 z_4}{z_2 - z_3}\\ &= z_1^2 z_4 \frac{z_2^5 z_3^3 - z_2^3 z_3^5}{z_2 - z_3}\\ &= z_1^2 z_2^3 z_3^3 z_4 \frac{z_2^2 - z_3^2}{z_2 - z_3}\\ &= z_1^2 z_2^3 z_3^3 z_4 (z_2 + z_3).\end{aligned}$$

We now define the operators $\pi_i(f) := \partial_i(z_i f)$ for $i = 1, \ldots, n-1$ whenever $f \in \mathbb{C}[z_1, \ldots, z_n]$. It is straightforward to verify the following properties of the π_i's:

- π_i preserves the degree,
- $\pi_i^2 = \pi_i$ for all i,
- $\pi_i \pi_j = \pi_j \pi_i$ whenever $|i - j| > 2$,
- $\pi_i \pi_{i+1} \pi_i = \pi_{i+1} \pi_i \pi_{i+1}$ for all i.

The last two properties allow us to make the following definition: Let $\sigma = s_{i_1} s_{i_2} \ldots s_{i_\ell}$ be a *reduced word* of a permutation $\sigma \in \mathfrak{S}_n$. Then let

$$\pi_\sigma := \pi_{i_1} \circ \pi_{i_2} \circ \cdots \circ \pi_{i_\ell}.$$

The action of π_σ is independent of the choice of reduced word, since we have the relations above.

We are now ready to define the key polynomials. Let λ be a partition with at most n parts, and let $\sigma \in \mathfrak{S}_n$ be a permutation. The *key polynomial* $\kappa_{\lambda,\sigma}(z)$ is defined as

$$\kappa_{\lambda,\sigma}(z) := \pi_\sigma \left(z_1^{\lambda_1} \cdots z_n^{\lambda_n} \right). \tag{10}$$

Example 3.3. Let $\lambda = (2, 1, 0, 0)$ and $\sigma = [2, 4, 3, 1] \in \mathfrak{S}_4$ in one-line notation. The permutation can be expressed as a reduced word as $\sigma = s_2 s_3 s_2 s_1$. We compute the key polynomial as follows:

$$\begin{aligned}\kappa_{\lambda,\sigma}(z) &= \pi_2\pi_3\pi_2\pi_1(z_1^2 z_2) = \pi_2\pi_3\pi_2\partial_1(z_1^3 z_2) = \pi_2\pi_3\pi_2\left(\frac{z_1^3 z_2 - z_1 z_2^3}{z_1 - z_2}\right) \\ &= \pi_2\pi_3\pi_2(z_1^2 z_2 + z_1 z_2^2).\end{aligned}$$

We continue the calculation by applying π_2 and get

$$\kappa_{\lambda,\sigma}(z) = \pi_2\pi_3(z_2 z_1^2 + z_3 z_1^2 + z_2^2 z_1 + z_3^2 z_1 + z_2 z_3 z_1).$$

Applying $\pi_2\pi_3$ then finally gives

$$\begin{aligned}\kappa_{\lambda,\sigma}(z) = {} & z_1^2 z_2 + z_1^2 z_3 + z_1^2 z_4 + z_1 z_2^2 + z_1 z_3^2 \\ & + z_1 z_4^2 + z_1 z_2 z_3 + z_1 z_2 z_4 + z_1 z_3 z_4.\end{aligned} \tag{11}$$

In general, some monomials may appear multiple times.

In [12], the following formula for key polynomials using Kogan faces was proved. This generalizes an earlier result by A. Postnikov and R. Stanley [24], who considered the case covered in Proposition 3.1.

Proposition 3.2. *Let $\mathcal{GT}(\lambda, \sigma)$ be defined as the polytopal complex*

$$\mathcal{GT}(\lambda, \sigma) := \bigcup_{\substack{\mathcal{F} \in \mathcal{GT}(\lambda) \\ \text{type}(\mathcal{F}) = w_0 \sigma}} \mathcal{F}.$$

That is, $\mathcal{GT}(\lambda,\sigma)$ is the union of all reduced Kogan faces of type $w_0\sigma$ in the polytope $\mathcal{GT}(\lambda)$. The key polynomial $\kappa_{\lambda,\sigma}(z)$ can be computed as

$$\kappa_{\lambda,\sigma}(z_1,\dots,z_n) = \sum_{G\in\mathcal{GT}(\lambda,\sigma)\cap\mathbb{Z}^{\frac{n(n+1)}{2}}} z_1^{w_1(G)}\cdots z_n^{w_n(G)} \tag{12}$$

where we use the same weight for integral Gelfand–Tsetlin patterns as in (4).

As an immediate corollary, it is clear that

$$s_\lambda(z_1,\dots,z_n) = \kappa_{\lambda,w_0}(z_1,\dots,z_n).$$

We now recalculate the key polynomial in Example 3.3 using (12).

Example 3.4. Let $\lambda = (2,1,0,0)$ and $\sigma = [2,4,3,1] \in \mathfrak{S}_4$. We have that $\omega_0\sigma = [1,3,4,2]$, and we have that

$$[1,3,4,2] = s_3 s_2.$$

There are no other reduced words that give rise to the same permutation. However, there are three reduced Kogan faces that give rise to this particular reduced word (and are hence of type $[1,3,4,2]$):

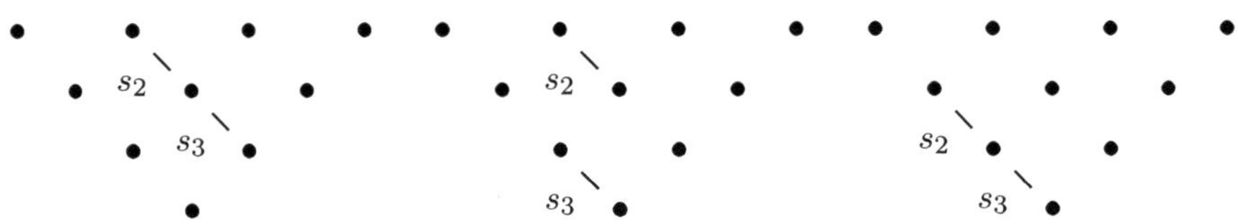

(A) (B) (C)

(13)

We expect nine lattice points in the union of these faces, as there are nine monomials in (11). These lattice points are given by the following Gelfand–Tsetlin patterns:

$$\begin{array}{ccccccc} 2 && 1 && 0 && 0 \\ & 1 && 0 && 0 & \\ && 1 && 0 && \\ &&& 1 &&& \\ &&& C &&& \\ &&& z_1z_4^2 &&& \end{array} \qquad \begin{array}{ccccccc} 2 && 1 && 0 && 0 \\ & 1 && 1 && 0 & \\ && 1 && 0 && \\ &&& 1 &&& \\ &&& BC &&& \\ &&& z_1z_3z_4 &&& \end{array} \qquad \begin{array}{ccccccc} 2 && 1 && 0 && 0 \\ & 1 && 1 && 0 & \\ && 1 && 1 && \\ &&& 1 &&& \\ &&& ABC &&& \\ &&& z_1z_2z_4 &&& \end{array}$$

$$\begin{array}{ccccccc} 2 && 1 && 0 && 0 \\ & 2 && 0 && 0 & \\ && 2 && 0 && \\ &&& 2 &&& \\ &&& C &&& \\ &&& z_1^2z_4 &&& \end{array} \qquad \begin{array}{ccccccc} 2 && 1 && 0 && 0 \\ & 2 && 1 && 0 & \\ && 1 && 0 && \\ &&& 1 &&& \\ &&& B &&& \\ &&& z_1z_3^2 &&& \end{array} \qquad \begin{array}{ccccccc} 2 && 1 && 0 && 0 \\ & 2 && 1 && 0 & \\ && 1 && 1 && \\ &&& 1 &&& \\ &&& AB &&& \\ &&& z_1z_2z_3 &&& \end{array}$$

$$
\begin{array}{ccccccc}
2 & & 1 & & 0 & & 0 \\
& 2 & & 1 & & 0 & \\
& & 2 & & 0 & & \\
& & & 2 & & & \\
& & & BC & & & \\
& & & z_1^2 z_3 & & &
\end{array}
\qquad
\begin{array}{ccccccc}
2 & & 1 & & 0 & & 0 \\
& 2 & & 1 & & 0 & \\
& & 2 & & 1 & & \\
& & & 1 & & & \\
& & & A & & & \\
& & & z_1 z_2^2 & & &
\end{array}
\qquad
\begin{array}{ccccccc}
2 & & 1 & & 0 & & 0 \\
& 2 & & 1 & & 0 & \\
& & 2 & & 1 & & \\
& & & 2 & & & \\
& & & AC & & & \\
& & & z_1^2 z_2 & & &
\end{array}
$$

The letters below each pattern indicate which reduced Kogan faces in (13) the pattern is a member of and the monomials represent $z^{w(G)}$ as defined in (4).

4. Ehrhart polynomials

From (6), it follows that the Ehrhart polynomial of $\mathcal{GT}(\lambda)$ is given by

$$i(\mathcal{GT}(\lambda), k) = \prod_{1 \leq i < j \leq n} \frac{k(\lambda_i - \lambda_j) + j - i}{j - i} \tag{14}$$

and it is clear that all coefficients of k are non-negative.

For a skew shape λ/μ, the lattice point enumerator is indeed a polynomial $i(\mathcal{GT}(\lambda/\mu), k) = \mathrm{s}_{k\lambda/k\mu}(\underbrace{1, 1, \ldots, 1}_{n})$. Table 1 below shows a sample computation of such polynomials. Computer evidence suggests the following conjecture:

Conjecture 4.1. *Let λ/μ be a skew shape. Then the polynomial $i(\mathcal{GT}(\lambda/\mu), k)$ has non-negative coefficients.*

We do not expect a closed-form formula for $i(\mathcal{GT}(\lambda/\mu), k)$ — the volume of $\mathcal{GT}(\lambda/\mu)$ is related to the number of skew standard Young tableaux for which there are no known closed formulas in general, see [23] and subsequent papers.

Table 1. Ehrhart polynomials for $\mathcal{GT}(\lambda/\mu)$.

λ/μ	$\mathrm{s}_{k\lambda/k\mu}(1,1,1)$
221	$\frac{1}{2}\left(k^2 + 3k + 2\right)$
221/1 221/21 221/11	$\frac{1}{4}\left(k^4 + 6k^3 + 13k^2 + 12k + 4\right)$
221/2 321	$k^3 + 3k^2 + 3k + 1$
321/1 321/2	$\frac{1}{2}(k^5 + 6k^4 + 14k^3 + 16k^2 + 9k + 2)$
321/21	$\frac{1}{8}(k^6 + 9k^5 + 33k^4 + 63k^3 + 66k^2 + 36k + 8)$

4.1. *Kostka coefficients*

Kostka coefficients are important numbers appearing in several branches of mathematics such as algebraic combinatorics, symmetric functions, representation theory and algebraic geometry among others. From a representation theoretical point of view, they are defined as the dimension of the weight subspace of the irreducible representation of the Lie algebra $\mathfrak{gl}(\mathbb{C})$. From a combinatorial point of view, they enumerate the number of semi-standard Young tableaux of fixed shape and weight.

Given two partitions λ and μ of n, Kostka coefficients arise in the expansion of the Schur polynomial as a linear combination of monomial symmetric polynomials

$$\mathrm{s}_\lambda(z_1, \dots, z_n) = \sum_\mu K_{\lambda\mu} m_\mu(z_1, \dots, z_n). \tag{15}$$

More generally, Kostka coefficients can be viewed as a special case of so called Littlewood–Richardson coefficients $K_{\lambda\mu} = c^{\tau}_{\sigma\lambda}$, where σ and τ are defined in terms of the partition μ. For more material about the latter coefficients and some standard results, the reader is referred to the book [7].

The partitions λ and μ are usually represented as integer vectors, so it makes sense to talk about scaling these vectors by an integer factor k. This operation gives rise to what is called "stretched" Kostka coefficients $K_{\lambda\mu}(k) := K_{k\lambda,k\mu}$.

Surprisingly at first, $K_{\lambda\mu}(k)$ turns out to be a polynomial function in k with rational coefficients depending on λ and μ — a fact initially shown by A. Kirillov and N. Reshetikhin [13].

Expressing Kostka coefficients using the Gelfand–Tsetlin polytopes from Section 2.1 provides a natural geometric interpretation of $K_{\lambda\mu}(k)$. To each Kostka coefficient $K_{\lambda\mu}$, we have the corresponding Gelfand–Tsetlin polytope $\mathcal{GT}(\lambda, \mu)$ in $\mathbb{R}^{\frac{n(n+1)}{2}}$ such that

$$K_{\lambda\mu} = |\mathcal{GT}(\lambda, \mu) \cap \mathbb{Z}^{\frac{n(n+1)}{2}}| \text{ and } K_{\lambda\mu}(k) = |k \cdot \mathcal{GT}(\lambda, \mu) \cap \mathbb{Z}^{\frac{n(n+1)}{2}}|. \tag{16}$$

The vertices of $\mathcal{GT}(\lambda, \mu)$ have rational coordinates in general ($n \geq 5$). From E. Ehrhart's fundamental work, it is well-known that $K_{\lambda\mu}(k)$ must be a quasipolynomial. This means that there exist an integer M and polynomials $g_0, g_1, \dots, g_{M-1}$ such that $K_{\lambda\mu}(k) = g_i(k)$ whenever $k \equiv i \pmod{M}$ — see details in [30]. The "surprising" fact here is that the function $K_{\lambda\mu}(k)$ is indeed a polynomial, which exhibits an example of *period collapse* in rational polytopes.

In their paper [11], the authors conjectured a formula for the degree of $K_{\lambda\mu}(k)$ which was later proved by T. McAllister [22]. In that same paper, they also conjectured that the coefficients of the polynomial $K_{\lambda\mu}(k)$ are non-negative. To the best of our knowledge, this conjecture remains open and provides an instance of a well-known phenomenon called Ehrhart positivity. For a recent survey about this topic, we recommend the article by F. Liu [18].

The polytopes $\mathcal{GT}(\lambda/\mu, w)$ also exhibit a period collapse, and have a polynomial Ehrhart funtion. This follows from the work of E. Rassart in [26]. Integrality and the integer decomposition property of the family $\mathcal{GT}(\lambda/\mu, w)$ is studied in the work of P. Alexandersson [1], where several questions are left unanswered. The *skew Kostka coefficients* $K_{\lambda/\mu,\nu}$ are defined analogously to the usual Kostka coefficients, via the identity

$$s_{\lambda/\mu}(z_1, \ldots, z_n) = \sum_{\nu} K_{\lambda/\mu,\nu} m_\nu(z_1, \ldots, z_n). \tag{17}$$

Scaling by an integer factor k gives rise to the polynomial $i(\mathcal{GT}(\lambda/\mu, \nu), k) = K_{k\lambda/k\mu,k\nu} \in \mathbb{Q}[k]$. The conjecture by R.C. King et al. seems to extend to the skew case:

Conjecture 4.2. *Let λ/μ be a skew shape and ν be a weight. Then the polynomial $i(\mathcal{GT}(\lambda/\mu, \nu), k)$ has non-negative coefficients.*

5. A conjecture on key polynomials

Given the positivity phenomena above, it is reasonable to ask if we have positive Ehrhart coefficients for the polytopal complexes in Proposition 3.2. In other words, for a partition λ and permutation $\sigma \in \mathfrak{S}_n$, do

$$i(\mathcal{GT}(\lambda, \sigma), k) = \kappa_{k\lambda,\sigma}(\underbrace{1, 1, \ldots, 1}_{n}) \in \mathbb{Q}[k]$$

have non-negative coefficients?

By using the divided difference operators in Equation (10), it is possible to compute the Ehrhart polynomials for small values of n. Note first that for all λ, σ and $m \geq 0$, we have that

$$\kappa_{m+\lambda,\sigma}(z_1, \ldots, z_n) = (z_1 z_2 \cdots z_n)^m \kappa_{\lambda,\sigma}(z_1, \ldots, z_n), \tag{18}$$

where $m + \lambda := (m + \lambda_1, m + \lambda_2, \ldots, m + \lambda_n)$. It follows that

$$\kappa_{m+\lambda,\sigma}(1, \ldots, 1) = \kappa_{\lambda,\sigma}(1, \ldots, 1) \text{ and } i(\mathcal{GT}(m + \lambda, \sigma), k) = i(\mathcal{GT}(\lambda, \sigma), k).$$

Example 5.1. Let $\lambda = (a+b, a, 0)$, with $a, b \geq 0$, since we can assume that the last part is equal to zero, due to (18). We have the following table for $\kappa_{\lambda,\sigma}(z_1, z_2, z_3)$, for different values of $\sigma \in \mathfrak{S}_3$:

By taking the limit $z_i \to 1$ in Table 2, and multiplying a and b with k, we get the Ehrhart polynomials $i(\mathcal{GT}(\lambda, \sigma), k)$, which we present in Table 3. The last polynomial for $\sigma = [3, 2, 1]$ agrees with Equation (14) with $\lambda_1 = a + b$, $\lambda_2 = a$ and $\lambda_3 = 0$, since we get

$$\kappa_{k\lambda,[3,2,1]}(1, 1, 1) = \frac{k(a + b - a) + (2 - 1)}{2 - 1} \frac{k(a - 0) + (3 - 2)}{3 - 2} \frac{k(a + b - 0) + (3 - 1)}{3 - 1}.$$

We are then fairly confident in the following conjecture:

Table 2. Key polynomials computed for all $\sigma \in \mathfrak{S}_3$.

σ	$\kappa_{\lambda,\sigma}(z_1, z_2, z_3)$
$[1,2,3]$	$z_1^{a+b} z_2^a$
$[2,1,3]$	$\frac{z_1^a z_2^a \left(z_1^{b+1} - z_2^{b+1}\right)}{z_1 - z_2}$
$[1,3,2]$	$\frac{\left(z_2^{a+1} - z_3^{a+1}\right) z_1^{a+b}}{z_2 - z_3}$
$[3,1,2]$	$\frac{z_1^a \left((z_1-z_3) z_2^{a+1}\left(z_1^{b+1} - z_2^{b+1}\right) - (z_1-z_2) z_3^{a+1}\left(z_1^{b+1} - z_3^{b+1}\right)\right)}{(z_1-z_2)(z_1-z_3)(z_2-z_3)}$
$[2,3,1]$	$\frac{(z_3-z_1)\left(z_3^{a+1} - z_2^{a+1}\right) z_1^{a+b+1} + (z_2-z_3)\left(z_3^{a+1} - z_1^{a+1}\right) z_2^{a+b+1}}{(z_1-z_2)(z_1-z_3)(z_2-z_3)}$
$[3,2,1]$	$\frac{z_1^{a+1}\left(z_3^{a+b+2} - z_2^{a+b+2}\right) + \left(z_2^{a+1} - z_3^{a+1}\right) z_1^{a+b+2} + z_2^{a+1} z_3^{a+1}\left(z_2^{b+1} - z_3^{b+1}\right)}{(z_1-z_2)(z_1-z_3)(z_2-z_3)}$

Table 3. Values of $\kappa_{k\lambda,\sigma}(z_1, z_2, z_3)$ after taking the limit $z_i \to 1$. Note that a and b have been multiplied by k.

σ	$\kappa_{k\lambda,\sigma}(1,1,1) = i(\mathcal{GT}(\lambda,\sigma),k)$
$[1,2,3]$	1
$[2,1,3]$	$1 + bk$
$[1,3,2]$	$1 + ak$
$[3,1,2]$	$\frac{1}{2}(bk+1)(2ak+bk+2)$
$[2,3,1]$	$\frac{1}{2}(ak+1)(2bk+ak+2)$
$[3,2,1]$	$\frac{1}{2}(ak+1)(bk+1)(2+ak+bk)$

Conjecture 5.1. *The Ehrhart polynomial $i(\mathcal{GT}(\lambda,\sigma),k) \in \mathbb{Q}[k]$ has only non-negative coefficients for all λ, σ. Furthermore, in the new variables variables $\lambda_i = a_1 + a_2 + \cdots + a_i$, we have*

$$i(\mathcal{GT}(\lambda,\sigma),k) \in \mathbb{Q}[k, a_1, a_2, \ldots, a_n] \tag{19}$$

with non-negative coefficients.

Above, we verified this for $\sigma \in \mathfrak{S}_3$, and we have also verified this for $\sigma \in \mathfrak{S}_4$. The symbolic computations done in Mathematica become tedious for larger values of n.

5.1. *A determinant formula*

An interesting special case was considered in [24], where σ is 231-avoiding and $i(\mathcal{GT}(\lambda,\sigma),k)$ reduces to the Ehrhart polynomial of a single reduced Kogan face of $\mathcal{GT}(\lambda)$, with type $w_0\sigma$. The number of 231-avoiding permutations in $\mathfrak{S}_n$ is given by the Catalan numbers, $\frac{1}{n+1}\binom{2n}{n}$. For an excellent survey on Catalan numbers, see the book by R. Stanley [31].

From [24, Corollary 14.6], it follows that the Ehrhart polynomial $i(\mathcal{GT}(\lambda, \sigma), k)$ for a 231-avoiding permutation $\sigma \in \mathfrak{S}_n$ can be obtained as the determinant

$$i(\mathcal{GT}(\lambda, \sigma), k) = \det\left(\binom{k\lambda_i + b_i - i}{b_i - j}\right)_{1 \leq i,j \leq n} \tag{20}$$

where $b_1, b_2, \dots, b_n$ form a sequence of non-negative integers (determined by σ) satisfying

$$b_1 \leq b_2 \leq \cdots \leq b_n \leq n \text{ and } b_i \geq i \text{ for } i = 1, 2, \dots, n.$$

Such sequences are in bijection with 231-avoiding permutations in $\mathfrak{S}_n$. Conjecture 5.1 thus states that for every integer partition λ, the determinant in (20) has non-negative coefficients (as a polynomial in k).

The reason (20) exists is due to a Jacobi–Trudi identity for flagged Schur polynomials, as key polynomials are certain flagged Schur polynomials in this special setting.

6. Faces of Gelfand–Tsetlin polytopes

As a consequence of Conjecture 5.1 and Proposition 3.1, it would follow that certain faces of Gelfand–Tsetlin polytopes have non-negative coefficients in the Ehrhart polynomial. It is therefore appropriate to investigate Ehrhart polynomials of general faces of $\mathcal{GT}(\lambda)$.

However, in this generality there are certain faces where negative coefficients appear. The following example was constructed in [2]:

Example 6.1. Consider the following face $\mathcal{F}_\ell$ of $\mathcal{GT}(\lambda)$ where $\lambda = (1^\ell, 0^{\ell+1})$:

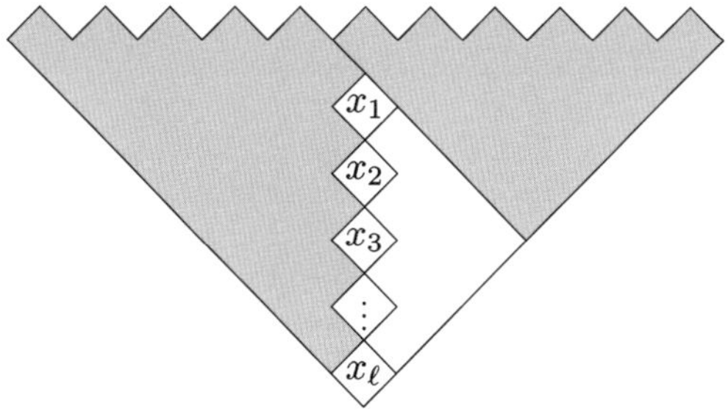

All entries in each respective region are forced to be equal by additional constraints, thus giving rise to a face of a Gelfand–Tsetlin polytope. One can construct a bijection between lattice points in this face, and lattice points in a certain order polytope. It is then straightforward to prove that

$$i(\mathcal{F}_\ell, k) = \sum_{j=1}^{k+1} j^\ell \in \mathbb{Q}[k]$$

and this polynomial has a negative coefficient for $\ell = 20$.

For more background on negative Ehrhart coefficients of order polytopes, we refer to [19].

Acknowledgements

The authors would like to thank Osaka University for their hospitality, in particular T. Hibi and A. Tsuchiya. The first author is funded by the *Knut and Alice Wallenberg Foundation* (2013.03.07).

References

[1] Per Alexandersson. Gelfand–Tsetlin polytopes and the integer decomposition property. *European Journal of Combinatorics*, 54:1–20, 2016.

[2] Per Alexandersson. Polytopes and large counterexamples. *Experimental Mathematics*, pages 1–6, sep 2017.

[3] Per Alexandersson and Mehtaab Sawhney. Properties of non-symmetric Macdonald polynomials at $q = 0$ and $q = 1$. *ArXiv e-prints*, 2018.

[4] Michel Demazure. Désingularisation des variétés de schubert généralisées. *Annales scientifiques de l´École normale supérieure*, 7(1):53–88, 1974.

[5] Michel Demazure. Une nouvelle formule des caractères. *Bulletin des Sciences Mathématiques*, 98(3):163–172, 1974.

[6] Boris Feigin and Igor Makhlin. A combinatorial formula for affine Hall–Littlewood functions via a weighted Brion theorem. *Selecta Mathematica*, 22(3):1703–1747, mar 2016.

[7] William Fulton and Joe Harris. *Representation theory, a first course.* Springer-Verlag New York Inc., 1991.

[8] Pavel Gusev, Valentina Kiritchenko, and Vladlen Timorin. Counting vertices in Gelfand–Zetlin polytopes. *J. Comb. Theory, Ser. A*, 120(4):960–969, 2013.

[9] James Haglund, Mark Haiman, and Nick Loehr. A Combinatorial Formula for Non-symmetric Macdonald Polynomials. *American Journal of Mathematics*, 130(2):359–383, 2008.

[10] Katharina Jochemko and Raman Sanyal. Arithmetic of marked order polytopes, monotone triangle reciprocity, and partial colorings. *SIAM Journal on Discrete Mathematics*, 28(3):1540–1558, jan 2014.

[11] R.C. King, C. Tollu, and F. Toumazet. Stretched Littlewood–Richardson coefficients and Kostka coefficients. *CRM Proceedings and Lecture Notes*, (34):99–112, 2004.

[12] Valentina A. Kirichenko, Evgeny Yu Smirnov, and Vladlen A. Timorin. Schubert calculus and Gelfand–Zetlin polytopes. *Russian Mathematical Surveys*, 67(4):685, 2012.

[13] A. N. Kirillov and N. Yu. Reshetikhin. The Bethe ansatz and the combinatorics of Young tableaux. *Journal of Mathematical Sciences*, 41:925–955, 1988.

[14] Donald E. Knuth. *The Art of Computer Programming, Volume 3: (2nd Ed.) Sorting and Searching.* Addison Wesley Longman Publishing Co., Inc., Redwood City, CA, USA, 1998.

[15] Mikhail Kogan. *Schubert Geometry of Flag Varieties and Gelfand–Cetlin Theory.* PhD thesis, Massachusetts Institute of Technology, June 2000.

[16] Daniel Kurland. An exploration of key polynomials. B.s thesis, University of Pennsylvania, 2016.

[17] Alain Lascoux and Marcel-Paul Schützenberger. Keys & standard bases. In *Invariant theory and tableaux (Minneapolis, MN, 1988)*, volume 19 of *IMA Vol. Math. Appl.*, pages 125–144. Springer, New York, 1990.

[18] Fu Liu. On positivity of Ehrhart polynomials. In *Recent Trends in Algebraic Combinatorics*, volume 16 of *Association for Women in Mathematics Series*, pages 189–237, 2019.

[19] Fu Liu and Akiyoshi Tsuchiya. Stanley's non-Ehrhart-positive order polytopes. *Adv. in Appl. Math.* 108:1–10, 2019.

[20] Jesús A. De Loera and Tyrrell B. McAllister. Vertices of Gelfand–Tsetlin polytopes. *Discrete & Computational Geometry*, 32(4):459–470, sep 2004.

[21] Sarah K. Mason. An explicit construction of type A Demazure atoms. *Journal of Algebraic Combinatorics*, 29(3):295–313, 2009.

[22] Tyrrell B. McAllister. Degrees of stretched Kostka coefficients. *J. Algebraic Comb.*, 27(3):263–273, May 2008.

[23] Alejandro H. Morales, Igor Pak, and Greta Panova. Hook formulas for skew shapes i. q-analogues and bijections. *Journal of Combinatorial Theory, Series A*, 154:350–405, feb 2018.

[24] Alexander Postnikov and Richard P. Stanley. Chains in the Bruhat order. *Journal of Algebraic Combinatorics*, 29(2):133–174, mar 2008.

[25] Anna Pun. *On Decomposition of the Product of Demazure Atoms and Demazure Characters*. PhD thesis, University of Pennsylvania, PA, USA, nov 2016.

[26] Etienne Rassart. A polynomiality property for Littlewood–Richardson coefficients. *J. Comb. Theory Ser. A*, 107(2):161–179, August 2004.

[27] Victor Reiner and Mark Shimozono. Key polynomials and a flagged Littlewood–Richardson rule. *J. Combin. Theory Ser. A*, 70(1):107–143, 1995.

[28] Richard P. Stanley. Two poset polytopes. *Discrete & Computational Geometry*, 1(1):9–23, 1986.

[29] Richard P. Stanley. *Enumerative Combinatorics: Volume 2*. Cambridge University Press, 1st edition, 2001.

[30] Richard P. Stanley. *Enumerative Combinatorics: Volume 1*. Cambridge University Press, 2nd edition, 2011.

[31] Richard P. Stanley. *Catalan Numbers*. Cambridge University Press, 2015.

Families of 3-dimensional polytopes of mixed degree one

Gabriele Balletti

Department of Mathematics, Stockholm University
SE-106 91, Stockholm, Sweden
E-mail: balletti@math.su.se

Christopher Borger

Fakultät für Mathematik, Institut für Algebra und Geometrie
Otto-von-Guericke-Universität Magdeburg
Universitätsplatz 2, 39106, Magdeburg, Germany
E-mail: christopher.borger@ovgu.de

The mixed degree of a family of n-dimensional lattice polytopes $P_1, \ldots, P_n$ is a generalization of the lattice degree of a single lattice polytope introduced as a special case by Soprunov and in full generality by Nill. A classification for the mixed degree zero case follows from a result by Cattani et al., while Soprunov posed the question to classify families of mixed degree one. In dimension three, these are families P_1, P_2, P_3 for which the Minkowski sum of any pair is a hollow polytope. As a preliminary study for tackling the question for a general classification we present a classification of families of 3-dimensional lattice polytopes of mixed degree one that are exceptional, in the sense that the Minkowski sum of at least one of the pairs is hollow without having a lattice projection onto a lower-dimensional hollow polytope.

Keywords: Lattice polytopes; mixed degree; Minkowski sum.

1. Introduction

The *mixed codegree* $\operatorname{mcd}(P_1, \ldots, P_n)$ of a family of lattice polytopes $P_1, \ldots, P_n \subset \mathbb{R}^n$ is defined as the smallest natural number c such that there exists a set $I \subseteq [n]$ with $|I| = c$ and with the polytope $P_I = \sum_{i \in I} P_i$ containing an interior lattice point. If no such I exists (that is, if $P_1 + \cdots + P_n$ is *hollow*) we set $\operatorname{mcd}(P_1, \ldots, P_n) := n+1$. The *mixed degree* is then defined as

$$\operatorname{md}(P_1, \ldots, P_n) := n + 1 - \operatorname{mcd}(P_1, \ldots, P_n).$$

This definition first appears in a recent paper of Nill [10], where it is also shown that the mixed degree is a nonnegative integer. Note that, if $P_1 = \cdots = P_n$, we recover the classical definition of degree as in Ehrhart theory (as defined e.g. in [5]). Here we are going to focus on families $P_1, \ldots, P_n$ of lattice polytopes of mixed degree at most one where all $P_i \subset \mathbb{R}^n$ are n-dimensional. As it was proven in [12] these are the ones minimizing the number of interior points in their Minkowski sum $P_1 + \cdots + P_n$ for a given mixed volume $V(P_1, \ldots, P_n)$. More explicitly, they are the families that fulfill the following inequality, that holds in general for n-dimensional polytopes, with equality:

$$|\operatorname{int}_{\mathbb{Z}}(P_1 + \cdots + P_n)| \geq V(P_1, \ldots, P_n) - 1,$$

where $V(P_1, \dots, P_n) := \sum_{I \subseteq [n]} (-1)^{n-|I|} v(P_I)$ denotes the lattice mixed volume of $P_1, \dots, P_n$ and $v(P_I)$ is the euclidean volume of P_I. Again, this generalizes the classical setting of Ehrhart Theory: if $P_1 = \cdots = P_n$, then the above inequality is fulfilled with equality if and only if the degree of the polytope is at most one. In [6] Soprunov already defined such families as the families of mixed degree one before the general notion of the mixed degree was introduced in [10]. Additionally, he posed the question for a classification of all such families. Here we begin the study of such a classification, starting from the 3-dimensional case.

Note that the mixed degree zero case has been solved by Cattani et al. [7].

Proposition 1.1 ([7, Proposition 2.7]). *Let $P_1, \dots, P_n$ be n-dimensional lattice polytopes. Then the mixed degree $\operatorname{md}(P_1, \dots, P_n)$ equals zero if and only if $P_1, \dots, P_n$ equal (up to translation) the same unimodular n-simplex.*

In the classical setting, a characterization of lattice polytopes of degree one has been given by Batyrev–Nill [5]. Given an n-dimensional lattice polytope $Q \subset \mathbb{R}^n$ with respect to $\mathbb{Z}^n$, we define the *lattice pyramid* $\operatorname{Pyr}(Q)$ as the $(n+1)$-dimensional polytope

$$\operatorname{Pyr}(Q) := \operatorname{conv}(Q \times \{0\} \cup \{\mathbf{e}_{n+1}\}) \subset \mathbb{R}^{n+1}.$$

The lattice pyramid construction preserves the degree (it actually preserves the h^*-vector too). We say that an n-dimensional lattice polytope Q is an *exceptional simplex* if Q can be obtained via $n-2$ iterations of the lattice pyramid construction over the second dilation of a unimodular triangle, that is,

$$Q = \operatorname{Pyr}(\cdots(\operatorname{Pyr}(\operatorname{conv}(\{\mathbf{0}, 2\mathbf{e}_1, 2\mathbf{e}_2\})))\cdots).$$

We say that an n-dimensional lattice polytope Q is a *Lawrence prism* if there exist nonnegative integers $a_0, \dots, a_{n-1}$ such that

$$Q = \operatorname{conv}(\{\mathbf{0}, a_0\mathbf{e}_n, \mathbf{e}_1, \mathbf{e}_1 + a_1\mathbf{e}_n, \dots, \mathbf{e}_{n-1}, \mathbf{e}_{n-1} + a_{n-1}\mathbf{e}_n\}).$$

We say that two lattice polytopes $P, Q \subset \mathbb{R}^n$ are *equivalent* if there exists a lattice preserving unimodular transformation $\varphi\colon \mathbb{R}^n \to \mathbb{R}^n$ such that $\varphi(P) = Q$.

Theorem 1.1 ([5, Theorem 2.5]). *Let P be a lattice polytope. Then $\deg(P) \leq 1$ if and only if P is equivalent to an exceptional simplex or a Lawrence prism.*

One can trivially generate infinitely many families $P_1, \dots, P_n$ having mixed degree one by setting $P_i = P$ for all i, for a degree one n-dimensional polytope. One can also be slightly more sophisticated and obtain a family having mixed degree one such that all the P_i are different polytopes, as in Example 1.1. In order to formalize this construction we introduce lattice projections. By *lattice projection*, we denote a surjective linear map $\varphi\colon \mathbb{R}^n \to H$ where $H \subseteq \mathbb{R}^n$ is an m-dimensional subspace such that the kernel of φ is linearly generated by lattice points of $\mathbb{Z}^n$ and φ restricted to H is the identity. The lattice projection φ is uniquely determined by its kernel. The image of one element of the lattice $\mathbb{Z}^n$ will be considered as an element of the rank m lattice $\varphi(\mathbb{Z}^n) \subset H$. Sometimes it can be convenient to identify H with $\mathbb{R}^m$ and $\varphi(\mathbb{Z}^n)$ with $\mathbb{Z}^m$.

Example 1.1. For all $1 \leq i \leq n$, let P_i be an n-dimensional lattice polytope inside $\Delta_{n-1} \times \mathbb{R}$. Then $P_1, \ldots, P_n$ has mixed degree at most one, indeed $P_I = \sum_{i \in I} P_i$ is hollow for any proper subsets $I \subsetneq [n]$. To check this it is enough to notice that P_I has a lattice projection to $c\Delta_{n-1}$, with $c = |I|$, which is hollow.

The following is a nontrivial example of a family P_1, P_2, P_3 of mixed degree one for which there are no pairwise common projections onto (translates) of the same unimodular triangle.

Example 1.2. The family P_1, P_2, P_3 defined as

$$\begin{aligned}
P_1 &= \operatorname{conv}(\{\mathbf{0}, \mathbf{e}_1 + 2\mathbf{e}_2, \mathbf{e}_1 + \mathbf{e}_2, -\mathbf{e}_1 + \mathbf{e}_3, -3\mathbf{e}_1 + 2\mathbf{e}_3)\},\\
P_2 &= \operatorname{conv}(\{\mathbf{0}, \mathbf{e}_1, -\mathbf{e}_1 + \mathbf{e}_3, -\mathbf{e}_1 + \mathbf{e}_2 + \mathbf{e}_3, \mathbf{e}_1 + 2\mathbf{e}_2)\},\\
P_3 &= \operatorname{conv}(\{\mathbf{0}, \mathbf{e}_1, \mathbf{e}_1 + \mathbf{e}_2, -\mathbf{e}_1 + \mathbf{e}_2 + \mathbf{e}_3, -3\mathbf{e}_1 + 2\mathbf{e}_3)\},
\end{aligned}$$

has mixed degree one. Indeed the Minkowski sums $P_1 + P_2$, $P_1 + P_3$ and $P_2 + P_3$ are hollow, while $P_1 + P_2 + P_3$ is not (see Figure 1). In particular $P_1 + P_2 + P_3$ has three interior lattice points, so the mixed volume of P_1, P_2, P_3 is four.

In order to avoid having to deal with duplicates, we will say that two families $P_1, \ldots, P_n \subset \mathbb{R}^n$ and $Q_1, \ldots, Q_n \subset \mathbb{R}^n$ are *equivalent* if there is a permutation $\sigma \in S_n$, a lattice preserving unimodular transformation $U \colon \mathbb{R}^n \to \mathbb{R}^n$ and vectors $t_1, \ldots, t_n \in \mathbb{Z}^n$ such that $(P_1, \ldots, P_n) = (U(\sigma(Q_1) + t_1), \ldots, U(\sigma(Q_n) + t_n))$. In Section 2 we describe a criterion to easily check equivalence using Cayley sums.

1.1. *Summary and main result*

In Section 2 we are going to introduce some necessary tools for the classification. Furthermore we define the notion of a certain simplicial complex, denoted the projection constellation associated with a family of lattice polytopes. Based on this notion we can structure the classification of the mixed families of degree one in dimension three into five different cases, one for every possible simplicial complex with three vertices (see Figure 2). Cases (A), (B) and (C) will be completely enumerated in Sections 3 and 4. In a subsequent paper ([3]) we show that these are exactly the cases in which the number of families is finite up to equivalence. While case (E) is trivial, as all the families are of the form given in Example 1.1, case (D) surprisingly unveils infinite non-equivalent families without any pair having a common projection onto a unimodular triangle.

The following is the main result of this note.

Theorem 1.2. *Let $P_1, P_2, P_3 \subset \mathbb{R}^3$ be a family of 3-dimensional lattice polytopes with* $\operatorname{md}(P_1, P_2, P_3) = 1$. *Then*

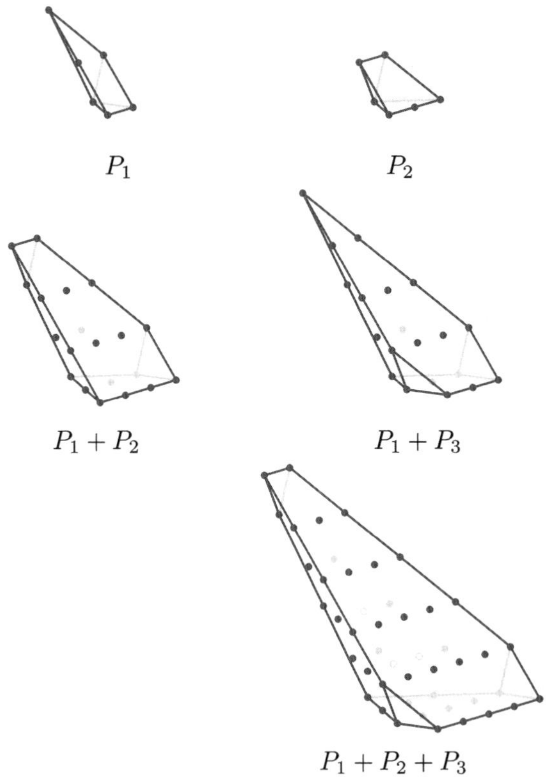

Fig. 1. An example of a family P_1, P_2, P_3 having mixed degree 1.

(A) if there is no pair in P_1, P_2, P_3 that has a common lattice projection onto translates of the same unimodular triangle, then P_1, P_2, P_3 is equivalent to one out of 29 *possible families,*
(B) if there is exactly one pair in P_1, P_2, P_3 having a common lattice projection onto translates of the same unimodular triangle, then P_1, P_2, P_3 is equivalent to one out of 141 *possible families,*
(C) if there are exactly two pairs in P_1, P_2, P_3 having a common lattice projections onto translates of the same unimodular triangle, then P_1, P_2, P_3 is equivalent to one out of 82 *possible families.*

The three parts of the statement of this theorem correspond to the cases (A), (B) and (C) mentioned earlier.

Finally in Section 5 we give an additional structure to this result by providing a list of 6 inclusion-maximal families of mixed degree one such that all families of the above result are contained in one of the maximals. All computations have been carried out using Magma [4] and the code can be found at `https://github.com/gabrieleballetti/mixed_degree_one`.

Acknowledgments

We would like to thank Takayuki Hibi and Akiyoshi Tsuchiya for the organization of the Summer Workshop on Lattice Polytopes in Osaka, providing the framework in which these notes have been written. Furthermore we would like to thank Benjamin Nill for pointing us to the question and for helpful discussions. The first author is partially supported by the Vetenskapsrådet grant NT:2014-3991 and the second author is funded by the Deutsche Forschungsgemeinschaft (DFG, German Research Foundation) - 314838170, GRK 2297 MathCoRe.

2. Toolbox

In this section we are going to introduce some preliminaries and notation that will be important during our classification.

2.1. *Cayley sums and equivalence of families*

While there is a standard way of determining whether two lattice polytopes are equivalent by comparing their Hermite normal forms (see e.g. [9]) the task of checking two families for equivalence seems a priori more complicated. However, we will show that the construction of the Cayley sum of a family actually allows us to reduce the problem of checking equivalence of families to checking equivalence of two higher-dimensional lattice polytopes. In order to phrase and prove this statement let us quickly recall the definition and some basic properties of the Cayley sum of a family of lattice polytopes.

Definition 2.1. Let $P_1, \dots, P_k \subset \mathbb{R}^n$ be (possibly empty) lattice polytopes. We define the *Cayley sum* $P_1 * \cdots * P_k$ as

$$P_1 * \cdots * P_k := \operatorname{conv}((P_1 \times \{0\}) \cup (P_2 \times \{e_1\}) \cup \cdots \cup (P_k \times \{e_{k-1}\})) \subset \mathbb{R}^{n+k-1}.$$

We call a Cayley sum $P_1 * \cdots * P_k$ *proper* if all P_i are non-empty. In this case one has $\dim(P_1 * \cdots * P_k) = \dim(P_1 + \cdots + P_k) + k - 1$.

The following characterization is straightforward.

Proposition 2.1. *Let $P \subset \mathbb{R}^n$ be a lattice polytope. Then the following are equivalent.*

(1) There exists a lattice projection $\pi\colon \mathbb{R}^n \to \mathbb{R}^{k-1}$ with $\pi(P) = \Delta_{k-1}$,
(2) there exist lattice polytopes $P_1, \dots, P_k \subset \mathbb{R}^{n-k+1}$ such that $P \cong P_1 * \cdots * P_k$.

Remark 2.1. Let $P = P_1 * \cdots * P_k$. As a consequence of the well-known connection between the faces of the Cayley sum and faces of certain Minkowski sums (see e.g. [8]), all faces of P are of the form $F = F_1 * \cdots * F_k$ for (possibly empty) faces $F_i \subseteq P_i$. One has $\dim(F) = \dim(F_1 + \cdots + F_k) + l - 1$ where l is the number of F_i that are nonempty. In particular any of the P_i corresponds to a face of P which we will denote by $\hat{P}_i$. Furthermore also $\hat{P}_i^c := \text{conv}(v \in V(P) \setminus V(\hat{P}_i))$, where $V(P)$ denotes the set of vertices of P, is always a proper face of P.

Let us now formulate the lemma which allows us to determine equivalence of families simply by checking equivalence of their Cayley sums given the families do not have a certain very special structure.

Lemma 2.1. *Let $P_1, \ldots, P_k \subset \mathbb{R}^n$ be a family of n-dimensional lattice polytopes such that $P_1, \ldots, P_k$ do not have a common lattice projection onto a unimodular $(k-1)$-simplex and let $Q_1, \ldots, Q_k \subset \mathbb{R}^n$ be any family of n-dimensional lattice polytopes. Then the families $P_1, \ldots, P_k$ and $Q_1, \ldots, Q_k$ are equivalent if and only if their Cayley sums $P_1 * \cdots * P_k$ and $Q_1 * \cdots * Q_k$ are equivalent.*

Proof. For the first implication we need to show $U(\sigma(P_1) + t_1) * \cdots * U(\sigma(P_k) + t_k) \cong P_1 * \cdots * P_k$ for $\sigma \in S_k$ a permutation, $U : \mathbb{R}^n \to \mathbb{R}^n$ a lattice preserving unimodular transformation and $t_1, \ldots, t_k \in \mathbb{Z}^n$. This is true as translating one of the summands P_i corresponds to shearing the Cayley sum and applying a common unimodular transformation corresponds to applying the unimodular transformation $U \times \text{id} : \mathbb{R}^n \times \mathbb{R}^{k-1} \to \mathbb{R}^n \times \mathbb{R}^{k-1}$ to the Cayley sum. Finally, applying a permutation σ to the P_i corresponds to permuting the Cayley coordinates $\mathbf{e}_{n+1}, \ldots, \mathbf{e}_{n+k-1}$ and $\mathbf{0}$. All of these transformations are unimodular.

For the converse suppose now that $P_1, \ldots, P_k$ do not have a common projection onto translates of the same unimodular $(k-1)$-simplex and suppose $P_1 * \cdots * P_k \cong Q_1 * \cdots * Q_k$. Let $P_1', \ldots, P_k'$ be the images of the $\hat{P}_i$ under that isomorphism. We will show that up to renumbering $P_i' = \hat{Q}_i$ for all $i \in [k]$. Suppose without loss of generality $P_1' \neq \hat{Q}_i$ for all $i \in [k]$. The face P_1' cannot properly contain any $\hat{Q}_i$ (by Remark 2.1 any face of $Q_1 * \cdots * Q_k$ that properly contains a $\hat{Q}_i$ is of dimension greater or equal to $n+1$). Together with the assumption that $P_1' \neq \hat{Q}_i$ for all $i \in [k]$ this implies that $(P_1')^c$ has nonempty intersection with all $\hat{Q}_i$. Therefore also P_1' cannot be disjoint to any $\hat{Q}_i$ as in that case $(P_1')^c$ would be a Cayley sum of length k with one Cayley summand being n-dimensional and therefore $\dim((P_1')^c) = n+k-1$. This is a contradiction to $(P_1')^c$ being a proper face of the $(n+k-1)$-dimensional Cayley sum (see Remark 2.1). As P_1' was chosen arbitrarily this argumentation implies that all P_i' have non-empty intersection with all $\hat{Q}_j$. Therefore the natural projection of $Q_1 * \cdots * Q_k$ onto Δ_{k-1} commonly projects all of the P_i' onto Δ_{k-1} in contradiction to our assumption.

This shows that, if $P_1, \ldots, P_k$ do not have a common projection onto a unimodular $(k-1)$-simplex, any lattice-preserving unimodular transformation $U : \mathbb{R}^{n+k-1} \to \mathbb{R}^{n+k-1}$ mapping $P_1 * \cdots * P_k$ to $Q_1 * \cdots * Q_k$ yields (up to renumbering) a bijection mapping the face $\hat{P}_i$ to the face $\hat{Q}_i$. We may without loss of generality assume that U preserves the origin. Restricting U to a map from the affine hull of P_1 to the affine hull of Q_1 we obtain a lattice-preserving unimodular transformation $L : \mathbb{R}^n \to \mathbb{R}^n$. It is straightforward to verify that restricting U to the affine hull of any other P_i results in a map $x \mapsto L(x) + t_i$ from $\mathbb{R}^n \to \mathbb{R}^n$ for some $t_i \in \mathbb{Z}^n$.

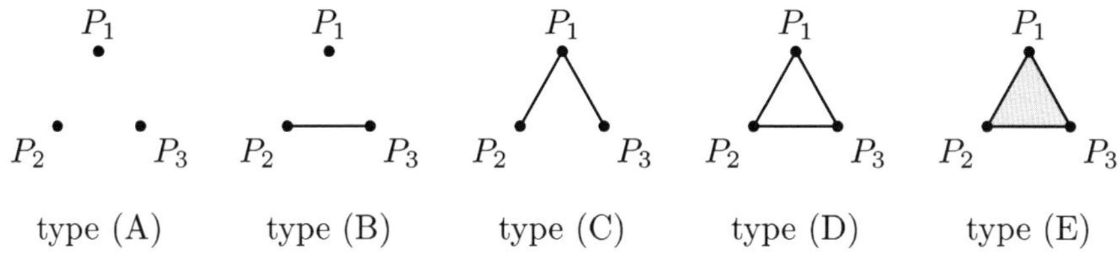

Fig. 2. The possible projection constellations for $n = 3$.

□

2.2. *Simplicial complexes and case distinction*

Let $P_1, \dots, P_n \subset \mathbb{R}^n$ be a family of lattice polytopes of mixed degree one. Then, by definition, the Minkowski sum P_I is hollow for all $I \subset [n]$ with $|I| = n-1$. By [11] we have the following structural result for lattice polytopes of that type.

Proposition 2.2. *Let $P_1, \dots, P_n \subset \mathbb{R}^n$ be lattice polytopes such that $P_1 + \dots + P_n$ is hollow. Then either*

(1) $P_1, \dots, P_n$ have a common lattice projection onto the same unimodular $(n-1)$-simplex, or
(2) $P_1 + \dots + P_n$ is a subpolytope of one of finitely many hollow n-polytopes which are inclusion-maximal.

We will call $(n-1)$-families of n-polytopes of type 2 *exceptional*. We want to distinguish between families $P_1, \dots, P_n \subset \mathbb{R}^n$ depending on which of their $(n-1)$-subfamilies are of which type.

Given a family of n-dimensional lattice polytopes $P_1, \dots, P_n$, we associate to it a simplicial complex $\Delta_{P_1,\dots,P_n}$ on the vertex set $\{P_1, \dots, P_n\}$ counting the common projections of the polytopes to a $(n-1)$-dimensional unimodular simplex. Specifically, $\sigma \subseteq [n]$ is a face of $\Delta_{P_1,\dots,P_n}$ if there is a unimodular $(n-1)$-simplex S and a lattice-preserving projection $\varphi : \mathbb{R}^n \to \mathbb{R}^{n-1}$ such that $\varphi(P_i) = S + t_i$ for all $i \in \sigma$ and some $t_i \in \mathbb{Z}^n$. We will call $\Delta_{P_1,\dots,P_n}$ the *projection constellation* associated to $P_1, \dots, P_n$.

In this note we are interested in the case $n = 3$. In this setting there are five different possibilities for Δ_{P_1,P_2,P_3}, as shown in Figure 2. Our goal is to classify all families $P_1, P_2, P_3 \subset \mathbb{R}^3$ whose projection constellation is of type (A), (B) or (C). In a subsequent paper we are going to show that these are exactly the cases in which the number of families is finite up to equivalence.

2.3. *Hollow 3-polytopes with width larger than one*

As an intermediate step towards the classification we are going to calculate all (equivalence classes of) exceptional pairs of 3-polytopes. In order to do that we consider the list of maximal hollow 3-polytopes classified by Averkov–Wagner–Weismantel [2] (see also [1]), and compute all subpolytopes of the maximal hollows that have lattice width greater than one.

Proposition 2.3 ([2, Corollary 1]). *Let P be a hollow 3-dimensional lattice polytope of width at least two. Then, up to a lattice transformation, P is contained either in the unbounded polyhedron $2\Delta_2 \times \mathbb{R}$ or in one of the 12 maximal hollow lattice polytopes of Figure 3.*

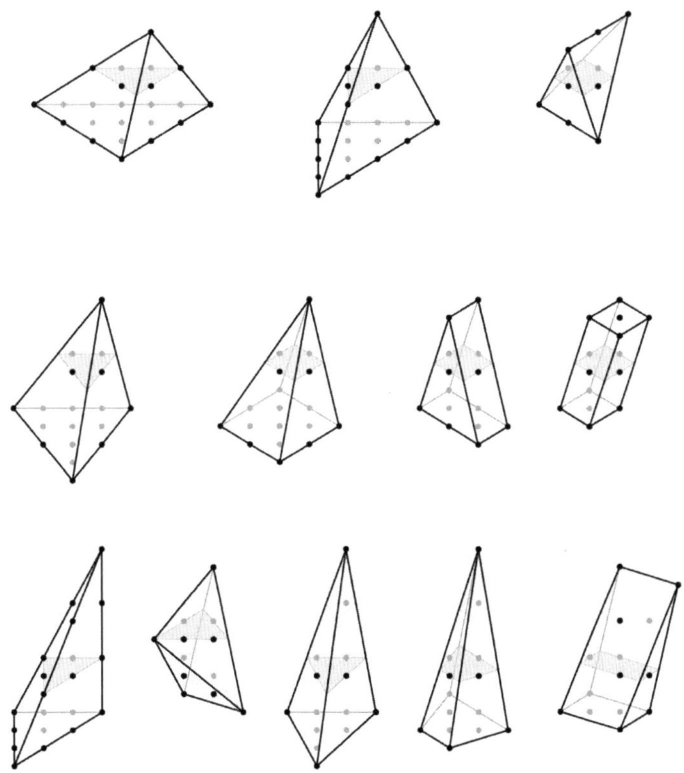

Fig. 3. All the hollow 3-dimensional lattice polytopes which are maximal up to inclusion. The polytopes in the first two rows have lattice-width two, while the last row contains polytopes of lattice-width three (picture from [1]).

As we are interested in obtaining a list of exceptional pairs $P, Q \subset \mathbb{R}^3$ we use an implementation in magma in order to compute the decompositions of all subpolytopes of the 12 maximal hollow polytopes into Minkowski sums of two 3-dimensional polytopes. Afterwards we determine those couples that do actually not have a common projection onto the same unimodular triangle and then determine equivalent couples using Lemma 2.1.

Proposition 2.4. *There are (up to equivalence) 32 pairs of lattice 3-polytopes P, Q whose Minkowski sum is hollow and does not have a projection onto a lower-dimensional hollow polytope.*

3. Cases (A) and (B)

In this section we discuss cases (A) and (B). In this setting P_1, P_2, P_3 is a family of 3-dimensional polytopes such that $\mathrm{md}(P_1, P_2, P_3) = 1$ and at most one of the couples (P_1, P_2), (P_1, P_3), (P_2, P_3) has a common projection onto Δ_2. As usual, by "have a common projection on Δ_2", we mean that there exists a lattice projection $\varphi : \mathbb{R}^3 \to \mathbb{R}^2$, such that $\varphi(P_i)$ and $\varphi(P_j)$ are translations of the same unimodular triangle.

In case (A) the Minkowski sums $P_1 + P_2$, $P_1 + P_3$ and $P_2 + P_3$ are hollow 3-dimensional polytopes which do not project to $2\Delta_2$. In particular, by Proposition 2.3, any of the Minkowski sum $P_1 + P_2$, $P_1 + P_3$ and $P_2 + P_3$ is contained in one of the 12 maximal hollow 3-dimensional polytopes of Proposition 2.3.

Similarly, in case (B), any of the Minkowski sums P_1+P_3 and P_2+P_3 is contained in one of the 12 maximal hollow 3-dimensional polytopes of Proposition 2.3.

Thus, we immediately deduce a finiteness result for the families of these cases.

Proposition 3.1. *Up to equivalence, there are finitely many families P_1, P_2, P_3 of 3-dimensional lattice polytopes such that $\mathrm{md}(P_1, P_2, P_3) = 1$ and the projection constellation Δ_{P_1,P_2,P_3} is either of type (A) or (B).*

In order to perform an actual enumeration of such families we go through the list of 32 couples of lattice polytopes of Proposition 2.4 and use them to generate the triples.

Theorem 3.1. *Up to equivalence, there are 170 possible families P_1, P_2, P_3 of 3-dimensional lattice polytopes such that $md(P_1, P_2, P_3) = 1$ and the projection constellation Δ_{P_1,P_2,P_3} is either of type (A) or type (B). In the first case there are 29 triples, in the latter they are 141.*

Proof. We can assume that (P_1, P_2) is one of the couples of Proposition 2.4. Once (P_1, P_2) is fixed, then (P_1, P_3) also has to be (up to equivalence and reordering) one of the couples mentioned above. At this point we have finitely many candidates for all the possible families P_1, P_2, P_3 as in the statement. If $P_2 + P_3$ is not hollow, then $\mathrm{md}(P_1, P_2, P_3) > 1$, so all such cases can be excluded. If on the other hand $P_2 + P_3$ is hollow, then $\mathrm{md}(P_1, P_2, P_3) = 1$ and we can make a distinction of the two cases just by checking if P_2 and P_3 have a common projection (up to translations) to a unimodular triangle. As these triples do not have a common projection onto translates of the same unimodular triangle by construction we may use Lemma 2.1 in order to determine equivalent triples. □

4. Case (C)

In this section we discuss case (C). Therefore $P_1, P_2, P_3 \subset \mathbb{R}^3$ is a family of 3-dimensional lattice polytopes with $\mathrm{md}(P_1, P_2, P_3) = 1$ and (without loss of generality) there are two lattice projections $\varphi_{12}\colon \mathbb{R}^3 \to \mathbb{R}^2$ and $\varphi_{13}\colon \mathbb{R}^3 \to \mathbb{R}^2$ where φ_{ij} maps P_i and P_j to translates of the same unimodular triangle. Furthermore we assume that φ_{12} and φ_{13} are different (as we assume not to be in case (E)). In this setting P_1 is a lattice polytope with two different lattice projections onto a unimodular triangle and we have the following.

Lemma 4.1. *Let P be a 3-dimensional lattice polytope projecting to a unimodular triangle via two different lattice projections. Then P is equivalent to Δ_3 or $\square_3 :=$ $\mathrm{Pyr}(\mathrm{conv}(\mathbf{0}, \mathbf{e}_1, \mathbf{e}_2, \mathbf{e}_1 + \mathbf{e}_2))$.*

Proof. As P has a lattice projection onto a unimodular triangle by Proposition 2.1 we may assume $P = I_1 * I_2 * I_3$ for some segments $I_i = [0, a_i]$ with $a_i \in \mathbb{Z}_{\geq 0}$. If one of the a_i was at least 2, any projection of P onto a unimodular triangle would have to be along the edge $\hat{I}_i$ corresponding to the segment I_i. This contradicts the assumption that P has two different lattice projections onto a unimodular triangle. Therefore $a_i \in \{0, 1\}$ for all $i \in [n]$. It is easy to check (by checking the projections along the edge directions) that the prism one obtains for $a_1 = a_2 = a_3 = 1$ does not have two different lattice projections onto a unimodular triangle and this proves the statement. $\square$

Let $P \subset \mathbb{R}^n$ be a (not necessarily n-dimensional) lattice polytope and $v \in \mathbb{Z}^n$ a lattice point. We call the polyhedron $P + \mathbb{R}v$ the *unbounded cylinder* in direction v over P.

Proposition 4.1. *Let $C_1 = \mathrm{conv}(v_1, v_2, v_3) + \mathbb{R}v \subset \mathbb{R}^3$ and $C_2 = \mathrm{conv}(w_1, w_2, w_3) + \mathbb{R}w \subset \mathbb{R}^3$ be cylinders over lattice triangles in directions $v, w \in \mathbb{Z}^3$ such that $\mathrm{conv}(v_1, v_2, v_3, v_1 + v)$ and $\mathrm{conv}(w_1, w_2, w_3, w_1 + w)$ both are unimodular 3-simplices and v and w are linearly independent. Define the lattice polytope $P_u := \mathrm{conv}(C_1 \cap (C_2 + u) \cap \mathbb{Z}^3)$ for $u \in \mathbb{Z}^3$ and let $u_1, u_2 \in \mathbb{Z}^3$. If P_{u_1} and P_{u_2} are both 3-dimensional, then they are bounded and P_{u_1} and P_{u_2} are the same up to translation.*

Proof. By Lemma 4.1 we know that, if P_u is 3-dimensional, then P_u is unimodularly equivalent either to Δ_3 or to $\square_3$ for any u and in particular P_u is bounded.

Assume now that there is a $u_1 \in \mathbb{Z}^3$ such that P_{u_1} is 3-dimensional. Again by Lemma 4.1 the vectors v and w are non-parallel edge-directions of a unimodular copy either of Δ_3 or $\square_3$. One easily verifies that this implies $\mathbb{Z}v + \mathbb{Z}w = (\mathbb{R}z + \mathbb{R}w) \cap \mathbb{Z}^3$. Furthermore P_{u_1} has another edge-direction $d \in \mathbb{Z}^3$ such that the polytope P_{u_1} only has lattice points in two different translates of the hyperplane defined by the normal vector d and such that v, w, d form a lattice basis of $\mathbb{Z}^3$. Let us therefore without loss of generality assume $d = \mathbf{e}_1$, $v = \mathbf{e}_2$ and $w = \mathbf{e}_3$ and that P_{u_1} only has lattice points of height 1 and 0 with respect to $\mathbf{e}_1$. As P_{u_1} is full-dimensional and therefore has lattice points on all edges of both C_1 and $C_2 + u_1$ this implies that any lattice point $p = (p_1, p_2, p_3) \in C_1$ and any lattice point $q = (q_1, q_2, q_3) \in C_2 + u_1$ satisfy $p_1, q_1 \in \{0, 1\}$. It is clear that shifting $C_2 + u_1$ by a lattice vector $s \in \mathbb{R}\mathbf{e}_2 + \mathbb{R}\mathbf{e}_3$ results in a shifted copy P_{u_1+s} of P_{u_1}. So the claim is true for $u_2 = u_1 + s$ for $s \in \mathbb{R}\mathbf{e}_2 + \mathbb{R}\mathbf{e}_3$.

On the other hand, shifting $C_2 + u_1$ by a lattice vector $s = (s_1, s_2, s_3) \in \mathbb{Z}^3$ with $s_1 \neq 0$ results in P_{u_1+s} having lattice points in at most the $\{\mathbf{e}_1 = 0\}$ or the $\{\mathbf{e}_1 = 1\}$ hyperplane (but not in both). Therefore P_{u_2} is lower-dimensional for all $u_2 \notin t_u + \mathbb{R}\mathbf{e}_2 + \mathbb{R}\mathbf{e}_3$.

□

As a corollary we obtain the finiteness of families in case (C).

Corollary 4.1. *Up to equivalence, there are finitely many families P_1, P_2, P_3 of 3-dimensional lattice polytopes such that* $\operatorname{md}(P_1, P_2, P_3) = 1$ *and the projection constellation Δ_{P_1,P_2,P_3} is of type (C).*

Proof. Let P_2, P_3 be the pair that is exceptional. Therefore we may assume that it equals one of the 32 pairs of Proposition 2.4. As we are in case (C) there exist different lattice projections $\varphi_{12}\colon \mathbb{R}^3 \to \mathbb{R}^2$ and $\varphi_{13}\colon \mathbb{R}^3 \to \mathbb{R}^2$ such that φ_{ij} maps P_i and P_j onto translates of the same unimodular 2-simplex. There are only finitely many choices for a pair $\varphi_{12}, \varphi_{13}$. As P_1 projects onto a unimodular 2-simplex both with P_2 and P_3 we have $P_1 \subseteq M_t := (P_2 + \ker\varphi_{12}) \cap (P_3 + \ker\varphi_{13} + t)$ for some $t \in \mathbb{Z}^3$ (up to translation of P_1). By Proposition 2.1 the cylinders $P_2 + \ker\varphi_{12}$ and $P_3 + \ker\varphi_{13}$ satisfy the assumptions of Proposition 4.1 and therefore all choices of $t \in \mathbb{Z}^3$ such that $\operatorname{conv}(M_t \cap \mathbb{Z}^3)$ is 3-dimensional result in translations of the same lattice polytope. This leaves us with finitely many possibilities (up to translations) to choose P_1 given P_2 and P_3. □

Using the fact that P_1 lies inside the intersection of certain cylinders over exceptional couples P_2, P_3 we can produce a list of all families in case (C) from the list of Proposition 2.4. As we may additionally assume that P_1 has $\mathbf{0}$ as a vertex we only need to check a finite number of translations of the cylinders. We fix a translation of, say, P_2 having $\mathbf{0}$ as a vertex and go through all translations of P_3 sending a vertex of P_3 to $\mathbf{0}$. If one of the intersections of the cylinders over these translations contains a 3-dimensional lattice polytope, we know that P_1 has to be contained in it.

Theorem 4.1. *Up to equivalence, there are 82 families of lattice polytopes P_1, P_2, P_3 such that* $\operatorname{md}(P_1, P_2, P_3) = 1$ *and the projection constellation Δ_{P_1,P_2,P_3} is of type (C).*

5. Maximal families

Unlike the classical setting of Theorem 1.1 where there is one exceptional case, namely a lattice pyramid over the second dilation of the unimodular triangle, the mixed setting presents a vast multitude of "exceptional cases". Indeed all the families of type (A), (B) or (C) do not trivially project to the same unimodular triangle. A behavior like that was to be expected as we have the following effect. The exceptional triple we obtain from the classical setting is $\mathrm{Pyr}(2\Delta_2), \mathrm{Pyr}(2\Delta_2), \mathrm{Pyr}(2\Delta_2)$. However, if for example we set P to be the lattice polytope obtained from $\mathrm{Pyr}(2\Delta_2)$ by chopping off one of the vertices of the base triangle, both the families $P, \mathrm{Pyr}(2\Delta_2), \mathrm{Pyr}(2\Delta_2)$ and $P, P, \mathrm{Pyr}(2\Delta_2)$ are of mixed degree one (as they are not copies of the same unimodular simplex and therefore of degree greater than 0). Furthermore $\mathrm{Pyr}(2\Delta_2)$ does not have a projection onto a unimodular triangle and therefore both triples are exceptional. However, such families are by construction contained in larger triples of mixed degree one and a natural way of deriving structural results about exceptional families is to isolate the effect described above by restricting to those families that are maximal with respect to inclusion. Indeed, of the 252 families of type (A), (B) or (C) that we have classified only 6 are inclusion-maximal.

Theorem 5.1. *All the mixed degree one triples P_1, P_2, P_3 of type (A), (B) or (C) are, up to equivalence, contained in one of the following* 6 *maximal families:*

- *the maximal family* $\mathrm{Pyr}(2\Delta_2), \mathrm{Pyr}(2\Delta_2), \mathrm{Pyr}(2\Delta_2)$,
- *the maximal family* $2\Delta_3, \Delta_3, \Delta_3$,
- *the maximal family* $\{\mathrm{conv}(\mathbf{0}, 2\mathbf{e}_i, \mathbf{e}_j, \mathbf{e}_k) \colon i, j, k \in [3] \text{ pairwise different}\}$,
- *the maximal family*

$$\begin{aligned}&\mathrm{conv}(\mathbf{e}_1, \mathbf{e}_2, -\mathbf{e}_2) * \mathrm{conv}(\mathbf{0}, \mathbf{e}_1),\\ &\mathrm{conv}(\mathbf{0}, \mathbf{e}_1, -\mathbf{e}_2) * \{-\mathbf{e}_2\},\\ &\mathrm{conv}(\mathbf{0}, \mathbf{e}_1, \mathbf{e}_2) * \{\mathbf{e}_2\},\end{aligned}$$

- *one of the two maximal families*

$$\begin{aligned}&\mathrm{conv}(\mathbf{0}, 2\mathbf{e}_2) * \mathrm{conv}(\mathbf{0}, \mathbf{e}_1),\\ &\mathrm{conv}(\mathbf{0}, -\mathbf{e}_1, -\mathbf{e}_1 - \mathbf{e}_2) * \{-\mathbf{e}_1 - 2\mathbf{e}_2\},\\ &\mathrm{conv}(\mathbf{0}, \mathbf{e}_2, -\mathbf{e}_1) * \{\mathbf{e}_1\} \textit{ or } \mathrm{conv}(\mathbf{0}, -\mathbf{e}_2, -\mathbf{e}_1) * \{\mathbf{e}_1 - 2\mathbf{e}_2\}.\end{aligned}$$

Note that, as in Example 1.1, a family P_1, P_2, P_3 of type (E) satisfies $P_i \subset \Delta_2 \times \mathbb{R}$. In a subsequent paper ([3]) we show that case (D) does not yield infinite families for any dimension greater than 3. Therefore it would be interesting to obtain a list like in Theorem 5.1 for any dimension to obtain a complete classification of families of mixed degree one.

References

[1] G. Averkov, J. Krümpelmann and S. Weltge, Notions of maximality for integral lattice-free polyhedra: the case of dimension three, *Math. Oper. Res.* **42**, 1035 (2017).

[2] G. Averkov, C. Wagner and R. Weismantel, Maximal lattice-free polyhedra: finiteness and an explicit description in dimension three, *Math. Oper. Res.* **36**, 721 (2011).

[3] G. Balletti and C. Borger, Families of full-dimensional polytopes of mixed degree one, manuscript.

[4] W. Bosma, J. Cannon and C. Playoust, The Magma algebra system. I. The user language, *J. Symbolic Comput.* **24**, 235 (1997), Computational algebra and number theory (London, 1993).

[5] V. Batyrev and B. Nill, Multiples of lattice polytopes without interior lattice points, *Mosc. Math. J.* **7**, 195 (2007).

[6] M. Beck, B. Nill, B. Reznick, C. Savage, I. Soprunov and Z. Xu, Let me tell you my favorite lattice-point problem . . . , in *Integer points in polyhedra—geometry, number theory, representation theory, algebra, optimization, statistics*, Contemp. Math. Vol. 452 (Amer. Math. Soc., Providence, RI, 2008) pp. 179–187.

[7] E. Cattani, M. A. Cueto, A. Dickenstein, S. Di Rocco and B. Sturmfels, Mixed discriminants, *Math. Z.* **274**, 761 (2013).

[8] B. Huber, J. Rambau and F. Santos, The Cayley trick, lifting subdivisions and the Bohne-Dress theorem on zonotopal tilings, *J. Eur. Math. Soc. (JEMS)* **2**, 179 (2000).

[9] M. Kreuzer and H. Skarke, Classification of reflexive polyhedra in three dimensions, *Adv. Theor. Math. Phys.* **2**, 853 (1998).

[10] B. Nill, The mixed degree of families of lattice polytopes, http://arxiv.org/abs/1708.03250, (2017).

[11] B. Nill and G. M. Ziegler, Projecting lattice polytopes without interior lattice points, *Math. Oper. Res.* **36**, 462 (2011).

[12] I. Soprunov, Global residues for sparse polynomial systems, *J. Pure Appl. Algebra* **209**, 383 (2007).

Some lattice parallelepipeds with unimodular covers

Mónica Blanco

Santander, Spain
E-mail: m.blanco@gmail.com

In this paper I will study a 3-parameter family of lattice parallelepipeds. I will study the behavior of the lattice with respect to the parameters and prove that three 2-parameter subfamilies have unimodular covers.

Keywords: Lattice parallelepiped; unimodular cover; dimension 3.

1. Introduction

During the workshop corresponding to these proceedings, Johannes Hofscheier, one of the lecturers of the mini-courses, presented the following problem:

Problem 1.1. *Let D be a unit square embedded in $\mathbb{R}^3$ and let $e \subset \mathbb{R}^3$ be a lattice segment. Does $D + e$ always have a unimodular cover?*

Remember that a *unimodular cover* of a lattice d-polytope P is a collection $\mathcal{T}$ of unimodular d-simplices with vertices in $P \cap \mathbb{Z}^d$ such that $P = \bigcup_{T \in \mathcal{T}} T$.

The following are related questions:

Question 1.1 (Question 4.1, [2]). *Do parallelepipeds $D + e$ that are* centrally symmetric *have a unimodular cover? Do 3-dimensional parallelepipeds have a unimodular cover?*

Question 1.2 (Question 9.3, [1]). *Is there a three-dimensional IDP polytope that does not have a unimodular cover?*

Recall that IDP is the *integer decomposition property.* A lattice d-polytope P is IDP if, for every $r \in \mathbb{N}$ and for every $m \in rP \cap \mathbb{Z}^d$, there exist $m_1, \dots, m_r \in P \cap \mathbb{Z}^d$ such that $m = m_1 + m_2 + \cdots + m_r$. That is, if the lattice points of rP can be written as the sum of exactly r lattice points of P, maybe with repetition.

The answers to the questions in Question 1.1, if positive, would mean another step towards giving a negative answer to Question 1.2, that is, to prove that IDP implies unimodular cover. On one hand, parallelepipeds are IDP, and on the other hand, by the results in [2], if parallelepipeds $D + e$ have a unimodular cover, this would also imply that every smooth centrally symmetric 3-polytope, which they prove to be IDP, admits a unimodular cover.

For the rest of the paper we will consider $D = [0,1]^2 \times \{0\}$ and $e =$

conv$\{(0,0,0),(a,b,h)\}$, for $a, b, h \in \mathbb{Z}$. That is, let

$$P(a,b,h) = \text{conv}\left\{\begin{array}{cccccccc} 0 & 1 & 0 & 1 & a & a+1 & a & a+1 \\ 0 & 0 & 1 & 1 & b & b & b+1 & b+1 \\ 0 & 0 & 0 & 0 & h & h & h & h \end{array}\right\}$$

A complete solution to Problem 1.1 is yet unknown, but in this contribution I collect some of the efforts made into studying their structure and finding a unimodular cover. Sections 2 and 3 are devoted to the understanding of $P(a,b,h)$ and the structure of the lattice points inside, and Section 4 contains the proof that the cases $P(a,0,h)$, $P(a,1,h)$ and $P(a,a,h)$ do indeed have a unimodular cover.

Remark 1.1. After this article was finished, and through discussions with Francisco Santos, he pointed out to me that the paths described in Lemma 4.1, and the use that is made of them in the proofs of Theorems 4.2 and 4.3 to describe unimodular covers, were already used in [4] and [3] in a very similar fashion. Moreover, if one uses some other arguments of [4] it is also possible to prove that the polytopes considered in those theorems have a unimodular triangulation.

2. Preliminaries on the polytopes $P(a, b, h)$

Let us study our polytopes $P(a,b,h) = [0,1]^2 \times \{0\} + \text{conv}\{o, v_h\}$, for o the origin and $v_h := (a,b,h)$.

Definition 2.1. We will call *slices* the horizontal integer cuts of $P(a,b,h)$. That is, $S_k := P(a,b,h) \cap \{z = k\}$ is the *k-th slice* of $P(a,b,h)$, for each $k \in \{0,1,\dots,h\}$.

Notice that each slice $S_k = [0,1]^2 \times \{0\} + v_k$, where $v_k := (\frac{ak}{h}, \frac{bk}{h}, k)$. See Fig. 1.

We will generally abuse notation and drop the parameters (a,b,h). For example, $S_k \equiv S_k(a,b,h)$ and $v_k \equiv v_k(a,b,h)$, but this will be done throughout the rest of the article to simplify notation. We also simplify notation for planes as follows: we write $\{\ell(x,y,z) = 0\}$ for the set $\{(x,y,z) \in \mathbb{R}^3 \text{ s. t. } \ell(x,y,z) = 0\}$, for $\ell : \mathbb{R}^3 \to \mathbb{R}$ an affine function.

The volume of $P(a,b,h)$ is always $|6h|$, so clearly $|h|$ is an invariant of each equivalence class. For fixed $|h|$, the following are unimodular transformations within polytopes of the family $\mathcal{P}$:

(1) $(x,y,z) \mapsto (x,y,-z)$ maps $P(a,b,h)$ to $P(a,b,-h)$.
(2) $(x,y,z) \mapsto (x+\alpha z, y+\beta z, z)$ maps $P(a,b,h)$ to $P(a+\alpha h, b+\beta h, h)$, for any $\alpha, \beta \in \mathbb{Z}$.
(3) $(x,y,z) \mapsto (y,x,z)$ maps $P(a,b,h)$ to $P(b,a,h)$.
(4) $(x,y,z) \mapsto (-x+z+1, y, z)$ maps $P(a,b,h)$ to $P(h-a,b,h)$.
(5) $(x,y,z) \mapsto (x,-y+z+1,z)$ maps $P(a,b,h)$ to $P(a,h-b,h)$.

Lemma 2.1. *If we consider the polytopes $P(a,b,h)$ modulo unimodular transformation, we can restrict ourselves to the cases with $0 \le b \le a \le \frac{h}{2}$.*

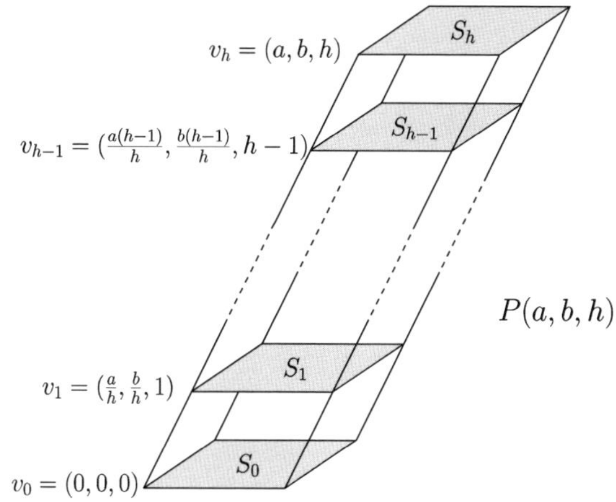

Fig. 1. The polytope $P(a, b, h)$ with the slices S_k and the points v_k, for each $k \in \{0, 1, \ldots, h\}$.

Proof. By (1), we can assume that $h > 0$. By (2) we can choose $a, b \in [0, h)$, that is, $a, b \in \{0, 1, \ldots, h-1\}$. Similarly, (3) gives $b \leq a$ and with (4) we can restrict ourselves to $a \leq h - a$, that is, $a \leq \frac{h}{2}$. □

Lemma 2.2. *Let $\gamma := \gcd(a, b, h)$. If $\gamma > 1$, then $P(a, b, h)$ can be subdivided into γ translated copies of $P(a', b', h')$, where $(a', b', h') := \frac{1}{\gamma}(a, b, h)$.*

Proof. It follows from the fact that, for each $k \in \{0, 1, \ldots \gamma - 1\}$, the polytope

$$P(a, b, h) \cap (\mathbb{R}^2 \times [kh', (k+1)h'])$$

is a translation of $P(a', b', h')$ by the vector (ka', kb', kh'). □

Notice that, in particular, if $\gcd(a, b, h) > 1$, then $S_i \cong S_{i+h'}$ under the translation by vector (a', b', h').

Remark 2.1. It is trivial that, if $P(a', b', h')$ has a unimodular cover, so does $P(a, b, h)$.

Example 2.1. Let $(a, b, h) = (18, 4, 12)$, which has $\gamma = \gcd(18, 4, 12) = 2$. Then $P(18, 4, 12)$ can be subdivided into two copies of $P(9, 2, 6)$, as shown in Fig. 2. Notice that $P(18, 4, 12) \cong P(6, 4, 12)$, which is the representative we have chosen (in Lemma 2.1). In this case we have shown a different element of the class in order for the slices not to overlap in the picture.

Notice that, in the example, we have chosen $(a', b', h') = (9, 2, 6)$ so that $\gcd(a', h'), \gcd(b', h') > 1$, so that all the possible *combinatorial types* of slices appear. By this we mean that we do not care about specific coordinates, but only about the degrees of freedom of the lattice points in the slice with respect to the border. These are the four possible combinatorial types:

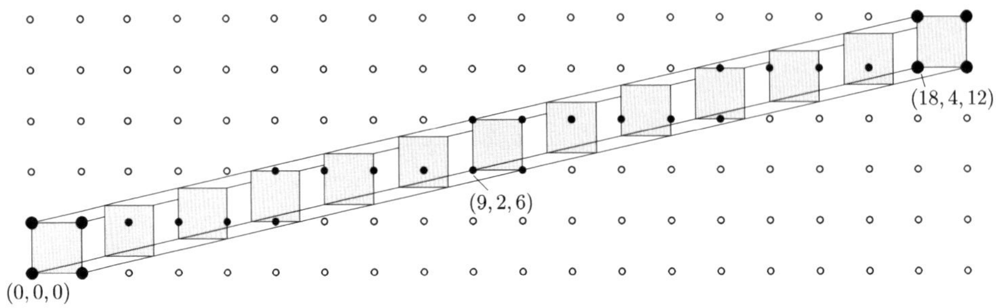

Fig. 2. The figure shows $P(18, 4, 12)$, and its 13 slices, in the 2-dimensional space $\mathbb{R}^3/\langle(0, 0, 1)\rangle$.

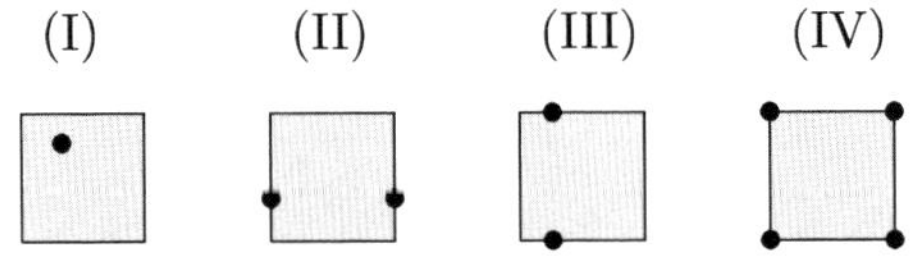

Lemma 2.3. *Let $P(a, b, h)$ and let $k \in \{0, 1, \ldots, h\}$. Then:*

- *If k is a multiple of $\frac{h}{\gcd(a,b,h)}$, then S_k is of type (IV).*
- *Otherwise:*
  - *If k is a multiple of $\frac{h}{\gcd(a,h)}$, then S_k is of type (II).*
  - *If k is a multiple of $\frac{h}{\gcd(b,h)}$, then S_k is of type (III).*
  - *If none of the above, then S_k is of type (I).*

Proof. The statement is trivial considering that the *left-down* corner of the slice S_k is the vertex v_k of $P(a, b, h)$, which has coordinates $v_k = (\frac{ak}{h}, \frac{bk}{h}, k)$. □

By Remark 2.1, from now on we will restrict ourselves to the cases where $\gcd(a, b, h) = 1$. In particular, the only slices of type (IV) are S_0 and S_h and the four edges $\{0, 1\}^2 \times \{0\} + (a, b, h)$ are primitive. We will also restrict to the representatives of Lemma 2.1. That is, we have

$$0 \leq b \leq a \leq \frac{h}{2}, \quad \gamma = \gcd(a, b, h) = 1. \tag{1}$$

3. The lattice $\mathbb{Z}^3$ with respect to $P(a, b, h)$

Now we want to understand how the lattice points are distributed in the slices. For this let us define the following automorphism of $P(a, b, h)$, which exchanges *opposite* vertices (in the natural sense) for any value of a, b, h:

$$\Gamma : \mathbb{R}^3 \to \mathbb{R}^3, \qquad \Gamma(x, y, z) = -(x, y, z) + (a + 1, b + 1, h) \tag{2}$$

Observe that Γ exchanges S_i with S_{h-i} for all $i \in \{0, \ldots, h\}$.

Definition 3.1. For each point $p \in P(a,b,h)$, we say that x is the *opposite* of $q \in P(a,b,h)$ (and vice versa), if p is the image of q under the automorphism Γ in (2).

We will apply the same definition for subpolytopes of P (facets, edges, simplices,...).

Example 3.1. Let us look at the slices of $P(3,1,7)$ (in $\mathbb{R}^3/\langle(0,0,1)\rangle$).

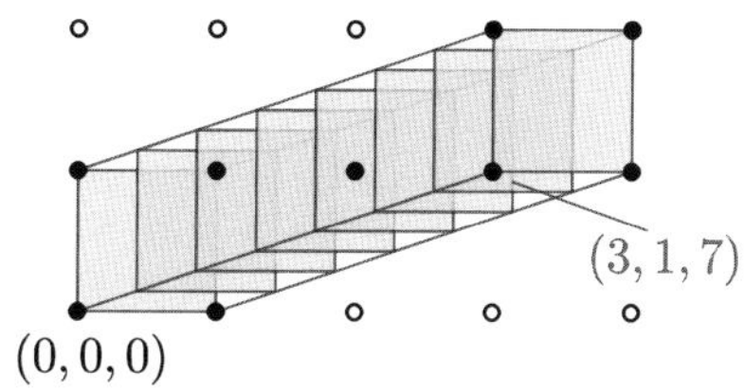

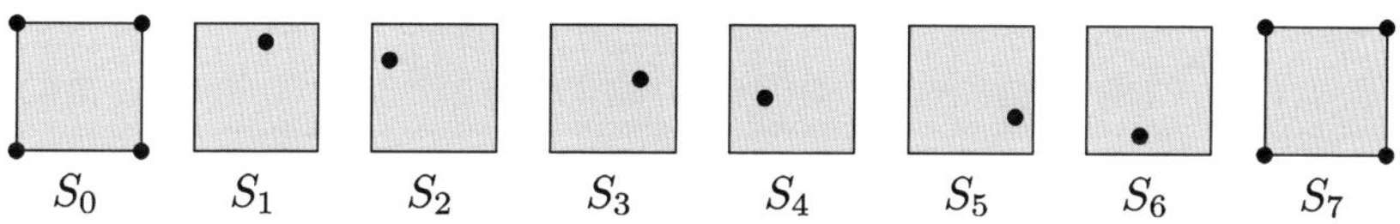

This does not give much information, unless we *overlap* the slices on top of each other:

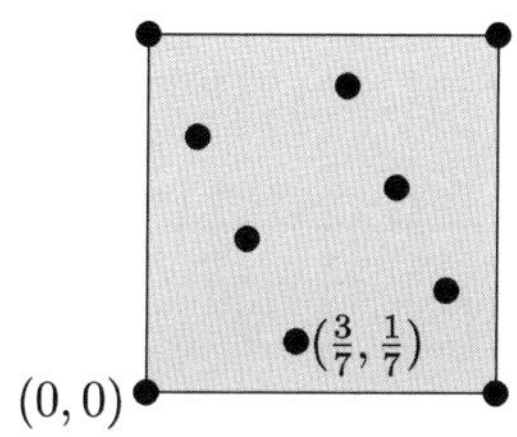

Notice that the point $\left(\frac{3}{7}, \frac{1}{7}\right)$ comes from the lattice point in slice S_6 (S_{h-1}).

This *overlapping* of the slices can be translated into the following map:

$$\Pi : \mathbb{R}^3 \to \mathbb{R}^2, \quad \Pi(x,y,z) = \left(x - \frac{az}{h}, y - \frac{bz}{h}\right)$$

Notice that Π sends the slices to the unit square $[0,1]^2$ but also the entire polytope $P(a,b,h)$. Let us see what is the image of the lattice $\mathbb{Z}^3$ under Π.

Lemma 3.1. *The map Π sends $P(a,b,h)$ to the unit square $[0,1]^2$ and it sends the lattice $\mathbb{Z}^3$ to the lattice*

$$\Lambda := \mathbb{Z}^2 + \langle(\frac{a}{h}, \frac{b}{h})\rangle_{\mathbb{Z}}$$

where $\langle(\frac{a}{h}, \frac{b}{h})\rangle_{\mathbb{Z}}/\mathbb{Z}^2$ is a group of order h.

Proof. To see that $\Pi(\mathbb{Z}^3) = \Lambda$ it suffices to compute the image of a lattice basis:

$$\Pi \begin{pmatrix} 0 & 1 & 0 & 0 \\ 0 & 0 & 1 & 0 \\ 0 & 0 & 0 & 1 \end{pmatrix} = \begin{pmatrix} 0 & 1 & 0 & -\frac{a}{h} \\ 0 & 0 & 1 & -\frac{b}{h} \end{pmatrix}$$

Notice that $\langle(\frac{a}{h}, \frac{b}{h})\rangle_{\mathbb{Z}} = \{k \cdot (\frac{a}{h}, \frac{b}{h}) \mid k \in \mathbb{Z}\} = \langle(\frac{-a}{h}, \frac{-b}{h})\rangle_{\mathbb{Z}}$. Let δ be the order of $\langle(\frac{a}{h}, \frac{b}{h})\rangle_{\mathbb{Z}}/\mathbb{Z}^2$. By definition $(\frac{\delta a}{h}, \frac{\delta b}{h}) \in \mathbb{Z}^2$. Since $h \cdot (\frac{a}{h}, \frac{b}{h}) = (a, b) \in \mathbb{Z}^2$, δ divides h. This implies that h/δ is an integer dividing a, b and h. By assumption, $\gcd(a, b, h) = 1$, hence $\delta = h$. □

Let

$$\Lambda_0 := \Lambda \cap [0, 1]^2. \tag{3}$$

Since $\Lambda = \Lambda_0 + \mathbb{Z}^2$, we can restrict ourselves to Λ_0 in order to understand Λ.

Example 3.2. Let us look at two different examples of lattices Λ (or, more concretely, Λ_0) that cover the cases when $\gamma_a := \gcd(a, h)$ and $\gamma_b := \gcd(b, h)$ are 1 or larger. In both cases, we have drawn the complete segment $\mathrm{conv}\{(0,0), (a,b)\}$, which in the lattice Λ is of length h:

- Λ and Λ_0, for $(a, b, h) = (3, 1, 7)$.

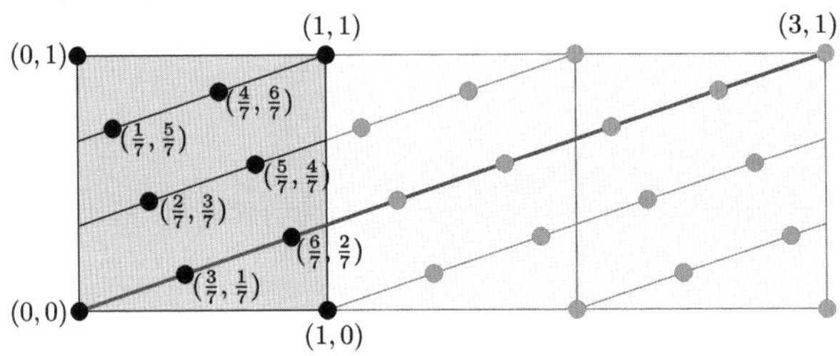

- Λ and Λ_0, for $(a, b, h) = (3, 2, 6)$.

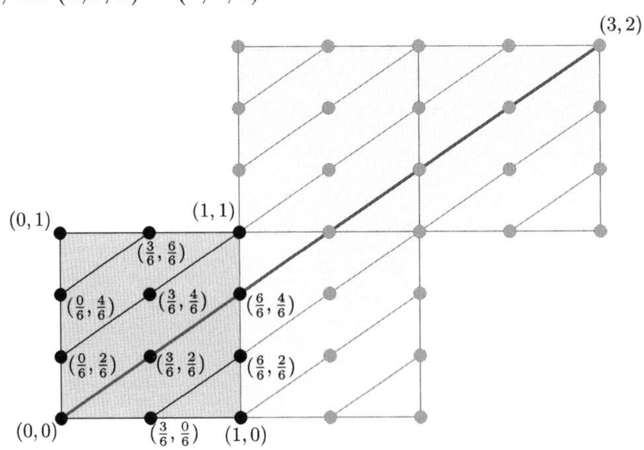

Lemma 3.2. *Let $(a, b, h) \in \mathbb{Z}^3$ with $\gcd(a, b, h) = 1$. The following holds:*

(1) Π *maps the lattice points in* $\mathbb{Z}^2 \times \{0\} + \langle(a,b,h)\rangle$ *to* $\mathbb{Z}^2$.
(2) Π *is a bijection between the lattice points of* $P(a,b,h)$ *that are not vertices and the lattice points of* $\Lambda_0 \setminus \{0,1\}^2$.

Proof. Part (1) is trivial.

For part (2), let us now see that Π is a bijection between the points of $P(a,b,h) \cap \mathbb{Z}^3 \setminus \mathrm{vert}(P(a,b,h))$ and the points of $\Lambda_0 \setminus \{0,1\}^2$ (notice that $\mathrm{vert}(P(a,b,h))$ denotes the vertices of $P(a,b,h)$). Let $p_0 \in \Lambda_0 \subset \Lambda$. Then there exists a point $p = (x_p, y_p, z_p) \in \mathbb{Z}^3$ such that $\Pi(p) = p_0$. First of all, $p \in \Pi^{-1}([0,1]^2) = ([0,1]^2 \times \{0\}) + \langle(a,b,h)\rangle_{\mathbb{R}}$. Suppose that $p \notin P(a,b,h)$, that is, $z_p \notin [0,h]$. Let $m \in \mathbb{Z}$, $n \in \{0,1,\ldots,h-1\}$ such that $z_p = mh + n$. Then the lattice point

$$p' := p - m \cdot (a,b,h) = (x_p - am, y_p - bm, n) \in P(a,b,h) \cap \mathbb{Z}^3$$

is such that

$$\begin{aligned}\Pi(p') &= \left(x_p - am - \frac{an}{h}, y_p - bm - \frac{bn}{h}\right) = \\ &= \left(x_p - \frac{a(mh+n)}{h}, y_p - \frac{b(mh+n)}{h}\right) = \Pi(p) = p_0.\end{aligned}$$

Notice that p' is a vertex of $P(a,b,h)$ if, and only if, p_0 is in $\{0,1\}^2$.

Take now $p_1 = (x_1, y_1, z_1), p_2 = (x_2, y_2, z_2) \in P(a,b,h) \cap \mathbb{Z}^3$ and suppose that $\Pi(p_1) = \Pi(p_2)$. That is:

$$x_1 - \frac{az_1}{h} = x_2 - \frac{az_2}{h}, \qquad y_1 - \frac{bz_1}{h} = y_2 - \frac{bz_2}{h}.$$

This implies that

$$x_2 = x_1 + \frac{a(z_2 - z_1)}{h}, \; y_2 = y_1 + \frac{b(z_2 - z_1)}{h} \in \mathbb{Z}$$

Now, the only way that $\frac{a(z_2-z_1)}{h}$ and $\frac{b(z_2-z_1)}{h}$ are both in $\mathbb{Z}$, given that $\gcd(a,b,h) = 1$, is that h divides $z_2 - z_1$. Since $z_1, z_2 \in [0,h]$, either $z_1 = z_2$, in which case $p_1 = p_2$, or $z_1 = 0$ and $z_2 = h$, in which case p_1 and p_2 are vertices of $P(a,b,h)$, and those are not considered. □

Example 3.3. In the following figures, the label(s) of a point $p_0 \in \Lambda_0$ is a number k such that there exists a point $p \in \mathbb{Z}^3 \cap S_k$ that is mapped to p_0 under Π:

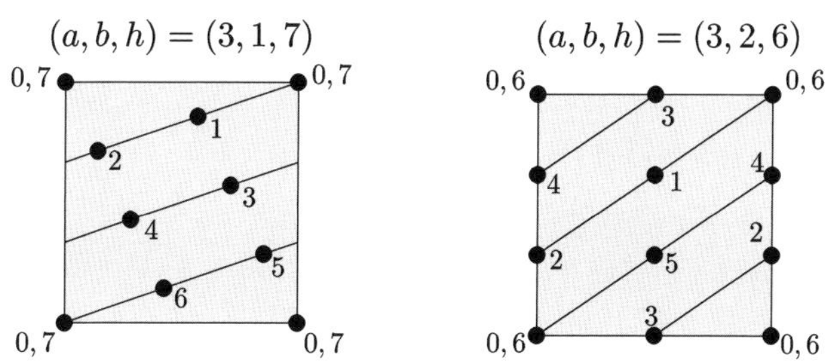

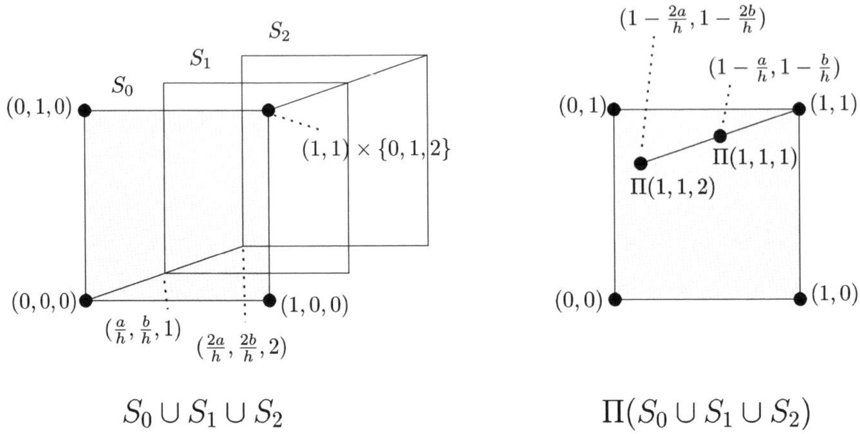

Fig. 3. The figure in the left shows the slices S_0, S_1 and S_2, and their lattice points, in the space $\mathbb{R}^3/\langle(0,0,1)\rangle$. The figure in the right shows the image of them under Π. Notice that in this case we are assuming $0 \leq b \leq a < \frac{h}{2}$, so that the slices S_1 and S_2 are of type (I) and they contain the lattice point $(1,1,\_)$.

To understand the general shape of Λ_0, look at Fig. 3.

Remark 3.1. In the unit square $[0,1]^2$ we call *opposite* points the pairs $\{p_0,(1,1)-p_0\}$. Notice that Π maps opposite pairs of points in $P(a,b,h)$ to opposite points in $[0,1]^2$. In particular, it maps opposite points of $P(a,b,h)\cap\mathbb{Z}^3$ to opposite points of Λ_0.

Moreover, by the bijection of Lemma 3.2, if $p_0, q_0 \in \Lambda_0 \setminus \{0,1\}^2$ are opposite points in $[0,1]^2$, then $p,q \in P(a,b,h)\cap\mathbb{Z}^3$ are opposite lattice points in $P(a,b,h)$, for p and q the unique lattice points of $P(a,b,h)$ that Π maps to p_0 and q_0, respectively.

4. Particular cases with unimodular covers

In this section we will consider three 2-parameter subfamilies of the $P(a,b,h)$ that have the property that we can place the lattice points of $P(a,b,h)$ in at most three consecutive parallel lattice hyperplanes, which helps us identify unimodular simplices within the polytope.

Theorem 4.1. *Let $0 \leq a \leq \frac{h}{2}$ with $\gcd(a,h)=1$. The lattice polytope $P(a,0,h)$ has a unimodular triangulation.*

Proof. First of all, notice that all lattice points of $P(a,0,h)$ are in the planes $\{y=0\}$ and $\{y=1\}$ (see Fig. 4). Moreover, if

$$B_0 := P(a,0,h)\cap\{y=0\} = \operatorname{conv}\{(0,0,0),(1,0,0),(a,0,h),(a+1,0,h)\},$$

then

$$B_1 := P(a,0,h)\cap\{y=1\} = B_0 + (0,1,0).$$

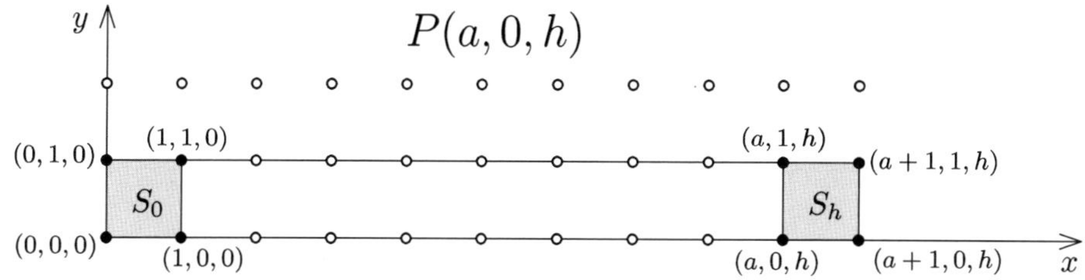

Fig. 4. The figure shows the polytope $P(a,0,h)$ in the space $\mathbb{R}^3/\langle(0,0,1)\rangle$. Notice that all its lattice points are contained in the planes $\{y=0\}$ and $\{y=1\}$.

That is, we have two copies of the same lattice polygon in consecutive lattice hyperplanes. Let $\mathcal{T}_0$ be a unimodular triangulation of B_0. Then $\mathcal{T}_1 := \{T+(0,1,0) \mid T \in \mathcal{T}_0\}$ is a unimodular triangulation of B_1. Then $\mathcal{S} := \{S_T,\ T \in \mathcal{T}_0\}$, where $S_T := \operatorname{conv}\{T \cup (T+(0,1,0))\}$, is a subdivision of $P(a,0,h)$ into prisms of height 1 over unimodular triangles. By *of height one* we mean that the two copies of the base are at lattice distance one. In particular, any of these prisms S_T is equivalent to: $\Delta_z := \{(0,0),(1,0),(0,1)\} \times \{0,1\}$, which has a unimodular triangulation as shown in the following picture:

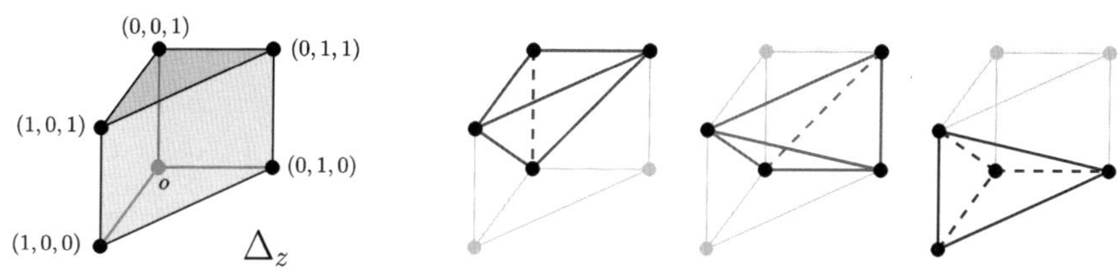

If we take the equivalent unimodular triangulation of each of the prisms S_T, that is a unimodular triangulation of $P(a,0,h)$. □

For the second and third cases, we first need to prove the following lemma:

Lemma 4.1. *Let* $B(a,h) := \operatorname{conv}\{(0,0),(1,0),(a,h),(a+1,h)\} \subset \mathbb{R}^2$, *for* $0 \le a \le h$. *Then the following holds:*

(i) There exists a path $p_0p_1p_2\dots p_h$ *in* $B(a,h)$ *from the origin* $p_0=(0,0)$ *to the facet* $\operatorname{conv}\{(a,h),(a+1,h)\}$ *such that, for each* $k \in \{0,\dots,h-1\}$, *either*

$$p_{k+1}-p_k=(0,1) \qquad \text{or} \qquad p_{k+1}-p_k=(1,1)$$

(ii) There exists a path $q_0q_1q_2\dots q_{h_a}$ *in* $B(a,h)$ *from the origin* $q_0=(0,0)$ *to the facet* $\operatorname{conv}\{(1,0),(a+1,h)\}$ *such that, for each* $k \in \{0,\dots,h_a-1\}$, *either*

$$q_{k+1}-q_k=q_1-(0,0) \qquad \text{or} \qquad q_{k+1}-q_k=q_1-(a,h),$$

for $q_1 \in B(a,b)\cap\mathbb{Z}^2$ *at lattice distance* 1 *from* $\operatorname{conv}\{(0,0),(a,h)\}$. *Here,* $h_a := \frac{h}{\gamma_a}$.

In particular, all the p_k *and* q_k *are lattice points.*

Observe that, by the description of the lemma, those paths use two types of lattice edges: parallel to $\operatorname{conv}\{o,(1,1)\}$ or $\operatorname{conv}\{(1,0),(1,1)\}$ in case (i), and parallel to $\operatorname{conv}\{o,q_1\}$ or $\operatorname{conv}\{(a,h),q_1\}$ in case (ii).

Proof. For convenience for later use, we will use the coordinates (x,z) in $\mathbb{R}^2$. A point $(x,z)\in\mathbb{R}^2$ belongs to $B(a,h)$ if, and only if:

$$0\le f(x,z):=xh-az\le h \qquad \text{and} \qquad 0\le g(x,z):=z\le h.$$

(i) By the assumption $0\le a\le h$, we have that $(1,1)\in B(a,h)$. Since $(1,1)-(0,0)=(1,1)$, then we can choose $p_1=(1,1)$. To verify the existence of the path it suffices to show that, for every lattice point $(x_0,z_0)\in B(a,h)\cap\mathbb{Z}^2$ with $z_0\le h-1$, we have that one of $(x_0,z_0)+(1,1)$ or $(x_0,z_0)+(0,1)$ also belong to $B(a,h)$.
Notice that, in particular, $(1,1)$ is a point of $B(a,h)$ at lattice distance one from the facet $\operatorname{conv}\{(0,0),(1,0)\}$. Also, for each new step of the path, we have $g(p_{k+1})=g(p_k)+1$ (the z-coordinate increases by one) and, if $z_0=h$, we are already at the end of the path.
So, we know that $0\le f(x_0,z_0)\le h$. Then,

$$f(x_0+1,z_0+1)=(x_0+1)h-a(z_0+1)=f(x_0,z_0)+h-a\ge f(x_0,y_0)\ge 0$$

$$f(x_0,z_0+1)=x_0h-a(z_0+1)=f(x_0,z_0)-a\le f(x_0,y_0)\le h$$

Suppose that neither (x_0+1,z_0+1) nor (x_0,z_0+1) are in $B(a,h)$, then we have that $f(x_0+1,z_0+1)>h$ and $f(x_0,z_0+1)<0$, but this yields a contradiction.

(1) For the second part, we do an analogous process, although now we do not know the coordinates of q_1. Because we are in dimension 2, we can choose such a q_1 at lattice distance one from $\operatorname{conv}\{(0,0),(a,h)\}$. This gets translated into $f(q_1)=\gamma_a$ (which we have defined to be $\gcd(a,h)$). In this case, in each step of the path we get $f(q_{k+1})=f(q_k)+\gamma_a$, which is why, after $\frac{h}{\gamma_a}$ steps we are already at the end of the path.
So, as before, it suffices to check that, for $(x_0,z_0)\in B(a,h)\cap\mathbb{Z}^2$ with $f(x_0,y_0)\le h-\gamma_a$, that either $(x_0,z_0)+q_1-(0,0)$ or $(x_0,y_0)+q_1-(a,h)$ is also in $B(a,h)$.
We know that $0\le g(x_0,z_0),g(q_1)\le h$. Then,

$$g((x_0,z_0)+q_1-(0,0))=z_0+g(q_1)\ge 0,$$

$$g((x_0,z_0)+q_1-(a,h))=z_0+g(q_1)-h\le h+h-h=h$$

Suppose that neither $(x_0,z_0)+q_1$ nor $(x_0,z_0)+q_1-(a,h)$ are in $B(a,h)$, then we must have $g((x_0,z_0)+q_1-(0,0))>h$ and $g((x_0,z_0)+q_1-(a,h))<0$, but this yields a contradiction.

□

In Fig. 5 one can see a couple of examples of the paths mentioned in Lemma 4.1. Moreover, one can see that there exist paths for all the following options of end-points: $p_0 \in \{(0,0),(1,0)\}$ and $p_h \in \{(a,h),(a+1,h)\}$ in part (i); $q_0 \in \{(0,0),(a,h)\}$ and $q_{h_a} \in \{(1,0),(a+1,h)\}$ in part (ii).

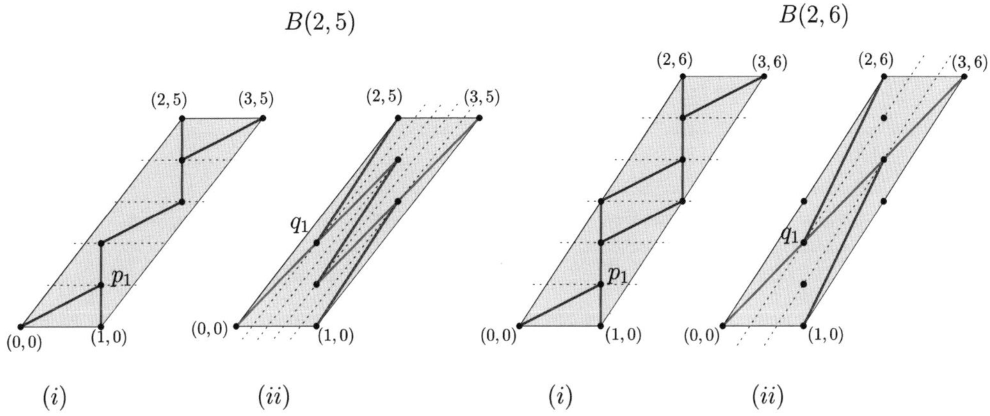

Fig. 5. The figure shows the polygons $B(2,5)$ and $B(2,6)$ with the possible paths of Lemma 4.1.

We can now prove the following:

Theorem 4.2. *Let $0 \le a \le \frac{h}{2}$. The lattice polytope $P(a,1,h)$ has a unimodular cover.*

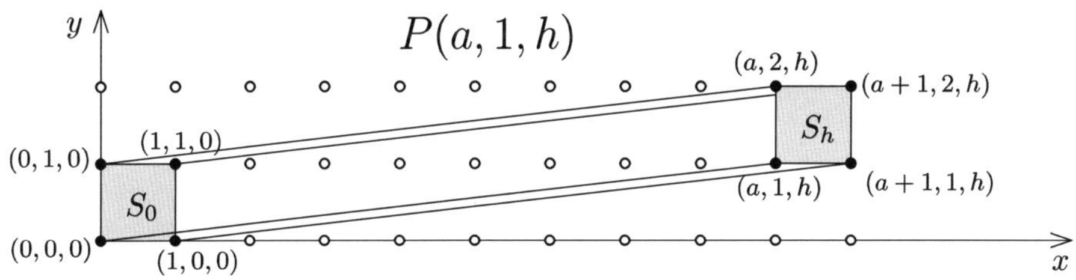

Fig. 6. The figure shows the polytope $P(a,1,h)$ in the space $\mathbb{R}^3/\langle(0,0,1)\rangle$. Notice that all its lattice points are contained in the planes $\{y=0\}$, $\{y=1\}$ and $\{y=2\}$.

Proof. All lattice points of $P(a,1,h)$ are in the planes $\{y=0\}$, $\{y=1\}$ and $\{y=2\}$ (see Fig. 6). In the plane $\{y=0\}$ there is only the lattice points $(0,0,0)$ and $(1,0,0)$, in the plane $\{y=2\}$ we have $(a,2,h)$ and $(a+1,2,h)$, and in the plane $\{y=1\}$ we have

$$B_1 := P(a,1,h) \cap \{y=1\} = \operatorname{conv}\{(0,1,0),(1,1,0),(a,1,h),(a+1,1,h)\},$$

which is equivalent to the $B(a,h)$ of Lemma 4.1, embedded in $\{y=1\}$ by $(x,z) \mapsto (x,1,z)$.

Let us see that $P := \operatorname{conv}\big(B_1 \cup \{(0,0,0),(1,0,0)\}\big) \subset P(a,1,h)$ has a unimodular cover. Observe that, by the automorphism Γ in (2), this suffices in order to prove that the entire polytope $P(a,1,h)$ has a unimodular cover.

Let $\mathcal{T}$ be a unimodular triangulation of B_1. Since the points $(0,0,0)$ and $(1,0,0)$ are at lattice distance one from B_1 (taking the functional y), then every tetrahedron of the form $\operatorname{conv}\big(T \cup \{(0,0,0)\}\big)$ or $\operatorname{conv}\big(T \cup \{(1,0,0)\}\big)$, for $T \in \mathcal{T}$, are unimodular. That is, the quadrangular pyramids

$$A_{(0,0,0)} := \operatorname{conv}\big(B_1 \cup \{(0,0,0)\}\big) \quad \text{and} \quad A_{(1,0,0)} := \operatorname{conv}\big(B_1 \cup \{(1,0,0)\}\big)$$

have a unimodular triangulation. Hence, to prove that P has a unimodular cover it remains to check that the region

$$R := P \setminus \big(A_{(0,0,0)} \cup A_{(1,0,0)}\big) = \operatorname{conv}\left\{(0,0,0),(1,0,0),\left(\frac{1}{2},\frac{1}{2},0\right),\left(\frac{a+1}{2},\frac{1}{2},\frac{h}{2}\right)\right\},$$

which is a rational tetrahedron, is covered by unimodular tetrahedra of $P(a,1,h)$. See Fig. 7.

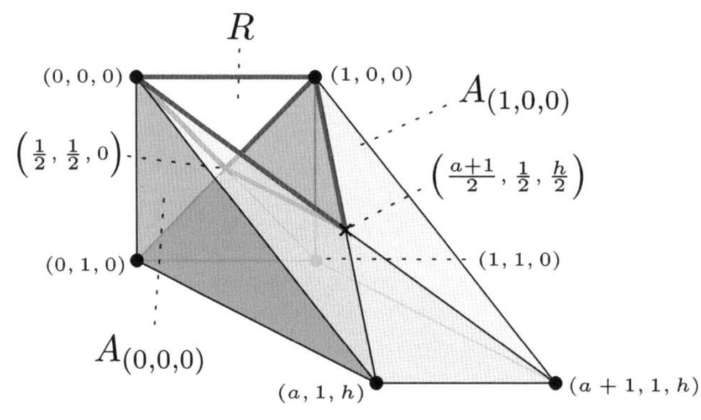

Fig. 7. The figure shows the quadrangular pyramids $A_{(0,0,0)}$ and $A_{(1,0,0)}$, and the *uncovered* region R, within the polytope $P \subset P(a,1,h)$. Notice that they are not drawn in the natural perspective. This particular orientation of the drawing has been chosen so that R and the pyramids are easier to visualize.

Let $p_0 p_1 \ldots p_h$ be the path of $B(a,h)$ given by Lemma 4.1, part (i), embedded in $\{y=1\}$ $((x,z) \in B(a,h) \mapsto (x,1,z) \in B_1)$. Then $p_0 = (0,1,0)$, $p_1 = (1,1,1)$ and, for each $k \in \{0,1,\ldots,h-1\}$, we have

$$p_{k+1} - p_k = (1,0,1) \quad \text{or} \quad p_{k+1} - p_k = (0,0,1).$$

Take, for $k \in \{0,1,\ldots,h-1\}$, the following tetrahedra:

$$T_k := \operatorname{conv}\{(0,0,0),(1,0,0),p_k,p_{k+1}\} \subset P$$

Let $p_k = (x_k, 1, z_k)$.

- If k is such that $p_{k+1} - p_k = (1,0,1)$, then T_k is equivalent to $\Delta_1 := \mathrm{conv}\{(0,0,0),(1,0,0),(0,1,0),(1,1,1)\}$ under the unimodular transformation $(x,y,z) \mapsto (x - x_k y, y, z - z_k y)$.
- If k is such that $p_{k+1} - p_k = (0,0,1)$, then T_k is equivalent to $\Delta_2 := \mathrm{conv}\{(0,0,0),(1,0,0),(1,1,0),(1,1,1)\}$ under the unimodular transformation $(x,y,z) \mapsto (x + (1-x_k)y, y, z - z_k y)$.

That is, all the tetrahedra T_k are unimodular.

Finally, it remains to check that $R \subset \cup_{k=0,1,\ldots,h-1} T_k$. For this, notice that $R_k := R \cap T_k = \mathrm{conv}\{(0,0,0),(1,0,0),p'_k,p'_{k+1}\}$, for:

$$
\begin{aligned}
p'_k &:= \mathrm{conv}\{(0,0,0),(1,0,0),p_k\} \cap \mathrm{conv}\left\{\left(\frac{1}{2},\frac{1}{2},0\right),\left(\frac{a+1}{2},\frac{1}{2},\frac{h}{2}\right)\right\} \\
&= \left(1-\frac{k}{h}\right)\left(\frac{1}{2},\frac{1}{2},0\right) + \frac{k}{h}\left(\frac{a+1}{2},\frac{1}{2},\frac{h}{2}\right).
\end{aligned}
$$

See Fig. 8 for a detailed drawing of this setup. □

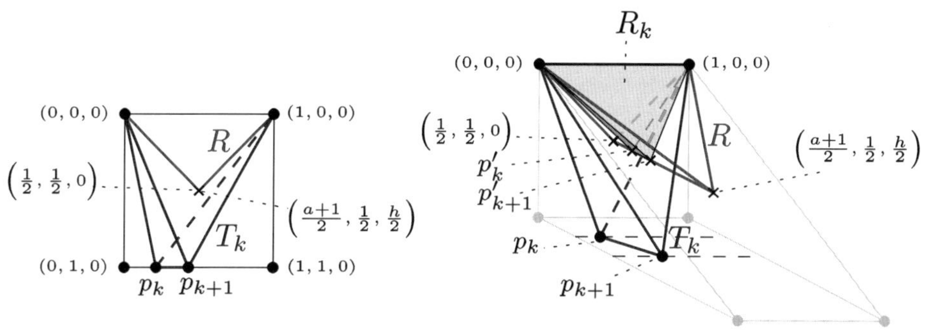

Fig. 8. The left figure shows the rational tetrahedron R and the lattice tetrahedron T_k in the space $\mathbb{R}^3/\langle(a,0,h)\rangle$. The right figure shows R, T_k and their intersection R_k, with the same perspective as in Fig. 7.

Remark 4.1. Notice that the path $p_0 p_1 \ldots p_h$ used in the proof of Theorem 4.2 uses a point of each slice of $P(a,1,h)$: $p_k \in S_k$ for each $k \in \{0,1,\ldots,h\}$.

Theorem 4.3. *Let* $0 \le a \le \frac{h}{2}$ *with* $\gcd(a,h) = 1$. *The lattice polytope* $P(a,a,h)$ *has a unimodular cover.*

Proof. The proof of this case is very similar to that of Theorem 4.2. All lattice points of $P(a,a,h)$ are in the planes $\{y = x\}$, $\{y = x+1\}$ and $\{y = x-1\}$ (see Fig. 9). Since $\gcd(a,h) = 1$, the edges $\mathrm{conv}\{(0,1,0),(a,a+1,h)\}$ and $\mathrm{conv}\{(1,0,0),(a+1,a,h)\}$ are primitive, and the only lattice points in the plane $\{y = x+1\}$ are

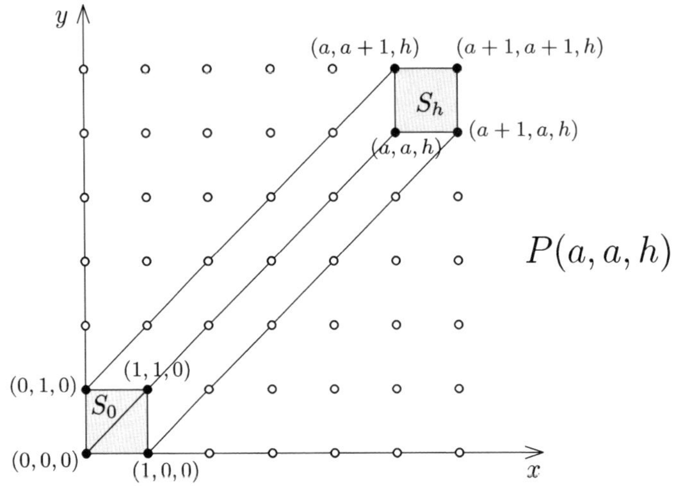

Fig. 9. The figure shows the polytope $P(a, a, h)$ in the space $\mathbb{R}^3/\langle(0,0,1)\rangle$. Notice that all its lattice points are contained in the planes $\{y = x\}$, $\{y = x+1\}$ and $\{y = x-1\}$.

$(0,1,0)$ and $(a, a+1, h)$, and in the plane $\{y = x-1\}$ we only have $(1,0,0)$ and $(a+1, a, h)$. In the plane $\{y = x\}$ we have

$$B_{xy} := P(a, a, h) \cap \{y = x\} = \operatorname{conv}\big\{(0,0,0), (1,1,0), (a, a, h), (a+1, a+1, h)\big\},$$

which is equivalent to the $B(a, h)$ of Lemma 4.1, embedded in $\{y = x\}$ by $(x, z) \mapsto (x, x, z)$.

We will now see that $P' := \operatorname{conv}\big(B_{xy} \cup \{(1,0,0), (a+1, a, h)\}\big) \subset P(a, a, h)$ has a unimodular cover.

The same as in the previous proof, since both $(1,0,0)$ and $(a+1, a, h)$ are at distance one from the plane $\{y = x\}$, then the following pyramids admit unimodular triangulations:

$$A'_{(1,0,0)} := \operatorname{conv}\big(B_{xy} \cup \{(1,0,0)\}\big) \quad \text{and} \quad A'_{(a+1,a,h)} := \operatorname{conv}\big(B_{xy} \cup \{(a+1, a, h)\}\big)$$

Hence, to prove that P' has a unimodular cover it remains to check that the region

$$\begin{aligned} R' &:= P' \setminus \big(A'_{(1,0,0)} \cup A'_{(a+1,a,h)}\big) \\ &= \operatorname{conv}\left\{(1,0,0), (a+1, a, h), \left(\frac{a+1}{2}, \frac{a}{2}, \frac{h}{2}\right), \left(\frac{a+2}{2}, \frac{a+1}{2}, \frac{h}{2}\right)\right\}, \end{aligned}$$

which is a rational tetrahedron, is covered by unimodular tetrahedra of $P(a, a, h)$. See Fig. 10.

Observe that B_{xy}, P', R' and the edge $\operatorname{conv}\{(1,0,0), (a+1, a, h)\}$ play the roles of B_1, P, R and the edge $\operatorname{conv}\{(0,0,0), (1,0,0)\}$, respectively, in the proof of Theorem 4.2. The only difference is that $\operatorname{conv}\{(0,0,0), (1,0,0)\}$ in $P(a, 1, h)$ is parallel to the edge of B_1 equivalent to $\operatorname{conv}\{(0,0), (1,0)\}$ in the polygon $B(a, b)$ of

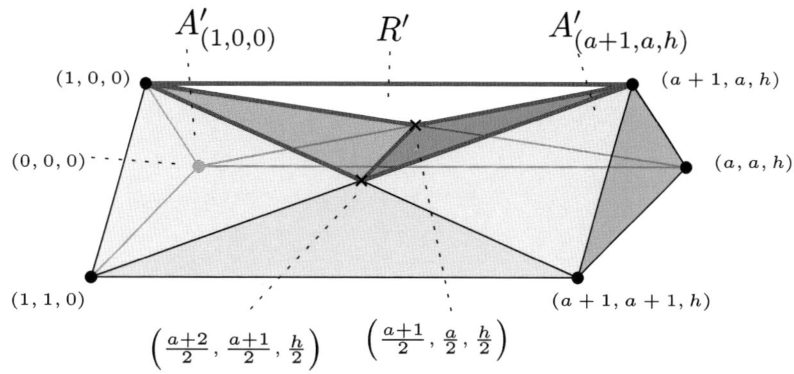

Fig. 10. The figure shows the quadrangular pyramids $A'_{(1,0,0)}$ and $A'_{(a+1,a,h)}$, and the *uncovered* region R', within the polytope $P' \subset P(a, 1, h)$.

Lemma 4.1, and in this case, $\text{conv}\{(1,0,0),(a+1,a,h)\}$ is parallel to the edge of B_{xy} equivalent to $\text{conv}\{(0,0),(a,h)\}$ in $B(a,b)$.

So now we use $q_0q_1\dots q_{h_a}$ the path of $B(a,h)$ given by Lemma 4.1, part (*ii*), embedded in $\{y = x\}$ $((x,z) \in B(a,h) \mapsto (x,x,z) \in B_{xy})$. Then $q_0 = (0,0,0)$, $q_1 = (1,1,1)$ and, for each $k \in \{0,1,\dots,h_a - 1\}$, we have

$$q_{k+1} - q_k = q_1 \quad \text{or} \quad q_{k+1} - q_k = q_1 - (a,a,h).$$

Take, for $k \in \{0,1,\dots,h_a - 1\}$, the following tetrahedra:

$$T'_k := \text{conv}\,\{(1,0,0),(a+1,a,h),q_k,q_{k+1}\} \subset P'$$

Let $q_k = (x_k, x_k, z_k)$.

- If k is such that $q_{k+1} - q_k = q_1$, then T'_k is equivalent to $\Delta'_1 := \text{conv}\{(1,0,0),(a+1,a,h),(0,0,0),p_1\}$ under the unimodular transformation $(x,y,z) \mapsto \big(x_k(x-y-1)+x,\ x_k(x-y-1)+y,\ z_k(x-y-1)+z\big)$.
- If k is such that $q_{k+1} - q_k = q_1 - (a,a,h)$, then T'_k is equivalent to $\Delta'_2 := \text{conv}\{(1,0,0),(a+1,a,h),(a,a,h),p_1\}$ under the unimodular transformation $(x,y,z) \mapsto \big((x_k-a)(x-y-1)+x,\ (x_k-a)(x-y-1)+y,\ (z_k-h)(x-y-1)+z\big)$.

In this case, we know that both Δ'_1 and Δ'_2 are unimodular because q_1 was chosen in Lemma 4.1 to be at lattice distance one from the edge $\text{conv}\{(0,0),(a,h)\}$. The integer functional of $\mathbb{R}^2$ that was 0 in that edge and 1 in q_1 is $\frac{1}{\gamma_a}(xh - az)$. Taking the functional $\frac{1}{\gamma_a}(yh - az)$ of $\mathbb{R}^3$, which in $B_{xy} \subset \{y = x\}$ takes the same values, gives that q_1 is at lattice distance one from the plane containing the four points $\{(0,0,0),(1,0,0),(a,a,h),(a+1,a,h)\}$.

Finally, it remains to check that $R' \subset \bigcup_{k=0,1,\dots,h_a-1} T'_k$. For this, notice that

$R'_k := R' \cap T'_k = \operatorname{conv}\{(1,0,0), (a+1,a,h), q'_k, q'_{k+1}\}$, for:

$$q'_k := \operatorname{conv}\{(1,0,0),(a+1,a,h),q_k\} \cap \operatorname{conv}\left\{\left(\frac{a+1}{2},\frac{a}{2},\frac{h}{2}\right),\left(\frac{a+2}{2},\frac{a+1}{2},\frac{h}{2}\right)\right\}$$
$$= \left(1-\frac{k}{h_a}\right)\left(\frac{a+1}{2},\frac{a}{2},\frac{h}{2}\right)+\frac{k}{h_a}\left(\frac{a+2}{2},\frac{a+1}{2},\frac{h}{2}\right)$$

□

Acknowledgments

I would like to thank Johannes Hofscheier for proposing this interesting problem, and all the people that initially discussed the problem with me, specially Gabriele Balletti.

References

[1] G. Balletti, *Enumeration of lattice polytopes by their volume*, To appear.
[2] M. Beck, C. Haase, A. Higashitani, J. Hofscheier, K. Jochemko, L. Katthän and M. Michałek, *Smooth centrally symmetric polytopes in dimension* 3 *are IDP*, Preprint, July 2018. arXiv:1802.01046.
[3] J. M. Kantor and K. S. Sarkaria, *On primitive subdivisions of an elementary tetrahedron*, Pacific J. Math., 211 (1) (2003) 123–155.
[4] F. Santos and G. M. Ziegler, *Unimodular triangulations of dilated 3-polytopes*, Trans. Moscow Math. Soc., 74 (2013), 293–311.

A brief survey on lattice zonotopes

Benjamin Braun

Department of Mathematics
University of Kentucky
Lexington, KY 40506–0027, USA
E-mail: benjamin.braun@uky.edu
https://sites.google.com/view/braunmath/

Andrés R. Vindas-Meléndez

Department of Mathematics
University of Kentucky
Lexington, KY 40506–0027, USA
E-mail: andres.vindas@uky.edu
https://math.as.uky.edu/users/arvi222

Zonotopes are a rich and fascinating family of polytopes, with connections to many areas of mathematics. In this article we provide a brief survey of classical and recent results related to lattice zonotopes. Our emphasis is on connections to combinatorics, both in the sense of enumeration (e.g. Ehrhart theory) and combinatorial structures (e.g. graphs and permutations).

Keywords: Zonotope, Ehrhart polynomial, lattice polytope.

1. Introduction

Zonotopes are a rich and fascinating family of polytopes, with connections to many areas of mathematics. In this article we provide a brief survey of classical and recent results related to *lattice* zonotopes, i.e. Minkowski sums of line segments with endpoints in the lattice $\mathbb{Z}^n$. Our emphasis is on connections to combinatorics, both in the sense of enumeration (e.g. Ehrhart theory) and combinatorial structures (e.g. graphs and permutations). Our primary goal in this article is to maintain a level of presentation accessible to beginning graduate students working in algebraic and geometric combinatorics or related fields. We make no effort here to be comprehensive; for example, we omit from our discussion the deep connections between zonotopes, hyperplane arrangements, and oriented matroids. Our selection of topics is based purely on personal taste, and includes various topics we find interesting for one reason or another. Our hope is that this survey will serve as a stepping stone to a deeper investigation of lattice zonotopes for interested readers.

2. Zonotope Basics

Zonotopes can be defined using either Minkowski sums or projections of cubes.

Definition 2.1. Consider polytopes, $P_1, P_2, \ldots, P_m \subset \mathbb{R}^n$. We define the

Minkowski sum of the m polytopes as

$$P_1 + P_2 + \cdots + P_m := \{x_1 + x_2 + \cdots + x_m : x_j \in P_j \text{ for } 1 \leq j \leq m\}.$$

Given $\mathbf{v}, \mathbf{w} \in \mathbb{R}^n$, we write $[\mathbf{v}, \mathbf{w}]$ for the line segment from $\mathbf{v}$ to $\mathbf{w}$.

Example 2.1. Consider the Minkowski sum of $[(0,0),(1,0)]$, $[(0,0),(0,1)]$, and $[(0,0),(1,1)]$. The Minkowski sum of the first two segments is a unit square. Taking the Minkowski sum of this square with the line segment $[(0,0),(1,1)]$ can be visualized as sliding the square up and to the right along the line segment, with the resulting polytope consisting of all points touched by the square during the sliding movement, see Figure 1.

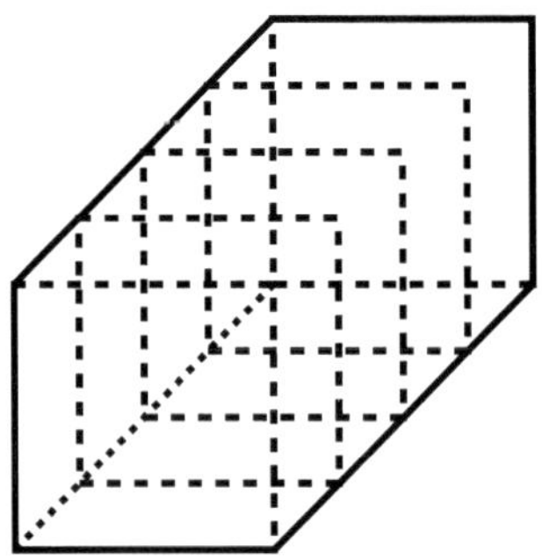

Fig. 1. The Minkowski sum of $[(0,0),(1,0)]$, $[(0,0),(0,1)]$, and $[(0,0),(1,1)]$.

Definition 2.2. Consider m vectors $\mathbf{v}_1, \ldots, \mathbf{v}_m$ in $\mathbb{R}^n$ and their corresponding line segments $[\mathbf{0}, \mathbf{v}_j]$. The *zonotope* corresponding to $\mathbf{v}_1, \ldots, \mathbf{v}_m$ is defined to be the Minkowski sum of these line segments:

$$Z(\mathbf{v}_1, \mathbf{v}_2, \ldots, \mathbf{v}_m) := \{\lambda_1\mathbf{v}_1 + \lambda_2\mathbf{v}_2 + \cdots + \lambda_m\mathbf{v}_m : 0 \leq \lambda_j \leq 1\} = [\mathbf{0}, \mathbf{v}_1] + \cdots + [\mathbf{0}, \mathbf{v}_m].$$

We call any polytope that is translation-equivalent to such a polytope of this type a *zonotope*. When each $\mathbf{v}_j \in \mathbb{Z}^n$, we say that $Z(\mathbf{v}_1, \mathbf{v}_2, \ldots, \mathbf{v}_m)$ is a *lattice zonotope*.

Example 2.2. The Minkowski sum in Figure 1 is $Z((1,0),(0,1),(1,1))$. The Minkowski sum in Figure 2 is $Z((1,0,0),(0,1,0),(0,0,1),(1,1,1))$.

Definition 2.3. The zonotope $Z_0(\mathbf{v}_1, \mathbf{v}_2, \ldots, \mathbf{v}_m) := Z(\pm\mathbf{v}_1, \pm\mathbf{v}_2, \ldots, \pm\mathbf{v}_m)$ is symmetric about the origin, that is, it has the property that $\mathbf{x} \in Z_0$ if and only if $-\mathbf{x} \in Z_0$; we call Z_0 a *centrally symmetric zonotope* defined by $\mathbf{v}_1, \ldots, \mathbf{v}_m$. Note that Z_0 can be obtained as $Z(2\mathbf{v}_1, 2\mathbf{v}_2, \ldots, 2\mathbf{v}_m) - (\mathbf{v}_1 + \cdots + \mathbf{v}_m)$.

An alternative definition of a zonotope is as a projection of the unit cube. In this case, let A denote the matrix with columns given by $\mathbf{v}_1, \ldots, \mathbf{v}_m$. Then by definition

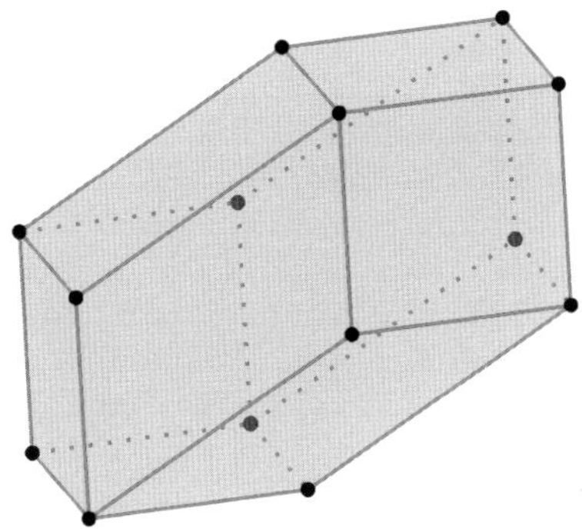

Fig. 2. $Z((1,0,0),(0,1,0),(0,0,1),(1,1,1))$

it is immediate that $Z(\mathbf{v}_1,\ldots,\mathbf{v}_m)$ is equal to $A\cdot[0,1]^m$. In some circumstances it is more convenient to work with this projection-based definition of zonotopes.

Remark 2.1. Zonotopes are deceptively simple to define, yet even the most elementary zonotopes are mathematically rich. For example, the unit cube $[0,1]^n$ is itself a zonotope, and the survey paper by Zong [21] shows that the mathematical properties of this object are both broad and deep.

In polyhedral geometry, it is typically useful when objects can be decomposed as unions of simpler objects, e.g. the theory of subdivisions and triangulations. Zonotopes admit a particularly nice decomposition into parallelepipeds; parts of the boundaries of these parallelepipeds can be removed resulting in a disjoint decomposition. More precisely, suppose that $\mathbf{w}_1,\mathbf{w}_2,\ldots,\mathbf{w}_k\in\mathbb{R}^n$ are linearly independent, and let $\sigma_1,\sigma_2,\ldots,\sigma_k\in\{\pm1\}$. Then we define

$$\Pi^{\sigma_1,\sigma_2,\ldots,\sigma_k}_{\mathbf{w}_1,\mathbf{w}_2,\ldots,\mathbf{w}_k} := \left\{\lambda_1\mathbf{w}_1+\lambda_2\mathbf{w}_2+\cdots+\lambda_k\mathbf{w}_k : \begin{array}{l} 0\le\lambda_j<1 \text{ if } \sigma_j=-1 \\ 0<\lambda_j\le 1 \text{ if } \sigma_j=1 \end{array}\right\}$$

to be the *half-open parallelepiped* generated by $\mathbf{w}_1,\mathbf{w}_2,\ldots,\mathbf{w}_k$. The signs $\sigma_1,\sigma_2,\ldots,\sigma_k$ keep track of the facets of the parallelepiped that are either included or excluded from the closure of the parallelepiped.

Theorem 2.1 (Shephard [15], Theorem 54). *The zonotope $Z(\mathbf{v}_1,\mathbf{v}_2,\ldots,\mathbf{v}_m)$ can be expressed as a disjoint union of translates of $\Pi^{\sigma_1,\sigma_2,\ldots,\sigma_k}_{\mathbf{w}_1,\mathbf{w}_2,\ldots,\mathbf{w}_k}$, where $\{\mathbf{w}_1,\mathbf{w}_2,\ldots,\mathbf{w}_k\}$ ranges over all linearly independent subsets of $\{\mathbf{v}_1,\mathbf{v}_2,\ldots,\mathbf{v}_m\}$, each equipped with an appropriate choice of signs $\sigma_1,\sigma_2,\ldots,\sigma_k$.*

For a complete proof of Theorem 2.1, see Lemma 9.1 of the textbook by Beck and Robins [6]. Figure 3 is an illustration of the decomposition of the zonotope $Z((0,1),(1,0),(1,1))\subset\mathbb{Z}^2$ as suggested by Shephard's theorem.

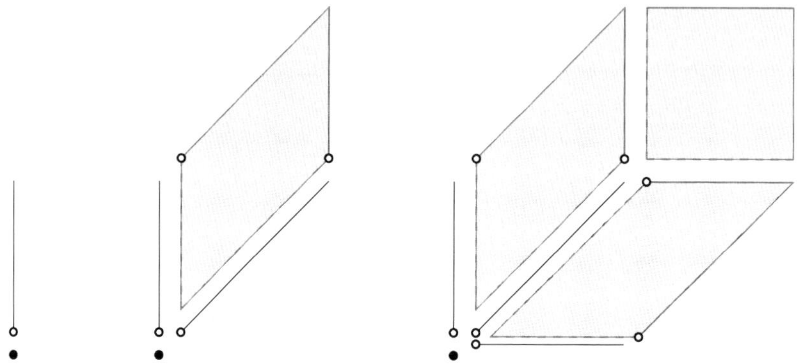

Fig. 3. A zonotopal decomposition of $Z((0,1),(1,0),(1,1))$

3. Ehrhart Theory

A *lattice polytope* $P \subset \mathbb{R}^n$ of dimension d is the convex hull of finitely many points in $\mathbb{Z}^n$ that together affinely span a d-dimensional hyperplane. For $t \in \mathbb{Z}_{>0}$, set $tP := \{tp : p \in P\}$, and let $L_P(t) = |\mathbb{Z}^n \cap tP|$. Ehrhart [10] proved a statement equivalent to the following. Recall that the set

$$\left\{\binom{t+d-i}{d} : i = 0, 1, \ldots, d\right\}$$

is a basis for polynomials of degree d, where

$$\binom{t+d-i}{d} = (1/d!)(t+d-i)(t+d-i-1)(t+d-i-2)\cdots(t-i+1)$$

is clearly a polynomial in t of degree d. For any lattice polytope P, there exist rational numbers $c_0, c_1, \ldots, c_d$ and $h_0^*, h_1^*, \ldots, h_d^*$ such that

$$L_P(t) = \sum_{i=0}^{d} h_i^* \binom{t+d-i}{d} = \sum_{i=0}^{d} c_i t^i .$$

The polynomial $L_P(t)$ is called the *Ehrhart polynomial of* P and has connections to commutative algebra, algebraic geometry, combinatorics, and discrete and convex geometry. Stanley [16] proved that $h_i^* \in \mathbb{Z}_{\geq 0}$ for all i. We call the polynomial $h^*(P;x) := h_0^* + h_1^* x + \cdots + h_d^* x^d$ encoding the h^*-coefficients the *h^*-polynomial* (or *δ-polynomial*) of P. The coefficients of $h^*(P;x)$ form the *h^*-vector of* P.

Example 3.1. The unit square as depicted in Figure 4 has Ehrhart polynomial $L_{[0,1]^2}(t) = t^2 + 2t + 1$ and $h^*([0,1]^2;x) = 1 + x$.

Various properties of P are reflected in its h^*-polynomial. For example, $\mathrm{vol}(P) = (\sum_i h_i^*)/d!$, where $\mathrm{vol}(P)$ denotes the Euclidean volume of P with respect to the integer lattice contained in the hyperplane spanned by P. We therefore define the *normalized volume* of P to be $\mathrm{Vol}(P) = \sum_i h_i^*$. Furthermore, it is known that

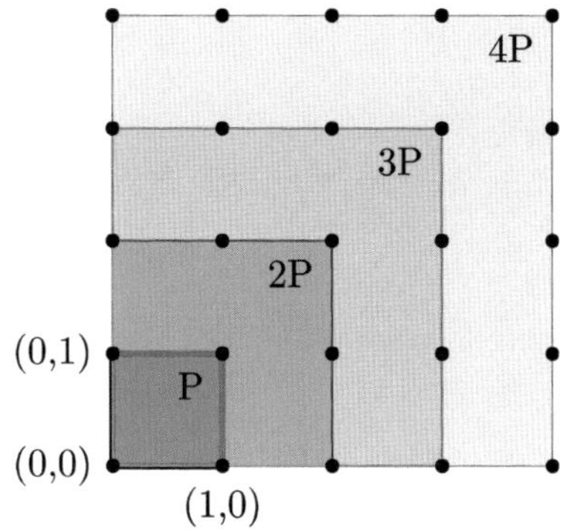

Fig. 4. The unit square $P = [0,1]^2 \subset \mathbb{Z}^2$ and some of its dilates.

$h_0^* = 1$ for all P, and that h_d^* is equal to the number of lattice points in the relative (topological) interior of P within the affine span of P. Another interesting combinatorial property displayed by $h^*(P;x)$ for some lattice polytopes is *unimodality.* A polynomial $a_0 + a_1 x + \cdots + a_d x^d$ is called *unimodal* if there exists an index j, $0 \le j \le d$, such that $a_{i-1} \le a_i$ for $i \le j$, and $a_i \ge a_{i+1}$ for $i \ge j$. Unimodality of h^*-polynomials is an area of active research [7].

4. Ehrhart Polynomials for Zonotopes

For a lattice zonotope Z, Stanley proved the following description of the coefficients of $L_Z(t)$.

Theorem 4.1 (Stanley [19], Theorem 2.2). *Let $Z := Z(\mathbf{v}_1, \ldots, \mathbf{v}_m)$ be a zonotope generated by the integer vectors $\mathbf{v}_1, \ldots, \mathbf{v}_m$. Then the Ehrhart polynomial of Z is given by*

$$L_Z(t) = \sum_S m(S) t^{|S|},$$

where S ranges over all linearly independent subsets of $\{\mathbf{v}_1, \ldots, \mathbf{v}_m\}$, and $m(S)$ is the greatest common divisor of all minors of size $|S|$ of the matrix whose columns are the elements of S.

The proof of Theorem 4.1 relies on Theorem 2.1, and is our first example of the usefulness of half-open decompositions of zonotopes. A generalization of Stanley's theorem was recently given by Hopkins and Postnikov [12].

Example 4.1. Consider the zonotope $Z((0,1),(1,0),(1,1),(1,-1)) \subset \mathbb{Z}^2$ depicted in Figure 5. We compute

$$\begin{aligned} L_{Z((0,1),(1,0),(1,1),(1,-1))}(t) = &\left|\det\begin{pmatrix}0&1\\1&0\end{pmatrix}\right|t^2 + \left|\det\begin{pmatrix}0&1\\1&1\end{pmatrix}\right|t^2 + \left|\det\begin{pmatrix}0&1\\1&-1\end{pmatrix}\right|t^2 + \\ &\left|\det\begin{pmatrix}1&1\\0&1\end{pmatrix}\right|t^2 + \left|\det\begin{pmatrix}1&1\\0&-1\end{pmatrix}\right|t^2 + \left|\det\begin{pmatrix}1&1\\1&-1\end{pmatrix}\right|t^2 + \\ &\gcd(0,1)t + \gcd(1,0)t + \gcd(1,1)t + \gcd(1,-1)t + 1 \\ = &\, 7t^2 + 4t + 1 \end{aligned}$$

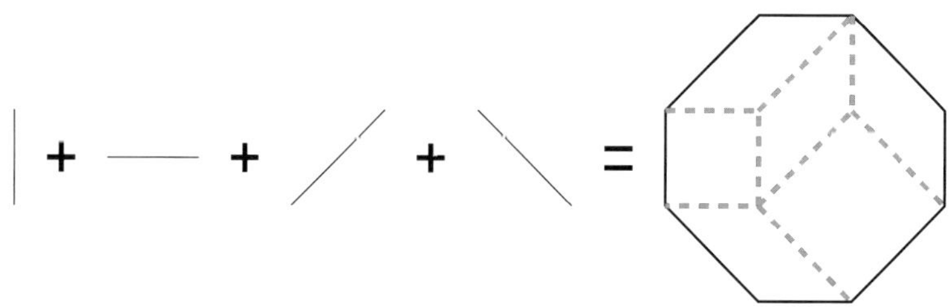

Fig. 5. The zonotope $Z((0,1),(1,0),(1,1),(1,-1))$ is a Minkowski sum of four line segments.

Recent work by D'Adderio and Moci has established a connection between $L_Z(t)$ and the arithmetic Tutte polynomial defined as follows by Moci [13].

Definition 4.1. Consider a collection $A \subseteq \mathbb{R}^n$ of integer vectors. The *arithmetic Tutte polynomial* is

$$M_A(x,y) := \sum_{B\subseteq A} m(B)(x-1)^{r(A)-r(B)}(y-1)^{|B|-r(B)},$$

where for each $B \subseteq A$, the multiplicity $m(B)$ is the index of $\mathbb{Z}B$ as a sublattice of $\text{span}(B) \cap \mathbb{Z}^n$ and $r(A)$ and $r(B)$ are the size of the largest independent subset of A and B, respectively. If we use the vectors in B as the columns of a matrix, then $m(B)$ equals the greatest common divisor of the minors of full rank.

The result of D'Adderio and Moci is to obtain the Ehrhart polynomial of Z as a specialization of the arithmetic Tutte polynomial associated to the lattice points generating Z.

Theorem 4.2 (D'Adderio and Moci [8], Theorem 3.2). *Let $Z := Z(\mathbf{v}_1, \ldots, \mathbf{v}_m)$ and let the columns of A be given by $\mathbf{v}_1, \ldots, \mathbf{v}_m$. Then*

$$L_Z(t) = t^{r(A)} M_A\left(1 + \frac{1}{t}, 1\right).$$

Example 4.2. Consider the zonotope $Z((0,1),(1,0),(1,1),(1,-1))$ from Example 4.1. Then

$$\begin{aligned}M_Z(x,y) &= (x-1)^2 + (1+1+1+1)(x-1) + (1+1+1+1+1+2) + (y-1)\\ &= x^2 + 2x + 3 + y.\end{aligned}$$

The corresponding Ehrhart polynomial is

$$L_Z(t) = t^2 M_Z\left(1 + \frac{1}{t}, 1\right) = 7t^2 + 4t + 1,$$

which agrees with Example 4.1.

From basic properties of Ehrhart polynomials, e.g. Ehrhart–Macdonald reciprocity, we obtain the following corollary observed by D'Adderio and Moci [8].

Corollary 4.1.

(1) For the interior Z° of the zonotope Z , the number of interior integer points in tZ° is $L_{Z^\circ}(t) = (-t)^{r(A)} M_A\left(1 - \frac{1}{t}, 1\right)$.
(2) The volume of the zonotope $Z(A)$ is $M_A(1,1)$.
(3) The zonotope $Z(A)$ contains $M_A(2,1)$ lattice points.
(4) The zonotope $Z(A)$ contains $M_A(0,1)$ interior lattice points.

These techniques have been applied in interesting ways; for example, Ardila, Castillo, and Henley [1] have computed the arithmetic Tutte polynomials and Ehrhart polynomials for zonotopes defined by the classical root systems.

5. Ehrhart h^*-Polynomials for Zonotopes

The h^*-vectors of lattice zonotopes have been the subject of significant recent investigation. Two properties of h^*-polynomials that have been studied in recent research in Ehrhart theory are unimodality and real-rootedness, where we refer to a polynomial as real-rooted if all of its roots are real. It is a well-known consequence of the general theory of real-rooted polynomials [17] that if $h^*(Z,x)$ has only real roots, its coefficient sequence is unimodal, but the converse does not hold. Recent work by Beck, Jochemko, and McCullough has shown that h^*-polynomials for zonotopes are as well-behaved as possible from this perspective.

Theorem 5.1 (Beck, Jochemko, and McCullough [5], Theorem 1.2). *Let Z be a n-dimensional lattice zonotope. Then the h^*-polynomial $h^*_Z(t) = h_0 + h_1 t + \cdots + h_n t^n$ has only real roots. Moreover,*

$$h_0 \le \cdots \le h_{\frac{n}{2}} \ge \cdots \ge h_n \text{ if } n \text{ is even}$$

and

$$h_0 \le \cdots \le h_{\frac{n-1}{2}} \text{ and } h_{\frac{n+1}{2}} \ge \cdots \ge h_n \text{ if } n \text{ is odd.}$$

Example 5.1. Consider the zonotope Z from Example 4.1. We saw that $L_Z(t) = 7t^2 + 4t + 1$, and it is a nice exercise to change basis and show that $h^*(Z;x) = 1 + 9x + 4x^2$. It follows easily that $h^*(Z;x)$ has only real roots and is clearly unimodal.

The key idea in the proof of Theorem 5.1 is again the half-open decomposition of Z. The paper by Beck, Jochemko, and McCullough [5] is well-written, and we recommend the interested reader look there for details. However, we do want to discuss a connection with permutations and descents that is a core ingredient of their proof.

Let $\mathfrak{S}_n$ denote the set of all permutations on $[n] := \{1, 2, 3, \dots, n\}$. For a permutation word $\sigma = \sigma_1\sigma_2\cdots\sigma_n$ in $\mathfrak{S}_n$ the *descent set* is defined as

$$\mathrm{Des}(\sigma) := \{i \in [n-1] : \sigma_i > \sigma_{i+1}\}\,.$$

The number of descents of σ is denoted by $\mathrm{des}(\sigma) := |\mathrm{Des}(\sigma)|$. The *Eulerian number* $a(n,k)$ counts the number of permutations in $\mathfrak{S}_n$ with exactly k descents:

$$a(n,k) := |\{\sigma \in \mathfrak{S}_n : \mathrm{des}(\sigma) = k\}|\,.$$

The (A,j)*-Eulerian number*

$$a_j(n,k) := |\{\sigma \in \mathfrak{S}_n : \sigma_n = n+1-j \text{ and } \mathrm{des}(\sigma) = k\}|$$

is a refinement of the descent statistic which gives the number of permutations $\sigma \in \mathfrak{S}_n$ with last letter $n+1-j$ and exactly k descents. The associated polynomial is known as the (A,j)*-Eulerian polynomial*, sometimes called the *restricted Eulerian polynomial*, and is defined as

$$A_j(n,x) := \sum_{k=0}^{n-1} a_j(n,k)x^k\,.$$

The half-open decomposition of a lattice zonotope leads to the following result.

Theorem 5.2 (Beck, Jochemko, and McCullough [5], Theorem 1.3). *Let $n \geq 1$. The convex hull of the h^*-vectors of all n-dimensional lattice zonotopes (viewed as points in $\mathbb{R}^{n+1}$) and the convex hull of the h^*-vectors of all n-dimensional lattice parallelepipeds are both equal to the n-dimensional simplicial cone*

$$A_1(n+1,x) + \mathbb{R}_{\geq 0}A_2(n+1,x) + \cdots + \mathbb{R}_{\geq 0}A_{n+1}(n+1,x).$$

It is fascinating that the convex hull of zonotope h^*-vectors admits such a beautiful combinatorial description, while the convex hull of the set of all h^*-vectors for lattice polytopes of a fixed dimension appears to be much more complicated [4, 20]. Given our knowledge of Theorem 5.2, it would be interesting if the following problem turns out to be tractable, at least in low dimensions.

Problem 5.1. *For fixed n, characterize the set of all h^*-vectors of n-dimensional parallelepipeds/zonotopes.*

6. Graphical and Laplacian Zonotopes

There are several interesting zonotopes associated with a finite simple graph G. In this section we discuss two such constructions, each of which is related to the number of spanning trees of G. For the first construction, we recall that for a polynomial $f = \sum_{\mathbf{a}\in\mathbb{Z}^n} \beta_{\mathbf{a}} t_1^{a_1} \cdots t_n^{a_n}$, the *Newton polytope* Newton(f) is the convex hull of integer points $\mathbf{a} \in \mathbb{Z}^n$ such that $\beta_{\mathbf{a}} \neq 0$. It is known that Newton$(f \cdot g)$ is the Minkowski sum Newton(f) + Newton(g), which is the fundamental ingredient for the following definition.

Definition 6.1. For a graph G on the vertex set $[n]$, the *graphical zonotope* Z_G is defined to be

$$Z_G := \sum_{(i,j)\in G} [e_i, e_j] = \text{Newton}\left(\prod_{(i,j)\in G} (t_i - t_j)\right),$$

where the Minkowski sum and the product are over edges (i,j), $i<j$, of the graph G, and $e_1, \ldots, e_n$ are the coordinate vectors in $\mathbb{R}^n$.

Example 6.1. The *n-permutahedron* is the polytope in $\mathbb{R}^n$ whose vertices are the $n!$ permutations of $[n]$:

$$\mathcal{P}_n := \text{conv}\{(\pi(1), \pi(2), \ldots, \pi(n)) : \pi \in \mathfrak{S}_n\}.$$

Using the Vandermonde determinant, it is straightforward to show that $\mathcal{P}_n$ is the graphical zonotope Z_{K_n} for the complete graph K_n. Figure 6 shows $\mathcal{P}_4$.

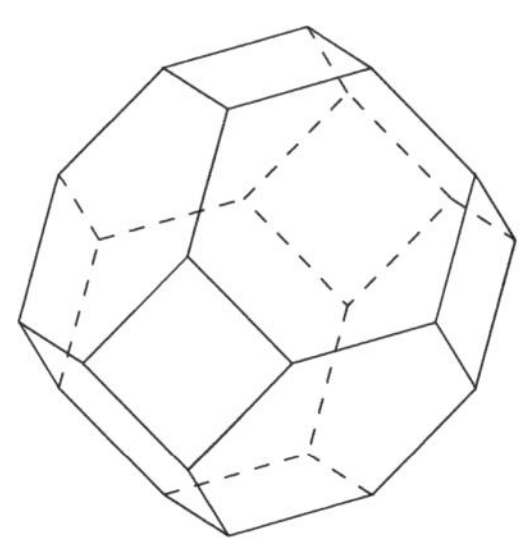

Fig. 6. The permutahedron $\mathcal{P}_4$.

The following beautiful theorem shows that the volume and lattice points of Z_G encode the number of spanning trees and forests in G.

Theorem 6.1 (Stanley [18], Exercise 4.32; Postnikov [14], Proposition 2.4). *For a connected graph G on n vertices, the volume of the graphical zonotope Z_G equals the number of spanning trees of G. The number of lattice points of Z_G equals the number of forests in the graph G.*

Example 6.2. The number of spanning trees of the connected graph K_n is n^{n-2} and by Theorem 6.1 it is also the volume of $\mathcal{P}_n$. Furthermore, the number of lattice points of $\mathcal{P}_n$ equals the number of forests on n labeled vertices.

Remark 6.1. Grujić [11] has shown that the f-polynomial of Z_G, which encodes the number of faces of Z_G in each dimension, can be obtained as the principal specialization of the q-analogue of the chromatic symmetric function of G.

Another zonotope associated with a finite simple graph arises from the Laplacian matrix, defined as follows.

Definition 6.2. Let G be an undirected graph with vertex set $V(G) = \{v_1, \ldots, v_p\}$. The *Laplacian matrix* $\mathbf{L} := \mathbf{L}(G)$ is the $p \times p$ matrix

$$\mathbf{L}_{ij} = \begin{cases} -1, & \text{if } i \neq j \text{ and there is an edge between} \\ & \text{vertices } v_i \text{ and } v_j \\ \deg(v_i), & \text{if } i = j \end{cases},$$

where $\deg(v_i)$ denotes the degree (number of incident edges) of v_i.

The Laplacian matrix of a graph admits a standard factorization that is relevant to our discussion. Let G have an arbitrary edge orientation. Let ∂ denote the $|V| \times |E|$ matrix where

$$\partial_{(v,e)} = \begin{cases} -1, & \text{if } v \text{ is the negative endpoint of } e \\ 1 & \text{if } v \text{ is the positive endpoint of } e \\ 0 & \text{otherwise} \end{cases}.$$

We call ∂ the *signed incidence matrix* for G. It is a straightforward exercise to verify that $\mathbf{L}(G) = \partial\partial^T$.

The following recent theorem due to Dall and Pfeifle connects the geometry of zonotopes arising from ∂ and $\mathbf{L}(G)$.

Theorem 6.2 (Dall and Pfeifle [9], Theorem 17). *Let ∂ and $\mathbf{L}(G)$ be the signed incidence matrix and Laplacian, respectively, for G. Let $\overline{\mathbf{L}}$ be the matrix obtained by taking any basis for $\mathbb{Z}^{|V|} \cap \operatorname{span}_{\mathbb{R}}(\mathbf{L}(G))$ from among the columns of $\mathbf{L}(G)$. Then the volume of the zonotope defined by the columns of ∂ is equal to the volume of the zonotope defined by the columns of $\overline{\mathbf{L}}$.*

The reason why Theorem 6.2 is interesting is that it allowed Dall and Pfeifle to give a polyhedral proof of the Matrix Tree Theorem, one of the most important theorems related to graph Laplacians, stated as follows.

Theorem 6.3. *Let G be a finite connected p-vertex graph without loops. Let $1 \leq i \leq p$, and let $\mathbf{L}_0$ denote $\mathbf{L}(G)$ with the i-th row and column removed, for any arbitrary choice of i. Let $\kappa(G)$ denote the number of spanning trees of G. Then*

$$\kappa(G) = |\det(\mathbf{L}_0)|.$$

Remark 6.2. Another interesting zonotope related to graphs is the zonotope of degree sequences of length n. This family of zonotopes was studied by Stanley [19] from an Ehrhart-theoretic perspective.

7. Fixed Polytopes of Permutahedra

Because of the close connection between permutahedra (defined in Section 6) and symmetric groups (defined in Section 5), it is of interest to investigate the geometric properties of the action of the symmetric group on the permutahedron. Specifically, the symmetric group $\mathfrak{S}_n$ acts on $\mathcal{P}_n \subset \mathbb{R}^n$ by permuting coordinates, i.e., $\sigma \in \mathfrak{S}_n$ acts on a point $\mathbf{x} = (x_1, x_2, \ldots, x_n) \in \mathcal{P}_n$, by $\sigma \cdot \mathbf{x} := (x_{\sigma^{-1}(1)}, x_{\sigma^{-1}(2)}, \ldots, x_{\sigma^{-1}(n)})$. Recent research [2] has focused on the fixed points of permutations under this action.

Definition 7.1. Let $\sigma \in \mathfrak{S}_n$. The *fixed polytope of the permutahedron* relative to σ is

$$\mathcal{P}_n^{\sigma} = \{\mathbf{x} \in \mathcal{P}_n : \sigma \cdot \mathbf{x} = \mathbf{x}\},$$

the polytope of the permutahedron $\mathcal{P}_n$ fixed by σ.

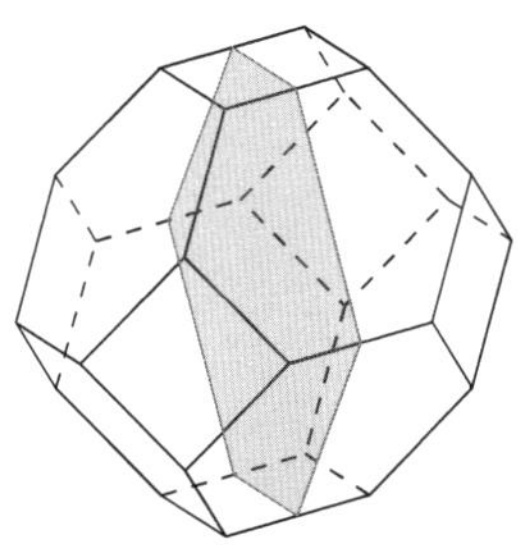

Fig. 7. The polytope $\mathcal{P}_4^{(12)}$ of the permutahedron $\mathcal{P}_4$ fixed by $(12) \in \mathfrak{S}_4$.

Given the combinatorial nature of the normalized volume of $\mathcal{P}_n$ described in Example 6.1, it is of interest to compute the normalized volumes of fixed polytopes of $\mathcal{P}_n$. Before doing so, we first describe the vertices, supporting hyperplanes, and Minkowski sum representation of $\mathcal{P}_n^{\sigma}$. Let e_i denote the i-th standard basis vector in $\mathbb{R}^n$.

Theorem 7.1 (Ardila, Schindler, Vindas-Meléndez [2], Theorem 2.12). *Let σ be a permutation of $[n]$ having disjoint cycles $\sigma_1, \ldots, \sigma_m$, which have respective lengths $l_1, \ldots, l_m$. The fixed polytope $\mathcal{P}_n^{\sigma}$ can be described in the following four ways:*

1. *It is the set of points x in the permutahedron $\mathcal{P}_n$ such that $\sigma \cdot \mathbf{x} = \mathbf{x}$.*
2. *It is the set of points $x \in \mathbb{R}^n$ satisfying*

(a) $x_1 + x_2 + \cdots + x_n = 1 + 2 + \cdots + n$,
(b) for any proper subset $\{i_1, i_2, \ldots, i_k\} \subset \{1, 2, \ldots, n\}$,

$$x_{i_1} + x_{i_2} + \cdots + x_{i_k} \leq 1 + 2 + \cdots + k, \text{ and}$$

(c) for any i *and* j *which are in the same cycle of* σ, $x_i = x_j$.

3. *It is the convex hull of the set of vertices* $\mathrm{Vert}(\sigma)$ *consisting of the* $m!$ *points*

$$\overline{\mathbf{v}_{\prec}} := \sum_{k=1}^{m} \left(\frac{l_k + 1}{2} + \sum_{j\,:\,\sigma_j \prec \sigma_k} l_j \right) e_{\sigma_k}$$

as $\prec$ *ranges over the* $m!$ *possible linear orderings of the disjoint cycles* $\sigma_1, \sigma_2, \ldots, \sigma_m$.

4. *It is the Minkowski sum*

$$M_\sigma := \sum_{1 \leq j < k \leq m} [l_j e_{\sigma_k}, l_k e_{\sigma_j}] + \sum_{k=1}^{m} \frac{l_k + 1}{2} e_{\upsilon_k}.$$

Consequently, the fixed polytope $\mathcal{P}_n^\sigma$ *is an* $(m-1)$*-dimensional zonotope that is combinatorially isomorphic to the permutahedron* $\mathcal{P}_m$.

Example 7.1. In Figure 7, we see that $\mathcal{P}_4^{(12)}$ has six vertices and Theorem 7.1 tells us that the vertices correspond to the 6 orderings of $\sigma_1 = (12), \sigma_2 = (3), \sigma_3 = (4)$:

$$\begin{array}{ll} \frac{1+2}{2} e_{12} + 3e_3 + 4e_4, & \frac{1+2}{2} e_{12} + 4e_3 + 3e_4, \\ \frac{2+3}{2} e_{12} + 1e_3 + 4e_4, & \frac{2+3}{2} e_{12} + 4e_3 + 1e_4, \\ \frac{3+4}{2} e_{12} + 1e_3 + 2e_4, & \frac{3+4}{2} e_{12} + 2e_3 + 1e_4. \end{array}$$

Furthermore, $\mathcal{P}_4^{(12)}$ is the Minkowski sum

$$[e_{12}, 2e_3] + [e_{12}, 2e_4] + [e_3, e_4] + \frac{3}{2} e_{12} + e_3 + e_4.$$

Using Theorem 7.1, it is possible to determine an arithmetical formula for the normalized volume of $\mathcal{P}_n$ related to only the disjoint cycle decomposition of σ.

Theorem 7.2 (Ardila, Schindler, Vindas-Meléndez [2], Theorem 1.2). *If* σ *is a permutation of* $[n]$ *whose cycles have lengths* $l_1, \ldots, l_m$, *then the normalized volume of the fixed polytope of* $\mathcal{P}_n$ *is*

$$\mathrm{Vol}\, \mathcal{P}_n^\sigma = n^{m-2} \gcd(l_1, \ldots, l_m).$$

8. Cyclic Zonotopes

Let $\{(1, t, t^2, t^3, \ldots, t^{n-1}) : t \in \mathbb{R}\}$ be the *moment curve* in $\mathbb{R}^n$. The moment curve is often used in discrete geometry to select points in general position, and the well-studied family of cyclic polytopes arise by taking the convex hull of a finite set of points on the moment curve. There is a zonotopal analogue of cyclic polytopes, defined as follows using centrally symmetric zonotopes.

Definition 8.1. The *cyclic zonotope* $Z(d,n)$ defined by $t_1 < t_2 < \cdots < t_d$ is the n-zonotope $Z_0(\mathbf{v}_1,\ldots,\mathbf{v}_d) = Z(\pm\mathbf{v}_1,\ldots,\pm\mathbf{v}_d)$ generated by the d vectors $\mathbf{v}_i = (1,t_i,\ldots,t_i^{n-1})$ in $\mathbb{R}^n$.

It is known that while $Z(d,n)$ depends on the set of parameters $\{t_1,\ldots,t_d\}$, its face poset depends only on d and n, hence the suppression of the values of t in our notation. When $\{t_1,\ldots,t_d\} \subset \mathbb{Z}$, we say that $Z(d,n)$ is an *integral* cyclic zonotope, though our discussion in this section will apply to cyclic zonotopes more generally.

Example 8.1. Figure 8 illustrates the cyclic zonotope $Z(3,2)$ generated by $t \in \{1,2,3\}$.

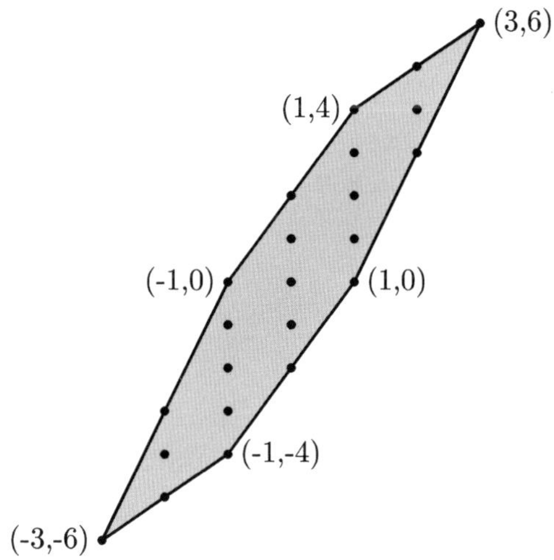

Fig. 8. The cyclic zonotope $Z(3,2)$ generated by $t \in \{1,2,3\}$.

Cyclic zonotopes have particularly nice geometric properties, especially with regard to subdivisions. We have seen throughout this paper the importance of the half-open decompositions of zonotopes established in Theorem 2.1. A relaxation of this idea is the notion of a zonotopal subdivision, which we now review in the setting of $Z(d,n)$. Denote by Λ_d the set $\{-,0,+\}^d$ of sign vectors of length d and write $X = (X_1, X_2, \ldots, X_d)$ for $X \in \Lambda_d$. The set Λ_d is partially ordered by extending coordinatewise the partial order on $\{-,0,+\}$ defined by the relations $0 < -$ and $0 < +$. A *subzonotope* Z_X of $Z(d,n)$ is a pair (X, Z_X) of a sign vector $X \in \Lambda_d$ and the translated zonotope

$$Z_X = \sum_{X_i=+} \mathbf{v}_i - \sum_{X_i=-} \mathbf{v}_i + \sum_{X_i=0} [-\mathbf{v}_i, \mathbf{v}_i]\,.$$

We say that Z_C is a *face* of Z_X if there exists a face F of the polytope Z_X such that C is minimum among all sign vectors $Y \geq X$ with $Z_Y = F$. Two subzonotopes

Z_X and Z_Y of $Z(d,n)$ are said to *intersect properly* if either $Z_X \cap Z_Y = \emptyset$ or there exists a sign vector C such that

(1) $Z_X \cap Z_Y = Z_C$ and
(2) Z_C is a face of both Z_X and Z_Y.

Definition 8.2. A *zonotopal subdivision* S of $Z(d,n)$ is a set of subzonotopes of $Z(d,n)$ with the following properties:

(1) if Z_C is a face of $Z_X \in S$ then $Z_C \in S$,
(2) any two elements of S intersect properly and
(3) $\bigcup_{Z_X \in S} Z_X = Z(d,n)$.

As with triangulations of polytopes, there is a notion of shellability for zonotopal subdivisions.

Definition 8.3. A zonotopal subdivision S of $Z(d,n)$ is *shellable* if it is pure of dimension n, i.e., each subzonotope of $Z(d,n)$ has dimension n, and there exists a linear ordering of the subzonotopes $Z_{X_1}, \ldots, Z_{X_k}$, called a *shelling*, such that for $1 < j \leq k$,

$$Z_{X_j} \cap \left(\bigcup_{i \leq i < j} Z_{X_i} \right)$$

is not empty and equals the union of the $(n-1)$-polytopes in an initial segment of a shelling of the boundary of Z_{X_j}.

Example 8.2. Consider the zonotope $Z_O((0,1),(1,0),(1,1),(1,-1))$ illustrated in Figure 5. The zonotope is shellable and Figure 9 shows a shelling, that is, the subdivision given by the numbered facets together with their faces is shellable.

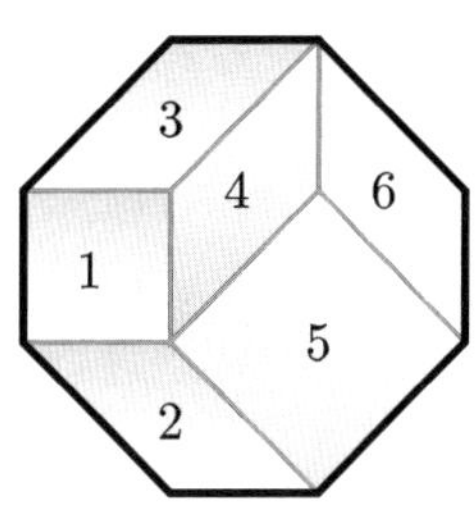

Fig. 9. A zonotopal subdivision of $Z_O((0,1),(1,0),(1,1),(1,-1))$ with a shelling.

A lovely result about cyclic zonotopes is the following theorem due to Athanasiadis.

Theorem 8.1 (Athanasiadis [3], Theorem 1.2). *All zonotopal subdivisions of a cyclic zonotope are shellable.*

In addition to Theorem 8.1, Athanasiadis [3] also proved that posets of proper zonotopal subdivisions of $Z(d, n)$ induced by canonical projections of cyclic zonotopes are homotopy equivalent to spheres. Thus, there are interesting connections between the structure of cyclic zonotopes and poset topology.

Acknowledgments

This work was partially supported by NSF Graduate Research Fellowship DGE-1247392 (ARVM). Thanks to Federico Ardila, Matt Beck, and Sam Hopkins for helpful comments.

References

[1] F. Ardila, F. Castillo, M. Henley, The arithmetic Tutte polynomials of the classical root systems, *Int. Math. Res. Not. IMRN*, 3830 (2015).

[2] F. Ardila, A. Schindler, A. R. Vindas-Meléndez, The equivariant volumes of the permutahedron, preprint at `https://arxiv.org/abs/1803.02377`.

[3] C. A. Athanasiadis, Zonotopal subdivisions of cyclic zonotopes, *Geom. Dedicata* 86, 37 (2001).

[4] M. Beck, J. A. De Loera, M. Develin, J. Pfeifle, R. P. Stanley, Coefficients and roots of Ehrhart polynomials, in *Integer points in polyhedra - geometry, number theory, algebra, optimization*, Contemp. Math. Vol. 374 (Amer. Math. Soc., Providence, RI, 2005) pp. 15 - 36.

[5] M. Beck, K. Jochemko, E. McCullough, h*-polynomials of zonotopes, to appear in Transactions of the American Mathematical Society.

[6] M. Beck, S. Robins, *Computing the continuous discretely*, Undergraduate Texts in Mathematics, second edn. (Springer, New York, 2015), Integer-point enumeration in polyhedra, With illustrations by David Austin.

[7] B. Braun, Unimodality problems in Ehrhart theory, in *Recent trends in combinatorics*, IMA Vol. Math. Appl. Vol. 159 (Springer, [Cham], 2016) pp. 687 - 711.

[8] M. D'Adderio, L. Moci, Ehrhart polynomial and arithmetic Tutte polynomial, *European J. Combin.* 33, 1479 (2012).

[9] A. Dall, J. Pfeifle, A polyhedral proof of the matrix tree theorem, preprint at `https://arxiv.org/abs/1404.3876`.

[10] E. Ehrhart, Sur les polyèdres rationnels homothétiques à n dimensions, *C. R. Acad. Sci. Paris* 254, 616 (1962).

[11] V. Grujić, Counting faces of graphical zonotopes, *Ars Math. Contemp.* 13, 227 (2017).

[12] S. Hopkins, A. Postnikov, A positive formula for the Ehrhart-like polynomials from root system chip-firing, preprint at `https://arxiv.org/abs/1803.08472`

[13] L. Moci, A Tutte polynomial for toric arrangements, *Trans. Amer. Math. Soc.* 364, 1067 (2012).

[14] A. Postnikov, Permutohedra, associahedra, and beyond, *Int. Math. Res. Not. IMRN*, 1026 (2009).

[15] G. C. Shephard, Combinatorial properties of associated zonotopes, *Canad. J. Math.* 26, 302 (1974).

[16] R. P. Stanley, Decompositions of rational convex polytopes, *Ann. Discrete Math.* 6, 333 (1980), Combinatorial mathematics, optimal designs and their applications (Proc. Sympos. Combin. Math. and Optimal Design, Colorado State Univ., Fort Collins, Colo., 1978).

[17] R. P. Stanley, Log-concave and unimodal sequences in algebra, combinatorics, and geometry, in *Graph theory and its applications: East and West (Jinan, 1986)*, Ann. New York Acad. Sci. Vol. 576 (New York Acad. Sci., New York, 1989) pp. 500 - 535.

[18] R. P. Stanley, *Enumerative combinatorics. Vol. 1*, Cambridge Studies in Advanced Mathematics, Vol. 49 (Cambridge University Press, Cambridge, 1997), With a foreword by Gian-Carlo Rota, Corrected reprint of the 1986 original.

[19] R. P. Stanley, A zonotope associated with graphical degree sequences, in *Applied geometry and discrete mathematics*, DIMACS Ser. Discrete Math. Theoret. Comput. Sci. Vol. 4 (Amer. Math. Soc., Providence, RI, 1991) pp. 555 - 570.

[20] A. Stapledon, Inequalities and Ehrhart δ-vectors, *Trans. Amer. Math. Soc.* 361, 5615 (2009).

[21] C. Zong, What is known about unit cubes, *Bull. Amer. Math. Soc. (N.S.) 42*, 181 (2005).

A pithy look at the polytope algebra

Federico Castillo
Mathematics Department, Kansas University
Lawrence, KS 66049, USA
E-mail: fcastillo@ku.edu

This is a hands-on introduction to McMullen's Polytope Algebra. More than interesting on its own, this algebra was McMullen's tool to give a combinatorial prove of the g-theorem.

Keywords: Polytopes; translation invariant valuations, Minkowski weights.

1. Introduction

One of the biggest achievements in polytope theory was the complete characterization of f-vectors of simplicial polytopes, known as the g-theorem (for a precise statement see [12, Chapter 3]). In the early 70's Peter McMullen conjectured a set of necessary and sufficient conditions and less than a decade later Lou Billera and Carl Lee proved sufficiency, whereas Richard Stanley proved necessity. For an interesting survey about the history of this result we refer the reader to [3].

Stanley's proof in [13] imports a result from complex geometry, namely the Hard-Lefschetz theorem, so for a while people attempted to find a more combinatorial proof, one that used more elementary arguments. McMullen succeeded initially in [10] and then corrected and simplified parts of it in [11]. The heart of the g-theorem lies in associating, to each simplicial polytope P, a finite dimensional algebra with particular properties. By now there are many different ways of doing this. For instance, in [2] Billera uses piecewise polynomial functions on a fan (an approach also explored by Brion [4]), and in [5] Fulton-Sturmfels use the Chow ring of the associated toric variety which they show is isomorphic to the Minkowski weights, which is the same ring that McMullen employs in [11] although he constructs it from his earlier work in [9].

Here we give an overview of the polytope algebra as developed in [9]. Of all the approaches mentioned above, this one is of the easiest to define and in a way the most "polytopal". Apart from the intrinsic interest, we also wanted to introduce, in a natural way, the notion of Minkowski weights which continues to play an important role in research nowadays, for example in Tropical geometry (See [8]).

Acknowledgments

The author is deeply grateful to Raman Sanyal from whom he learned most of the material. These notes combines the original material straight from [9] together with Sanyal lectures in the MSRI Summer School *Positivity Questions in Geometric*

Combinatorics in summer 2017.

2. Polytopes and their faces

Definition 2.1. A *polytope* $P \subset \mathbb{R}^d$ is the convex hull of finitely many points $\{\mathbf{v}_1. \cdots, \mathbf{v}_m\}$. More precisely,

$$P = \operatorname{conv}(\mathbf{v}_1, \cdots, \mathbf{v}_m) = \left\{ \sum_{i=1}^{m} \lambda_i \mathbf{v}_i : \lambda_1 + \cdots + \lambda_m = 1, \lambda_1, \cdots, \lambda_m \geq 0 \right\}.$$

An inclusion minimal set of points whose convex hull is P is called the vertex set of P, $\operatorname{vert}(P)$. Such a minimal set exists and is unique.

Theorem 2.1 (Minkowski-Weyl). *Every polytope P is the bounded intersection of finitely many halfspaces. More precisely, there exist $\boldsymbol{a}_1, \cdots, \boldsymbol{a}_n \in \mathbb{R}^d \backslash \{0\}$ and $b_1, \cdots, b_n \in \mathbb{R}$ such that*

$$P = \left\{ \boldsymbol{x} \in \mathbb{R}^d : \boldsymbol{a}_1^t \boldsymbol{x} \leq b_1, \cdots, \boldsymbol{a}_n^t \boldsymbol{x} \leq b_n \right\}.$$

As before, there may be redundant inequalities. An inclusion minimal set of inequalities that define P is called the set of facet inequalities. Such a minimal set exists and is unique.

Definition 2.2. Let $P \subset \mathbb{R}^d$ be a polytope. We define the following notions.

(1) The *affine hull* of P: $\operatorname{aff}(P) = \bigcap \{L \subset \mathbb{R}^d$ affine space such that $P \subset L\}$.
(2) The *linear span* of P: $\operatorname{lin}(P) =$ the linear subspace that is parallel to $\operatorname{aff}(P)$.
(3) The *relative interior* of P: $\operatorname{relint}(P) =$ the interior of P relative to $\operatorname{aff}(P)$.
(4) The *dimension* of P: $\dim(P) = \dim \operatorname{lin}(P)$.

Basic information: From now on we are going to repeatedly use the following notation. P is a d-dimensional polytope, or d-polytope, with n facet inequalities (or just facets for short) and m vertices. The set of all polytopes in $\mathbb{R}^d$ (not necessarily full dimensional) is denoted $\mathcal{P}_d$ and $\hat{\mathcal{P}_d} := \mathcal{P}_d \backslash \{\emptyset\}$.

2.1. *Faces of polytopes*

For any $\mathbf{c} \in \mathbb{R}^d$ (which we will think of as a linear functional) define

$$\begin{aligned} P^{\mathbf{c}} &:= \{\mathbf{x} \in P : \mathbf{c}^t\mathbf{x} \geq \mathbf{c}^t\mathbf{y}, \forall \mathbf{y} \in P\}, \\ &= \{\mathbf{x} \in P : \mathbf{c}^t\mathbf{x} = \delta\}, \quad \delta = \max_{\mathbf{x} \in P} \mathbf{c}^t\mathbf{x}. \end{aligned}$$

Subsets of P of the above form are called *faces* of P. If $P = \operatorname{conv}(\mathbf{v}_1, \cdots, \mathbf{v}_m)$ is the vertex description of P, then $P^{\mathbf{c}} = \operatorname{conv}(\mathbf{v}_i : \mathbf{c}^t v_i = \delta)$, so faces are polytopes themselves. Moreover, there can be only finitely many faces. By convention, $\emptyset$ and P itself are faces.

Remark 2.1. In what follows we will only consider face directions $\mathbf{c}$ such that $|\mathbf{c}| = 1$.

Definition 2.3. Let P be a polytope and let $\mathcal{F}_k(P)$ be the set of k-dimensional faces. Without a subscript $\mathcal{F}(P)$ is the set of all faces of P. Equipped with the partial order given by containment, $\mathcal{F}(P)$ is called the *face lattice* of P. This captures the combinatorial information of P. Two polytopes are said to be *combinatorially equivalent* if they have isomorphic face lattices.

Definition 2.4. A weaker invariant is the *f-vector*: $f(P) = (f_0, f_1, \cdots, f_{d-1}, f_d)$. Where f_i is the number of i-dimensional faces of P (recall that faces are themselves polytopes, hence they have dimension). Often people add $f_{-1} = 1$ for the empty set, but in this note we are going to obviate it sometimes. Also note that $f_d = 1$ always. The 0-faces are the vertices, 1-faces the edges, and $(d-1)$-faces are the facets.

We say a d-polytope is *simple* if each vertex is contained in exactly d facets.

3. Valuations

For any set $S \subset \mathbb{R}^d$ denote by $[S]$ the *indicator function* from $\mathbb{R}^d$ to $\{0,1\}$ defined as

$$[S](p) = \begin{cases} 1, & p \in S \\ 0, & p \notin S \end{cases}$$

Definition 3.1. A function $f : \mathbb{R}^d \longrightarrow \mathbb{Z}$ is a *polytopal simple function* if it can be written as

$$f = \alpha_1[Q_1] + \cdots + \alpha_k[Q_k], \quad \alpha_i \in \mathbb{Z},$$

with $Q_1, \cdots, Q_k$ polytopes.

Notation 3.1. Recall that $\mathcal{P}_d = \{$ polytopes $\subset \mathbb{R}^d\}$ and now we define $\mathcal{SP}_d :=$ $\{$ polytopal simple functions $\}$. Often we won't distinguish between a polytope and its indicator function.

By definition, indicator functions of $\mathcal{P}_d$ span $\mathcal{SP}_d$ but they are far from a basis, they satisfy relations like $[P \cup Q] = [P] + [Q] - [P \cap Q]$.

Definition 3.2. A *valuation* is any group homomorphism $\phi : \mathcal{SP}_d \longrightarrow G$ with G an abelian group.

The first and most fundamental valuation is the *Euler characteristic* (See [1, Theorem 2.4]).

Theorem 3.1. *There exists a unique valuation $\chi : \mathcal{P}_d \longrightarrow \mathbb{Z}$, called the* Euler characteristic *with*

- $\chi(\emptyset) = 0$.

- $\chi(P) = \begin{cases} 1, & P \in \hat{\mathcal{P}}_d \\ 0, & \textit{else.} \end{cases}$

Even though $\mathcal{SP}_d$ is spanned by indicator functions of polytopes, it also contains indicator functions of other non polytopal sets. For instance, the Euler formula is equivalent to the relation

Lemma 3.1. *Let P be a d-polytope then*

$$[P^\circ] = \sum_{F \subseteq P} (-1)^{dim(P) - dim(F)} [F].$$

Furthermore $\chi([P^\circ]) = (-1)^{dim(P)}$.

Even though indicators functions come with a natural product, namely pointwise multiplication, we consider the following ring structure.

Proposition 3.1. *The map* $* : \mathcal{SP}_d \times \mathcal{SP}_d \to \mathcal{SP}_d$ *defined by* $[P] * [Q] := [P + Q]$ *gives* $\mathcal{SP}_d$ *the structure of a commutative ring. The multiplicative unit is* $[\mathbf{0}]$, *the indicator function at the origin.*

With this operation we gain some interesting inverses.

Proposition 3.2. *Let* $P \in \hat{\mathcal{P}}_d$, *then* $[P]$ *is invertible:*

$$[P]^{-1} = (-1)^{\dim P} [-relint(P)]. \tag{1}$$

The next proposition is fundamental for what follows.

Proposition 3.3. *If* $P = conv\{v_1, \cdots, v_m\}$ *is a polytope, then*

$$([P] - [v_1]) * ([P] - [v_2]) * \cdots * ([P] - [v_m]) = 0. \tag{2}$$

Let's first illustrate an example.

Example 3.1. Let P be a one dimensional polytope $[a, b] \subset \mathbb{R}$. Then $([P] - [a]) * ([P] - [b]) = [2P] - [P + a] - [P + b] + [a + b] = 0$, as can be seen by inspection in Figure 1.

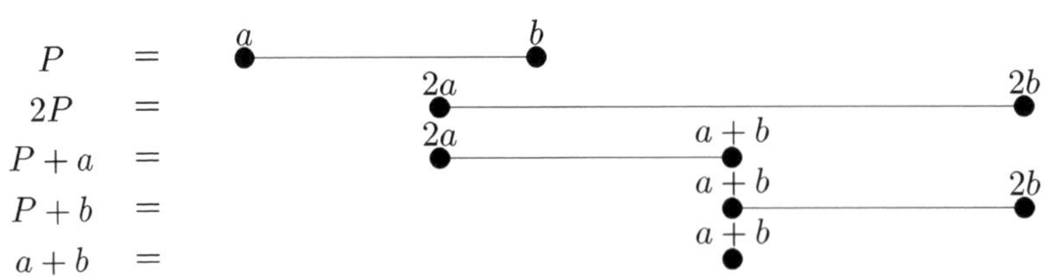

Fig. 1. An illustration of Proposition 3.3

Sketch of proof. We first prove it for $P = \Delta_{m-1} = \text{conv}\{e_1, \cdots, e_m\} = \{\mathbf{x} \in \mathbb{R}^m : x_i \geq 0, \quad \sum x_i = 1\}$. We define $e_I := \sum_{i \in I} e_i$ for $I \subset [m]$. We must show that

$$\sum_{I \subset [m]} (-1)^{|I|} [(m - |I|)\Delta_{m-1} + e_I] = 0. \tag{3}$$

For $q \in \mathbb{R}^m$ how do we decide if $q \in [(m - |I|)\Delta_{m-1} + e_I]$? This is equivalent to $q - e_I \in (m - |I|)\Delta_{m-1}$. This happens if and only if

$$q - e_I \geq 0 \qquad \sum q_i = m,$$

where the first inequality is coordinate wise. Now define $I_0 := \{i : q_i \geq 1\}$. For a fixed q with $\sum q_i = m$ the value of the function in the left hand side of Equation (3) is

$$\sum_{I \subset I_0} (-1)^{|I|} = \sum_{j=0}^{|I_0|} \binom{I_0}{j} (-1)^j = (1-1)^{|I_0|} = \begin{cases} 0 & I_0 \neq \emptyset \\ 1 & I_0 = \emptyset \end{cases}$$

but I_0 will never be empty since we are asumming $\sum q_i = m$. This finishes the proof for the standard simplex.

For the general case we can write $P = \pi(\Delta_{m-1})$, where $\pi : \mathbb{R}^m \longrightarrow \mathbb{R}^d$, defined by mapping $e_i \longrightarrow v_i$. Then one can argue that π induces a map $\pi_* : \mathcal{SP}_m \to \mathcal{SP}_d$ sending $\pi_*[P] = [\pi(P)]$ for $P \in \mathcal{P}_m$ and such that the linear relations are preserved. □

4. The polytope algebra

We define $\mathcal{T} := \mathbb{Z}\{[P+t] - [P] \quad \forall P \in \mathcal{P}_d, t \in \mathbb{R}^d\}$. This is actually an *ideal* of $\mathcal{SP}_d$, called the translation ideal.

Definition 4.1. The polytope algebra is

$$\Pi^d := \mathcal{SP}_d/\mathcal{T}.$$

We denote by $[\![P]\!]$ the class of $[P]$ in Π^d.

The multiplicative identity is $1 := [\![\{\mathbf{0}\}]\!]$, the class of the origin as a zero dimensional polytope.

Definition 4.2. We define also *dilation* maps which are in fact endormorphism. Let $D_\lambda : \Pi^d \to \Pi^d$ be defined in the generators as $D_\lambda[\![P]\!] := [\![\lambda P]\!]$ for $\lambda \in \mathbb{R}_{\geq 0}$. Notice that D_λ and $D_{\lambda^{-1}}$ are inverses to each other for $\lambda > 0$.

Example 4.1. Let's begin by understanding Π^1. This is generated by *integer* combinations of segments. For each line segment $[a, b]$ we have $[\![[a, b]]\!] = [\![[a, b)]\!] + [\![\{b\}]\!] = [\![[0, b-a)]\!] + [\![\{0\}]\!]$, since we can decompose and translate the pieces. Notice that the sum of two classes of half open segments can be represented again by a half open segment; $[\![[0, r)]\!] + [\![[0, s)]\!] = [\![[0, r)]\!] + [\![[r, s+r)]\!] = [\![[0, s+r)]\!]$.

It is not true that $[\![[0,s)]\!] * [\![[0,r)]\!] = [\![[0,s+r)]\!]$, since the definition of Minkowski sum only is intended for the generators, the indicator functions of polytopes. Indeed we have

$$[\![[0,s)]\!] * [\![[0,r)]\!] = ([\![[0,s]]\!] - [\![\{0\}]\!]) * ([\![[0,r]]\!] - [\![\{0\}]\!]),$$

which we already saw in Example 3.1 to be zero.

With this is mind, $\Pi^1 \cong \mathbb{Z} \oplus \mathbb{R}$ with multiplication defined by $(a,b) \cdot (a',b') = (aa', ab' + a'b)$.

Now that we understand Π^1, we move on to the general case. The following is the main structural result.

Theorem 4.1. *The polytope algebra is a graded ring, generated in degree 1.* $\Pi^d = \Pi_0 \oplus \Pi_1 \oplus \cdots \oplus \Pi_d$. *Furthermore*

(i) $\Pi_0 \cong \mathbb{Z}$.
(ii) Π_i *is an* $\mathbb{R}$*-vector space for* $i > 0$.
(iii) $\Pi_d \cong \mathbb{R}$.

The theorem means that Π^d is almost an $\mathbb{R}$-algebra, except for the fact that $\Pi_0 \cong \mathbb{Z}$. Notice that Π^d comes from $\mathcal{SP}_d$ which is made of *integer* combinations of indicator functions, so the fact that we end up with a $\mathbb{R}$ action should be surprising, in fact we will see below that scaling can be complicated. As a first step in proving Theorem 4.1, we have

Proposition 4.1. *The map* $\Pi^d \longrightarrow \mathbb{Z}$ *induced by* χ *allows us to decompose* $\Pi^d = \mathbb{Z} \oplus \Pi_+$ *as a direct sum, where* $\Pi_+ := \ker \chi$.

Proof. Decompose $x = \sum \alpha_i [\![P_i]\!] \in \Pi^d$ as

$$x = \sum \alpha_i \chi(P_i) \cdot 1 + \sum \alpha_i ([\![P_i]\!] - 1) = \chi(x) \cdot 1 + (x - \chi(x) \cdot 1). \tag{4}$$

□

From Equation (4) we have that $\Pi_+ = \mathbb{Z}\{[\![P]\!] - 1 : \ P \in \hat{\mathcal{P}}_d\}$. Proposition 3.3 in the polytope algebra means that all those generators are nilpotent. More precisely:

Corollary 4.1. *For* $P \in \hat{\mathcal{P}}_d$ *we have* $\Big([\![P]\!] - 1\Big)^r = 0$ *in* Π^d *for* $r > d$.

Proof. From Theorem 3.3 we know that $\Big([\![P]\!] - 1\Big)^{f_0(P)} = 0$. Now we argue that we can lower the exponent. Notice that $[\![nP]\!] = [\![P + P \cdots + P]\!] = [\![P]\!]^n$ so we can write

$$[\![nP]\!] = (1 + ([\![P]\!] - 1))^n = \sum_{k=0}^{n} \binom{n}{k} ([\![P]\!] - 1)^k. \tag{5}$$

Every polytope can be triangulated which means we can always write, through inclusion-exclusion, $[\![P]\!] = \sum \alpha_i [\![T_i]\!]$, where each T_i is a simplex of dimension $\leq d$

(and hence with at most $d+1$ vertices) and $\alpha_i \in \{-1,+1\}$. When dilating P we can dilate each piece so that

$$[\![nP]\!] = \sum \alpha_i [\![nT_i]\!]. \tag{6}$$

Expanding the right hand side with an analogous relation to Equation (5) for each T_i, we get a polynomial in n with coefficients in Π^d of degree at most d, since all terms $([\![T_i]\!]-1)^r$ vanish for $r > d$. This means that the right hand side of Equation (5) is also a polynomial in n of degree $\leq d$ and hence $([\![P]\!]-1)^r$ vanish for $r > d$. □

Theorem 4.2. *The abelian group Π_+ is a $\mathbb{Q}$-vector space.*

This is a nontrivial statement. We need to make sense of $\frac{1}{m}x$ for $x \in \Pi_+$, more precisely, we need that for every $m \in \mathbb{Z}_{>0}$ there exists a unique $h \in \Pi_+$ with $x = m \cdot h$. We prove existence and uniqueness in two separate lemmas. Together they prove Theorem 4.2.

Lemma 4.1. *The abelian group Π_+ is divisible. For every $x \in \Pi_+$ and $m \in \mathbb{Z}_{>0}$, there exist $h \in \Pi_+$ such that $m \cdot h = f$.*

Remark 4.1. It is important to keep in mind that if P is a polytope, then $2[\![P]\!]$ is **not** equal to $[\![2P]\!]$, the indicator of the second dilation of P. One quick way to remember this is to apply Euler characteristic: $\chi(2[\![P]\!]) = 2$ whereas $\chi([\![2P]\!]) = 1$.

Proof. The following proof is indirect and dry. To see how to actually divide see Example 4.3. It is enough to show the result for $m > 1$ prime. Consider $N = m^e > d+1$ a large power of m. Then we have

$$[\![P]\!] - 1 = [\![\frac{1}{N}P]\!]^N - 1 = \sum_{i=1}^{d} \binom{N}{i} \left([\![\frac{1}{N}P]\!] - 1\right)^i.$$

Now since m is prime and N is a power of m, we have that m divides all the binomial coefficients. □

Example 4.2. Notice that in the real line it is straighforward to divide half open segments, since $m \cdot [0, r/m) = [0, r)$ up to translation.

Example 4.3. Now sketch an example in two dimensions. Again we will not divide a whole polytope, but a polytope minus a point. Since all points are equivalent under translation, we choose to remove a vertex.
The idea now is that we are going to decompose that simplex in a convenient way. Along the way we will get 4 copies of its $\frac{1}{4}$-dilation, as may be expected, but we also get a number of products of simplices of strictly smaller dimensions. Then by induction in dimension we can divide each of them. See Figure 3.

Lemma 4.2. *The abelian group Π_+ has no torsion elements.*

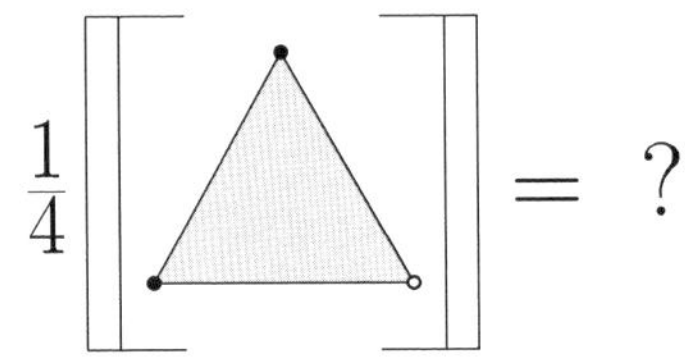

Fig. 2. We need to find an integer combination of polytopes h such that $[\![P]\!] = 4h$

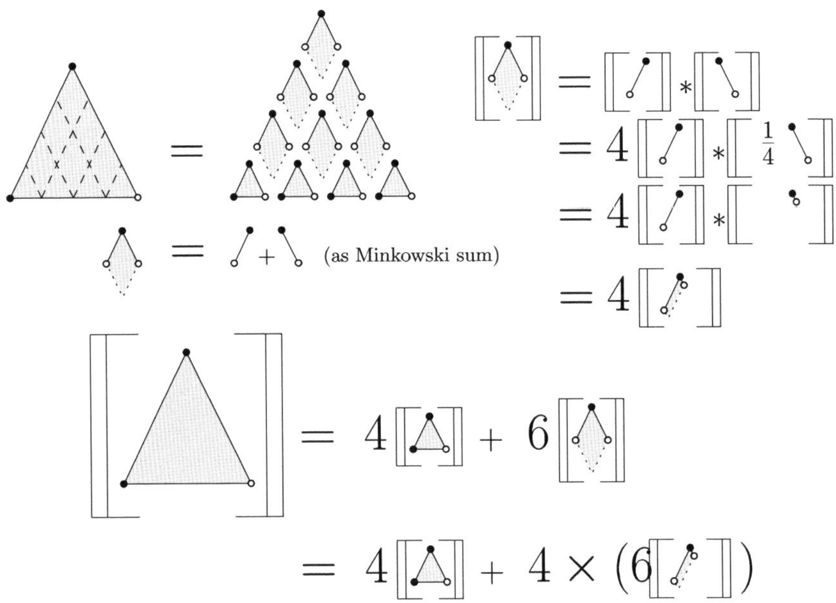

Fig. 3. An example of division by 4.

Proof. Consider the following filtration $\Pi_+ = Z_1 \supset Z_2 \supset \cdots \supset Z_d \supset Z_{d+1}$ where Z_r is generated by elements of the form $([\![P]\!] - 1)^j$ for $j \geq r$. The proof of the lemma follows from two observations.

The first one is that $D_\lambda\left([\![P]\!] - 1\right)^r = ([\![\lambda P]\!] - 1)^r$, since D_λ is a ring endomorphism, so it commutes with taking powers. This implies that $D_\lambda Z_r \subset Z_r$ and $Z_r \subset D_{\lambda^{-1}} Z_r$.

The second observation is that if $x \in Z_r$, then

$$D_n x - n^r x \in Z_{r+1}, \tag{7}$$

for n a natural number. It is enough to check it on the generators. We apply Equation (5)

$$D_n\left([\![P]\!] - 1\right) = ([\![nP]\!] - 1) = \binom{n}{1}([\![P]\!] - 1) + \binom{n}{2}([\![P]\!] - 1)^2 + \cdots. \tag{8}$$

Raising the above expression to the r power we get that $D_n\left([\![P]\!] - 1\right)^r -$

$n^r\left([\![P]\!]-1\right)^r \in Z_{r+1}$. Since it is true for the generators then Equation (7) holds for all Z_r.

Now we finish the proof of the lemma. Let $x \in Z_r$ for some $r \geq 1$, and $nx = 0$ for a natural n. Then $D_n x = D_n x - n^r x$ since $nx = 0$ and hence $D_n x \in Z_{r+1}$ by 7 and also $x \in D_{n^{-1}} Z_{r+1} \subset Z_{r+1}$. This implies that $x \in Z_j$ for $j >> 0$, but since they are eventually zero, x must be zero. □

Combining Lemmas 4.2 and 4.1, we get Theorem 4.2. Now that we can make sense of *rational multiples* of element in Π_+, we can define the exponential and the logarithm as formal power series with coefficients in $\mathbb{Q}$.

$$\exp(z) := \sum_{k\geq 0} \frac{1}{k!} z^k, \qquad \log(1+z) = \sum_{k\geq 1} \frac{(-1)^{k-1}}{k} z^k.$$

The usual identities formally apply

$$\begin{aligned} \log\exp(z) &= \exp\log(z) = z, \\ \exp(a+b) &= \exp(a)\cdot\exp(b), \\ \log(a\cdot b) &= \log(a) + \log(b). \end{aligned}$$

With this in mind, we can define

$$\log([\![P]\!]) = \log(1 + ([\![P]\!]-1)) = \sum_{k\geq 1}^{d} \frac{(-1)^{k-1}}{k}\left([\![P]\!]-1\right)^k,$$

since $\left([\![P]\!]-1\right)^i = 0, i > d$.

Theorem 4.3. *For $P \in \mathcal{P}_d$, define $p := \log\left([\![P]\!]\right)$, then*

$$[\![P]\!]^n = [\![nP]\!] = 1 + pn + \frac{1}{2}p^2n^2 + \cdots + \frac{1}{d!}p^d n^d. \tag{9}$$

Proof. We simply manipulate our expressions.

$$[\![nP]\!] = [\![P]\!]^n = \exp\left(\log[\![P]\!]^n\right) = \exp\left(np\right) = \sum_{i=0}^{d} \frac{1}{i!} p^i n^i \tag{10}$$

□

A consequence of Theorem 4.3 is that for $n = 1$:

$$[\![P]\!] = 1 + p + \frac{1}{2}p^2 + \cdots + \frac{1}{d!}p^d. \tag{11}$$

which is going to be our graded decomposition.

Definition 4.3. Define $\Pi_k := \mathbb{Z}\{\frac{1}{k!}\log\left([\![P]\!]\right)^k \quad P \in \mathcal{P}_d\}$.

With this we now have $\Pi = \Pi_0 + \Pi_1 + \cdots + \Pi_d$. However we want this sum to be direct. For this we shall use our dilation maps. We have:

$$D_r(\log[\![P]\!]) = \log D_r([\![P]\!]) = \log([\![rP]\!]) = \log([\![P]\!]^r) = r\log([\![P]\!]).$$

Moreover we have

$$x \in \Pi_k \iff D_r x = r^k x \quad r \in \mathbb{Q}_{>0}.$$

Proposition 4.2. *We have $\Pi = \Pi_0 \oplus \Pi_1 \oplus \cdots \oplus \Pi_d$ as a direct sum.*

Proof. Suppose there exist $x_i \in \Pi_i$ with $x_0 + x_1 + \cdots + x_d = 0$. From this, we conclude $D_N(x_0 + x_1 + \cdots + x_d) = 0$ which is the same as $x_0 + x_1 N^1 + x_2 N^2 + \cdots x_d N^d = 0$. This last equality is true for all $N > 0$ hence we can conclude that $x_i = 0$. □

Proposition 4.3. *The decomposition $\Pi = \Pi_0 \oplus \Pi_1 \oplus \cdots \oplus \Pi_d$ gives a standard grading.*

Proof. What we need to prove is that $x \in \Pi_i, y \in \Pi_j$ imply $x * y \in \Pi_{i+j}$. This follows from the fact that D_r is a ring map,

$$D_r(x * y) = D_r(x) * D_r(y) = r^i x * r^j y = r^{i+j}(x * y).$$

□

Corollary 4.2. *For any $\phi : \mathcal{P}_d \longrightarrow G$ translation invariant valuation, ϕ is homogeneous of degree k (i.e., $\phi(nP) = n^k$ for $P \in \mathcal{P}_d$) if and only if $\phi(\Pi_j) = 0$ for $j \neq k$. Also, for any translation invariant valuation ϕ, homogeneous or not, we can uniquely decompose it $\phi = \phi_0 + \phi_1 + \cdots + \phi_d$ in homogeneous parts.*

The volume is the *unique* (up to a multiple) translation invariant valuation of degree d on $\mathcal{P}_d$ (See [9, Section 7]).

Corollary 4.3. *The volume valuation induces an isomorphism $Vol_d : \Pi_d \to \mathbb{R}$.*

We can convince ourselves that Π_d is not trivial. In fact, the class of each half open segment is in Π_1. And hence $[\![[\mathbf{0}, \mathbf{e}_1)]\!] * \cdots * [\![[\mathbf{0}, \mathbf{e}_d)]\!] \in \Pi_d$. Such a class can be represented by the half open cube $\{\mathbf{x} \in \mathbb{R}^d : 0 \leq x_i < 1 \text{ for all } i \in [d]\}$ which has volume one, so it can be taken a the generator of Π_d.

Example 4.4. Corollary 4.3 implies that any two elements in Π_d with the same volume are equivalent in Π. In Figure 4 we illustrate one example.

4.1. *Two applications*

4.1.1. *Mixed volumes*

For any polytope $P \in \mathcal{P}_d$ we have that the volume is an homogeneous valuation of degree d, i.e., $\mathrm{Vol}(tP) = t^d \mathrm{Vol}(P)$. Here is a more general version.

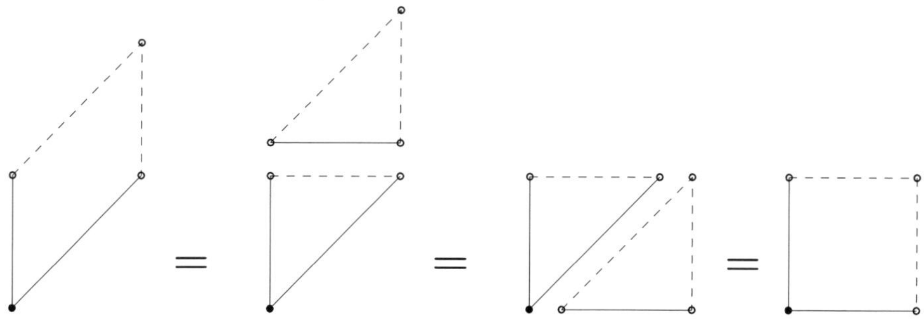

Fig. 4. An instace of two half open parallelograms with the same area.

Theorem 4.4. *For polytopes $P_1, \cdots, P_m \in \mathcal{P}_d$, we have*

$$Vol(\lambda_1 P_1 + \lambda_2 P_2 + \cdots + \lambda_m P_m) = \sum_{i_1,\cdots,i_d=1}^{d} V(P_{i_1}, \ldots, P_{i_d})\lambda_{i_1} \cdots \lambda_{i_d},$$

where each symmetric coefficient $V(P_{i_1}, \ldots, P_{i_d})$ depends only on the bodies $P_{i_1}, \ldots, P_{i_d}$.

Proof. In the polytope algebra, consider the element $[\![P_1]\!]^{\lambda_1} * [\![P_2]\!]^{\lambda_2} * \cdots * [\![P_m]\!]^{\lambda_m}$, where, for now, the λ's are integers. Using Equation (9) we get

$$[\![P_1]\!]^{\lambda_1} * [\![P_2]\!]^{\lambda_2} * \cdots * [\![P_m]\!]^{\lambda_m} = \prod_{i=1}^{m} \sum_{j=1}^{d} \frac{1}{d!} p_i^j \lambda_i^j, \tag{12}$$

where as usual $p_i = \log[\![P_i]\!]$. Taking Vol_d at both sides we get precisely

$$\mathrm{Vol}(\lambda_1 P_1 + \lambda_2 P_2 + \cdots + \lambda_m P_m) = \sum_{i_1,\cdots,i_d=1}^{d} \mathrm{Vol}_d(p_{i_1} * \cdots * p_{i_d})\lambda_{i_1} \cdots \lambda_{i_d}.$$

Notice that Vol_d is homogeneous of degree d so we only need to keep track of the degree d part of the right hand side. Hence we can define $V(P_{i_1}, \ldots, P_{i_d}) = \mathrm{Vol}_d(p_{i_1} * \cdots * p_{i_d})$ to finish the proof. □

Definition 4.4. The function $V(P_{i_1}, \ldots, P_{i_d})$ is the *mixed volume* of the tuple of polytopes $P_{i_1}, \ldots, P_{i_d} \in \mathcal{P}_d$.

4.1.2. *Ehrhart polynomial*

Let's focus briefly on *lattice* polytopes and *lattice* invariant valuations for some lattice $\Lambda \subset \mathbb{R}$. The main example of a valuation invariant under lattice translation (but not under all translations) is the counting valuation, $E(P) := |P \cap \Lambda|$. This case is substantially different and we cannot directly apply our results. However we can prove the following classic theorem.

Theorem 4.5 (Ehrhart). *The function $E_P(n) := E(nP)$ agrees with a polynomial in n of degree d whenever $n \in \mathbb{Z}_{\geq 0}$.*

Proof. Notice that at least Equation (8) doesn't involve any scaling (since the binomial coefficients are integers), it is invariant under lattice translation, and all polytopes appearing are lattice polytopes. Then one can apply the counting valuation $E(P) := |P \cap \Lambda|$ on both sides to obtain:

$$E_P(n) := E(nP) = \sum_{i=0}^{d} \binom{n}{i} \tilde{f}_i^*(P), \tag{13}$$

where $\tilde{f}_i^*(P)$ are some integers depending on P. Notice that for $n = 0$ we indeed get $E_P(0) = 1$. □

4.2. *Minkowski weights*

Corollary 4.3 does not say that $\mathrm{Vol}_d(P) = \mathrm{Vol}_d(Q)$ implies $[\![P]\!] = [\![Q]\!]$, since such elements do not belong to Π_d. Nevertheless, we have the following theorem. For an modern elementary proof see [7].

Theorem 4.6 (Minkowski). *Let $P, Q \in \mathcal{P}_d$ be two* full *dimensional polytopes. Then $Vol_{d-1}(P^c) = Vol_{d-1}(Q^c)$ for all c implies that P and Q are equal up to translation.*

A priori we need to check infinitely many directions, but it could be finite if we know where to look (there are finite c that give facets, so most of the time both quantities are zero). We need to generalize a bit previous theorem so that we have a criterion for lower dimensional polytopes.

Definition 4.5. A $(d-k)$ *frame* is a $(d-k)$ tuple of vectors $U = (u_1, u_2, \cdots, u_{d-k})$ such that $u_i^t u_j = \delta_{ij}$. Given a $(d-k)$ frame u we define the face $P^U := (\cdots((P^{u_1})^{u_2})\cdots)^{u_{d-k}}$. Because orthogonality, dimension is reduced by at least 1 on each step, so $\dim(P^u) \leq k$.

We also define the *frame functionals* to be $V_U(P) := \mathrm{Vol}_k(P^U)$. These are homogeneous valuations of degree k.

Theorem 4.7 (Generalized Minkowski). *Let $P, Q \in \mathcal{P}_d$ be two k-polytopes. Then $V_U(P) = V_U(Q)$ for all $(d-k)$-frame functionals U implies that P and Q are equal up to translation.*

5. A finitely generated subalgebra

So far Π^d is **not** finitely generated as an algebra. We will restrict to a subalgebra.

Definition 5.1. For $P, Q \in \mathcal{P}_d$ we say Q is a *Minkowski summand* of P, and we write $P \leq Q$ if there exists R such that $P = Q + R$. We say Q is a *weak Minkowski summand* of P, and we write $P \preceq Q$ if there exists $\lambda \in \mathbb{R}_{>0}$ such that $Q \leq \lambda P$.

Definition 5.2. Fixing $P \in \mathcal{P}_d$, we define $\Pi(P) = \mathbb{Z}\{[\![Q]\!] : Q \preceq P\}$.

Proposition 5.1.

(i) $\Pi(P)$ *is a* finitely generated *graded subalgebra of* Π.
(ii) $Q \preceq P \Longrightarrow \Pi(Q) \subset \Pi(P)$.
(iii) $\Pi(P)$ *is generated by Minkowski summands of* P.
(iv) $\Pi(P) + \Pi(Q) \subset \Pi(P + Q)$.

Proof. It is a subalgebra because the Minkowski sum of two weak summands is still a weak summand. For finite generation see Remark 5.2 below. It is graded since $Q \in \Pi(P)$ implies $\log(Q) \in \Pi(P)$. The other conditions are not hard to check. □

There is a criterion for determining when a polytope is a summand of another (See [6, Chapter 15]).

Theorem 5.1 (Shepard). *The polytope* Q *is a Minkowski summand of* P *if and only if the following conditions are satisfied:*

(i) $\dim P^c \geq \dim Q^c$ *for any* $c \in \mathbb{R}^d$.
(ii) If for some $c \in \mathbb{R}^d$ *we have* $\dim P^c = 1$, *then* $vol_1(P^c) \geq vol_1(Q^c)$.

Corollary 5.1. *The polytope* Q *is a weak Minkowski summand of* P *if and only if* $\dim P^c \geq \dim Q^c$ *for any* $c \in \mathbb{R}^d$.

Remark 5.1. The condition in Corollary 5.1 is equivalent to saying that $Q \preceq P$ if and only if the normal fan of P refines the normal fan of Q. From that, it is not hard to show that for any polytope P one can find a *simple* polytope P' with $P \preceq P'$, and since $\Pi(P) \subset \Pi(P')$ we can always assume that P is a simple polytope.

Corollary 5.1 is crucial in turning the infinite conditions of Theorem 4.7 into a finite set of conditions.

Definition 5.3. Fix a simple polytope P together with frame functionals $U(F)$ for $F \in \mathcal{F}(P)$ such that $P^U = F$. Define the *Minkowski* map ϕ as:

$$\phi : \Pi(P) \longrightarrow \bigoplus_{i=0}^{d} \mathbb{R}^{f_i(P)}, \tag{14}$$

$$[\![Q]\!] \longrightarrow \left(V_{U(F)}(Q)\right)_{F \in \mathcal{F}(P)}. \tag{15}$$

Theorem 4.7 guarantees that this map is an injection. However it is not surjective. The frame functionals satisfy linear relations.

Theorem 5.2 (Minkowski Relations). *Given a polytope* P *with unit facet normals* $u_1, \cdots, u_n$, *the following linear equation holds:*

$$\sum_{i=1}^{n} u_i \, Vol_{d-1}(P^{u_i}) = 0. \tag{16}$$

Proof. We choose a generic direction u to project the polytope into $u^{\perp}$. The volume of the image of each facet P^{u_i} is proportional to $|\langle u, u_i\rangle|\mathrm{Vol}_{d-1}(P^{u_i})$. The projection of the lower facets of P with respect to u, by which we mean the facets whose normals have negative inner product with u, cover the projection $\pi(P)$. And the same is true for the upper facets. We can compute the volume of $\pi(P)$ using upper or lower facets, which yields

$$\sum_{i:\langle u,-u_i\rangle<0} \langle u, u_i\rangle \mathrm{Vol}_{d-1}(P^{u_i}) = \sum_{i:\langle u,u_i\rangle>0} \langle u, u_i\rangle \mathrm{Vol}_{d-1}(P^{u_i}),$$

which means that $\langle u, \sum_{i=1}^{n} u_i \mathrm{Vol}_{d-1}(P^{u_i})\rangle = 0$ for all u. This implies the result

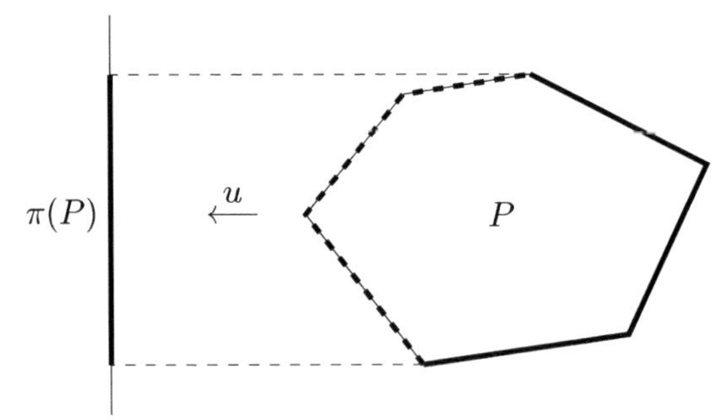

Fig. 5. Projection of a polygon onto a line segment.

□

In light of these relations we now give the following definition.

Definition 5.4. A k-balanced Minkowski weight on a polytope P is a function $\omega_k : \mathcal{F}_k(P) \to \mathbb{R}$ such that for every $F \in \mathcal{F}_{k+1}(P)$ we have the following equality in the subspace $\mathrm{lin}(F)$:

$$\sum_{\substack{G\subset F\\ \text{a facet}}} \omega_k(G) \cdot u_{G/F} = 0.$$

Here $u_{G/F}$ is the unit outer normal in direction G in the subspace $\mathrm{lin}(F)$. The set of all k-balanced Minkowski weights on P is denoted $\Omega_k(P)$, and $\Omega(P) := \bigoplus_k \Omega_k(P)$.

Remark 5.2. The set of all weak Minkowski summands of P can be identified with the set of *positive* 1-weights, which can be described as

$$\left\{ y \in \mathbb{R}^{f_1(P)} : \sum_{i=1}^{m} y(e_i) \cdot e_i = 0 \text{ if } e_1, \cdots, e_m \text{ form a 2-face}, y(e) \geq 0 \text{ for all edges } e \right\}.$$

This is a pointed cone in $\mathbb{R}^{f_1(P)}$, so it has finitely many rays. The rays correspond to *indecomposable* polytopes Q, i.e., polytopes whose only weak Minkowski summands are its scalar multiples (an three dimensional example is a pyramid over a square). These correspond to the finitely many generators of the subalgebra $\Pi(P)$.

Finally, McMullen uses the Minkowski map to obtain a different presentation of $\Pi(P)$.

Theorem 5.3. *The Minkowski map ϕ induces an graded isomorphism $\Pi(P) \cong \Omega(P)$ as graded vector spaces.*

McMullen's simplifies his original proof of the g-theorem in [11] by working over the ring of Minkowski weights. One technical difficulty is that we have to actually give a ring structure to $\Omega(P)$, to define, in coherence with ϕ, a way to multiply weights. This is a delicate part of [11], but ultimately McMullen gives a simplification of the proof of the g-theorem by replacing $\Pi(P)$ with $\Omega(P)$. He writes "The reader who wishes to work through the proof of the g-theorem in the light of the weight algebra can effectively discard much of [10]".

References

[1] A. Barvinok, Integer points in polyhedra. Zurich Lectures in Advanced Mathematics. European Mathematical Society (EMS), Zürich, 2008.
[2] L. Billera, The algebra of continuous piecewise polynomials. Adv. Math. 76 (1989), no. 2, 170–183.
[3] L. Billera, "Even more intriguing, if rather less plausible..." Face numbers of convex polytopes. The mathematical legacy of Richard P. Stanley, 65–81, Amer. Math. Soc., Providence, RI, 2016.
[4] Brion, Michel The structure of the polytope algebra. Tohoku Math. J. (2) 49 (1997), no. 1, 1–32.
[5] W. Fulton; B. Sturmfels, Intersection theory on toric varieties. Topology 36 (1997), no. 2, 335–353.
[6] B. Grünbaum, Convex polytopes. Second edition.
[7] D. Klain, The Minkowski problem for polytopes. Adv. Math. 185 (2004), no. 2, 270–288.
[8] D. Maclagan; B. Sturmfels, Introduction to Tropical Geometry, Graduate Studies in Mathematics, Vol 161, American Mathematical Society, 2015.
[9] P. McMullen, The polytope algebra. Adv. Math. 78 (1989), no. 1, 76–130.
[10] P. McMullen, On simple polytopes. Invent. Math. 113 (1993), no. 2, 419–444.
[11] P. McMullen, Weights on polytopes. Discrete Comput. Geom. 15 (1996), no. 4, 363–388.
[12] R. Stanley, Combinatorics and commutative algebra. Second edition.
[13] R. Stanley, The number of faces of a simplicial convex polytope. Adv. in Math. 35 (1980), no. 3, 236–238.

Restrictions on the singularity content of a Fano polygon

Daniel Cavey

School of Mathematical Sciences
University of Nottingham
Nottingham, NG7 2RD, UK
E-mail: danielcavey27@gmail.com

We determine restrictions on the singularity content of a Fano polygon, or equivalently of certain orbifold del Pezzo surfaces. We establish bounds on the maximum number of $\frac{1}{R}(1,1)$ singularities in the basket of residual singularities. In particular, there are no Fano polygons without T-singularities and with a basket given by (i) $\{k \times \frac{1}{R}(1,1)\}$ for $k \in \mathbb{Z}_{>0}$ and $R \geq 5$, or (ii) $\{\frac{1}{R_1}(1,1), \frac{1}{R_2}(1,1), \frac{1}{R_3}(1,1)\}$.

Keywords: del Pezzo surfaces; Fano polygons; Mirror Symmetry; singularity content.

1. Introduction

The motivation for this work comes from an approach to classifying del Pezzo surfaces via Mirror Symmetry that has been introduced in recent years by Coates–Corti–Kasprzyk et al. [1, 2]. Mirror Symmetry establishes a conjectural relation between certain Laurent polynomials $f \in \mathbb{C}[x_1^{\pm 1}, \cdots, x_n^{\pm 1}]$ and n-dimensional Fano varieties X. If f is associated to X under this correspondence then we say that f is *mirror dual* to X.

Example 1.1. The Laurent polynomial $x + y + 1/xy$ is known to be mirror dual to $\mathbb{P}^2$. The corresponding Newton polytope is:

$$P := \mathrm{Newt}\left(x + y + \frac{1}{xy}\right) =$$

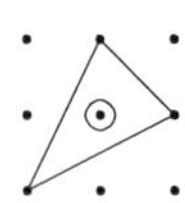

From the polygon P we construct a toric variety X_P by taking the spanning fan. In this case X_P is again $\mathbb{P}^2$.

In general the toric variety X_P associated to the Newton polytope P of f mirror dual to X will be Fano, and it is conjectured that X_P admits a $\mathbb{Q}$-*Gorenstein (qG-)* deformation, see [3, 4], to X. The toric variety X_P may be more singular than X, but this is compensated for by being able to use the language of toric geometry to describe the variety. There is an additional complication that the choice of mirror dual is not unique. To this end, Akhtar–Coates–Galkin–Kasprzyk [5] introduced the notion of *mutation* of a Laurent polynomial: a birational transformation transforming one mirror dual to X to another mirror dual to X [5, Lemma 1]. This notion of mutation for Laurent polynomials can be translated to a combinatorial operation on lattice polytopes. See [5, 6] for the details. An important open question is to begin to classify mutation-equivalence classes of polytopes.

From the viewpoint of Mirror Symmetry, it is natural to restrict ourselves to the study of *Fano* polytopes. Recall that a full-dimensional lattice polytope P is Fano if the vertices $\mathcal{V}(P)$ are all primitive and if the origin lies in the strict interior of P. The Newton polytope of any Laurent polynomial mirror is necessarily Fano. Furthermore, when considering the spanning fan, it makes sense to restrict oneself to Fano polytopes. For an overview of Fano polytopes see [7].

An important mutation invariant of a Fano polygon is its singularity content, introduced by Akhtar–Kasprzyk [8]. In order to describe this invariant, we first recall the definition of a cyclic quotient singularity. Consider the action of the cyclic group of order R, denoted μ_R, on $\mathbb{C}^2$ via $(x, y) \mapsto (\epsilon^a x, \epsilon^b y)$. Here ϵ is a primitive R-th root of unity. A *quotient singularity* $\frac{1}{R}(a, b)$ is defined by the germ of the origin of $\operatorname{Spec}(\mathbb{C}[x, y]^{\mu_R})$. See [9] for further details.

Example 1.2. Consider a $\frac{1}{2}(1, 1)$ singularity. Let $G = \mathbb{Z}/2\mathbb{Z}$ and $\epsilon = -1$. The action of G on $\mathbb{C}^2$ is given by $-1 \cdot (x, y) = (-x, -y)$, and

$$\begin{aligned}\operatorname{Spec}\left(\mathbb{C}[x,y]^G\right) &= \operatorname{Spec}\left(\mathbb{C}[x^2, xy, y^2]\right)\\ &= \operatorname{Spec}\left(\mathbb{C}[u,v,w]/(uw - v^2)\right)\\ &= \mathbb{V}(uw - v^2) \subset \mathbb{C}^3.\end{aligned}$$

A quotient singularity $\frac{1}{R}(a, b)$ is cyclic if $\gcd(R, a) = \gcd(R, b) = 1$. Set $k = \gcd(a + b, R)$, so $R = kr$ and $a + b = k\tilde{c}$ for some $r, \tilde{c} \in \mathbb{Z}_{>0}$. The cyclic quotient singularity can be written as $\frac{1}{kr}(1, kc - 1)$, where $ca \cong \tilde{c} \pmod R$.

Two important classes of cyclic quotient singularities are described by Kollar–Shepherd-Barron [3] and Akhtar–Kasprzyk [8]. A cyclic quotient singularity $\frac{1}{kr}(1, kc - 1)$ is

(i) a *T-singularity* if $r \mid k$;
(ii) an *R-singularity* if $k < r$.

In addition, a T-singularity is *primitive* if $r = k$. The significance of these definitions comes when one attempts to smooth the cyclic quotient singularities via a qG-deformation, that is, a deformation that preserves the numerics of the anti-canonical divisor. A cyclic quotient singularity is qG-smoothable if and only if it is a T-singularity, whereas R-singularities are rigid under qG-deformation. The deformation theory of cyclic quotient singularities has been well studied by Altmann, Christopherson, Ilten and Stevens [10–17]. In particular Altmann provides a description of the deformations at a combinatorial level.

Example 1.3. A cyclic quotient singularity $\frac{1}{R}(1, 1)$ is a T-singularity if and only if $R \in \{1, 2, 4\}$.

Consider an arbitrary cyclic quotient singularity $\sigma = \frac{1}{kr}(1, kc-1)$ not necessarily satisfying either $r \mid k$ or $k < r$. There exists unique non-negative integers n and k_0

such that $k = nr + k_0$ and $k_0 < r$. If $k_0 > 0$ then σ qG-deforms to a $\frac{1}{k_0 r}(1, k_0 c - 1)$ cyclic quotient singularity. Informally σ decomposes as n primitive T-singularities and an R-singularity; the T-part can be smoothed away leaving the R-singularity, which we call the *residue*. More precisely, the residue of σ is given by:

$$\operatorname{res}(\sigma) = \begin{cases} \varnothing, & \text{if } k_0 = 0; \\ \frac{1}{k_0 r}(1, k_0 c - 1), & \text{otherwise.} \end{cases}$$

With notation as above, the *singularity content* of σ is denoted by the pair:

$$\mathrm{SC}(\sigma) = (n, \operatorname{res}(\sigma)) .$$

Associated to a cyclic quotient singularity $\sigma = \frac{1}{R}(a, b)$ is a cone $C_\sigma = \operatorname{cone}(e_1, e_2)$ in the lattice $\mathbb{Z}^2 + (a/r, b/r) \cdot \mathbb{Z}$, defined up to a change of basis. By abuse of notation we often confuse the distinction between cones and singularities; namely we refer to T-cones, primitive T cones, R-cones and $\frac{1}{R}(a, b)$-cones.

Given a cone $C_\sigma \subset N_\mathbb{R} = N \otimes_\mathbb{Z} \mathbb{R}$ where $N \cong \mathbb{Z}^2$, let ρ_1, ρ_2 be the primitive lattice points generating the rays of C_σ. Choose a point $v \in C_\sigma$ such that all other points of C_σ can be written as a linear combination of ρ_1, ρ_2 and v over the integers. Suppose $v = \frac{1}{R}(a\rho_1 + b\rho_2)$. Then C_σ corresponds to the cyclic quotient singularity $\frac{1}{R}(a, b)$. Additionally there is a unique hyperplane H through ρ_1, ρ_2, and $E = C_\sigma \cap H$ is the edge over which C_σ is defined. The decomposition of σ has a description in the combinatorics of E.

The *lattice length*, denoted $l(E)$, of $E \subset N_\mathbb{R}$ is given by the value $|E \cap N| - 1$. The *lattice height* $h(E)$ of E is given by the lattice distance from the origin: that is, given the unique primitive inward pointing normal $n_E \in M = \operatorname{Hom}(N, \mathbb{Z})$ of E, the height is given by $|\langle v, n_E\rangle|$, for any $v \in E$. There exist $n, l_0 \in \mathbb{Z}_{\geq 0}$ such that $l = nh + l_0$ and $l_0 < h$. Divide C into separate sub-cones $C_0, \ldots, C_n$, where $C_1, \ldots, C_n$ have lattice length h, and C_0 has lattice length l_0. The cones C_i, for $1 \leq i \leq n$, are primitive T-cones and C_0 is an R-cone.

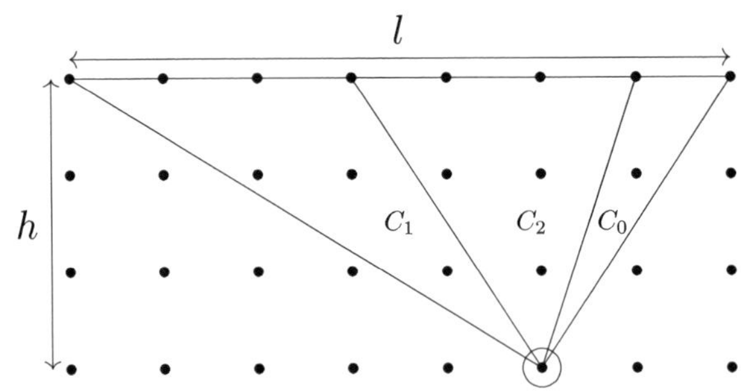

Fig. 1. Division of a cone of lattice length 7 and lattice height 3. C_0 is a $\frac{1}{3}(1,1)$ cone and C_1, C_2 are $\frac{1}{9}(1,2)$ cones.

Definition 1.1 ([8, Definition 3.1]). Let $P \subset N_{\mathbb{R}}$ be a Fano polygon. Label the edges of P clockwise $E_1, \ldots, E_k$. Let C_{σ_i} be the cone over the edge E_i. Set

$$\mathrm{SC}(\sigma_i) = (n_i, \mathrm{res}(\sigma_i)).$$

Define the *singularity content* of P to be:

$$\mathrm{SC}(P) = \left(\sum_{i=1}^{k} n_i, \mathcal{B} \right),$$

where $\mathcal{B} = \{\mathrm{res}(\sigma_1), \ldots, \mathrm{res}(\sigma_k)\}$ is a cyclically ordered set known as the *basket of residual singularities.*

The singularity content of P, a combinatorial property, describes the singularities on X_P, a geometrical property.

Definition 1.2 ([1, Definition 1]). A del Pezzo surface with cyclic quotient singularities is of class TG if it admits a qG-degeneration with reduced fibres to a normal toric del Pezzo surface.

Del Pezzo surfaces of class TG are exactly those which (conjecturally) can be described by this application of Mirror Symmetry:

Conjecture 1.1 ([1, Conjecture A]). *There exists a bijective correspondence between the set of mutation-equivalence classes of Fano polygons and the set of qG-deformation equivalence classes of locally qG-rigid TG del Pezzo surfaces with cyclic quotient singularities.*

Recent results support this conjecture [6, 18, 19], and understanding the possible values taken by the singularity content is an important open question. As a first step towards addressing this question, the two main results of this paper are:

Theorem 1.1. *There are no Fano polygons with singularity content*

$$\left(0, \left\{k \times \frac{1}{R}(1,1)\right\}\right), \qquad \text{where } k \in \mathbb{Z}_{>0}, R \in \mathbb{Z}_{\geq 5}.$$

Theorem 1.2. *There are no Fano polygons with singularity content*

$$\left(0, \left\{\frac{1}{R_1}(1,1), \frac{1}{R_2}(1,1), \frac{1}{R_3}(1,1)\right\}\right), \qquad \text{where } R_i \in \{3\} \cup \mathbb{Z}_{\geq 5}.$$

These two theorems are conjecturally equivalent to:

- There are no del Pezzo surfaces admitting a toric degeneration whose topological Euler number is 0 and singular locus consists of only isolated $\frac{1}{R}(1,1)$ cyclic quotient singularities, where $R \in \mathbb{Z}_{\geq 5}$;
- There are no del Pezzo surfaces admitting a toric degeneration whose topological Euler number is 0 and singular locus consists of exactly three qG-rigid isolated cyclic quotient singularities $\frac{1}{R_1}(1,1)$, $\frac{1}{R_2}(1,1)$ and $\frac{1}{R_3}(1,1)$.

2. Restrictions via Matrices

Let $P \subset N_{\mathbb{R}}$ be a Fano polygon with vertices $v_1, v_2, \ldots, v_k$ labelled anticlockwise. By convention our subscripts are considered modulo k to be in the range $\{1, \ldots, k\}$. Consider the set of uniquely determined matrices $\{M_i \in GL_2(\mathbb{Z})\}_{1 \leq i \leq k}$ satisfying

$$M_i v_i = v_{i+1}, \qquad M_i v_{i+1} = v_{i+2}.$$

It follows that

$$M_k M_{k-1} \cdots M_1 = Id.$$

By understanding the matrix M_i when the cones $\mathrm{span}_{\mathbb{R}_{\geq 0}}(v_i, v_{i+1})$ and $\mathrm{span}_{\mathbb{R}_{\geq 0}}(v_{i+1}, v_{i+2})$ describe particular cyclic quotient singularities, we create restrictions on when this equality can hold.

2.1. *Matrices in* $\frac{1}{3}(1,1)$ *Case*

We start with the simple case of a polygon P consisting entirely of $\frac{1}{3}(1,1)$ cones (we already know exactly one such polygon exists by Kasprzyk–Nill–Prince [6]).

Let E be an edge of a Fano polygon such that the cone over E is a $\frac{1}{3}(1,1)$ cone. Assume without loss of generality that E has vertices $(-1\ 3)^{\mathsf{T}}$ and $(-2\ 3)^{\mathsf{T}}$. Further suppose that the edge adjacent to E sharing the vertex $(-2\ 3)^{\mathsf{T}}$ also has a corresponding $\frac{1}{3}(1,1)$ cone.

The corresponding matrix $M = \begin{pmatrix} a & b \\ c & d \end{pmatrix} \in GL_2(\mathbb{Z})$ satisfies:

$$M \begin{pmatrix} -1 \\ 3 \end{pmatrix} = \begin{pmatrix} -2 \\ 3 \end{pmatrix} \qquad \text{and} \qquad \det(M) = 1.$$

The second condition follows from the fact that M maps a $\frac{1}{3}(1,1)$ cone onto a $\frac{1}{3}(1,1)$ cone, and so lattice length and lattice height must be preserved. Hence M is of the form

$$M = \begin{pmatrix} 3 - 2a & \frac{1-2a}{3} \\ 3a - 3 & a \end{pmatrix}, \qquad \text{for } a \in \mathbb{Z}.$$

The only remaining restriction is that $(1-2a)/3$ belongs to $\mathbb{Z}$ and so $a \equiv 2 \pmod 3$. Substituting $a = 3n + 2$, we obtain the set of matrices:

$$A_n = \begin{pmatrix} -6n - 1 & -2n - 1 \\ 9n + 3 & 3n + 2 \end{pmatrix}, \qquad \text{for } n \in \mathbb{Z}.$$

The image of the point $(-2\ 3)^{\mathsf{T}}$ under A_n, that is the second vertex of the second $\frac{1}{3}(1,1)$ cone, is given by:

$$A_n \begin{pmatrix} -2 \\ 3 \end{pmatrix} = \begin{pmatrix} 6n - 1 \\ -9n \end{pmatrix}.$$

Note if $n < 0$, convexity of the Fano polygon is broken. Therefore we have a 1-dimensional family of suitable matrices parametrised by $\mathbb{Z}_{\geq 0}$ each giving a point v such that $\mathrm{span}_{\mathbb{R}_{\geq 0}}\left(\left(-2\ 3\right)^{\mathsf{T}}, v\right)$ is a cone representing a $\frac{1}{3}(1,1)$ singularity.

$$n = 0 \longleftrightarrow \begin{pmatrix} -1 \\ 0 \end{pmatrix}$$

$$n = 1 \longleftrightarrow \begin{pmatrix} 5 \\ -9 \end{pmatrix}$$

$$n = 2 \longleftrightarrow \begin{pmatrix} 11 \\ -18 \end{pmatrix}$$

$$\vdots$$

Lemma 2.1. *A Fano polygon consisting only of* $\frac{1}{3}(1,1)$ *R-cones satisfies*

$$M_k M_{k-1} \cdots M_1 = A_{n_1} A_{n_2} \cdots A_{n_k}.$$

Proof. We have that

$$\begin{aligned} M_1 &= A_{n_1} \\ M_2 &= M_1 A_{n_2} M_1^{-1} = A_{n_1} A_{n_2} A_{n_1}^{-1} \\ M_3 &= M_2 M_1 A_{n_2} M_1^{-1} M_2^{-1} = A_{n_1} A_{n_2} A_{n_3} A_{n_2}^{-1} A_{n_1}^{-1} \\ &\vdots \\ M_i &= A_{n_1} A_{n_2} \cdots A_{n_{i-1}} A_{n_i} A_{n_{i-1}}^{-1} \cdots A_{n_2}^{-1} A_{n_1}^{-1} \end{aligned}$$

The identity follows by substitution. □

The problem remains to test when the identity $A_{n_1} A_{n_2} \cdots A_{n_k} = Id$ holds. First consider the A_{n_i} modulo 3:

$$\begin{aligned} A_{n_1} &\equiv \begin{pmatrix} 2 & 2n_1 + 1 \\ 0 & 2 \end{pmatrix} \pmod 3, \\ A_{n_1} A_{n_2} &\equiv \begin{pmatrix} 2 & 2n_1 + 1 \\ 0 & 2 \end{pmatrix} \begin{pmatrix} 2 & 2n_2 + 1 \\ 0 & 2 \end{pmatrix} \equiv \begin{pmatrix} 1 & n_1 + n_2 + 1 \\ 0 & 1 \end{pmatrix} \pmod 3, \\ A_{n_1} A_{n_2} A_{n_3} &\equiv \begin{pmatrix} 1 & n_1 + n_2 + 1 \\ 0 & 1 \end{pmatrix} \begin{pmatrix} 2 & 2n_3 + 1 \\ 0 & 2 \end{pmatrix} \equiv \begin{pmatrix} 2 & 2(n_1 + n_2 + n_3) \\ 0 & 1 \end{pmatrix} \pmod 3, \\ &\vdots \end{aligned}$$

Note that the multiplication of an odd number of matrices can never equal the identity matrix modulo 3, since the upper left entry is $2 \not\equiv 1 \pmod 3$. Indeed this follows by noting $A_{n_7} A_{n_6} \cdots A_{n_1} = A_{n_1 + \cdots + n_7}$ and then induction. Therefore if

$A_{n_1}A_{n_2}\cdots A_{n_k} = Id$, it follows that k is even. Looking modulo 9 further narrows down the possibilities. We have that:

$$A_{n_1}A_{n_2} \equiv \begin{pmatrix} * & * \\ 6 & * \end{pmatrix} \not\equiv Id \pmod 9,$$

$$A_{n_1}A_{n_2}A_{n_3}A_{n_4} \equiv \begin{pmatrix} * & * \\ 3 & * \end{pmatrix} \not\equiv Id \pmod 9.$$

Therefore the smallest possible value of k satisfying $A_{n_1}A_{n_2}\cdots A_{n_k} = Id$ is 6.

2.2. *Winding Number in $\frac{1}{3}(1,1)$ Case*

To finish the argument, we use the fact that for a Fano polygon, the boundary is a closed loop that wraps around the origin once. We shall use the winding number defined in [20], which we now describe.

Considering $SL_2(\mathbb{R})$ as a topological space, the fundamental group is given by $\pi_1(SL_2(\mathbb{R})) = \mathbb{Z}$. The universal cover, denoted $\widetilde{SL_2(\mathbb{R})}$, is the connected topological group fitting into the exact sequence:

$$0 \longrightarrow \mathbb{Z} \longrightarrow \widetilde{SL_2(\mathbb{R})} \longrightarrow SL_2(\mathbb{R}) \longrightarrow 0.$$

There is no description of $\widetilde{SL_2(\mathbb{R})}$ as a group of matrices subject to some algebraic conditions. The commonly used description is that of pairs $(M, [\gamma])$, where

$$M = \begin{pmatrix} a & b \\ c & d \end{pmatrix} \in SL_2(\mathbb{Z})$$

and $[\gamma]$ is a homotopy equivalence class of paths in $\mathbb{R}^2\backslash\{\mathbf{0}\}$ from $\begin{pmatrix}0 & 1\end{pmatrix}^{\mathsf{T}}$ to $\begin{pmatrix}c & d\end{pmatrix}^{\mathsf{T}}$. Hence $\widetilde{SL_2(\mathbb{R})}$ has the structure of a group via the composition law

$$(M_1, [\gamma_1]) \cdot (M_2, [\gamma_2]) = (M_1M_2, [\gamma_2 \star \gamma_1]),$$

where $\star$ denotes concatenation.

Define $\widetilde{SL_2(\mathbb{Z})}$ to be the inverse image of $SL_2(\mathbb{Z})$ under the map $\widetilde{SL_2(\mathbb{R})} \to SL_2(\mathbb{R})$. Note this is not a covering space of $SL_2(\mathbb{Z})$ since it is not a connected topological space. Lift each matrix A_{n_i} to $\widetilde{SL_2(\mathbb{Z})}$ by equipping with an appropriate straight line path denoted γ_i. The algebraic condition on the A_{n_i} then becomes

$$(A_{n_1}, [\gamma_1]) \cdot (A_{n_2}, [\gamma_2]) \cdot \cdots \cdot (A_{n_k}, [\gamma_k]) = (Id, [\text{anticlockwise loop}]). \qquad (1)$$

In Poonen–Rodriguez-Villegas [20], a homomorphism $\Phi : \widetilde{SL_2(\mathbb{Z})} \to \mathbb{Z}$ is introduced to act as a winding number. The aim is to apply Φ to both sides of the above equality to obtain an extra condition on k.

Similarly to how $SL_2(\mathbb{Z})$ is generated by

$$S = \begin{pmatrix} 0 & -1 \\ 1 & 0 \end{pmatrix} \quad \text{and} \quad T = \begin{pmatrix} 1 & 1 \\ 0 & 1 \end{pmatrix},$$

it is known that $\widetilde{SL_2(\mathbb{Z})}$ is generated by the two elements $\tilde{S}$ and $\tilde{T}$ obtained from lifting S and T to $\widetilde{SL_2(\mathbb{R})}$ by equipping them with the straight line path from $(0\ 1)^\mathsf{T}$ to $(1\ 0)^\mathsf{T}$ and the trivial path respectively. Furthermore it is shown in [20] that:

$$\Phi(\tilde{S}) = -3 \qquad \text{and} \qquad \Phi(\tilde{T}) = 1.$$

It is routine to check that $(\tilde{S})^4 = (Id, [\text{anticlockwise loop}])$ and so

$$\Phi\left(Id, [\text{anticlockwise loop}]\right) = -12.$$

It remains to calculate $\Phi\left(A_{n_i}, [\gamma_i]\right)$. By using an algorithm of Conrad [21], we obtain the expression:

$$A_{n_i} = TS^{-1}T^{-2}S^{-1}T^{-(n_i+1)}ST^{-3}.$$

After lifting to $\widetilde{SL_2(\mathbb{Z})}$ and applying the winding number homomorphism we obtain:

$$\Phi\left(A_{n_i}, [\gamma_i]\right) = -2 - n_i.$$

Applying Φ to both sides of (3) gives the expression:

$$\sum_{i=1}^{k} n_i = 12 - 2k.$$

If $k > 6$ this implies $\sum_{i=1}^{k} n_i < 0$, but convexity demands the n_i to be positive and so there are no solutions. The only remaining case is $k = 6$, for which the equation becomes

$$\sum_{i=1}^{k} n_i = 0.$$

Therefore there is a single possible solution given by $k = 6$ and $n_i = 0$. This recovers the known Fano polygon consisting of only $6 \times \frac{1}{3}(1,1)$ R-singularities.

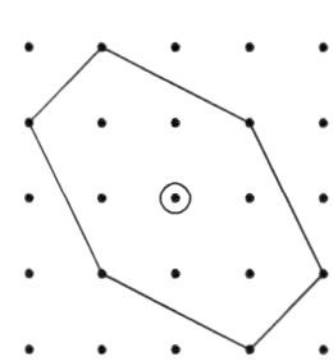

Fig. 2. Fano polygon with singularity content $\left(0, \left\{6 \times \frac{1}{3}(1,1)\right\}\right)$

2.3. *General* $\frac{1}{R}(1,1)$ *Case*

We now generalise our approach to prove Theorem 1.1.

Proof of Theorem 1.1. First assume R odd. Consider the standard position of the $\frac{1}{R}(1,1)$ cone to have vertices $(-(R+1)/2, R)$ and $(-(R-1)/2, R)$. Then we are looking for $A \in SL_2(Z)$ such that

$$A\begin{pmatrix} -\frac{R-1}{2} \\ R \end{pmatrix} = \begin{pmatrix} -\frac{R+1}{2} \\ R \end{pmatrix} \qquad \text{and} \qquad \det(A) = 1.$$

Such a matrix takes the form

$$A = \begin{pmatrix} \frac{-aR-a+2R}{R-1} & \frac{R-aR-a-1}{2R} \\ \frac{2R(a-1)}{R-1} & a \end{pmatrix}.$$

The entries of A belong to $\mathbb{Z}$ if and only if

$$a \equiv 1 \pmod{(R-1)/2}, \qquad \text{and} \qquad a \equiv -1 \pmod{R}.$$

This implies that $a = 2R - 1 + n((R-1)R)/2$ for some $n \in \mathbb{Z}$. Making this substitution into A gives:

$$A_n^{(R)} = \begin{pmatrix} -n\frac{R+1}{2}R - 2R - 1 & -n\frac{R^2-1}{4} - R \\ nR^2 + 4R & 2R - 1 + n\frac{R-1}{2}R \end{pmatrix}.$$

The problem is reduced to testing when the identity $A_{n_1}^{(R)}A_{n_2}^{(R)}\cdots A_{n_k}^{(R)} = Id$ can hold using a generalised version of Lemma 2.1. Studying $A_{n_i}^{(R)}$ modulo R:

$$A_{n_1}^{(R)} \equiv \begin{pmatrix} -1 & (\frac{R^2-1}{4})n_1 \\ 0 & -1 \end{pmatrix} \not\equiv Id \pmod{R},$$

$$A_{n_1}^{(R)}A_{n_2}^{(R)} \equiv \begin{pmatrix} 1 & \frac{R^2-1}{4}(-n_1 - n_2) \\ 0 & 1 \end{pmatrix} \equiv Id\ , \text{ if } n_1 + n_2 \equiv 0 \pmod{R},$$

$$A_{n_1}^{(R)}A_{n_2}^{(R)}A_{n_3}^{(R)} \equiv \begin{pmatrix} -1 & \frac{R^2-1}{4}(n_1 + n_2 + n_3) \\ 0 & -1 \end{pmatrix} \equiv A_{n_1+n_2+n_3}^{(R)} \not\equiv Id \pmod{R}.$$

Continuing inductively there cannot be a solution if k is odd. Furthermore for k even, the identity $A_{n_1}^{(R)}A_{n_2}^{(R)}\cdots A_{n_k}^{(R)} = Id$ holds only if $\sum_{i=1}^{k} n_i \equiv 0 \pmod{R}$. Studying the product of A_{n_i}'s modulo R^2, observe that:

$$\prod_{i=1}^{k} A_{n_i}^{(R)} \equiv \begin{pmatrix} * & * \\ (-1)^k 4kR & * \end{pmatrix} \pmod{R^2},$$

and so $A_{n_1}^{(R)}A_{n_2}^{(R)}\cdots A_{n_k}^{(R)} = Id$ holds only if k is a multiple of R. The smallest possible value for k is $2R$. Finally, appealing to the winding number argument

calculate that:

$$A_n^{(R)} = TS^{-1}(T^{-2}S^{-1})^{\frac{R-3}{2}}T^{-2}S^{-1}T^{-(n+2)}S^{-1}(T^{-2}S^{-1})^{\frac{R-5}{2}}T^{-2}ST^{-3},$$

and so $\Phi\left(A_n^{(R)}, [\gamma_i]\right) = -6 + R - n$. Applying Φ to (3) obtain

$$\sum_{i=1}^{k} n_i = 12 - (R-6)\,k.$$

Since k must be a multiple of $2R$, and $\sum_{i=1}^{k} n_i$ must be congruent to 0 modulo R,

$$12 \equiv 0 \pmod{R}.$$

This implies $R \mid 12$ and since R is odd and greater or equal 5, there are no solutions. The case R even follows similarly. □

Note that $R = 3$ satisfies the congruence $12 \equiv 0 \pmod{R}$ corresponding to the fact that there is a solution in this case.

3. Restrictions via Continued Fractions

In this section, we use results on continued fractions to prove Theorem 1.2. The geometry of continued fractions can be studied in Karpenkov [22].

3.1. *Continued Fractions*

Definition 3.1. For $a_0, a_1, \cdots, a_k \in \mathbb{R}$, consider the *continued fraction*:

$$[a_0 : a_1 : \cdots : a_k] = a_0 + \cfrac{1}{a_1 + \cfrac{1}{a_2 + \cfrac{1}{\cdots + \frac{1}{a_k}}}}.$$

The numbers a_i are called the *elements* of the continued fraction. A continued fraction is *odd/even* if there are an odd/even number of elements.

There uniquely exist polynomials P_k and Q_k in variables a_i satisfying:

$$[a_0 : a_1 : \cdots : a_k] = \frac{P_k(a_0, \ldots, a_k)}{Q_k(a_0, \ldots, a_k)}, \qquad \text{and} \qquad P_k(0, \ldots, 0) + Q_k(0, \ldots, 0) = 1.$$

The first few of these polynomials are:

$$[a_0] = \frac{P_0(a_0)}{Q_0(a_0)} = \frac{a_0}{1},$$

$$[a_0 : a_1] = \frac{P_1(a_0, a_1)}{Q_1(a_0, a_1)} = \frac{a_0 a_1 + 1}{a_1},$$

$$[a_0 : a_1 : a_2] = \frac{P_2(a_0, a_1, a_2)}{Q_2(a_0, a_1, a_2)} = \frac{a_0 a_1 a_2 + a_0 + a_2}{a_1 a_2 + 1}.$$

The polynomials P_k and Q_k satisfy the recursions:

$$P_k = a_k P_{k-1} + P_{k-2}, \qquad \text{and} \qquad Q_k = a_k Q_{k-1} + Q_{k-2}.$$

3.2. *Integer Geometry*

Definition 3.2. Consider an integer triangle ΔABC, that is a triangle whose vertices are the integer points A, B and C. The *integer area* of ΔABC, denoted $l\text{Area}(\Delta ABC)$, is given by the index of the sublattice generated by the line segments AB and AC thought of as vectors in the integer lattice.

Definition 3.3. Consider an integer angle $\angle ABC$, that is an angle between two integer lines based at an integer point. The *integer sine* of $\angle ABC$, denoted $l\sin(\angle ABC)$, is given by

$$l\sin(\angle ABC) = \frac{l\text{Area}(\Delta ABC)}{l(AB)l(BC)}.$$

Note that the integer sine is independent of the orientation of the integer angle, that is, $l\sin(\angle ABC) = l\sin(\angle CBA)$.

Definition 3.4. A *broken line* is defined by $L = A_0A_1 \cdots A_n = \bigcup_{i=0}^{n-1} L_i$, where L_i is the line segment between the integer points A_i and A_{i+1}. Let L be an integer broken line that does not contain the origin $\mathbf{0} \in \mathbb{Z}^2$. If all the L_i are at lattice height 1, then L is called an $\mathbf{0}$-*broken line.*

Definition 3.5. Let $A_0A_1 \cdots A_n$ be an $\mathbf{0}$-broken line. Associate to the broken line its *lattice-signed-length-sine (LSLS) sequence* given by $(a_0, a_1, \ldots, a_{2n-2})$, where

$$\begin{aligned}
a_0 &= \text{sign}(A_0\mathbf{0}A_1) \cdot l(A_0A_1),\\
a_1 &= \text{sign}(A_0\mathbf{0}A_1) \cdot \text{sign}(A_1\mathbf{0}A_2) \cdot \text{sign}(A_0A_1A_2) \cdot l\sin(\angle A_0A_1A_2),\\
a_2 &= \text{sign}(A_1\mathbf{0}A_2) \cdot l(A_1A_2),\\
&\ \ \vdots\\
a_{2n-3} &= \text{sign}(A_{n-2}\mathbf{0}A_{n-1}) \cdot \text{sign}(A_{n-1}\mathbf{0}A_n) \cdot \text{sign}(A_{n-2}A_{n-1}A_n) \cdot l\sin(\angle A_{n-2}A_{n-1}A_n),\\
a_{2n-2} &= \text{sign}(A_{n-1}\mathbf{0}A_n) \cdot l(A_{n-1}A_n),
\end{aligned}$$

and $\text{sign}(ABC)$ is defined for arbitrary integer points A, B, C by

$$\text{sign}(ABC) = \begin{cases} 1, & \text{if } (BA, BC) \text{ is orientated positively;}\\ 0, & \text{if } A, B, C \text{ are collinear;}\\ -1, & \text{if } (BA, BC) \text{ is orientated negatively.} \end{cases}$$

Given an $\mathbf{0}$-broken line the LSLS sequence measures the lattice lengths and lattice sine of the angles as we travel along the broken line, up to some change in sign.

Since lattice length and lattice sign are invariant under $GL_2(\mathbb{Z})$ transformations, so is the LSLS sequence of a broken line.

A strong link between LSLS sequences of $\mathbf{0}$-broken lines and continued fractions is made via the following theorem from [22]:

Theorem 3.1 ([22, Corollary 11.14]). *Consider an $\mathbf{0}$-broken line $A_0 \dots A_n$ with LSLS sequence $(a_0, a_1, \dots, a_{2n})$. Let also $A_0 = (1,0)$ and $A_1 = (1, a_0)$. Then*

$$A_n = (Q_{2n+1}(a_0, a_1, \dots, a_{2n}), P_{2n+1}(a_0, a_1 \dots, a_{2n}))\,.$$

3.3. *LSLS sequence of Fano Polygons with R-singularities*

Let C be the cone over the edge of a Fano polygon.

Definition 3.6. The *sail* $S(C)$ of C is given by $\operatorname{conv}\big(C \backslash \{\mathbf{0}\} \cap \mathbb{Z}^2\big)$.

Lemma 3.1. *The boundary $\delta S(C)$ of the sail of a cone C defines an $\mathbf{0}$-broken line.*

Proof. We need to show that each component L_i of the broken line $\delta S(C) = A_0 A_1 \cdots A_n$ is at lattice height 1. Consider the line segment L_i with vertices A_i and A_{i+1}. By the definition of $S(C)$, there are no interior points in $\operatorname{conv}(\mathbf{0}, A_i, A_{i+1})$ which is equivalent to the Euclidean area of $\operatorname{conv}(\mathbf{0}, A_i, A_{i+1})$ being 1/2 which is equivalent to the lattice height of L_i equalling 1. □

Using this lemma, associate to a cone an LSLS sequence.

Example 3.1. Consider a $\frac{1}{R}(1,1)$ R-singularity. First suppose R is even. Since $\frac{1}{2}(1,1)$ and $\frac{1}{4}(1,1)$ are T-singularities, assume $R \geq 6$. Without loss of generality the $\frac{1}{R}(1,1)$ cone has ray generators $(-1, R/2)$ and $(1, R/2)$. The corresponding broken line then has vertices $(-1, R/2)$, $(0,1)$ and $(1, R/2)$ giving a LSLS sequence of $[1 : R-2 : 1]$. The odd case with $R > 3$ is treated similarly by considering the cone $C_{\frac{1}{R}(1,1)}$ with ray generators $(-(R+1)/2, R)$ and $(-(R-1)/2, R)$. The LSLS sequence is again $[1 : R-2 : 1]$.

Observe that the sum of the elements of the LSLS sequence of a $\frac{1}{R}(1,1)$ singularity is equal to R; the Gorenstein index. This is not a property that generalises to arbitrary cyclic quotient singularities. Consider a $\frac{1}{9}(1,2)$ cone with rays generated by $(-1,3)$ and $(2,3)$. The LSLS sequence of the cone is $[1 : 3 : 2]$, and the sum of the elements is not equal to 9.

We use the following corollary of Theorem 3.1 alongside the LSLS sequence of a $\frac{1}{R}(1,1)$ cone to find combinations of R-singularities which cannot be glued together to form a Fano polygon.

Corollary 3.1 ([22, Corollary 11.14]). *Consider a broken line $A_0 A_1 \cdots A_n$ with the LSLS sequence $(a_0, a_1, \dots, a_{2n})$. Then the broken line is closed if and only if*

$$P_{2n+1}(a_0, a_1, \dots, a_{2n}) = 0 \qquad \textit{and} \qquad Q_{2n+1}(a_0, a_1, \dots, a_{2n}) = 1.$$

Example 3.2. Consider the unique Fano polygon P with six $\frac{1}{3}(1,1)$ cones. By glueing the broken line of each cone together along each vertex of P, obtain a broken line associated to P and through this an LSLS sequence. It is routine to check that the integer sine for each of the angles at a vertex of P is -1, and that the LSLS sequence satisfies

$$\left[1:1:1:-1:1:1:1:-1:1:1:1:-1:\cdots:1:1:1\right]=\frac{0}{1}.$$

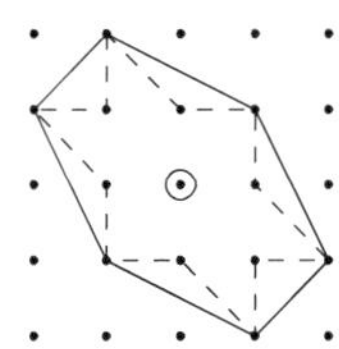

Fig. 3. Broken line associated to the Fano polygon with singularity content $\left(0,\left\{6\times\frac{1}{3}(1,1)\right\}\right)$.

Corollary 3.1 provides a test as to whether it is possible to glue together combinations of R-cones to form a Fano polygon. Namely if there exists a Fano polygon made of cones, cyclically ordered and corresponding to the cyclic quotient singularities $\frac{1}{R_1}(1,1)$, $\frac{1}{R_2}(1,1),\ldots,\frac{1}{R_k}(1,1)$ respectively, then there is a solution to the identity

$$[1:R_1-2:1:m_1:1:R_2-2:1:m_2:\cdots:m_{k-1}:1:R_k-2:1]=\frac{0}{1},$$

where $m_i\in\mathbb{Z}$ is the integer sine of the angle of the associated broken line lying between consecutive cones. Furthermore the convexity of the Fano polygon dictates that $m_i<0$. Note this variable m_i is analogous to the 1-dimensional family of matrices parametrised by $\mathbb{Z}_{\geq 0}$ obtained in Section 2.

The association of a broken line to a polygon is not unique. The choice of starting point for the broken line may change the continued fraction of the broken line since the integer sine of the angle at this point is omitted from the LSLS sequence. However the choice of starting point does not affect that the associated continued fraction should evaluate to $0/1$. Indeed this condition is required to hold at all choices of starting point.

We are now able to prove Theorem 1.2.

Proof of Theorem 1.2. If such a Fano polygon did exist, then by Corollary 3.1 there would be a solution $(m_1,m_2)\in\mathbb{Z}^2_{<0}$ to the following continued fraction:

$$[1:R_1-2:1;m_1:1:R_2-2:1:m_2:1:R_3-2:1]=\frac{0}{1}.$$

By calculating the polynomials $P_{10}(a_0,\cdots,a_{10})$ and $Q_{10}(a_0,\cdots,a_{10})$ and substituting appropriately for the a_i the condition of the continued fraction translates to the simultaneous equations:

$$\begin{aligned}
P_{10}(1,R_1-2,1,m_1,1,R_2-2,&1,m_2,1,R_3-2,1) = \\
&Am_1m_2+Bm_1+Cm_2+D=0,\\
Q_{10}(1,R_1-2,1,m_1,1,R_2-2,&1,m_2,1,R_3-2,1) = \\
&Em_1m_2+Fm_1+Gm_2+H=1,
\end{aligned}$$

where

$$\begin{aligned}
A &= R_1R_2R_3,\\
B &= 2R_1R_2R_3 - R_1R_3 - R_2R_3,\\
C &= 2R_1R_2R_3 - R_1R_2 - R_1R_3,\\
D &= 4R_1R_2R_3 - 2R_1R_2 - 4R_1R_3 - 2R_2R_3 + R_1 + R_2 + R_3 - 12,\\
E &= R_1R_2R_3 - R_2R_3,\\
F &= 2R_1R_2R_3 - R_1R_2 - R_1R_3 - 2R_2R_3 + R_2 + R_3,\\
G &= 2R_1R_2R_3 - R_1R_3 - 3R_2R_3 + R_3,\\
H &= 4R_1R_2R_3 - 2R_1R_2 - 4R_1R_3 - 6R_2R_3 + R_1 - 3R_2 + 5R_3 + 1.
\end{aligned}$$

Solving the simultaneous equations for m_2 gives

$$m_2 = \frac{-(CF+DE+A-G) \pm \sqrt{(CF+DE+A-G)^2 - 4CE(DF+B-H)}}{2CE}.$$

This expression is not integer for $R_i \in \{3\} \cup \mathbb{Z}_{\geq 5}$. □

Acknowledgements

I would like to thank Alexander Kasprzyk, my doctoral advisor, for his guidance. Additionally I am grateful to Takayuki Hibi and Akiyoshi Tsuchiya for their organisation of the Summer Workshop on Lattice Polytopes held in Osaka University 2018. Finally I would like to thank Irem Portakal for helpful discussion at the afore mentioned Workshop. This work was partially supported by Kasprzyk's EPRSC Fellowship EP/NO22513/1.

References

[1] M. Akhtar, T. Coates, A. Corti, L. Heuberger, A. Kasprzyk, A. Oneto, A. Petracci, T. Prince and K. Tveiten, Mirror symmetry and the classification of orbifold del Pezzo surfaces, *Proc. Amer. Math. Soc.* **144**, 513 (2016).

[2] T. Coates, A. Corti, S. Galkin, V. Golyshev and A. Kasprzyk, Mirror symmetry and Fano manifolds, in *European Congress of Mathematics*, (Eur. Math. Soc., Zürich, 2013) pp. 285–300.

[3] J. Kollár and N. I. Shepherd-Barron, Threefolds and deformations of surface singularities, *Invent. Math.* **91**, 299 (1988).

[4] J. Kollár, Flips, flops, minimal models, etc, in *Surveys in differential geometry (Cambridge, MA, 1990)*, (Lehigh Univ., Bethlehem, PA, 1991) pp. 113–199.

[5] M. Akhtar, T. Coates, S. Galkin and A. M. Kasprzyk, Minkowski polynomials and mutations, *SIGMA Symmetry Integrability Geom. Methods Appl.* **8**, Paper 094, 17 (2012).

[6] A. Kasprzyk, B. Nill and T. Prince, Minimality and mutation-equivalence of polygons, *Forum Math. Sigma* **5**, e18, 48 (2017).

[7] A. M. Kasprzyk and B. Nill, Fano polytopes, in *Strings, gauge fields, and the geometry behind*, (World Sci. Publ., Hackensack, NJ, 2013) pp. 349–364.

[8] M. Akhtar and A. Kasprzyk, Singularity content, `https://arxiv.org/abs/1401.5458`, (2014).

[9] M. Reid, Young person's guide to canonical singularities, in *Algebraic geometry, Bowdoin, 1985 (Brunswick, Maine, 1985)*, Proc. Sympos. Pure Math. Vol. 46 (Amer. Math. Soc., Providence, RI, 1987) pp. 345–414.

[10] K. Altmann, Deformations of affine torus varieties, *Beiträge Algebra Geom.* **34**, 119 (1993).

[11] K. Altmann, Computation of the vector space T^1 for affine toric varieties, *J. Pure Appl. Algebra* **95**, 239 (1994).

[12] K. Altmann, Minkowski sums and homogeneous deformations of toric varieties, *Tohoku Math. J. (2)* **47**, 151 (1995).

[13] K. Altmann, The versal deformation of an isolated toric Gorenstein singularity, *Invent. Math.* **128**, 443 (1997).

[14] K. Altmann, P-resolutions of cyclic quotients from the toric viewpoint, in *Singularities (Oberwolfach, 1996)*, Progr. Math. Vol. 162 (Birkhäuser, Basel, 1998) pp. 241–250.

[15] J. A. Christophersen, On the components and discriminant of the versal base space of cyclic quotient singularities, in *Singularity theory and its applications, Part I (Coventry, 1988/1989)*, Lecture Notes in Math. Vol. 1462 (Springer, Berlin, 1991) pp. 81–92.

[16] N. O. Ilten, One-parameter toric deformations of cyclic quotient singularities, *J. Pure Appl. Algebra* **213**, 1086 (2009).

[17] J. Stevens, On the versal deformation of cyclic quotient singularities, in *Singularity theory and its applications, Part I (Coventry, 1988/1989)*, Lecture Notes in Math. Vol. 1462 (Springer, Berlin, 1991) pp. 302–319.

[18] A. Corti and L. Heuberger, Del Pezzo surfaces with $\frac{1}{3}(1,1)$ points, *Manuscripta Math.* **153**, 71 (2017).

[19] D. Cavey and T. Prince, Del Pezzo surfaces with a single $1/k(1,1)$ singularity, `https://arxiv.org/abs/1707.09213` (July, 2018).

[20] B. Poonen and F. Rodriguez-Villegas, Lattice polygons and the number 12, *Amer. Math. Monthly* **107**, 238 (2000).

[21] K. Conrad, $SL_2(\mathbb{Z})$, available to read online at `http://www.math.uconn.edu/ kconrad/blurbs/grouptheory/SL(2,Z).pdf`.

[22] O. Karpenkov, *Geometry of continued fractions*, Algorithms and Computation in Mathematics, Vol. 26 (Springer, Heidelberg, 2013).

Finding a fully mixed cell in a mixed subdivision of polytopes

Giulia Codenotti
Institut für Mathematik, Freie Universität Berlin
Berlin, 14195, Germany
E-mail: codenotti@math.fu-berlin.de
www.fu-berlin.de

Lena Walter
Institut für Mathematik, Freie Universität Berlin
Berlin, 14195, Germany
E-mail: lenawalter@math.fu-berlin.de
www.fu-berlin.de

Motivated by the algorithmic problem of finding a mixed cell in a random mixed subdivision, we survey objects capturing the structure of mixed cells in mixed subdivisions. One such object, via the Cayley trick, is the secondary fan of the Cayley embedding. Another is the mixed secondary polytope introduced by Cools and Michiels.

Keywords: Cayley embedding, mixed subdivisions, fully mixed cells.

1. Introduction

In the study of solutions of systems of polynomials, the Bernstein-Khovanskii-Kushnirenko theorem [1] highlights an important connection to discrete geometry. It states that the number of solutions of a generic system of polynomials is equal to the mixed volume of the Newton polytopes of the polynomials:

Theorem 1.1. *For $f_1, \dots, f_d \in \mathbb{C}[x_1, \dots, x_d]$, the number of isolated solutions to $f_1 = \cdots = f_d = 0$ in $(\mathbb{C}^*)^d$ is bounded above by $MixVol(New(f_1), \dots, New(f_d))$.*

Indeed, if P_i is the Newton polytope of f_i and ω_i is the vector of coefficients of f_i, then a fully mixed cell C of the regular triangulation of the Cayley embedding $\mathrm{Cay}(P_1, \dots, P_d)$ given by the weights $\omega := (\omega_1, \dots, \omega_d)$ corresponds to $\mathrm{vol}(C)$ many solutions of the system. For further details on this correspondence between fully mixed cells and solutions of the polynomial system, we refer the reader to Chapters 3 and 9 of [6].

This motivates us to ask if it is possible to find one fully mixed cell of such a triangulation efficiently. More precisely, we are given, for each $i = 1, \dots, d$, a finite integer point configuration $A_i \subseteq \mathbb{Z}^d$ whose convex hull $P_i = \mathrm{conv}(A_i)$ is a full-dimensional lattice polytope in $\mathbb{R}^d$. We denote the number of points in A_i by n_i, so the total number of points is $n := \sum_{i=1}^{d} n_i$. Also given are vectors $\omega_i \in \mathbb{R}^{n_i}$ for each i, which we will call *weight vectors*. For convenience we also define $\omega := (\omega_1, \dots, \omega_d) \in \mathbb{R}^n$.

From the weight vector ω we can obtain a regular subdivision $\mathcal{S}_\omega$ of the Cayley embedding $\text{Cay}(A_1, \dots, A_d) := \text{Cay}(P_1, \dots, P_d)$ of the point configurations. For generic ω, this will be a triangulation, which we denote by $\mathcal{T}_\omega$.

Our goal is to find one fully mixed cell that is part of this triangulation $\mathcal{T}_\omega$. Computing the whole triangulation would be the naive approach for finding one such fully mixed cell. Since we are only interested in a small part of the information, we want to avoid doing the whole expensive computation. The question we discuss in this note is if it is possible to compute one fully mixed cell of $\mathcal{T}_\omega$ in expected polynomial time. Here, 'polynomial time' is understood to mean polynomial in n and d and 'expected' refers to the expected running time when varying the weight vector ω.

2. Background and notation

In this section, we give all the definitions and constructions necessary to understand the statement of the problem, and state a few main theorems that we will need in the following.

2.1. *Regular subdivisions and triangulations*

A *point configuration* for us will always be a finite set of points $A \subseteq \mathbb{R}^d$. A *face* of a point configuration A is the set of all points of A which maximize the value of some linear functional among all points of A. A face which is not the whole configuration A is called a *proper face.* A face of maximal dimension among proper faces is called a *facet.* A (sub)configuration A is a *simplex* if $|A| = \dim \text{aff}(\text{conv}(A)) + 1$.

We can now define a *subdivision* of A as a collection $\mathcal{S}$ of subsets, called *cells*, of A, such that

- the union of the convex hulls of all cells of $\mathcal{S}$ covers $\text{conv}(A)$,
- the intersection of any two cells of $\mathcal{S}$ is a face of both cells.

A *triangulation* of $\mathcal{T}$ of A is a subdivision where all of the cells are simplices.

We can also speak of a subdivision of a polytope P: given any point configuration A such that $\text{conv}(A) = P$ and any subdivision $\mathcal{S}$ of A, we say that the collection of convex hulls of all cells of $\mathcal{S}$ is a subdivision of P. A triangulation of P is a subdivision where all the elements are simplices. In example 2.1 we will see that there is a subtle difference between triangulations of the polytope and triangulations of the point configuration.

A *regular* subdivision of a point configuration $A \subseteq \mathbb{R}^d$ is a subdivision that can be obtained as the orthogonal projection of the lower facets of a 'lift' of A. More precisely, given a weight vector $\omega \in \mathbb{R}^{|A|}$, lift all points of A to $\mathbb{R}^{d+1}$ using the given weights as the last coordinate, to define

$$\tilde{A} = \left\{(a, \omega(a)) \in \mathbb{R}^{d+1} \mid a \in A\right\},$$

a point configuration in $\mathbb{R}^{d+1}$. Now consider the lower facets of $\tilde{A}$, that is, those facets whose outer normal has negative last coordinate. Their image in A via the projection which forgets the last coordinate is the *regular* subdivision $\mathcal{S}_\omega$ of A induced by the weight vector ω.

We can then define a regular subdivision of a polytope P via a subdivision of A such that $P = \mathrm{conv}(A)$ just as before.

Example 2.1. Consider five points on a line $A = \{1,2,3,4,5\}$, and let $P = \mathrm{conv}\,\{1,2,3,4,5\} \subset \mathbb{R}$. Let $\omega = (5,4,3,5,4)$. The regular subdivision of A induced by ω is $\mathcal{S}_\omega = \{\{1,2,3\},\{3,5\}\}$. Observe that although as a subdivision of the polytope P this is a triangulation, as a subdivision of the point configuration A it is not, since $\{1,2,3\}$ is not a simplex. The weight vector $\omega' = (5,5,3,5,4)$ instead does induce a triangulation $\mathcal{T}_{\omega'} = \{\{1,3\},\{3,5\}\}$.

As subdivisions of the polytope P both vectors induce the same triangulation, as can be seen in figure 1, which shows the subdivision induced by ω, consisting of the two full-dimensional cells $C_1 = \mathrm{conv}(\{1,2,3\})$ and $C_2 = (\{3,5\})$.

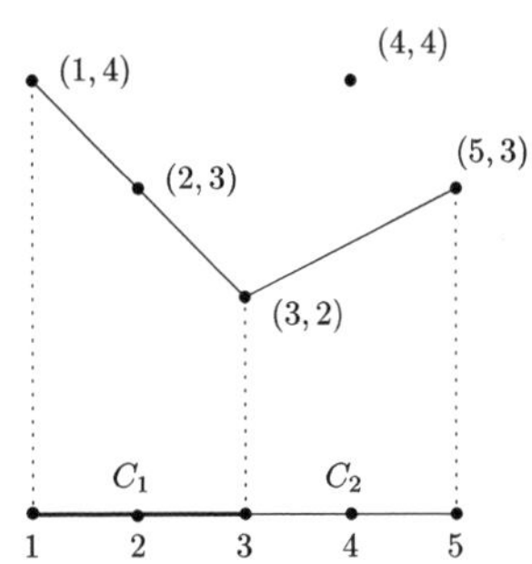

Fig. 1. A regular subdivision of a line segment

2.2. *Minkowski sums and mixed cells*

In the introduction, we mentioned the term 'fully mixed cell'; this only makes sense if several polytopes $P_1 = \mathrm{conv}(A_1), \ldots, P_d = \mathrm{conv}(A_d) \subset \mathbb{R}^d$ and weights $\omega_1, \ldots, \omega_d$ are considered.

The definition can be given by looking at two different constructions, the Minkowski sum and the Cayley embedding. We begin by introducing the former. The latter will be discussed in the next section. The *Minkowski sum* $\sum_{i=1}^d P_i$ of $P_1, \ldots, P_d$ is the polytope defined as

$$\sum_{i=1}^{d} P_i := \{x_1 + \ldots + x_d \mid x_i \in P_i\}.$$

An example of a Minkowski sum consisting of two summands is shown in figure 2.

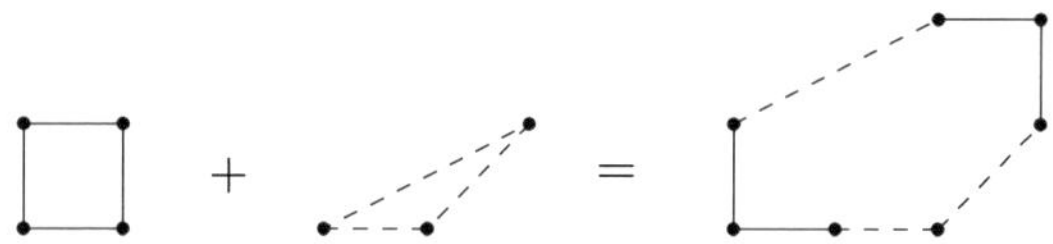

Fig. 2. The Minkowski sum of a square and a triangle

In the previous section we introduced subdivisions of a polytope. In the case of Minkowski sums, we can define a special kind of subdivisions, the so-called *mixed subdivisions*, which respect the structure of the summands. More precisely: a mixed subdivision of $\sum_{i=1}^{d} P_i$ is a subdivision where all cells are of the form $\sum_{i=1}^{d} F_i$, where each F_i is the convex hull of a subset $B_i \subseteq A_i$ and where all cells intersect properly, that is, if $\sum_{i=1}^{d} \operatorname{conv}(B_i)$ and $\sum_{i=1}^{d} \operatorname{conv}(B_i')$ are two cells then

$$\sum_{i=1}^{d} \operatorname{conv}(B_i) \cap \sum_{i=1}^{d} \operatorname{conv}(B_i') = \sum_{i=1}^{d} \operatorname{conv}(B_i \cap B_i').$$

We call a cell itself *mixed* if $\dim(F_i) \geq 1$ for at least two $i \in \{1, \dots, d\}$, and *fully mixed*, if $\dim(F_i) = 1$ for all i., i.e. if it is the Minkowski sum of d edges, one coming from each of the polytopes.

To obtain a regular subdivision on the Minkowski sum $\sum_{i=1}^{d} P_i$, we apply a variation of the construction described in the previous section: we lift each of the polytopes P_i to $\tilde{P}_i \in \mathbb{R}^{d+1}$ according to the given weights ω_i. Then we project the lower convex hull of the Minkowski sum $\sum_{i=1}^{d} \tilde{P}_i$. This yields a regular subdivision of the Minkowski sum $\sum_{i=1}^{d} P_i$, see Theorem 1.3.5 in [2]. Regular subdivisions of Minkowski sums obtained this way are by construction mixed.

Example 2.2. Let us continue with the example shown in figure 2 and let us therefore give coordinates in the plane to the square $P_1 = \operatorname{conv}(A_1)$ and the triangle $P_2 = \operatorname{conv}(A_2)$ where

$$A_1 = \begin{pmatrix} 0\,1\,0\,1 \\ 0\,0\,1\,1 \end{pmatrix} \qquad A_2 = \begin{pmatrix} 0\,1\,2 \\ 0\,0\,1 \end{pmatrix}.$$

We are slighty abusing notation here, since A_i is also used to denote the matrix having $a_1, \dots, a_{n_i}$ as columns. Choosing $\omega = (\omega_1, \omega_2) = (0,0,0,0,1,2,0)$ as a weights vector one obtains the following regular mixed subdivision $\mathcal{S}_\omega(P_1 + P_2)$. The induced subdivision contains two fully mixed cells, which are the unshaded ones in figure 3.

2.3. *The Cayley trick*

There is another way to look at mixed subdivisions and mixed cells, via the Cayeley embedding of the point configurations $A_1, \dots, A_d$. We embed the configurations in parallel affine subspaces in a higher dimensional space: each A_i is embedded in the

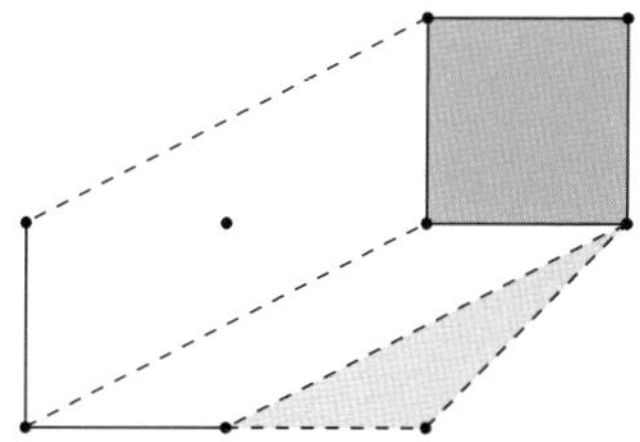

Fig. 3. A regular mixed subdivision of the Minkowski sum of a square and a triangle

subspace of $\mathbb{R}^{2d}$ defined by $x_{d+1} = 0, \ldots, x_{d+i} = 1, \ldots, x_{2d} = 0$. That is, the *Cayley embedding* $\mathrm{Cay}(A_1, \ldots, A_d) \subset \mathbb{R}^d \times \mathbb{R}^d$ is the point configuration

$$\mathrm{Cay}(A_1, \ldots, A_d) := (A_1 \times e_1) \cup \cdots \cup (A_d \times e_d)$$

where $e_1, \ldots, e_d$ denote the standard basis vectors of $\mathbb{R}^d$. For polytopes $P_1 = \mathrm{conv}(A_1), \ldots, P_d = \mathrm{conv}(A_d)$, the *Cayley embedding* $\mathrm{Cay}(P_1, \ldots, P_d)$ is given by $\mathrm{Cay}(P_1, \ldots, P_d) := \mathrm{conv}(\mathrm{Cay}(A_1, \ldots, A_d))$. In the case of full-dimensional polytopes $P_1, \ldots, P_d \subset \mathbb{R}^d$, the Cayeley embedding will be a $2d-1$-dimensional polytope in $\mathbb{R}^{2d}$, with the unique affine relation being that the last d coordinates add up to 1.

Example 2.3. Applying the Cayley embedding to our running example 2.2 yields the point configuration

$$\mathrm{Cay}(A_1, A_2) = \begin{pmatrix} 0 & 1 & 0 & 1 & 0 & 1 & 2 \\ 0 & 0 & 1 & 1 & 0 & 0 & 1 \\ 1 & 1 & 1 & 1 & 0 & 0 & 0 \\ 0 & 0 & 0 & 0 & 1 & 1 & 1 \end{pmatrix}$$

The corresponding Cayley embedding $\mathrm{Cay}(P_1, P_2) \subset \mathbb{R}^4$ can be pictured by making use of the fact that it is contained in the 3-dimensional subspace given by the equation $x_3 + x_4 = 1$.

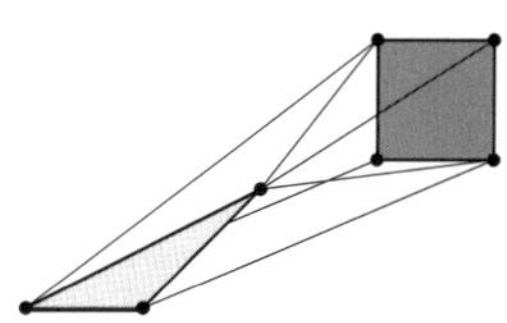

Fig. 4. The Cayley embedding of a square and a triangle

So far the two constructions, Minkowski sum and Cayley embedding, may seem unrelated. But they are far from it, since the Minkowski sum is the intersection of the Cayley embedding with the subspace $\{x \in \mathbb{R}^{2d} \mid x_{d+1} = \ldots = x_{2d} = 1/d\}$, up to a scaling factor:

$$\sum_{i=1}^{d} P_i = d \cdot \mathrm{Cay}(P_1, \ldots, P_d) \cap \{x \in \mathbb{R}^{2d} \mid x_{d+1} = \ldots = x_{2d} = 1/d\}$$

For our running example 2.2 this is illustrated in figure 5.

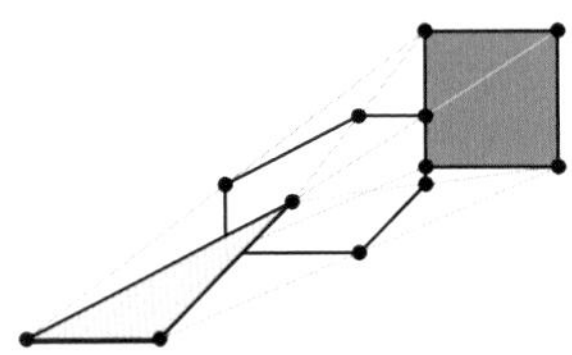

Fig. 5. The connection between the Cayley embedding and the Minkowski sum

By a *cell* C in the Cayley embedding we mean a cell in some polyhedral subdivision of it, and we will often denote it as $C = (C_1, \ldots, C_d)$, where $C_i = C \cap A_i$, to keep track of which points of the cell belong to which point configuration. Since the original polytopes $P_1, \ldots, P_d$ are embedded in parallel affine subspaces, a full-dimensional, i.e. $2d-1$-dimensional, cell must always involve vertices of all original polytopes $P_1, \ldots, P_d$. A *fully mixed* cell will be a full-dimensional cell which involves two vertices of each polytope. It is not a coincidence that this reminds us of the fully mixed cells appearing in regular subdivisions of Minkowski sums. Indeed, the following statement relating subdivisions of Cayley embedding and mixed subdivision of the Minkowski sum holds:

Theorem 2.1 (Theorem 9.2.16 in [2]). *Let $P_1, \ldots, P_d \subset \mathbb{R}^d$ be full-dimensional polytopes. Then polyhedral subdivisions of the Cayley embedding* $\mathrm{Cay}(P_1, \ldots, P_d)$ *are in one-to-one correspondence with mixed polyhedral subdivisions of the Minkowski sum $\sum_{i=1}^{d} P_i$.*

Explicitly, given a polyhedral subdivision of the Cayley embedding, we can recover (a scaled version of) the corresponding subdivision by intersecting the Cayley embedding as we did above with the subspace $\{x \in \mathbb{R}^{2d} \mid x_{d+1} = \ldots = x_{2d} = 1/d\}$. In particular, fully mixed cells of the mixed subdivision are sections of fully mixed cells of the subdivision of the Cayley embedding. For more details on the Cayley trick, the reader is referred to [2] or [3].

Example 2.4. We have already seen a mixed subdivision of the Minkowski sum $P_1 + P_2$ for our running example in figure 3. Each cell corresponds to a cell in the Cayley embedding $\mathrm{Cay}(P_1 + P_2)$ and all together they form a polyhedral subdivision of $\mathrm{Cay}(P_1 + P_2)$. For one of the cells this correspondence is illustrated in figure 6.

We call a mixed subdivision of the Minkowski sum a *fine mixed subdivision* if it corresponds to a triangulation of the Cayley embedding. The name 'fine' comes from the fact that it is impossible to have a mixed subdivision strictly refining it.

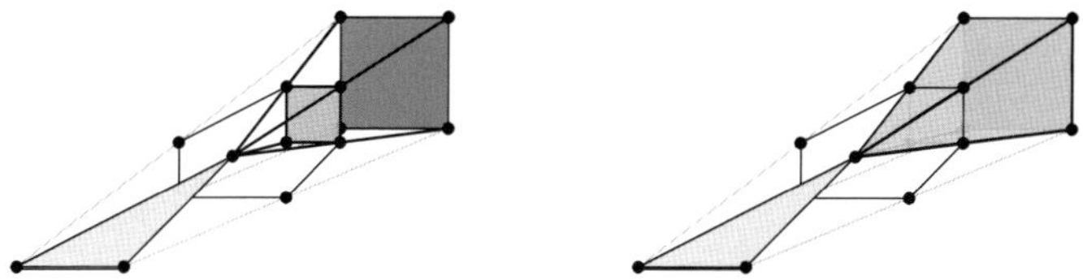

Fig. 6. The correspondence between Minkowski cells and Cayley cells

We now have two ways of thinking of fully mixed cells, and we will often switch between the two. There should be no ambiguity since we have given an explicit bijection.

2.4. *The secondary polytope and the secondary fan*

If we assume the weight vector ω to be generic, the regular polyhedral subdivision it induces will always be a triangulation. In this section we introduce the secondary polytope, which captures useful information about all possible triangulations of a given point configuration. Before we can give a precise definition we have to introduce some notation. We follow the construction of the secondary polytope presented in [2].

Consider a point configuration $A \subset \mathbb{R}^d$ consisting of n points. To each triangulation $\mathcal{T}$ of A we associate a vector $\Phi_A(\mathcal{T}) \in \mathbb{R}^n$. The entry of this vector corresponding to the point $a \in A$ is defined as the sum of all cells of the triangulation which contain the point a as a vertex, where the volume is understood to be the usual Euclidean volume. More precisely, if we denote e_i the i-th unit vector.

Definition 2.1. Let A be a point configuration with $n = |A|$ and $\mathcal{T}$ be a triangulation of A. The vector

$$\Phi_A(\mathcal{T}) := \sum_{i\in[n]} \sum_{C\in\mathcal{T}:a_i\in C} \mathrm{vol}(C)e_i \in \mathbb{R}^n$$

is called the *Gelfand-Kapranov-Zelevinsky* vector of $\mathcal{T}$, often referred as *GKZ* vector.

Example 2.5. Let us come back to the one-dimensional example of 5 points on a line which we studied in example 2.1. The GKZ-vector for the triangulation $\mathcal{T}_{\omega'}$ is given by

$$\Phi_{\{1,2,3,4,5\}}(\mathcal{T}_{\omega'}) = \begin{pmatrix} \mathrm{vol}(13) \\ 0 \\ \mathrm{vol}(13) + \mathrm{vol}(35) \\ 0 \\ \mathrm{vol}(35) \end{pmatrix} = \begin{pmatrix} 2 \\ 0 \\ 4 \\ 0 \\ 2 \end{pmatrix}$$

The convex hull of the GKZ-vectors of all triangulations is what is called the secondary polytope:

Definition 2.2. Let A be a point configuration with $n = |A|$ and $\mathcal{T}$ be a triangulation of A. The *secondary polytope* $\Sigma - \text{poly}(A)$ of A is defined as

$$\Sigma - \text{poly}(A) := \text{conv}\,\{\Phi_A(\mathcal{T}) \mid \mathcal{T} \text{ is a triangulation of } A\}\,.$$

Example 2.6. We continue with example 2.5. The given point configuration $A = \{1, 2, 3, 4, 5\} \subset \mathbb{R}$ has six different triangulations in total. For each of them one can compute the corresponding GKZ-vector as in the previous example. The resulting secondary polytope $\Sigma - \text{poly}(A)$ has the combinatorics of a cube as shown in figure 7.

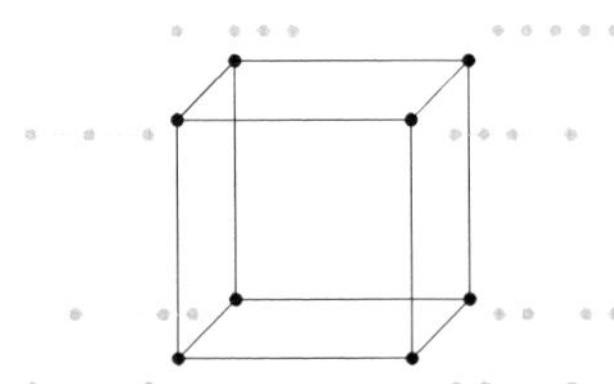

Fig. 7. The secondary polytope of five points on a line

An important result about the secondary polytope is that, although we took the convex hull of GKZ vectors of all triangulations, the vertices correspond only to the regular triangulations (see for example Theorem 5.1.9 in [2]).

Not only vertices, but the whole face lattice of the secondary polytope encodes information about regular subdivisions of the original polytope. In fact, if there is an edge connecting two vertices, the associated triangulations are similar in some sense. That is, they only differ by a local change, a so-called flip.

A point configuration $A \subset \mathbb{R}^d$ with $|A| = n$ is called a *corank 1* configuration if there is a unique affine dependence relation among its points, $\sum_{i\in[n]} \lambda_i a_i = 0$, with $\sum_{i\in[n]} \lambda_i = 0$. The *signed circuit*, denoted by $\mathcal{C} = (\mathcal{C}_+, \mathcal{C}_-)$ keeps track of the indices of the points that are involved in the dependence relation, i.e. $\mathcal{C}_+ = \{i \in [n] \mid \lambda_i > 0\}$ and $\mathcal{C}_- = \{i \in [n] \mid \lambda_i < 0\}$. We refer to $\mathcal{C}_+$ or $\mathcal{C}_-$ as one *side* of the signed circuit $\mathcal{C}$. Since this describes a minimal affine dependence, the two sides of the circuit intersect in a unique point, as can for instance be seen in figure 8.

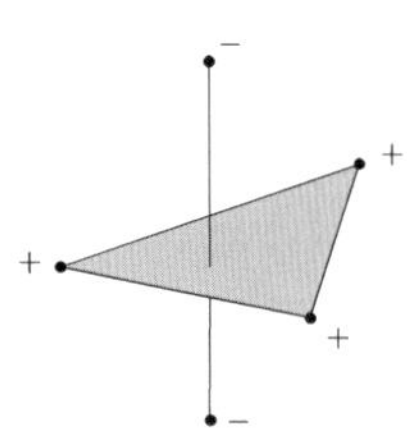

Fig. 8. A signed circuit

The important property of corank 1 configurations is that they only have two possible triangulations:

Lemma 2.1 (Lemma 2.4.2 in [2]). *Let A be a configuration of corank 1 and $\mathcal{C} = (\mathcal{C}_+, \mathcal{C}_-)$ the associated circuit. Then the following are the only two triangulations of A:*

$$\mathcal{T}_+ = \{C \subset [n] \mid \mathcal{C}_+ \not\subseteq C\} \quad \text{and} \quad \mathcal{T}_- = \{C \subset [n] \mid \mathcal{C}_- \not\subseteq C\}.$$

The triangulations are regular.

An *almost triangulation* is a subdivision such that it itself is not a triangulation, but all its proper refinements are. Any cell of an almost triangulation is composed of points which are either affinely independent or are of corank 1, and all its cells of corank 1 contain the same circuit. Therefore by lemma 2.1 it has exactly two proper refinements, which are both triangulations. We say that two triangulations of the same point configuration are *connected by a flip supported on the almost triangulation* $\mathcal{S}$ if they are the only two triangulations refining $\mathcal{S}$.

For the secondary polytope this means that two vertices are connected by an edge if the corresponding triangulations differ by a flip, and the edge corresponds the involved almost triangulation. Figure 9 shows all the possible flips in the plane.

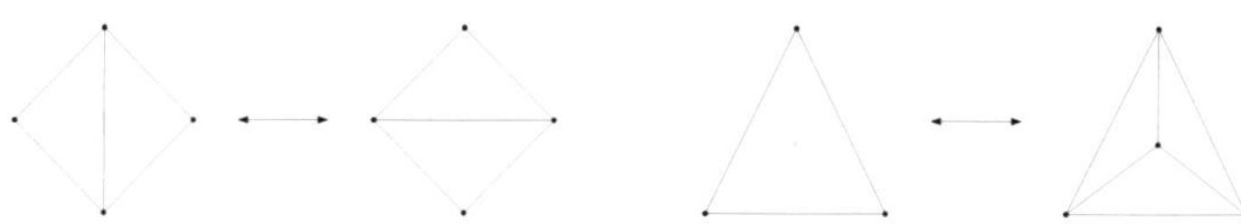

Fig. 9. The possible flips in the plane

In some contexts it is more useful to consider not the secondary polytope, but the secondary fan. The *secondary fan* is the normal fan of the secondary polytope and denoted by Σ – fan. Since each vertex of the secondary polytope corresponds to a regular triangulation, a full-dimensional cone in the secondary fan also corresponds to a regular triangulation. It actually is the cone of all the weight vectors inducing the regular triangulation.

Example 2.7. The secondary fan corresponding to the point configuration of example 2.6 has the combinatorics of an octahedron and can be visualized in figure 10.

3. The mixed secondary fan of the Cayley embedding

Since we are only interested in a single fully mixed cell, not in the whole triangulation, the secondary polytope captures more information than needed.

In [5] Cools and Michiels introduce the so-called 'mixed secondary polytopes', which is related to the secondary polytope but which captures less information

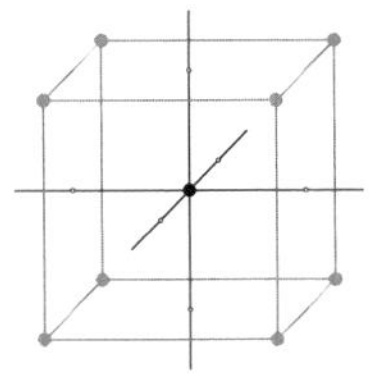

Fig. 10. The secondary fan of five points on a line

and is therefore less complicated to compute. Roughly speaking a vertex of such a polytope represents the collection of fully mixed cells that are part of a given mixed subdivision and ignores all other cells of the subdivision. Thus two mixed subdivisions that contain the same fully mixed cells are represented by the same vertex.

In the following we want to summarize parts of Cool's and Michiel's ideas and results that are useful in our context.

In a similar way to what is done for a single point configuration, they define a secondary polytope of a tuple of point configurations. We let as usual $(A_1, \dots, A_d)$ be a tuple of finite point configurations with $A_i \subset \mathbb{R}^d$, $|A_i| = n_i$ and $n = \sum_{i=1}^{d} n_i$.

Definition 3.1 (Definition 3.2 in [5]). The *secondary polytope* of a tuple of point configurations $(A_1, \dots, A_d)$, denoted by $\Sigma - \text{poly}(A_1, \dots, A_d)$, is

$$\text{conv}\left(\{\Phi_{(A_1,\dots,A_d)}(\mathcal{S}) \mid \mathcal{S} \text{ is a mixed subdivision of } (A_1, \dots, A_d)\}\right).$$

The GKZ-vector $\Phi_{(A_1,\dots,A_d)}(\mathcal{S}) \in \mathbb{R}^n$ mentioned in the theorem is given as definition 2.1 suggests, i.e., for a point $a \in A_i$, the value at its coordinate is given by

$$\left(\Phi_{(A_1,\dots,A_d)}(\mathcal{S})\right)_a = \sum_{C \in \mathcal{S}: a \in C} \text{vol}(C),$$

where vol(C) is short for the respective volume in the Cayley embedding.

The vertices of $\Sigma - \text{poly}(A_1, \dots, A_d)$ correspond to regular mixed subdivisions of $\text{conv}(A_1) + \cdots + \text{conv}(A_d)$ and the edges correspond to flips between these subdivisions, see Theorem 3.3 in [5]. By definition, the GKZ vectors $\Phi_{(A_1,\dots,A_d)}(\mathcal{S}) = \Phi_{(A_1,\dots,A_d)}(\mathcal{T})$ agree for a triangulation $\mathcal{T}$ of the Cayley embedding $\text{Cay}(A_1, \dots, A_d)$ that corresponds to a mixed subdivision $\mathcal{S}$ of the Minkowski sum $\text{conv}(A_1) + \cdots + \text{conv}(A_d)$.

Example 3.1. Let us consider as an example a mixed subdivision of the standard triangle and square in the plane, with the following coordinates (see figure 11):

$$A_1 = \begin{pmatrix} 0 & 1 & 0 \\ 0 & 0 & 1 \end{pmatrix} \qquad A_2 = \begin{pmatrix} 0 & 1 & 0 & 1 \\ 0 & 0 & 1 & 1 \end{pmatrix}$$

For convenience the vertices are labeled by a, b, c, d, e, f, g instead of using additional indices.

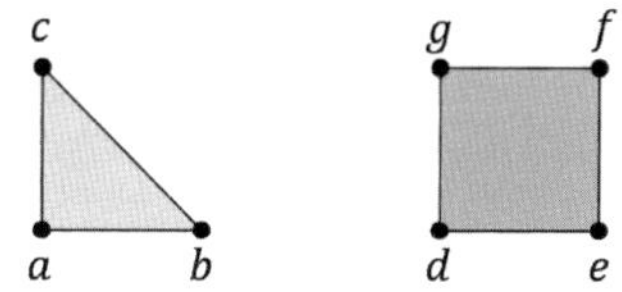

Fig. 11. The standard triangle and square

In figure 12, we can see one possible mixed subdivision of the Minkowski sum with the corresponding labels for each cell. The fully mixed cells are the unshaded ones. In the lower part of figure 12, we have drawn all possible mixed subdivisions of the point configuration. The labeling $\mathcal{T}_1, \ldots, \mathcal{T}_{16}$ refers to the corresponding triangulations of the Cayley embedding.

Each of these mixed subdivisions corresponds to a vertex of the secondary polytope, whose combinatorics is drawn in figure 13.

In [5], they show that the secondary polytope $\Sigma - \text{poly}(A_1, \ldots, A_d)$ can be Minkowski-decomposed into so-called 'mixed secondary polytopes' (Theorem 4.3 in [5]). We are only interested in one of these summands, so we do not give the general definition. We instead introduce the specific object needed in our context. For this construction, we consider a variant of the GKZ-vector which in its construction involves only cells that are close to being a fully mixed cell. To specify what we mean by close to fully mixed, we introduce the 'type' of a cell.

Definition 3.2. The *type* of a cell $C = (C_1, \ldots, C_d)$, denoted by $\tau(C)$, is the set of indices $\{i \in [d] \mid |C_i| > 1\}$.

Definition 3.3. Let $(A_1, \ldots, A_d)$ be a tuple of point configurations, with $A_i \subset \mathbb{R}^d$, $|A_i| = n_i$ and $n = \sum_{i=1}^d n_i$ and let $\mathcal{S}$ be a mixed subdivision. The vector $\Phi^*_{(A_1,\ldots,A_d)}(\mathcal{S}) \in \mathbb{R}^n$ is called the *fully mixed GKZ vector* of $\mathcal{S}$ and its coordinate for $a \in A_i$ is given by

$$\left(\Phi^*_{(A_1,\ldots,A_d)}(\mathcal{S})\right)_a = \sum_{C \in \mathcal{S}: a \in C_i, \tau(C) \cup \{i\} = [d]} \text{vol}(C),$$

where vol(C) is short for the respective volume in Caley embedding.

From this variant of the GKZ vector we can now build a polytope, as we did before for the secondary polytope.

Definition 3.4. The *mixed secondary polytope* of a tuple $(A_1, \ldots, A_d)$ of point configurations with $A_i \subset \mathbb{R}^d$, denoted by $\Sigma - \text{poly}^*(A_1, \ldots, A_d)$, is

$$\text{conv}\left(\left\{\Phi^*_{(A_1,\ldots,A_d)}(\mathcal{S}) \mid \mathcal{S} \text{ is a regular mixed subdivision of } (A_1, \ldots, A_d)\right\}\right)$$

The vertices of the mixed secondary polytope keep track of the fully mixed cells that are involved in a subdivision. We call this set $\mathcal{S}^* = \{C \in \mathcal{S} \mid \tau(C) = [d]\}$ a *fully mixed cell configuration*.

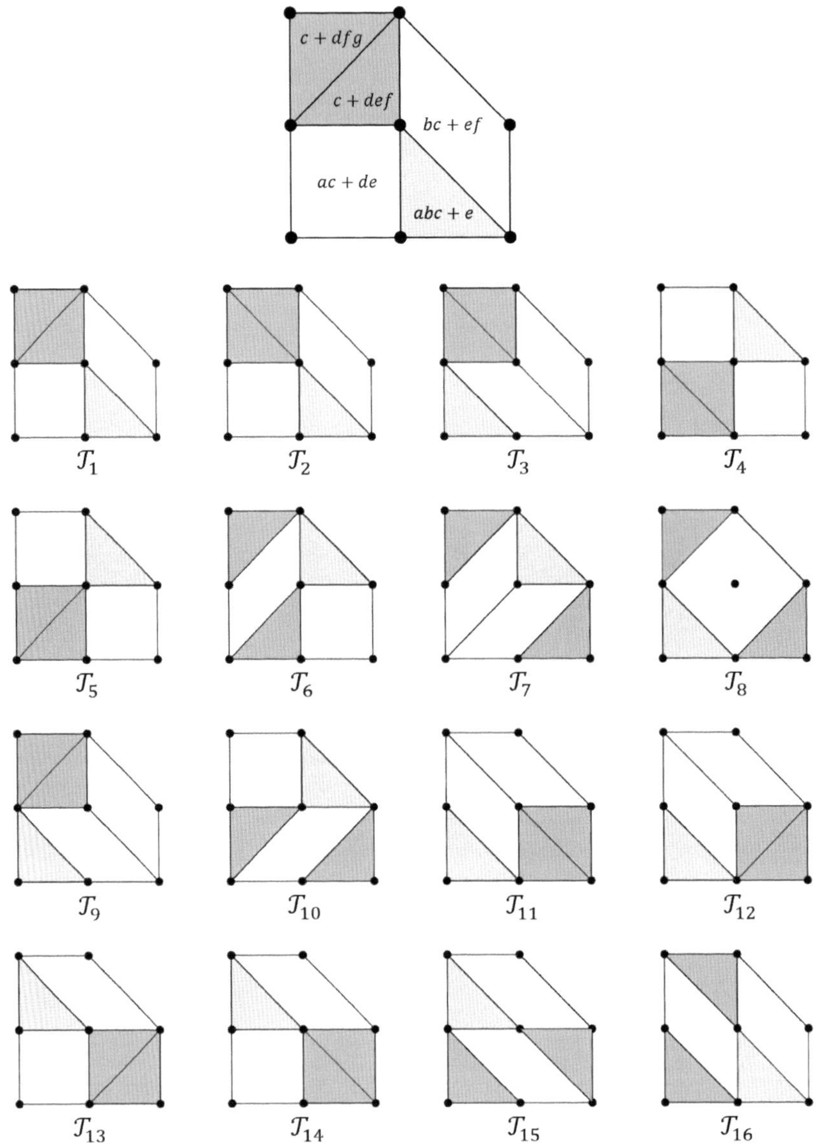

Fig. 12. All mixed subdivisions of the Minkowski sum of triangle and square

Theorem 3.1 (Theorem 6.6 in [5]). *For a tuple of point configurations* $(A_1, \ldots, A_d)$ *from* $\mathbb{R}^d$

(1) *The vertices of* $\Sigma - poly^*(A_1, \ldots, A_d)$ *correspond to the fully mixed cell configurations of* $(A_1, \ldots, A_d)$.

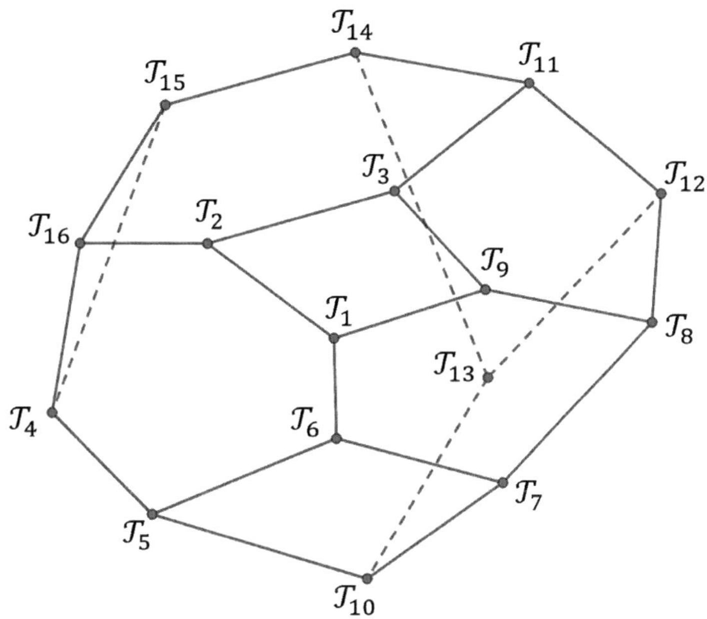

Fig. 13. The combinatorics of the secondary polytope of the Minkowski sum of triangle and square

(2) The edges of $\Sigma - poly^*(A_1, \ldots, A_d)$ correspond to flips involving fully mixed cells.

Example 3.2. Let us return to example 3.1 to illustrate the differences between the secondary polytope and the mixed secondary polytope of the standard triangle and square. In figure 14 we have an example of two distinct mixed subdivisions which contain the same fully mixed cells (in the figure, these are the unshaded cells).

These correspond to distinct vertices in the secondary polytope, vertices $\mathcal{T}_1$ and $\mathcal{T}_2$ in figure 13. Since they share the same fully mixed cell configurations, they will correspond to the same vertex in the mixed secondary polytope. In this example, there are four pairs of vertices which are identified, and the combinatorics of the mixed secondary polytope can be seen in figure 15.

4. The secondary fan as the chamber complex of the Gale dual

The goal of this section is to present an explicit construction of the secondary fan of a point configuration, and to propose a tentative approach to finding a fully mixed cell efficiently.

We call a *fully mixed cell candidate* a collection of pairs of indices $((a_1, b_1), \ldots, (a_d, b_d))$, with $a_i, b_i \in A_i$ for all i such that the matrix formed by those

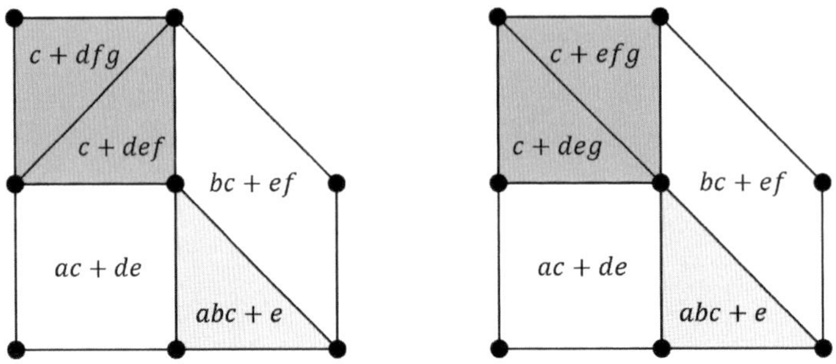

Fig. 14. Two mixed subdivisions with the same fully mixed cell configurations

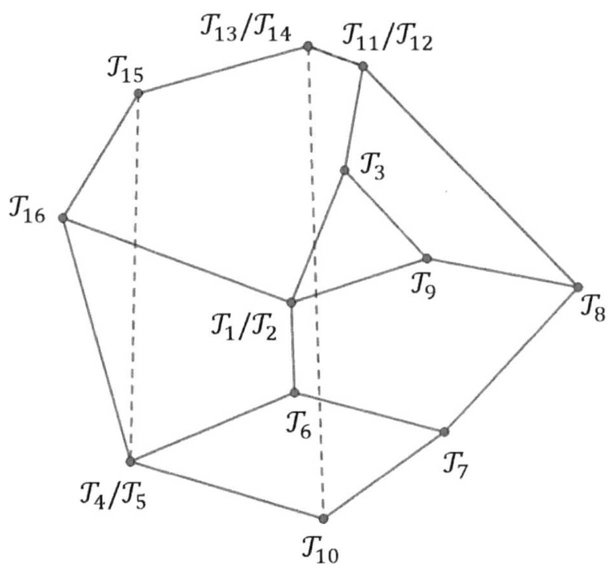

Fig. 15. The combinatorics of the mixed secondary polytope

columns of the A_i indexed by $a_1, b_1, \ldots, a_d, b_d$ has full rank. In the Cayley setting, this corresponds to a simplex contained in the Cayley embedding with one edge coming from each of the point configurations $A_1, \ldots, A_d$, while in the Minkowski setting, it corresponds to a sum of segments from each of the point configurations $A_1, \ldots, A_d$.

Given our point configuration $A = (A_1, \ldots, A_d)$, and weights $\omega = (\omega_1, \ldots, \omega_d)$, we are looking for a fully mixed cell of the mixed subdivision. One possible idea to approach our problem would be to find a collection $\mathcal{M}$ of fully mixed cell candidates such that any mixed subdivision contains exactly one cell of $\mathcal{M}$. Once we have such a collection, we can restrict our search to $\mathcal{M}$, rather than checking all potential mixed cell candidates. If we could find such an $\mathcal{M}$ which is sufficiently small, then this would yield an effective way of searching for a fully mixed cell.

So what does such a family $\mathcal{M}$ look like? To answer this question, we need to introduce the Gale dual (or Gale transform) of a vector configuration and exploit a theorem by Gelfand, Kapranov and Zelevisnki linking triangulations to the chamber fan of the Gale dual.

Given a point configuration $A = \{a_1, a_2, \ldots, a_n\}$ of n points in $\mathbb{R}^{d-1}$ whose affine hull has dimension $r-1$. If we think of them as a matrix

$$A = \begin{pmatrix} a_1 & a_2 & \ldots & a_n \\ 1 & 1 & \ldots & 1 \end{pmatrix}$$

in $\mathbb{R}^{d \times n}$ whose columns correspond to the points embedded into the affine subspace at height 1, the rank of A is r. Hence the kernel of A has dimension $n-r$. A *Gale dual* of A will be a vector configuration $b_1, b_2, \ldots, b_n$ of n vectors in $\mathbb{R}^{n-d}$ such that if we arrange the vectors in columns of an $(n-d) \times n$ matrix B, we have $AB^\intercal = 0$, or equivalently, the rows of B are a basis for the kernel of A.

The key property of the Gale dual is that dependences vectors of A are the evaluation vectors of B and vice versa. Equivalently, one can say that the oriented matroids of A and B are dual to each other.

It will be convenient to speak in the language of (oriented) matroids associated to the point (or vector) configuration. For the reader not familiar with matroids, it will be sufficient to know that a basis of A or B is a set indexing a maximal independent point configuration, and to observe that the complement of a basis of A is a basis of B.

Given a vector configuration B, a *relatively open chamber* is a minimal non-empty set, that can be obtained as intersections of relatively open cones spanned by subconfigurations in B. The closure of a relatively open chamber is a *closed chamber.* The *chamber complex*, or *chamber fan*, of B is the polyhedral fan consisting of the closed chambers of B. (Definition 5.4.6 in [2])

The main reason we introduce this object is that the chamber complex of the Gale dual B is a convenient way to obtain a full-dimensional description of the secondary fan of the original point configuration A, where we have eliminated its lineality space. In particular we have the following:

Theorem 4.1 (Gelfand-Kapranov-Zelevinski). *Let A and B be vector configurations which are Gale duals of each other (equivalently, their oriented matroids are dual to each other). Then there is a bijection between regular triangulations of A and chambers of B.*

Explicitly, the bijection is as follows. Take a triangulation $\mathcal{T}$ of A. If $I_1, \dots I_k \subseteq [n]$ are the bases associated to the simplices of $\mathcal{T}$, and we denote by $\overline{I}$ the complement of a set I, define

$$C_{\mathcal{T}} = \cap_{j=1}^{k} \text{cone}(\overline{I}_j).$$

This is a chamber in the chamber fan of B, and it gives the desired bijection.

What would a family $\mathcal{M}$ of bases of A with the property that any triangulation contains exactly one of the simplices of the family correspond to in the Gale dual? The set of all complements $\overline{\mathcal{M}}$ will be bases of B indexing a triangulation of B: the property that any triangulation of A contains at least one of the cells of $\mathcal{M}$ translates into the complements covering all of B, while the fact that any triangulation of A contains at most one of the cells of $\mathcal{M}$ gives the fact that none of the complements intersect in the relative interior.

Example 4.1. Let's go back to example 2.1. The point configuration can be represented in the matrix

$$A = \begin{pmatrix} 0 & 1 & 2 & 3 & 4 \\ 1 & 1 & 1 & 1 & 1 \end{pmatrix}$$

We think of A as a vector configuration in $\mathbb{R}^2$. This is necessary since a point configuration corresponds to an acyclic vector configuration, so its dual will be totally cyclic, that is it is represented by a collection of vertices who contain the origin in the interior of the convex hull. We can write its Gale dual B as

$$B = \begin{pmatrix} 1 & 0 & -2 & 0 & 1 \\ 2 & -3 & 0 & 1 & 0 \\ 1 & -2 & 1 & 0 & 0 \end{pmatrix},$$

which we can visualize as points on the 2-sphere, see figure 16.

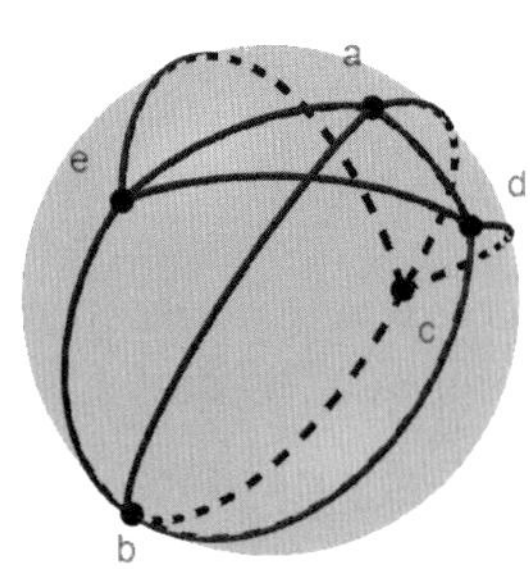

Fig. 16. Gale dual of the point configuration A

Now suppose we have a collection of bases of A: $\{a,b\},\{a,c\},\{a,d\},\{a,e\}$. These clearly have the property that any triangulation of A must contain exactly one of them, since they intersect and they are all the bases containing a. The collection of complements are the following bases of B: $\{b,c,d\},\{b,c,e\},\{b,d,e\},\{c,d,e\}$; these do indeed form a triangulation of the Gale dual (check on figure 16!).

Of course, in our mixed setting, $\overline{M}$ cannot be any triangulation of the Gale dual: if $A_1,\ldots,A_d$ are point configurations in $\mathbb{R}^d$, and A is the matrix of the Cayley embedding, the problem becomes to find a triangulation of the Gale dual such that the complement of all bases of the triangulation are fully mixed cell candidates. Not only this, but in order for this to yield a fast algorithm (on average), we want this triangulation to be small enough compared to the set of all fully mixed cell candidates.

We tried to find such a triangulation in the small case from example 2.2:

Example 4.2. We call A the matrix of $\mathrm{Cay}(A_1,A_2)$ we saw in example 2.2; the columns of the matrix B form a point configuration which is a Gale dual of A:

$$B=\begin{pmatrix} 2 & 0 & -1 & -1 & -1 & 0 & 1 \\ 1 & -1 & 0 & 0 & -1 & 1 & 0 \\ -1 & 1 & 1 & -1 & 0 & 0 & 0 \end{pmatrix}$$

We can visualize this on the unit sphere:

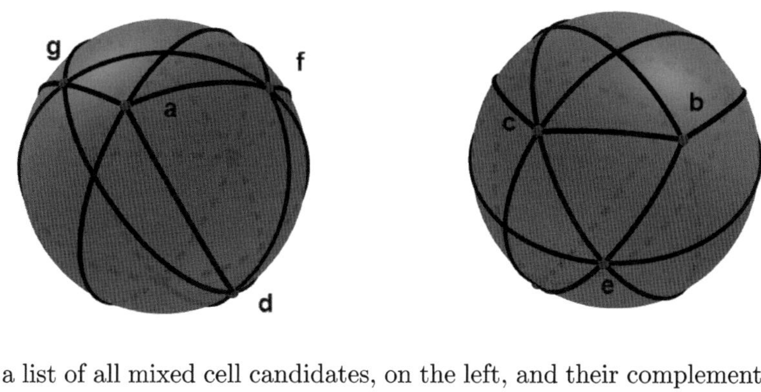

This is a list of all mixed cell candidates, on the left, and their complements, on the right:

$$
\begin{array}{ll}
\{a,b,d,e\} & \{c,f,g\} \\
\{a,b,d,f\} & \{c,e,g\} \\
\{a,b,e,f\} & \{c,d,g\} \\
\{a,b,d,g\} & \{c,e,f\} \\
\{a,b,e,g\} & \{c,d,f\} \\
\{a,b,f,g\} & \{c,d,e\} \\
\{a,c,d,e\} & \{b,f,g\} \\
\{a,c,d,f\} & \{b,e,g\} \\
\{a,c,e,f\} & \{b,d,g\} \\
\{a,c,d,g\} & \{b,e,f\} \\
\{a,c,e,g\} & \{b,d,f\} \\
\{a,c,f,g\} & \{b,d,e\} \\
\{b,c,d,e\} & \{a,f,g\} \\
\{b,c,d,f\} & \{a,e,g\} \\
\{b,c,e,f\} & \{a,d,g\} \\
\{b,c,d,g\} & \{a,e,f\} \\
\{b,c,e,g\} & \{a,d,f\} \\
\{b,c,f,g\} & \{a,d,e\}
\end{array}
$$

One possible triangulation of the dual configuration consisting of complements of mixed cell candidates is:

$$\{c,d,f\},\{c,d,g\},\{c,f,g\},\{a,f,g\},\{a,d,f\},\{a,d,g\}.$$

All such triangulations contained at least 6 cells, so this is a minimal example.

The example suggests that this approach may not be viable in general: if the triangulation is not much smaller than the collection of all mixed cell candidates, it will not aid us in our search. Still, this is only one small example and one can hope that in general it is possible to find a small such triangulation.

References

[1] D. N. Bernstein, *The number of roots of a system of equations*, Functional Analysis and its Applications, 9: 183-185 (1975)

[2] J. A. De Loera, J. Rambau and F. Santos, *Triangulations: Structures for Algorithms and Applications*, 1st edn. (Springer Publishing Company, Incorporated, 2010).

[3] B. Huber, J. Rambau and F. Santos, The Cayley trick, lifting subdivisions and the bohne-dress theorem on zonotopal tilings, *Journal of the European Mathematical Society* **2**, 179 (Jun 2000).

[4] A. Nedergaard Jensen, Tropical homotopy continuation (01 2016).

[5] T. Michiels and R. Cools, Decomposing the secondary Cayley polytope, *Discrete & Computational Geometry* **23**, 367 (Mar 2000).

[6] B. Sturmfels, *Solving Systems of Polynomial Equations*, volume 97 of *CBMS Regional Conference Series in Mathematics*. Published for the Conference Board of the Mathematical Sciences, Washington, DC; by the American Math- ematical Society, Providence, RI, (2002).

Predicting the integer decomposition property via machine learning

Brian Davis

Department of Mathematics
University of Kentucky
Lexington, KY 40506–0027, USA
E-mail: brian.davis@uky.edu

In this paper we investigate the ability of a neural network to approximate algebraic properties associated to lattice simplices. In particular we attempt to predict the distribution of Hilbert basis elements in the fundamental parallelepiped, from which we detect the integer decomposition property (IDP). We give a gentle introduction to neural networks and discuss the results of this prediction method when scanning very large test sets for examples of IDP simplices.

Keywords: Machine learning, neural network, Hilbert basis, integer decomposition property.

1. Introduction

Due to the maturity and ubiquity of machine learning techniques and applications, open–source software libraries such as **Tensorflow** have become available to non-specialists. These libraries are typically well–documented, and friendly technical references are freely available online, e.g., [4]. In this environment, it seems natural to ask: *How do we apply machine learning technology to algebraic combinatorics?*

It is not clear how to extract human–understandable meaning from the raw numerical data of, for example, a neural network (for a discussion of comprehensibility, see [7].) We therefore employ these techniques for their prediction and approximation power, rather than for use in theorems and their proofs. There is a long history of using neural networks in order to approximate solutions to combinatorial optimization problems, e.g. the traveling salesman problem [6], and in [5], Gryak, Haralick, and Kahrobaei use machine learning to predict if two elements of a group are conjugate. It seems reasonable, then, to hope that machine learning has some applicability to problems at the intersection of combinatorics and algebra.

We intend for this work to be an introduction to neural networks and a proof of concept for the use of machine learning, and neural networks in particular, in predicting properties relevant to lattice points in polyhedra. As a particular application, we attempt to predict the integer decomposition property (IDP) in a special class of lattice simplices.

In their paper [1], Braun, Davis, and Solus study the infinite family of lattice simplices of the form

$$\Delta_{(1,q)} = \operatorname{conv}\left\{ e_1, \dots, e_d, -\sum_{i=1}^{d} q_i e_i \right\} \subset \mathbb{R}^d,$$

where $q_i \in \mathbb{Z}_{\geq 0}$ for all i, and give necessary and sufficient conditions on the entries q_i (the **q-vector**) for a such a simplex to be IDP in the case that it is reflexive. In the present work we will "train" a neural network to predict if a given example of a $\Delta_{(1,q)}$ simplex is IDP without actually computing the Hilbert basis.

In Section 2, after presenting the basics of Hilbert bases, we interpret the integer decomposition property as a composition of functions to be approximated. In Section 3 we develop the general framework for training a neural network using the language of piecewise linear functions[a] and stochastic gradient descent. In Section 4 we discuss a piecewise linear approximation of the integer decomposition property and its accuracy.

2. Approximating the integer decomposition property as a function

We will now introduce the Hilbert basis and use it to define the integer decomposition property. We encode the integer decomposition property as a real-valued function $\mathbb{IDP}$ in Subsection 2.3, and define what it means to approximate the integer decomposition property.

2.1. *Hilbert basics*

Let $v_1, \ldots, v_{d+1}$ be affinely independent elements of the integer lattice $\mathbb{Z}^d$. Their convex hull is a d-simplex

$$\Delta := \left\{ \sum_{i=1}^{d+1} \gamma_i v_i \ : \ 0 \leq \gamma_i \ , \ \sum_{i=1}^{d+1} \gamma_i = 1 \right\} \subset \mathbb{R}^d,$$

and we define cone(Δ) to be the non-negative real span of the points in $(1, \Delta)$, i.e., Δ embedded into $\mathbb{R}^{d+1}$ at height 1 in the zeroth coordinate:

$$\text{cone}(\Delta) := \left\{ \sum_{i=1}^{d+1} \gamma_i (1, v_1) \ : \ 0 \leq \gamma_i \right\} \subset \mathbb{R}^{d+1}.$$

The set cone$(\Delta) \cap \mathbb{Z}^{d+1}$ is closed under addition, and the unique minimal collection of additive generators is called the **Hilbert basis** of cone(Δ).

We call the zeroth coordinate of a point z in cone$(\Delta) \cap \mathbb{Z}^{d+1}$ the **height** of z, denoted

$$\text{height}(z) := z_0,$$

and we say that the simplex Δ has the **integer decomposition property** (IDP) if, for each element of the Hilbert basis of cone(Δ), the height is equal to 1.

[a]This exposition agrees with the more common descriptions of neural networks when restricted to the case that the source of the training data is a well-defined function and we use **ReLU** activation functions.

Define the **fundamental parallelepiped** of Δ to be the weighted sum of the cone generators $(1, v_i)$ with non-negative weights strictly less than 1:

$$\Pi_\Delta := \left\{ \sum_{i=1}^{d+1} \gamma_i(1, v_i) \; : \; 0 \leq \gamma_i < 1 \right\} \subset \operatorname{cone}(\Delta)\,.$$

Because any element z of $\operatorname{cone}(\Delta) \cap \mathbb{Z}^{d+1}$ lies in $\operatorname{cone}(\Delta)$, it is a non-negative linear combination of the $(1, v_i)$'s, i.e., there exist non-negative real coefficients g_i such that

$$z = \sum_{i=1}^{d+1} g_i(1, v_i) = \left(\sum_{i=1}^{d+1} \lfloor g_i \rfloor (1, v_i) \right) + \left(\sum_{i=1}^{d+1} \{g_i\}(1, v_i) \right)$$

where $\{g_i\}$ means the fractional part of g_i. By setting γ_i equal to $\{g_i\}$, we see that any point z may be written as a non-negative *integral* combination of the $(1, v_i)$'s and an integer point in $\Pi_\Delta \cap \mathbb{Z}^{d+1}$. Consequently, the Hilbert basis consists of the cone generators $(1, v_i)$ together with the additively minimal elements of $\Pi_\Delta \cap \mathbb{Z}^{d+1}$.

2.2. *Partitioning* Π

We partition Π_Δ into disjoint subsets we call **bins** B_α for α in $\{0, \ldots, d\}^{d+1}$, with $z \in B_\alpha$ if and only if

$$(\lfloor (d+1)\gamma_1 \rfloor, \ldots, \lfloor (d+1)\gamma_{d+1} \rfloor) = \alpha,$$

where the γ_i's are the coefficients of the representation of z in terms of the generators $(1, v_i)$.

Proposition 2.1. *Let z be an integer point in B_α. Then*

$$\textit{height}(z) = \left\lceil \frac{\sum_{i=1}^{d+1} \alpha_i}{d+1} \right\rceil .$$

Proof. Considering the zeroth coordinate of z, it is clear that

$$\operatorname{height}(z) = \sum_{i=1}^{d+1} \gamma_i.$$

Note that since $\alpha = (\lfloor (d+1)\gamma_1 \rfloor, \ldots, \lfloor (d+1)\gamma_{d+1} \rfloor)$, the inequality

$$(d+1)\gamma_i - 1 < \lfloor (d+1)\gamma_i \rfloor \leq (d+1)\gamma_i$$

implies that

$$\sum_{i=1}^{d+1} ((d+1)\gamma_i - 1) < \sum_{i=1}^{d+1} \alpha_i \leq \sum_{i=1}^{d+1} (d+1)\gamma_i.$$

Thus

$$(d+1)\big(\operatorname{height}(z) - 1\big) < \sum_{i=1}^{d+1} \alpha_i \leq (d+1)\operatorname{height}(z),$$

and

$$\text{height}(z) - 1 < \frac{\sum_{i=1}^{d+1} \alpha_i}{d+1} \leq \text{height}(z),$$

from which the result follows. □

This leads to the following characterization of the integer decomposition property:

Corollary 2.1. *The simplex* Δ *is IDP if and only if for each* z *in the Hilbert basis of* $\text{cone}(\Delta)$, $z \in B_\alpha$ *implies that* $\sum_{i=1}^{d+1} \alpha_i$ *is at most* $d+1$.

2.3. *The function* $\mathbb{IDP}$

Definition 2.1. $\mathbb{IDP}$ is the 0/1 function which takes as input the vertices of a lattice d-simplex Δ and returns one if Δ is IDP and zero otherwise.

We were unsuccessful in approximating $\mathbb{IDP}$ directly using techniques presented in this paper. Instead, we find success approximating another function, $\mathbb{HB}$, from which the value of $\mathbb{IDP}$ can be inferred.

For a vector $x \in \mathbb{R}^{(d+1)^{d+1}}$, we consider x to be multi-indexed by the collection of $\alpha \in \{0, \ldots, d\}^{d+1}$.

Definition 2.2. $\mathbb{HB}$ is the function taking as input the vertices of a lattice d-simplex and returning an element of $\{0,1\}^{(d+1)^{d+1}}$, with coordinate α equal to one if and only if there exists a Hilbert basis element in bin B_α.

Example 2.1. Consider the $\Delta_{(1,q)}$ simplex in dimension $d = 2$ with q-vector $(2,1)$. The ray generators are $v_1 = (1,0,1)$, $v_2 = (0,1,1)$, and $v_3 = (-2,-1,1)$. Computation with **Normaliz** [3] yields that the set of lattice points in $\Pi_{\Delta_{(1,g)}}$ is equal to $\{(0,0,0),\ (-1,0,1),\ (0,0,1)\}$. The representation of these points in terms of the ray generators are $(0,0,0)$, $(0, 1/2, 1/2)$, and $(1/2, 1/4, 1/4)$. The bins B_α which contain these points have α equal to:

$$\begin{aligned}
&(\lfloor 3 \cdot 0 \rfloor, \lfloor 3 \cdot 0 \rfloor, \lfloor 3 \cdot 0 \rfloor) = (0,0,0),\\
&(\lfloor 3 \cdot 0 \rfloor, \lfloor 3 \cdot 1/2 \rfloor, \lfloor 3 \cdot 1/2 \rfloor) = (0,1,1), \text{ and}\\
&(\lfloor 3 \cdot 1/2 \rfloor, \lfloor 3 \cdot 1/4 \rfloor, \lfloor 3 \cdot 1/4 \rfloor) = (1,0,0),
\end{aligned}$$

respectively.

When the α are lexicographically ordered, we may write the image under $\mathbb{HB}$ as the vector

$$(1,0,0,0,1,0,0,0,0,1,0,0,0,0,0,0,0,0,0,0,0,0,0,0,0,0,0) \in \mathbb{R}^{3^3}.$$

Let **supp** be the 0/1 function on $\mathbb{R}^{(d+1)^{d+1}}$ which, for a vector x, returns zero if and only if there exists an index α such that $x_\alpha \neq 0$ and $\sum_{i=1}^{d+1} \alpha_i > d+1$. Then, using Corollary 2.1, we may write the functional equality

$$\mathbb{IDP} = \mathbf{supp} \circ \mathbb{HB}.$$

Note that for Example 2.1, the non-zero entries are at multi-indices $(0,0,0)$, $(0,1,1)$ and $(1,0,0)$, and that the sum of each individual multi-index is not more than 3. Thus the image of $\mathbb{IDP}$ is equal to 1, indicating that the example is IDP. We can verify this fact by noting that the height of each lement of the Hilbert basis

$$\{v_1, v_2, v_3\} \bigcup \{(-1,0,1),(0,0,1)\}$$

is equal to 1. We remark that it is not true in general that the Hilbert basis elements are the non-zero lattice points of Π_Δ.

We have developed a theoretical framework for approximating the integer decomposition property by approximating the real-valued function $\mathbb{HB}$. One difficulty in the implementation of this scheme is the fact that **supp** is not sensitive to how close to zero a value is. If the entry at some multi-index α in the approximation of $\mathbb{HB}$ is close to but not equal to zero, and $\sum_{i=1}^{d+1} \alpha_i > d+1$, then the image of **supp** will be 0, i.e., our approximation of $\mathbb{IDP}$ will almost always predict that an example is *not* IDP. A standard solution to this issue is to first map our approximation into the open interval $(0,1)$, then choose a value $0 \leq \eta \leq 1$, then interpret values less than or equal to η as 0 and greater than η as 1. For the first step, we use the Sigmoid function:

$$\sigma(x) := (1+e^{-x})^{-1},$$

mapping $\mathbb{R}$ one-to-one onto the open interval $(0,1)$. For some fixed $0 \leq \eta \leq 1$, define

$$\mathbf{cutoff}(\mathbf{x}) = \begin{cases} 0 & \text{if } x \leq \eta, \text{ and} \\ 1 & \text{otherwise.} \end{cases}$$

The composition of **cutoff** and σ allows us to turn any real-valued function of one variable into a 0/1 function, and by applying it coordinate-wise, we may turn any function $f : \mathbb{R}^u \longrightarrow \mathbb{R}^v$ into $\mathbf{cutoff} \circ \sigma \circ \mathbf{f} : \mathbb{R}^{\mathbf{u}} \longrightarrow \{\mathbf{0},\mathbf{1}\}^{\mathbf{v}}$. In particular, consider $\mathbb{R}^u$ to be the space parameterizing the vertex sets of lattice d-simplices: $\mathbb{R}^u = \mathbb{R}^d \times \cdots \times \mathbb{R}^d$ with $d+1$ factors $\mathbb{R}^d$ (not all points in the space give rise to full-dimensional simplices.) Further consider $\mathbb{R}^v$ to have basis multi-indexed by $\alpha \in \{0,\ldots,d\}^{d+1}$. Then for any map f from $\mathbb{R}^u$ to $\mathbb{R}^v$, $\mathbf{cutoff} \circ \sigma \circ \mathbf{f}$ may be considered as a map from lattice d-simplices to 0/1 vectors indexed by bins B_α.

Continuing Example 2.1, consider the function $f : \mathbb{R}^{2\times 3} \longrightarrow \mathbb{R}^{3^3}$ defined by

$$f(\boldsymbol{x}) = \frac{1}{\|\boldsymbol{x}\|}\left(\sum_{i=1}^{27}(-1)^i \cdot e_i\right).$$

Then $f(1,0,0,1,-2,-1) = \frac{1}{\sqrt{7}}\left(\sum_{i=1}^{27}(-1)^i \cdot e_i\right)$. We compute that $\sigma(1/\sqrt{7}) = 0.593$, and that $\sigma(-1/\sqrt{7}) = 0.407$. Thus if:

- $0 \leq \eta < 0.407$, then $\mathbf{cutoff} \circ \sigma \circ \mathbf{f}(\mathbf{1},\mathbf{0},\mathbf{0},\mathbf{1},-\mathbf{2},-\mathbf{1})$ is the all-ones vector,
- if $0.593 \leq \eta \leq 1$ then $\mathbf{cutoff} \circ \sigma \circ \mathbf{f}(\mathbf{1},\mathbf{0},\mathbf{0},\mathbf{1},-\mathbf{2},-\mathbf{1})$ is the zero vector, and
- if $0.407 \leq \eta < 0.593$, then $\mathbf{cutoff} \circ \sigma \circ \mathbf{f}(\mathbf{1},\mathbf{0},\mathbf{0},\mathbf{1},-\mathbf{2},-\mathbf{1})$ is the 0/1-vector $\sum_{i=1}^{27} \frac{(1+(-1)^i)}{2} \cdot e_i$.

As this example demonstrates, the quality of the approximation depends heavily on the choice of value for η, as for the fixed function f, $\mathbf{cutoff} \circ \sigma \circ \mathbf{f}$ can be correct on 11%, 89%, or 33% of the entries of $\mathbb{HB}$, depending on the choice of η.

Definition 2.3. Let f be any function from $\mathbb{R}^{d(d+1)}$ to $\mathbb{R}^{(d+1)^{d+1}}$. Then we call the (coordinate-wise) composite function

$$\widehat{\mathbb{IDP}} := \mathbf{supp} \circ \mathbf{cutoff} \circ \sigma \circ \mathbf{f}$$

an approximation of the integer decomposition property.

Note that when $\sigma \circ f$ closely approximates $\mathbb{HB}$ coordinate-wise, $\widehat{\mathbb{IDP}}$ agrees with $\mathbb{IDP}$. For a given f, we will use the shorthand notation $\widehat{\mathbb{HB}}$ for the composite function $\mathbf{cutoff} \circ \sigma \circ \mathbf{f}$.

3. A general approximation method

In this section we describe piecewise linear functions as compositions of affine transformations and a well-behaved piecewise linear function ρ. We next describe the use of a loss function L in quantifying the accuracy of an approximation $\widehat{f}$ of a function f. We then describe an algorithm called gradient descent, which deforms the piecewise linear function $\widehat{f}$ in order to minimize the loss function L with respect to the target function f.

Let f be any set map from $\mathbb{R}^u$ to $\mathbb{R}^v$. We will approximate f by constructing a *random* initial "approximation" $\widehat{f}$, which we will deform until we have a sufficiently accurate approximation.

For a positive integer m, fix m positive integers ℓ_1 through ℓ_m, as well as a small real $\epsilon > 0$. We will call this the collection of **hyper-parameters**. Choose matrices $W_k \in \mathbb{R}^{\ell_{k-1} \times \ell_k}$ for $1 \leq k \leq m+1$, where we set $\ell_0 = u$ and $\ell_{m+1} = v$ (the dimensions of the domain and codomain of f.) Additionally, for each k, choose vectors $b_k \in \mathbb{R}^{\ell_k}$. The entries $(W_k)_{i,j}$ are called **weights**, and the $(b_k)_i$ are called **biases**. Generally, the initial values are randomized by an algorithm we will not discuss here. We will consider each such collection of **parameters** to be a point

$$p = (W_1, b_1, \ldots, W_{m+1}, b_{m+1})$$

in the space of parameters $\mathbb{R}^{(\ell_0+1)\times\ell_1} \times \cdots \times \mathbb{R}^{(\ell_m+1)\times\ell_{m+1}}$. We define ρ to be the function which returns the coordinate-wise maximum of 0 and the identity, i.e.,

$\rho(x_i) = \max(0, x_i)$. The map ρ is an example of an **activation function** and is called **ReLU** (Rectified Linear Unit.) Let ω_k be the affine map $x \mapsto W_k(x) + b_k$ composed with ρ. Then the approximation $\widehat{f}$ is the function

$$\widehat{f}(x;p) = W_{m+1} \circ \omega_m \circ \cdots \circ \omega_1(x) + b_{m+1}.$$

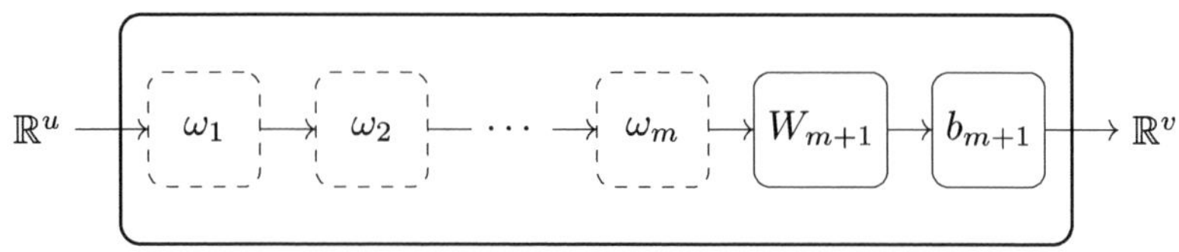

Fig. 1. The approximation $\widehat{f}$.

Example 3.1. Let $f(x) = \log(x)$. We will approximate f on the interval $[1, 3]$. Let $m = 1$ and $\ell_1 = 2$. We initially set the parameters $p = (\omega_1, b_1, \omega_2, b_2) \in \mathbb{R}^{(1+1)\times 2} \times \mathbb{R}^{(2+1)\times 1}$ by

$$W_1 = [0.75, -0.5]^T \quad b_1 = [-0.75, 1] \quad W_2 = [1, 1] \quad b_2 = [-0.5].$$

The resulting approximation, which we expect to be poor because it knows nothing about the function it is supposed to approximate, is given by the piecewise linear function (the dotted graph in Figure 2)

$$\begin{aligned}\widehat{f}(x;p) &= [1,1]\rho\left(\begin{bmatrix}0.75\\-0.5\end{bmatrix}[x] + \begin{bmatrix}-0.75\\1\end{bmatrix}\right) + [-0.5]\\ &= 1\cdot\rho(0.75x - 0.75) + 1\cdot\rho(-0.5x+1) - 0.5\\ &= \begin{cases}0.25x - 0.25 & 1 \le x \le 2\\ 0.75x - 1.25 & 2 < x \le 3.\end{cases}\end{aligned}$$

3.1. *Loss functions and gradient descent*

We measure the quality of the approximation via a **loss function** $L(x;p)$ which we attempt to minimize. By minimizing its value at many "training" points x distributed throughout the domain, we hope that the value of the approximation $\widehat{f}$ will be close to that of f at points outside of training set, i.e., that the magnitude of the loss function will be small at new points as well.

One example of a loss function is the Euclidean distance

$$D(x;p) = \left\|f(x) - \widehat{f}(x,p)\right\|.$$

Continuing Example 3.1,

$$\begin{aligned}D(x;p) = \big\|\log(x) - \big(&(W_2)_{1,1}\rho\big((W_1)_{1,1}(x) + b_{1,1}\big)\\ &+ (W_2)_{1,2}\rho\big((W_1)_{2,1}(x) + b_{1,2}\big) + b_{2,1}\big)\big\|,\end{aligned}$$

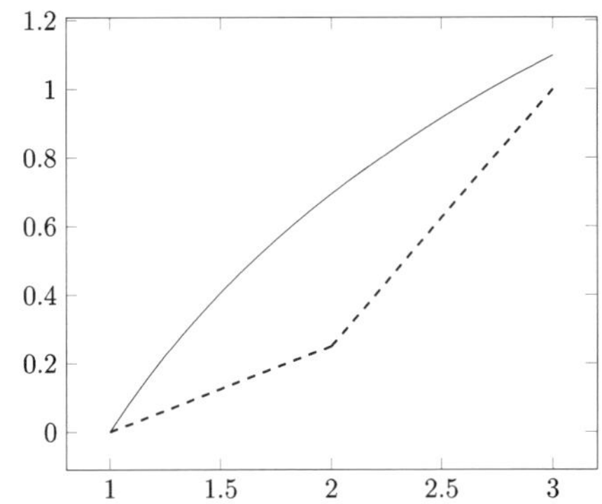

Fig. 2. The function f and the approximation $\widehat{f}$ (dashed).

and for our specific parameters p,

$$D(x;p) = \big\|\log(x) - \big(1\cdot\rho(0.75x - 0.75) + 1\cdot\rho(0.75x + 1) - 0.5\big)\big\| \,.$$

Although for fixed parameters p, the loss $L(x;p)$ is a function of x, the "learning" step of machine learning happens by interpreting it as a function of p, holding x fixed. We can imagine L as a surface above the parameterization space which is fixed by the choice of hyper-parameters and x. In order to improve our approximation $\widehat{f}$ at a particular point x in the domain, we modify its parameters in such a way that the value of the loss function L is reduced, i.e., "moving downhill" on the surface L.

We compute the gradient ∇L with respect to the parameters p at the point (x,p) and update the parameters by $p \mapsto p - \epsilon\nabla L$. The value of ϵ is chosen small enough that $L(x;p-\epsilon\nabla L) < L(x;p)$. When we repeatedly apply this process for points x sampled uniformly at random, this method is called **stochastic gradient descent** or SGD. In practice, for reasons of computational efficiency and stability, a batch of points are sampled and the mean of the gradients is used for the update. This is known as **mini batch** SGD.

Continuing our example, fix $x = 1.5$ and use the chain rule to compute that

$$\begin{aligned}\nabla D(1.5;p) &= \left\langle \frac{\partial D}{\partial\omega_1}\,, \frac{\partial D}{\partial b_1}\,, \frac{\partial D}{\partial\omega_2}\,, \frac{\partial D}{\partial b_2}\right\rangle_{x=1.5} \\ &= \langle -1.5\,, -1.5\,, -1\,, -1\,, -0.375\,, -0.25\,, -1\rangle\,.\end{aligned}$$

Then for $\epsilon = 0.02$, the update $p' = p - \epsilon\,\nabla D(1.5;p)$ is given by

$$\omega_1 = [0.78\,, -0.47]^T \quad b_1 = [-0.73\,, 1.02] \quad \omega_2 = [1.0075\,, 1.0075] \quad b_2 = [-0.48].$$

The resulting updated approximation is

$$\widehat{f}(x;p') = \begin{cases} 0.312x - 0.187 & 1 \le x \le 2.17 \\ 0.786x - 1.215 & 2.17 < x \le 3 \end{cases}.$$

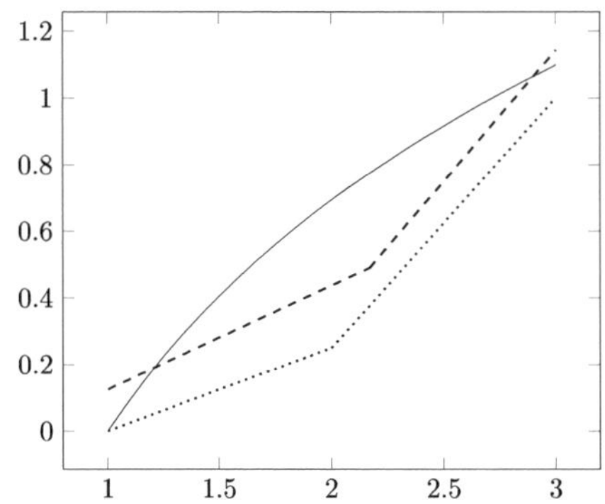

Fig. 3. The approximations $\widehat{f}(x;p)$ (dotted) and $\widehat{f}(x;p')$ (dashed).

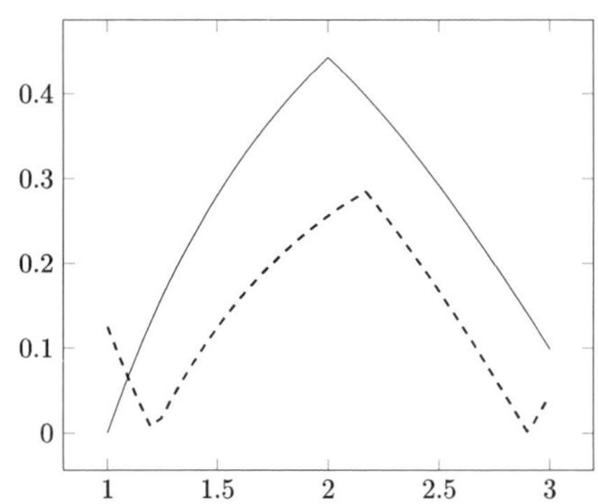

Fig. 4. The loss function D for $\widehat{f}(x;p)$ and $\widehat{f}(x;p')$ (dashed).

3.2. *Training and validation*

In practice, we perform the update step many thousands of times at x-values distributed throughout the domain. Often we gather a large collection of pairs $(x, f(x))$ called a **training set** to store for later use in the update process, rather than computing the value of f when needed. When generating this collection is costly, as in the case of the function $\mathbb{IDP}$, we use each pair from the collection multiple times over, in some cases as many as 100 times. By analogy with polynomial approximation, where we fit a polynomial to a finite set of points on the graph of a function, one may wonder if, when reusing sampled points in refining our approximation, we are simultaneously losing accuracy at other points in the domain. The short answer is yes.

This phenomenon of **overfitting** is a principal concern in the process of refining our approximation, and there are some standard techniques for mitigating its effect, including:

- creating two collections of pairs $(x, f(x))$ — one for training and one for validation. As we train $\widehat{f}$ we simultaneously monitor its accuracy on the **validation set**. If the performance on the validation set worsens while improving on the training set, we stop training.
- introducing a component to the loss function for the magnitudes of the parameters. Experience shows that this method, called **regularization**, reduces overfitting to the training data.
- using the simplest "structure" possible to achieve the desired performance. Complicated models require more training to achieve their optimal performance, and hence increase the number of times training data is reused. We balance the expressive capability of a complicated approximation with the need to minimize overfitting.

4. Implementation and results

4.1. *Implementation*

Our first goal is an approximation of the function $\mathbb{HB}$ restricted to the vertex sets of $\Delta_{(1,q)}$ simplices of dimension $d = 4$ and with q-vector entries bounded by 25. Recall also that, even though the target space of $\mathbb{HB}$ has dimension $(d+1)^{d+1}$, the relevant values are those at indices α whose coordinates sum to more than $d + 1$. We restrict to these 2,877 relevant indices. Hence the input to our function is the tuple $(q_1, q_2, q_3, q_4) \in [1, 25]^4$ and the output is in $\mathbb{R}^{2,877}$.

There is no general-purpose best design of hyper-parameters that works for every application of a neural network. In fact, it is possible to approximate with arbitrary accuracy any continuous function on a compact subset of $\mathbb{R}^u$ using only one "hidden layer" ($m = 1$.) The general rule is that higher values of m allow smaller values for the ℓ_i's while maintaining approximation flexibility. Optimizing hyper-parameters is a process that is outside the scope of this work, so we will simply report that, after experimenting with several values of m and ℓ_i's in order to minimize the loss function and computation time, we proceeded using the following choice of hyper-parameters:

Table 1. Hyperparameters

m	ℓ_1	ℓ_2	ℓ_3	ℓ_4	ϵ
4	100	400	800	3,000	0.001

Tensorflow produces a neural network with the specified dimensions and initializes the weights and biases automatically. In order to implement mini batch SGD, the user must make more decisions than just specifying the hyper-parameters:

(1) Amount of training/validation data:
We used Normaliz and a script to compute $\mathbb{HB}$ for a sample of size 50,000, 10% of which we reserved for validation.

(2) Batch size:
During training we computed the gradient ∇L for batches of 10 q-vectors at a time and used the mean for the update of the parameters p.

(3) Loss function:
Because the image of $\mathbb{HB}$ is contained by the set $\{0,1\}^{2,877}$, we may consider the approximation to be the composite $\sigma\widehat{f}$ and use the **Binary Cross Entropy** loss function BCE summed entry-wise over

$$BCE = (\mathbb{HB} - 1) \cdot \log\left(1 - \sigma\widehat{f}\right) - \mathbb{HB} \cdot \log \sigma\widehat{f}.$$

When the value of $\mathbb{HB}$ is one, the value of BCE is decreased by increasing the value of $\log \sigma\widehat{f}$, i.e., increasing the value of $\widehat{f}$. In this case, minimizing BCE coincides with minimizing the difference between $\sigma\widehat{f}$ and $\mathbb{HB}$. A similar analysis for the case when $\mathbb{HB}$ equals zero shows that BCE is a measure of the accuracy of $\sigma\widehat{f}$ as an approximation of $\mathbb{HB}$. We used a modification of BCE, which we discuss in Section 4.2.

(4) Training length:
We performed roughly 100,000 updates in the process of training the approximation.

The result of this training procedure was a piecewise linear function f. It was the well-defined and deterministic[b] result of the specific choices outlined above.

An approximation $\widehat{\mathbb{HB}}$ requires a choice of cutoff parameter η, and $\widehat{\mathbb{IDP}}$ requires the additional choice of a tolerance parameter τ (introduced in Subsection 4.2). These parameters control the functions **cutoff** and **supp**, respectively. Recall that the resulting approximations are given by

$$\widehat{\mathbb{HB}} := \mathbf{cutoff} \circ \sigma \circ \mathbf{f} \qquad \text{and} \qquad \widehat{\mathbb{IDP}} := \mathbf{supp} \circ \widehat{\mathbb{HB}}.$$

We present the results in terms of the values η and τ.

4.2. *The approximation* $\widehat{\mathbb{HB}}$

While it is tempting to present the accuracy of $\widehat{\mathbb{HB}}$ as the percentage of indices on which it agrees with $\mathbb{HB}$, this is problematic due to the scarcity of non-zero entries in any given image of $\mathbb{HB}$. Consider the q-vector $(4, 10, 14, 14)$; there are just 14 non-zero entries among the 2,877 relevant entries in its image under $\mathbb{HB}$. Consequently, an approximation which is uniformly equal to zero would be correct 99.5% of the time, while knowing essentially nothing about the function it is trying to approximate other than that it is typically equal to zero! We therefore present the accuracy in the form of a **confusion table**, which breaks down the indices α

[b]For purposes of analysis and reproducibility, we initialize the computer's randomness generator so that the stochastic processes are, in fact, deterministic, while still having good randomness properties.

along two criteria — firstly depending on whether $\mathbb{HB}_\alpha$ is equal to 1 (positive) or 0 (negative), and secondly whether $\widehat{\mathbb{HB}}_\alpha$ is positive or negative.

Example 4.1. Again using the q-vector $(4, 10, 14, 14)$, we set $\eta = 0.1$ and present the resulting confusion table below:

Table 2. The confusion table for $\widehat{\mathbb{HB}}(4, 10, 14, 14)$

	PREDICTED 0	PREDICTED 1
ACTUAL 0	2,808	55
ACTUAL 1	0	14

Observe that the sum of the table entries is, in fact, 2,877. We call entries appearing in the upper right cell of the table "**false positive**" because the approximation incorrectly predicted that a bin contained a Hilbert basis element. Similarly, entries in the bottom left cell are called "**false negative**".

We may summarize the table with the pair of ratios

$$\text{specificity} = \frac{\text{true negatives}}{\text{true negatives} + \text{false positives}}, \quad \text{and}$$

$$\text{sensitivity} = \frac{\text{true positives}}{\text{true positives} + \text{false negatives}}.$$

For the present example, they are 98% and 100%, respectively. The specificity and sensitivity vary with the cutoff value η, and are negatively correlated with each other, as demonstrated in Table 3:

Table 3. The effect of varying η

η	specificity	sensitivity
0.1	0.981	1.00
0.25	0.986	0.857
0.5	0.993	0.214

4.3. *Validation*

When we sampled 50,000 examples for training, we reserved 5,000 of them for validation purposes. We now report the performance on this validation set, which we denote $\mathcal{S}$. We aggregate (sum entry-wise) the confusion tables for $\eta = 0.1$ in Table 4.

The corresponding aggregated specificity is 89.0%, and sensitivity is 79.7%. One can account for the difference between specificity and sensitivity by recalling the scarcity of non-zero entries of $\mathbb{HB}$, i.e., the low total number of positives. If we use the loss function

$$BCE = (\mathbb{HB} - 1) \cdot \log\left(1 - \sigma\widehat{f}\right) - \mathbb{HB} \cdot \log \sigma\widehat{f}$$

Table 4. An aggregated confusion table for $\mathcal{S}$ ($\eta = 0.1$)

	PREDICTED 0	PREDICTED 1
ACTUAL 0	12,726,675	1,573,167
ACTUAL 1	22,569	88,482

as earlier described for our gradient descent, the resulting approximation will essentially be the constant zero function. In order for the model to learn to identify positives, we must balance the contributions to the loss function associated to positive and negative according to the inverse of their frequency. We accomplish this by introducing a positive term β which we call the **balance** term:

$$L = (\mathbb{HB} - 1) \cdot \log\left(1 - \sigma\widehat{f}\right) - \beta \cdot \mathbb{HB} \cdot \log \sigma\widehat{f}.$$

The results presented in this section correspond to a β value of 10. All other parameters remaining fixed, a higher value, roughly $\beta = 75$, is required in order achieve approximately equal sensitivity and specificity. However, it is not necessarily desirable to match the sensitivity and specificity, as we will discuss.

4.4. *The approximation* $\widehat{\mathbb{IDP}}$

Under the unrealistic assumption that Hilbert basis elements are distributed roughly uniformly among bins, consider an approximation with a specificity of 99.9% applied to the q-vector of an IDP $\Delta_{(1,q)}$ simplex. Because there are 2,877 bins, the probability that all bins will be correctly identified as negative (not containing a Hilbert basis element) can be estimated as $0.999^{2,877} \approx 5.6\%$. Since we expect the incidence of the integer decomposition property to be low, a true positive rate for IDP of 5.6% may result in few or even no examples being correctly predicted as IDP! We have several tools to combat this issue:

(1) manipulating the balance term β to produce high specificity (possibly at the expense of sensitivity)
(2) manipulating the cutoff value η to produce high specificity (again, at the expense of sensitivity)
(3) tolerating some number of positive entries in $\widehat{\mathbb{HB}}$ (under the assumption that many of them are false.)

For this last option we introduce the **tolerance** parameter τ, which sets an upper bound on the number of positive entries before the function $\widehat{\mathbb{IDP}}$ returns that an example is IDP negative. In our original description of $\widehat{\mathbb{IDP}}$, τ was implicitly set to zero.

Table 5 records the number of true positives over the total number of positives of $\widehat{\mathbb{IDP}}$ when applied to the sample $\mathcal{S}$ for select values of η and τ.

From Table 5, we see that there is not one optimal choice for the values of η and τ, since higher specificity is correlated with few examples being found; the

Table 5. The rate of true positives (specificity) for given values of η and τ

τ \ η	0.5	0.25	0.12	0.05
0	3/7 (42.9%)	3/4 (75.0%)	3/3 (100.0%)	3/3 (100.0%)
10	21/320 (6.6%)	11/38 (29.0%)	8/21 (38.1%)	6/12 (50.0%)
20	46/1026 (4.5%)	21/102 (20.6%)	11/45 (24.4%)	8/27 (29.6%)
30	65/1770 (3.7%)	35/196 (17.9%)	23/103 (22.3%)	16/64 (25.0%)

goals of specificity and sensitivity are in tension. Figure 5 shows the (log-scale) relationship between specificity and sensitivity induced by varying these values. When we actually checked for IDP using Normaliz, we found 112 positive examples among 5,000. The analogous "specificity" is 2.24%, but the "sensitivity" is 100% — we plot this point $(2.24, 100)$ for reference.

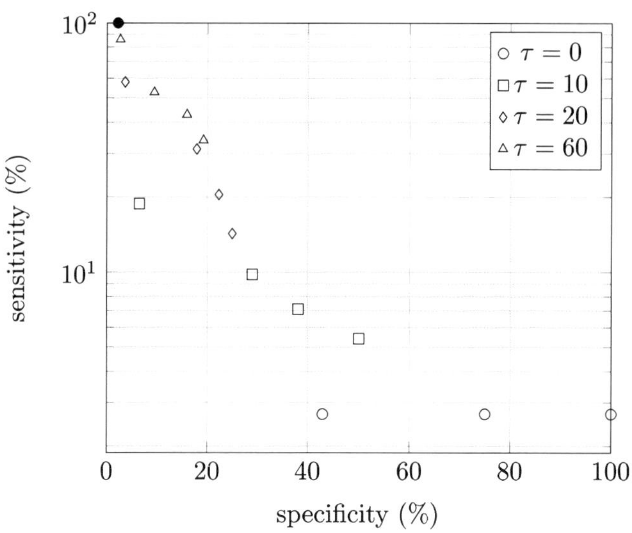

Fig. 5. The effect of varying η for fixed values of τ.

Table 6 lists all 112 q-vectors of $\mathcal{S}$ that correspond to IDP $\Delta_{(1,q)}$ simplices according to Normaliz. Recall that the rate of IDP in $\mathcal{S}$ is 2.24%. Table 7 lists the subset of $\mathcal{S}$ which are predicted to be IDP when $\eta = 0.5$ and $\tau = 0$, with the correct positive predictions highlighted. Observe that the incidence of IDP among the *predicted* IDP examples is about 43%, much higher than the rate in the sample at large.

We highlight the q-vectors in Table 6 that correspond to true IDP positive predictions made by setting $\eta = 0.1$ and $\tau = 65$ (the specificity was 15% and the sensitivity was 58%.)

Table 6. The 112 IDP examples in the sample $\mathcal{S}$

1,1,1,1	1,1,3,9	1,1,21,24	1,2,14,10	1,2,14,10
1,3,16,3	1,3,24,1	1,4,2,16	1,4,20,20	1,8,1,1
1,10,10,8	1,10,24,24	1,12,4,12	1,15,3,1	1,18,1,6
1,21,1,4	1,24,1,9	1,24,14,2	1,24,17,1	1,24,18,1
1,24,18,4	1,24,24,20	2,2,2,7	2,3,12,18	2,8,8,4
2,10,1,16	2,20,10,5	3,1,1,9	3,6,12,1	3,12,2,24
3,14,21,3	3,19,3,1	3,23,15,3	4,1,1,4	4,8,2,16
4,20,1,14	4,20,10,20	4,23,4,12	4,24,1,16	6,1,2,12
6,2,6,3	6,2,18,9	6,6,6,3	6,14,6,15	6,17,9,18
7,3,21,7	7,7,1,7	7,7,16,16	8,1,8,2	8,2,12,24
8,16,4,2	9,1,1,9	9,6,18,2	9,9,4,4	9,18,4,4
9,18,18,6	9,22,1,11	10,1,5,22	10,5,10,9	10,24,4,1
11,22,5,5	12,1,2,6	12,1,24,19	12,2,3,12	12,2,18,3
12,3,2,6	12,3,11,6	12,6,1,1	12,6,1,3	12,12,4,12
12,16,1,16	12,24,2,24	12,24,6,1	13,2,2,20	14,6,14,7
14,7,2,24	14,7,12,1	15,1,13,15	15,15,1,1	16,1,6,6
16,4,2,16	16,7,16,16	16,8,4,2	16,16,12,3	16,24,1,22
17,1,7,1	17,17,8,4	17,17,17,1	18,1,1,15	18,2,6,6
18,2,22,1	18,10,1,15	19,19,1,16	20,2,1,12	20,8,19,8
20,14,24,1	20,20,1,20	20,20,4,1	20,20,4,20	20,22,1,22
21,21,16,4	22,2,2,22	22,16,4,1	22,16,22,1	22,22,20,1
23,2,2,6	23,18,3,24	23,24,24,12	24,2,1,16	24,4,2,4
24,24,6,24	24,24,23,12			

Table 7. Predicted IDP examples ($\eta = 0.5$, $\tau = 0$)

1,1,1,1	1,2,10,2	1,3,24,1	2,2,2,7	2,3,4,7
4,3,2,5	11,6,9,6			

4.5. *Discussion*

As a demonstration of the utility of the approximation method presented here, we could attempt to advance the previously mentioned work in [1] on $\Delta_{(1,q)}$ simplices by producing a large and diverse collection of IDP examples from which to form conjectures to try to prove. A natural scheme for arriving at such a collection is to first generate a test set, say, all $\Delta_{(1,q)}$ simplices of dimension d with q-vector entries bounded by n, then verify the integer decomposition property with a program like Normaliz, collecting the positive examples. We could augment this scheme with machine learning by performing an initial sieving step prior to testing with Normaliz. By developing a computationally–cheap approximation to the integer decomposition property, we can reserve the relatively expensive Normaliz computations for those examples that, according to the approximation, are more likely to be IDP.

In the context of this application, the results outlined above point to a tradeoff between the computational efficiency (controlled by the specificity) and the number of examples that are ultimately produced (controlled by the sensitivity). It also seems that the approximation $\widehat{\mathbb{IDP}}$ is biased in favor of repeated entries (see the highlighted examples in Table 6), which brings into question how diverse a set of examples it is capable of producing.

We computed the value of $\widehat{\mathbb{IDP}}$ for all 390,625 $\Delta_{(1,q)}$ simplices with q-vector in $[1,25]^4$ using $\eta = 0.1$ and $\tau = 65$. The computation produced 2,520 predicted positives. We then computed $\mathbb{IDP}$ for these examples and found that 521 were IDP. This corresponds to a specificity of 20.7%. It is impractical to compute $\mathbb{IDP}$ over the entire collection of 390,625 examples in order to compute the sensitivity, so it is not known.

5. Concluding Remarks

It is very likely that other choices of hyper-parameters, or even entirely different machine learning techniques, will yield improved performance. However, the results, such as they are, do indicate that functions like $\mathbb{IDP}$ have the potential to be modeled by machine learning techniques. The following remarks point out directions in which this investigation might be continued.

Remark 5.1. Figure 5 shows the tradeoff between specificity and sensitivity for an approximation $\widehat{\mathbb{IDP}}$ that is a product of a choice of hyper-parameters, balance β, and training size. It would be useful to see the effect of different values of β in the plot. Does there exist a choice which achieves sensitivity and specificity of 50%?

Remark 5.2. The intermediate step of computing an approximation of $\mathbb{HB}$ has several potential applications which are not explicitly discussed in this paper. In particular we note that by computing the set of lattice points in each predicted–positive bin, we have an approximation of the Hilbert basis itself.

If the sensitivity of $\widehat{\mathbb{HB}}$ is high then it is very likely that the Hilbert basis is *contained* by the approximated Hilbert basis, and may be recovered by the reduction algorithm used by Normaliz (implemented by a python script, for example.) This could potentially be more efficient than Normaliz, which reduces the entire fundamental parallelepiped, if the specificity is high.

Remark 5.3. The Ehrhart h^*-vector records the number of lattice points at each height in Π_Δ. In the case that the h^*-vector is the concatenation of two vectors — the first increasing and the second decreasing — we call it **unimodal**. Unimodality is another interesting property to investigate for lattice simplices, see, e.g., [2]. If, rather than recording the presence of a Hilbert basis element, we were to record the number of fundamental parallelepiped points in each bin, we could approximate the h^*-vector using Proposition 2.1. Thus we have a framework for predicting both IDP *and* unimodality.

Acknowledgement

The author thanks Devin Willmott and Kyle Helfrich for many helpful conversations.

References

[1] B. Braun, R. Davis, and L. Solus, *Detecting the Integer Decomposition Property and Ehrhart Unimodality in Reflexive Simplices*, ArXiv e-prints (2016).

[2] Benjamin Braun, *Unimodality problems in Ehrhart theory*, Recent trends in combinatorics, IMA Vol. Math. Appl., vol. 159, Springer, [Cham], 2016, pp. 687–711. MR 3526428

[3] W. Bruns, B. Ichim, T. Römer, R. Sieg, and C. Söger, *Normaliz. algorithms for rational cones and affine monoids*, Available at https://www.normaliz.uni-osnabrueck.de.

[4] Ian Goodfellow, Yoshua Bengio, and Aaron Courville, *Deep learning*, MIT Press, 2016, `http://www.deeplearningbook.org`.

[5] Jonathan Gryak, Robert Haralick, and Delaram Kahrobaei, *Solving the conjugacy decision problem via machine learning*, (2017).

[6] Jean-Yves Potvin, *State-of-the-art survey?the traveling salesman problem: A neural network perspective*, ORSA Journal on Computing **5** (1993), no. 4, 328–348.

[7] Geoffrey G Towell and Jude W Shavlik, *Extracting refined rules from knowledge-based neural networks*, Machine learning **13** (1993), no. 1, 71–101.

Lattice polytopes in mathematical physics

Alexander Engström
Department of Mathematics and Systems Analysis
Aalto University
P.O. Box 11100, FI-00076 Aalto, Finland
E-mail: alexander.engstrom@aalto.fi

Florian Kohl
Department of Mathematics and Systems Analysis
Aalto University
P.O. Box 11100, FI-00076 Aalto, Finland
E-mail: florian.kohl@aalto.fi

The Tutte polynomial of graphs is deeply connected to the q-state Potts model in statistical mechanics. In this survey, we describe this connection and show how one can use lattice polytopes and transfer matrices to tackle a special case given by the zero-temperature anti-ferromagnetic Potts model, where computing the Tutte polynomial reduces to computing the chromatic polynomial. We apply the techniques earlier introduced by the authors to this special case. This article can be seen as a brief summary of this work with a view towards statistical mechanics. In particular, we illustrate this procedure by computing the chromatic polynomials of some (families of) graphs. As it turns out, counting integer points in lattice polytopes is one of the key ingredients of these computations.

Keywords: q-state Potts model, partition function, Tutte polynomial, chromatic polynomial, lattice polytope, transfer-matrix method, Ehrhart theory.

1. Introduction

Lattice polytopes are beautiful objects and they have a myriad of applications. For instance, they are connected to algebraic geometry and commutative algebra [BH93, CLS11, Sta96], optimization [BP03, Stu96], number theory [BBK$^+$15, BK14, Pom93], and combinatorics [BR15, Sta96]. However, lattice polytopes are also of interest in mathematical physics. The perhaps most prominent example of this fact is the relation between reflexive polytopes and mirror symmetry. This connection was first discovered by Batyrev [Bat94], and a beautiful survey can be found in [Cox15]. Roughly speaking, a reflexive lattice polytope together with its dual polytope give rise to two mirror-dual Calabi–Yau manifolds. This article is not about reflexive polytopes and mirror symmetry. This article is about the connections of the chromatic polynomial and the q-state Potts model. The chromatic polynomial counts the number of proper k-colorings of a given graph G. Lattice polytopes enter the picture due to a geometric interpretation of colorings given in [BZ06] by Beck and Zaslavsky. They show that proper colorings correspond to integer points in what they call inside-out polytopes. In [EK18], the authors use this connection to develop a transfer-matrix method to compute the chromatic polynomial of graphs $G \times P_n$ and to determine the asymptotic behavior of $G \times C_n$, where P_n is the path and C_n is the cycle graph on n nodes, respectively. In [BEMPS10], Beaudin, Ellis-Monaghan, Pangborn, and Shrock give an introduction to the q-state Potts model to people coming from graph theory. All the physics described in our article comes from their paper and we do not claim originality. We merely wanted to bring together lattice polytopes and statistical mechanics.

The structure of this article is as follows. In Section 2, we introduce the relevant basics for graph theory and for counting lattice points in (lattice) polytopes. In Section 3, we show that computing the Tutte polynomial gives the partition function for the q-state Potts model. This section is entirely based on [BEMPS10]. In Section 4, we introduce the techniques developed in [EK18] to compute the chromatic polynomial of $G \times P_n$ and describe the asymptotics of $G \times C_n$. In Section 5, we explicitly compute the chromatic polynomials of given graphs.

Acknowledgments

This project was supported by Academy of Finland project number 288318. Moreover, the second author would like to thank Akiyoshi Tsuchiya and Takayuki Hibi for organizing the summer workshop on lattice polytopes at Osaka university, which motivated this paper.

2. Background

The first part of this section briefly introduces the relevant basics from graph theory such as the Tutte polynomial, the chromatic polynomial, and the Cartesian graph product. It may safely be skipped by people familiar with graph theory. The second part concerns lattice polytopes and Ehrhart theory. We follow [BZ06] and show how proper graph colorings correspond to integer points in inside-out polytopes.

2.1. *Graph Theory*

Let $G = (V, E)$ be a simple, finite graph, i.e., a graph on finitely many nodes without any loops or multiple edges. The elements of V are called the *vertices* or *nodes* and the elements of E are called the *edges*. A *k-coloring* c of G is a map $c\colon V \to [k]$, where $[k] := \{1, 2, \ldots, k\}$. We say that a k-coloring c is *proper* if it satisfies $c(u) \neq c(v)$ for all $\{u, v\} \in E$, see Figure 1.

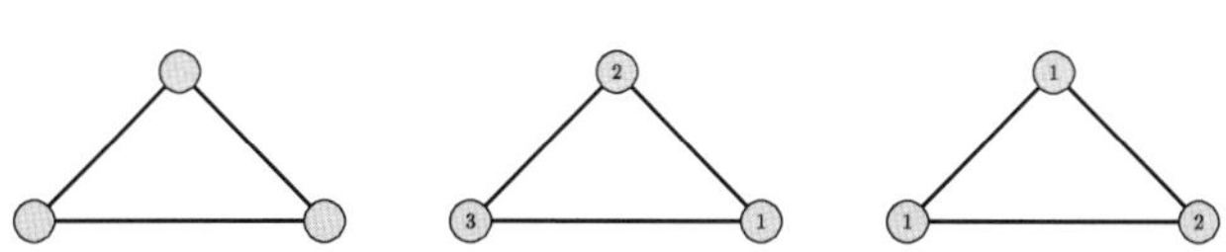

Fig. 1. C_3, a proper coloring of C_3, and a non-proper coloring of C_3.

Historically, proper colorings first arose in the context of the 4-color conjecture. Roughly speaking, the 4-color conjecture — now the 4-color theorem [AHK77, AH77a, AH77b] — states that every planar map can be colored using only 4 colors such that any two adjacent countries have different colors. This problem can be reformulated using proper graph colorings. For a given planar graph G, this led Birkhoff to introduce the *chromatic function* χ_G

$$\chi_G(k) := \#\text{proper } k\text{-colorings of } G.$$

Whitney later generalized this notion to arbitrary (simple) graphs and he showed that this function actually agrees with a polynomial in k:

Theorem 2.1 ([Whi32, Sec 6]). *Let G be a simple graph on n vertices. Then χ_G agrees with a monic polynomial in k of degree n.*

Due to this result, the chromatic function is known as the *chromatic polynomial*. A slick way of proving this theorem is using a process called *deletion-contraction*. Let us fix an edge $e = \{u, v\} \in E$. We define the *deletion* of G with respect to e as $G \backslash e := (V, E \setminus \{e\})$. Similarly, the *contraction G/e of G along e* is defined by identifying the nodes u and v and by removing all edges between them. This procedure is illustrated in Figure 2.

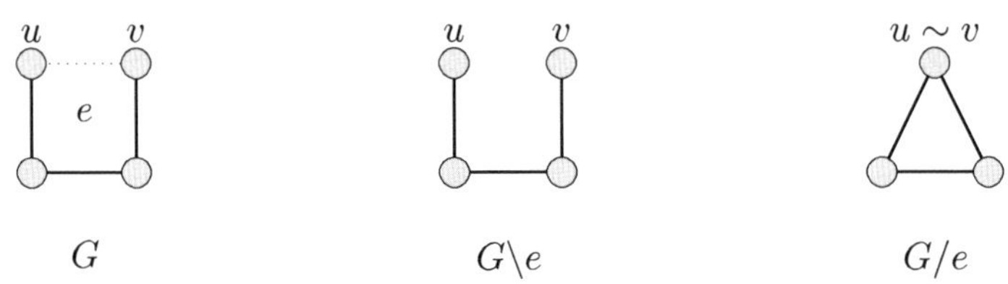

Fig. 2. Deletion-contraction along e, see [Koh18, Fig 2.3].

On the level of chromatic polynomials, we have that

$$\chi_{G\backslash e}(k) = \chi_G(k) + \chi_{G/e}(k).$$

One way to see this is noticing that there are two types of proper colorings of $G\backslash e$. If $e = \{u, v\}$, then either u and v are colored differently or u and v have the same color. In the first case, these correspond to proper colorings of G, whereas in the latter case the colorings are in bijection with colorings of G/e. To prove Whitney's theorem, one can use deletion-contraction and induct on the number of edges.

Inspired by a question of Read [Rea68], a lot of effort has gone into classifying chromatic polynomials. We list a few of the known properties:

- The coefficients of χ_G alternate in sign and they form a log-concave sequence, see [Huh12, Cor 27].
- The second highest coefficient up to a minus sign equals the number of edges.

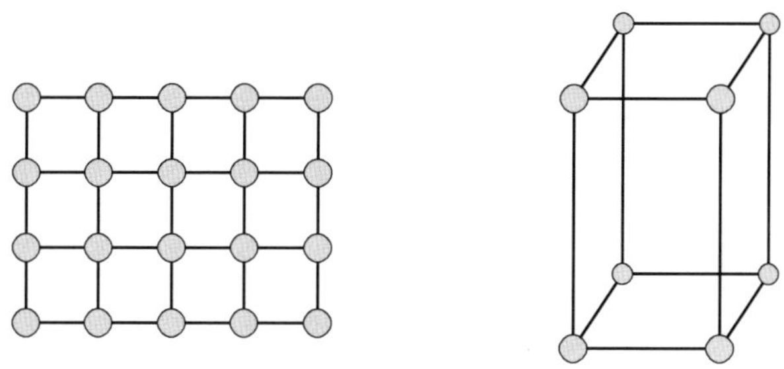

Fig. 3. $P_4 \times P_5$ and $C_4 \times P_2$.

- The chromatic polynomial of a graph with several connected components is the product of the chromatic polynomials of the connected components.
- The lowest degree of the monomials in χ_G is the number of connected components.

The clear highlight of this list is Huh's result about the log-concavity of the coefficients. Since log-concavity implies unimodality, we get that the absolute values of the coefficients form a unimodal sequence.

Remark 2.1. From now on we will always assume that G is connected, since by the previous properties we can just multiply the chromatic polynomials of the connected components to obtain the chromatic polynomial of the entire graph.

We will be interested in the chromatic polynomials of a special family of graphs. This family is given by the Cartesian product of a G with either a path graph P_n or a cycle graph C_n. The *path graph* P_n is the graph on the vertex set $[n]$ with edges $\{i, i+1\}$ for $i \in [n]$. Furthermore, the *cycle graph* C_n is the graph with vertex set $[n]$ and with edge $\{\{i, i+1\} : i \in [n-1]\} \cup \{1, n\}$.

Definition 2.1. Let $G_1 = (V(G_1), E(G_1))$ and $G_2 = (V(G_2), E(G_2))$ be two graphs. The *Cartesian product* $G_1 \times G_2$ (sometimes in the literature also denoted $G_1 \square G_2$) is the graph with vertex set $V(G_1) \times V(G_2)$, and the vertices (u_1, v_1) and (u_2, v_2) are connected by an edge if

- either $u_1 = u_2$ and $\{v_1, v_2\} \in E(G_2)$,
- or if $v_1 = v_2$ and $\{u_1, u_2\} \in E(G_1)$.

This definition is illustrated in Figure 3.

One of the most well-studied generalizations of the chromatic polynomial is the *Tutte polynomial.* The Tutte polynomial is a bivariate polynomial satisfying a deletion-contraction rule, see Theorem 2.2. It was introduced by Tutte [Tut54] and there are several cryptomorphic ways of defining it. For our purposes, the most convenient way is to define it via a deletion-contraction formula. Before we state the definition, we need two more terms. A *loop* is an edge that is incident to only one vertex. A *bridge* is an edge whose removal increases the number of connected components.

Definition 2.2. Let G be a finite graph. The *Tutte polynomial* $T(G;x,y)$ is the polynomial defined by the recursion

$$T(G;x,y) = T(G \setminus e;x,y) + T(G/e;x,y),$$

if e is neither a loop nor a bridge, and if G only consists of i bridges and j loops, then

$$T(G;x,y) = x^i y^j.$$

Remark 2.2. The reason we mention loops here is that starting from a simple graph the deletion-contraction process can create multiple edges or loops. With a view toward the chromatic polynomial, we remark that multiple edges don't change the proper colorings, but if a graph has a loop, there are simply no proper colorings.

Remark 2.3. The Tutte polynomial doesn't only generalize the chromatic polynomial, it encodes several more interesting graph properties and polynomials. We only list a few:

(1) The chromatic polynomial can be recovered by computing

$$\chi_G(k) = (-1)^{\#V - r(G)} k^{r(G)} T(G; 1-k, 0).$$

(2) The *Jones polynomial* can be recovered by evaluating the Tutte polynomial along $xy = 1$.
(3) The *flow polynomial* is obtained by specializing the Tutte polynomial at $x = 0$.
(4) $T(G;2,1)$ counts the number of acyclic edge subsets.
(5) $T(G;1,1)$ counts the number of spanning forests.
(6) $T(G;2,0)$ counts the number of acyclic orientations of edges.

The problem with this definition is that it is a priori not clear that it is independent of the choice and the order of the deleted edges. We will take this fact for granted. Let us illustrate this definition:

Example 2.1. Let $G = C_4$ and let e be any edge. Then

$$T(G;x,y) = T(G/e;x,y) + T(G \setminus e;x,y) = T(C_3;x,y) + T(P_4;x,y).$$

All edges of P_4 are bridges and thus $T(P_4;x,y) = x^3$. It remains to compute $T(C_3;x,y)$. Let's pick any other edge e'. Then

$$T(C_3;x,y) = T(C_3/e';x,y) + T(C_3 \setminus e';x,y) = T(\tilde{P}_2;x,y) + x^2,$$

where $\tilde{P}_2$ denotes the path graph on 2 nodes with a double edge, and where $C_3 \setminus e' \cong P_3$. Using deletion-contraction one more time gives us

$$T(\tilde{P}_2;x,y) = T(P_2;x,y) + T(\tilde{P}_1;x,y),$$

where $\tilde{P}_1$ is the path graph on 1 vertex with a loop. Putting everything together, we obtain

$$T(G;x,y) = x^3 + x^2 + x + y.$$

The following is known as the *recipe theorem* and it essentially says that the Tutte polynomial is *the* universal deletion-contraction invariant graph polynomial.

Theorem 2.2 (see e.g. [BEMPS10, Thm 3.1]). *Let $f\colon \mathcal{G} \to R$ be a function on graphs, where $\mathcal{G}$ is a minor closed class of graphs and let R be a commutative ring. If there are $a, b \in R$ with $ab \neq 0$ such that*

(1) $f(G) = 1$ if G consists of only one vertex and no edges,
(2) $f(G) = af(G \setminus e) + bf(G/e)$ whenever e is neither a loop nor a bridge,
(3) $f(GH) = f(G)f(H)$ where either GH is the disjoint union of G and H or G and H share at most one vertex,

then f is an evaluation of the Tutte polynomial of the form

$$f(G) = a^{\#E-(\#V-r(G))} b^{\#V-r(G)} T(G; x_0/a, y_0/b),$$

where $r(G)$ is the number of connected components of G.

As we will see in Section 3, Tutte polynomials make a surprising appearance in statistical mechanics as they determine the partition function of the q-state Potts model. In the zero-temperature case of the anti-ferromagnetic q-state Potts model, this reduces to the chromatic polynomial. We will be mostly be concerned with this case.

In physics, graphs are used to represent molecular structures. Since these structures tend to be extremely large, it makes sense to consider growing families of graphs. Motivated by the connection to the q-state Potts model, we now pose the following problem:

Problem 2.1. *Let G be a simple graph. How many proper k-colorings of $G \times P_n$ ($G \times C_n$) are there, where both k and n are variables?*

There are two special cases of Problem 2.1. If one fixes n and hence the size of the graph, Problem 2.1 reduces to the classical computation of the chromatic polynomial. If however one fixes the number of colors k, one can use a transfer-matrix method to compute $\chi_{G\times P_n}$. This can be done by constructing a graph M_G such that walks of length n on this graph correspond to the proper k-colorings of $G \times P_n$. This graph is explicitly constructed in [EK18, Sec 4.1]. The basic combinatorial idea behind this is that the adjacency matrix A (*or transfer matrix*) of an undirected graph can be used to count the number of walks of length n. Unfortunately, the size of M_G and thus the size of A depends on k, so we need to compactify A while keeping the biggest eigenvalue the same, see Remark 2.4.

The following result forms the basis for transfer-matrix methods:

Corollary 2.1 (Special case of [Sta12, Thm 4.7.1]). *Let G be an undirected, simple graph and let A be the adjacency matrix of G. Then $(A^n)_{ij}$ counts the number of walks of length n from v_i to v_j.*

Proper colorings of $G \times C_n$ correspond to *closed* walks on the graph M_G, and these walks are counted by the entries on the diagonal of A^n.

Remark 2.4 ([Koh18, Rem 2.4.7]). The number of *closed* walks is counted by trace of A^n. Since the trace is the sum of the eigenvalues, and since the (absolute value-wise) biggest eigenvalue $\lambda_{\max}$ is positive by the Perron–Frobenius theorem [Mey00, Ch 8], the trace is asymptotically dominated by $\lambda_{\max}^n$. where we implicitly use that G is connected.

We want to illustrate the previous result in a simple example:

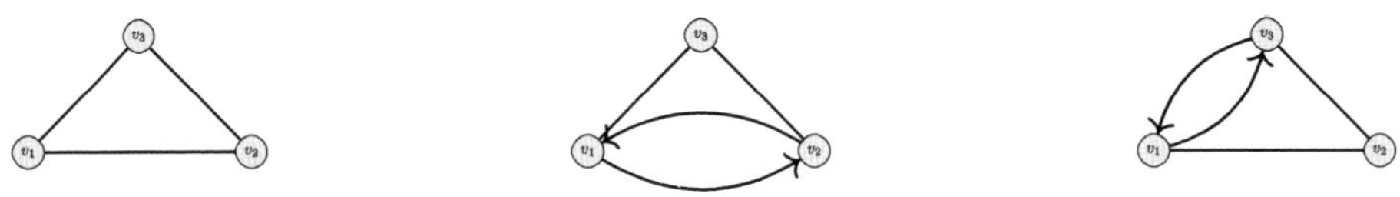

Fig. 4. $G = C_3$ and the two walks of length 2 from v_1 to v_1, see [Koh18, Fig 2.5].

Example 2.2 (see also [Koh18, Ex 2.4.8]). Let $G = C_3$, see Figure 4. The adjacency matrix of G is given by

$$A = \begin{pmatrix} 0 & 1 & 1 \\ 1 & 0 & 1 \\ 1 & 1 & 0 \end{pmatrix}$$

and

$$A^2 = \begin{pmatrix} 2 & 1 & 1 \\ 1 & 2 & 1 \\ 1 & 1 & 2 \end{pmatrix}.$$

Now we see that $(A^2)_{12} = a_{11}a_{11} + a_{12}a_{21} + a_{13}a_{31} = 2$ is the number of walks of length 2 from vertex v_1 back to itself, as predicted by Corollary 2.1. These walks are shown in Figure 4. We specifically wrote out matrix multiplication to illustrate that $a_{1i}a_{i1}$ counts the number of walks from v_1 to v_1 via the node v_i. This product is 1 if and only if all factors are 1 and it is 0 otherwise. The sum is over all possible intermediate nodes. The eigenvalues of A are $-1, -1$, and 2. So the trace of A^n is going to be dominated by 2^n. In particular, for $n = 5$ we have

$$A^5 = \begin{pmatrix} 10 & 11 & 11 \\ 11 & 10 & 11 \\ 11 & 11 & 10 \end{pmatrix}$$

and thus $\text{trace}(A^5) = 30$ and $\lambda_{\max}^5 = 32$.

2.2. *Lattice Polytopes and Ehrhart Theory*

In [BZ06], Beck and Zaslavsky showed that proper graph colorings actually have a beautiful geometric interpretation. This geometric interpretation is given in terms of integer points in lattice polytopes. This section recalls the basic notions about lattice polytopes and Ehrhart theory. We then reinterpret proper graph colorings in this language.

A *polytope* $P \subset \mathbb{R}^d$ is the convex hull of finitely many points $x_1, \ldots, x_s \in \mathbb{R}^d$, i.e.,

$$P = \operatorname{conv}\{x_1, \ldots, x_s\} := \left\{ \sum_{i=1}^{s} \lambda_i x_i \colon \lambda_i \geq 0, \sum_{i=1}^{s} \lambda_i = 1 \right\}.$$

The inclusion minimal set $\operatorname{Vert}(P) = \{v_1, \ldots, v_r\}$ such that $P = \operatorname{conv}\{v_1, \ldots, v_r\}$ is called the *vertex set of* P. If $\operatorname{Vert}(P) \subset \mathbb{Z}^d$, we say that P is a *lattice polytope.* The *dimension* $\dim P$ of a polytope P is defined as the dimension of its affine span. If $d = \dim P$, we say that P is a *d-polytope.* We will soon see that the chromatic polynomial counts integer points in dilates of polytopes. Hence, we should be interested in the number of lattice points in dilates of polytopes. This leads us directly to the notion of Ehrhart polynomials. Let P be a lattice polytope. The *Ehrhart polynomial of P* $i_P \colon \mathbb{Z}_{\geq 1} \to \mathbb{Z}_{\geq 1}$ is defined as

$$i_P(k) := \# kP \cap \mathbb{Z}^d,$$

where $kP = \{kx \colon x \in P\}$ is the kth dilate of P. As the name suggests, this function actually agrees with a polynomial.

Theorem 2.3 ([Ehr62]). *Let P be a lattice d-polytope. Then i_P agrees with a polynomial of degree d of leading coefficient* vol$P$ *and with constant coefficient* 1.

The study of integer points in polytopes is called *Ehrhart theory.* The book [BR15] offers a beautiful introduction to this active field of research. The Ehrhart polynomial extends the domain of the Ehrhart function to $\mathbb{C}$. It is natural to ask whether other evaluations have a geometric interpretation. This is indeed the case:

Theorem 2.4 (Ehrhart-Macdonald reciprocity). *Let P be a d-dimensional lattice polytope. Then*

$$i_P(-t) = (-1)^d i_{P^\circ}(t), \tag{1}$$

where $i_{P^\circ}(t)$ counts the number of integer points in the interior of tP.

This result was first conjectured by Ehrhart, who managed to prove it in several special cases. Macdonald later proved the general statement, see [Mac71, Prop 4.1].

We will now follow [BZ06] and outline how one can use Ehrhart polynomials of lattice polytopes to determine the chromatic polynomial of a given graph. Let

$G = (V, E)$ be a simple graph on n nodes. The edge set E gives rise to a set of hyperplanes $\mathcal{H}_G$ defined by

$$\mathcal{H}_G = \{x_i = x_j \colon \{i, j\} \in E\}.$$

A k-coloring c of G can be thought of as an integer point in $c \in [1, k]^n$. The coloring c is proper if and only if c does not lie on any of the hyperplanes in $\mathcal{H}_G$. Thus, c is a proper coloring of G if and only if it lies in the *interior* of $[0, k+1]^n$ and not on any hyperplane in $\mathcal{H}_G$. Let $P := [0, 1]^n$. The hyperplanes in $\mathcal{H}_G$ dissect P into closed regions $R_1, \ldots, R_m$. Every proper k-coloring c corresponds to an interior integer point in one of the dilated regions $(k+1)R_i$, and vice versa. The pair of $(P, \mathcal{H}_G)$ is called an *inside-out polytope.* This is illustrated in Figure 5.

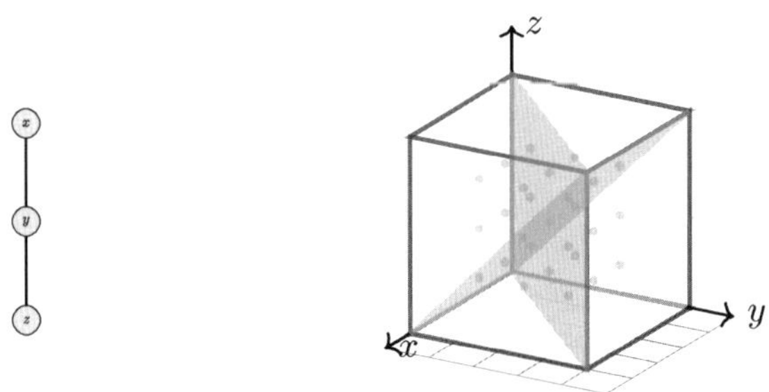

Fig. 5. P_3 and the corresponding inside-out polytope $(P, \mathcal{H})$.

Setting

$$i^{\circ}_{(P^{\circ}, \mathcal{H}_G)}(k) = \sum_{i=1}^{m} i_{R_i^{\circ}}(k),$$

where $i_{R_i^{\circ}}(k)$ is the Ehrhart polynomial of the open regions R_i, we get the following result.

Theorem 2.5 ([BZ06, Thm 5.1]). *Let G be an simple graph on n vertices and let $P = [0, 1]^n$. Moreover, we define*

$$\mathcal{H}(G) := \{x_i = x_j \colon \{x_i, x_j\} \in E\}.$$

Then

$$i^{\circ}_{(P^{\circ}, \mathcal{H}_G)}(t) = \chi_G(t-1). \tag{2}$$

This result directly implies the following properties of the chromatic polynomial:

- The chromatic function χ_G is a sum of the polynomials $i_{R_i^{\circ}}$ and it is thus a polynomial.
- The leading coefficient is 1, since the sum of the volumes of the R_i's equals the volume of the cube $[0, 1]^n$.

- The coefficients are alternating in sign, which can be seen by doing inclusion-exclusion on the closed regions $R_1, R_2, \ldots, R_m$.
- The second highest coefficient equals $-m$, which follows from the inclusion-exclusion argument since the only codimension 1 pieces come from the hyperplanes $x_i = x_j$ for $\{i,j\} \in E$.

3. Graph Theory Meets Statistical Mechanics

In this section, we introduce the q-state Potts model. This model shows how a micro-scale interactions between nearest neighbors determine the macro-scale behavior of the entire system. This section is entirely based on the excellent survey [BEMPS10]. The goal of this section is to very briefly describe why physicists are interested in the Tutte polynomial and why Problem 2.1 arises naturally in this context. None of the physics are relevant for the later sections.

Let $G = (V, E)$ be a graph and let S be a set of cardinality q. We call the elements of S *spins*. A *state of a graph* G is a map $\sigma \colon V \to S$. If $i \in V$, we will often write σ_i instead of $\sigma(i)$. In our previous language, a state ω is simply a q-coloring. Moreover, let us assume that every edge $\{i,j\} \in E$ has the same constant weight J. We are interested in the energy of a given state. In physics, this energy can be measured using the *Hamiltonian*.

Definition 3.1 ([BEMPS10, Def 2.1]). Two common formulations of the Hamiltonian are:

$$h_1(\omega) = -J \sum_{\{i,j\}\in E} \delta(\sigma_i, \sigma_j) \text{ and } h_2(\omega) = J \sum_{\{i,j\}\in E} (1 - \delta(\sigma_i, \sigma_j)),$$

where ω is a state of a graph G and δ is the Kronecker delta function.

Since (up to a factor) h_1 counts the number of edges $\{i,j\}$ with the same spins on i and j and since h_2 counts the number of edges $\{i,j\}$ with different spins on i and j, we immediately deduce

$$h_2(\omega) = J \cdot \#E + h_1(\omega). \tag{3}$$

If the weight J is positive, we call the model *ferromagnetic* and if J is negative, we say that the model is *anti-ferromagnetic*. A lot of the thermodynamic functions like the internal energy, the entropy, the (reduced) free energy, and the specific heat can be derived from what is called the *partition function*.

Definition 3.2 ([BEMPS10, Def 2.2]). Let G be a graph and let S be a set of q spins. Then for $i = 1, 2$ the q-state Potts model partition function is defined as

$$Z_i(G) = \sum_{\omega} \exp(-\beta(h_i(\omega))),$$

where $\beta = 1/(\kappa T)$, T is the temperature, and κ is the *Boltzmann constant.*

This turns Z_i into a bivariate function, where the two indeterminates are given by the number of states q and the temperature T. As a direct consequence of (3), we deduce that

$$Z_2(G; q, \beta) = \exp(-K \cdot \#E) \cdot Z_1(G; q, \beta),$$

where $K = \beta J$. This shows that both partition functions contain the same information. The reason we are interested in the partition function of the Potts model is its surprising connection to the Tutte polynomial of G.

Theorem 3.1 ([BEMPS10, Thm 3.2]). *Let G be a graph and let S be a set of q spins. Then*

$$Z_1(G; q, \beta) = q^{r(G)} v^{\#V - r(G)} T(G; (q+v)/v, v+1),$$

where $v = e^K - 1$, and where $r(G)$ is the number of connected components of G.

One way to prove this relation is showing that Z_1 satisfies a deletion-contraction recurrence. From (3), we immediately derive that (up to a prefactor) Z_2 also is an evaluation of the Tutte polynomial, see [BEMPS10, Cor 3.3].

As it turns out, phase transitions are manifested in the nonanalyticity of the reduced free energy. For a fixed graph G, the reduced free energy per vertex is given by

$$f(G; q, \beta) = \frac{\ln(Z_1(G; q, \beta))}{\#V}.$$

If G is a fixed graph, this function is a real analytic function in the variables q and β (or T). However, if one has an infinite family of graphs such as $G \times P_n$, this might not be the case, i.e.

$$f(G \times P_n; q, \beta) = \lim_{n \to \infty} \frac{1}{\#V(G \times P_n)} \ln(Z_1(G \times P_n; q, \beta))$$

might have points of nonanalyticity. Phase transitions are now given by the points of nonanalyticity of this limit of the reduced free energy. This shows that the structure of the Tutte polynomial of $G \times P_n$ is of special interest. For details, we refer the reader to [BEMPS10, Sec 4].

So far, we have only outlined why the Tutte polynomial is a relevant invariant of a graph G. However, in an important special case, understanding the chromatic polynomial becomes essential. This happens in the extremal case of zero-temperature in the anti-ferromagnetic Potts model. There, the partition function becomes the chromatic polynomial, where we use the h_1 Hamiltonian. For details about this connection, we refer the interested reader to [BEMPS10, Sec 5]. As we have seen before, points of nonanalyticity correspond to phase transitions. So in the zero-temperature anti-ferromagnetic model, this reduces to examining the non-analyticity of the chromatic polynomial. Hence, a lot of effort has gone in finding regions of the complex plane where the chromatic polynomial is non-zero. As a forward pointer, we would like to mention that one might be able to use the structure of the chromatic polynomial described in Theorem 2.5 to find regions of analyticity.

4. Transfer-Matrix Methods Meet Ehrhart Theory

In this section, we introduce the methods described in [EK18] to determine the chromatic polynomial of graphs of the form $G \times P_n$, where both the size of the path graph n and the number of colors k are treated as variables. Moreover, we will describe the doubly asymptotic behavior of graphs of the form $G \times C_n$. As the name of [EK18] suggests, this can be achieved by combining transfer-matrix methods with Ehrhart theory. We will set up a compactified transfer matrix L, whose size is independent of the number of colors k. The main ingredients will be two types of symmetry, namely permutation of the colors and symmetries of the underlying graph G. In order to properly distinguish between the two actions, we introduce some notation:

Definition 4.1 ([EK18, Def 5.1]). Let G be a simple graph on N vertices and let $\mathcal{C}$ be the set of proper k-colorings, where $k \geq N$. Let $\mathfrak{S}_k$ be the symmetric group on k elements and let $\mathcal{G}$ be a possibly trivial subgroup of the automorphism group of G. The group $\mathfrak{S}_k$ acts on $\mathcal{C}$ by permuting the colors and it gives rise to orbits $\tilde{o}_1, \dots, \tilde{o}_q$. The group $\mathcal{G}$ is acting on $\tilde{o}_1, \dots, \tilde{o}_q$ giving rise to orbits $o_1, \dots, o_p$.

We want to emphasize that it can happen that $o_i = \tilde{o}_i$ for some i. However, the numbers of orbits normally differ quite a lot. This is illustrated in the following example.

Example 4.1 ([EK18, Ex 5.2]). Let $G = C_5$. We first quotient the proper colorings by permutations of colors. This group action induces orbits $\tilde{o}_1$, $\tilde{o}_2$, $\dots$, $\tilde{o}_{11}$ (as computer generated in no particular order):

14 2 3 5	14 2 35	14 25 3	1 24 3 5	1 24 35	1 2 3 4 5
1 2 35 4	1 25 3 4	13 24 5	13 2 4 5	13 25 4	

Here 14 2 3 5 means that vertex 1 and vertex 4 have the same color, but they have a different color than vertices 2, 3, and 5. Similarly, vertex 2 is colored differently than all other vertices, etc.

The automorphism group of C_5 is the dihedral group generated by (12345) and (1)(25)(34). The 11 partitions of the vertex set end up in 3 orbits o_1, o_2, and o_3 after quotienting by the dihedral group. The classes are represented by:

1 2 3 4 5 | 1 24 35 | 1 2 4 35.

In [Ciu98], Ciucu used symmetry to define a compactified transfer matrix while working on the 3-dimensional dimer problem. The following definition is inspired by his approach:

Definition 4.2 ([EK18, Rem 5.9]). Let G be a graph on N nodes. Let o_1, o_2, $\dots$, o_p be orbits as defined in Definition 4.1 and let $k \geq N$ be any integer. We define a $p \times p$ matrix L whose entries are given by

$L_{i,j} = $ #prop. k-col. where $G \times \{1\}$ is col. by fixed rep of o_i and $G \times \{2\}$ is in o_j.

L is called the *compactified transfer matrix of* G.

The entries of L count *restricted*, proper colorings of a graph, i.e., proper colorings of graphs where as subset of the nodes is already colored properly. As it turns out, these entries are polynomial in the number of colors:

Theorem 4.1 (see [EK18, Thm 5.14]). *With the notation from above, we have that $L_{i,j}$ is a polynomial in k of degree equals to the number of colors in o_j and it is independent of the choice of the representative of o_i.*

Remark 4.1. The statement that the entries are polynomials also follows from [HM07, Thm 1]. Jochemko and Sanyal give a geometric proof of the polynomiality result, see [JS14, Thm 4.2].

This result enables us to compute the transfer matrix L using explicit computer calculations for small k.

Example 4.2 (see [Koh18, Ex 3.4.15], [EK18, Ex 5.15]). Let $G = P_3$ with orbits $o_1 = \{\{1,3\},\{2\}\}$ and $o_2 = \{\{1\},\{2\},\{3\}\}$. Then from explicit computations for $k = 3$, 4, 5, 6 we infer that

$$L = \begin{pmatrix} k^2 - 3k + 3 & k^3 - 6k^2 + 13k - 10 \\ k^2 - 4k + 5 & k^3 - 6k^2 + 14k - 13 \end{pmatrix}.$$

Let us illustrate the combinatorial interpretation for L_{11} and L_{12} in the case where $k = 3$, see Figure 6.

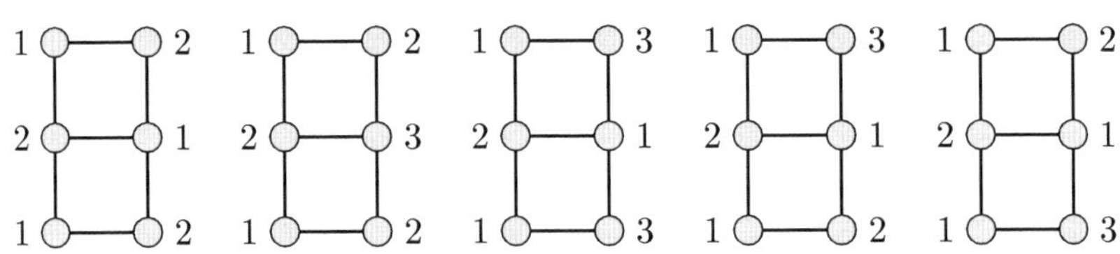

Fig. 6. An illustration of the proper 3-colorings counted by the first row of L, originally in [Koh18, Fig 3.7].

The reason we are interested in this matrix is the following result:

Theorem 4.2 ([EK18, Cor 5.20]). *Let $G \times P_{n+1}$ and L be as above. Then*

$$\chi_{G\times P_{n+1}}(k) = (w_1(k), \ldots, w_p(k))L^n\mathbf{1}, \tag{4}$$

where $w_i(k)$ is the size of o_i and $\mathbf{1} := (1, \ldots, 1)^t$.

Remark 4.2. Even though we have defined L only for $k \geq N$, where $N := \#V$, this result makes sense for *all* k. If there are not enough colors to get orbit o_i, then the weight w_i will be 0.

This shows that the difficulty of solving Problem 2.1 really lies in determining the orbits and determining the matrix L. A fact the we omitted is that L actually comes from the adjacency matrix A_{M_G} of a graph M_G. The size of the graph M_G depends on the number of colors k, but after quotienting by symmetries this can be reduced to a matrix of fixed size. It's important to note that the biggest eigenvalue of L agrees with the biggest eigenvalue of M_G, see [Ciu98, Lem 3.2]. For details of this construction, we refer to [EK18, Sec 4.1]. This suffices to determine the asymptotic behavior of the number of proper k-colorings of $G \times C_n$.

Proposition 4.1 ([EK18, Prop 5.31]). *Let G be a graph on N vertices, let $\delta(L)$ and $\Delta(L)$ be smallest and biggest row sum of L, respectively. Then the doubly asymptotic behavior of the number of proper k-colorings of $G \times C_n$ is dominated by $\lambda_{\max}^{n-1}$ and*

$$\delta(L) \leq \lambda_{\max} \leq \Delta(L),$$

where $\delta(L) = \sum_{i=0}^{N} a_i k^i$ is the smallest and $\Delta(L) = \sum_{i=0}^{N} b_i k^i$ is the biggest row sum of L, respectively. Furthermore, $a_N = b_N$, and $a_{N-1} = b_{N-1}$.

By Theorem 2.5 we know that once the matrix L is known, we can determine the chromatic polynomial of $G \times P_n$ and give asymptotics of the number of proper colorings of $G \times C_n$. Determining L is essentially a three-step process, which we will illustrate in Section 5:

(1) Determine as many graph automorphisms as possible.
(2) Determine the orbits $o_1, \ldots, o_p$.
(3) Compute the entries $L_{i,j} = L_{o_i,o_j}$.

We will now give an interpretation of the entries $L_{i,j}$ in terms of *induced inside-out polytopes*. We will give this description in the case where we do *not* consider graph automorphisms. In spirit of Definition 4.1, this gives orbits $\tilde{o}_i$ which are then mapped to orbits o_k. This constructs a compactified matrix $\tilde{L}$ from which we can derive the compactified transfer matrix L by fixing a row $\tilde{o}_i \in o_i$ and adding the entries $\tilde{L}_{ik}$ for $\tilde{o}_k \in o_j$. To keep the notation as simple as possible, we call $\tilde{L} = L$ and we will write o instead of $\tilde{o}$.

In Section 2.2, we have already seen that there is a geometric interpretation of proper graph colorings. If G is a graph on N vertices, we intersected the cube $P = [0,1]^N$ with hyperplanes coming from the edge set E. Let us determine the entry L_{ij}. According to Definition 4.2, we need to create a partially colored graph, where the partial coloring of $G \times \{1\}$ is given by orbit o_i. We need to extend it to a coloring of $G \times P_2$ such that the coloring of $G \times \{2\}$ lies in orbit o_j. This relates to the inside-out polytope defined in Section 2.2 in the following way. If G has N vertices, then $G \times P_2$ has $2N$ vertices. Let $P = [0,1]^{2N}$ and let

$$\mathcal{H}_{G \times P_2} = \{x_i = x_j \colon \{i,j\} \in E\,(G \times P_2)\}.$$

be the hyperplane arrangement coming from the edges of $G \times P_2$. Geometrically, coloring $G \times \{1\}$ means we intersect $(P, \mathcal{H}_{G\times P_2})$ with the hyperplanes $x_1 = c(1)$, $x_2 = c(2)$, ..., $x_N = c(N)$, where the vertices of $G \times \{1\}$ correspond to the coordinates x_1, ..., x_N. One important difference to the classical case is that this induces *forbidden* hyperplanes of the form $x_i =$ const. This means that even though the full-dimensional regions are lattice polytopes, their dilation factor is different, see Figure 7. Furthermore, we need to make sure that the coloring of $G \times \{2\}$ actually lies in o_j. If we do not take graph automorphisms into account, an orbit o_j can be thought of as a partition of the vertex set into independent sets. A set of vertices is called *independent*, if no two vertices in this set are adjacent. If two vertices k, l are in the same independent set of o_j, this forces the condition $x_k = x_l$, i.e., we additionally intersect our induced lattice polytope with these hyperplanes. If k and l are in different independent sets of o_j, we get the *forbidden hyperplanes* of the form $x_k = x_l$, since the vertices l and k have to be colored by different colors. All of these conditions together induce a lower-dimensional inside-out polytope, whose closed regions are all lattice polytopes.

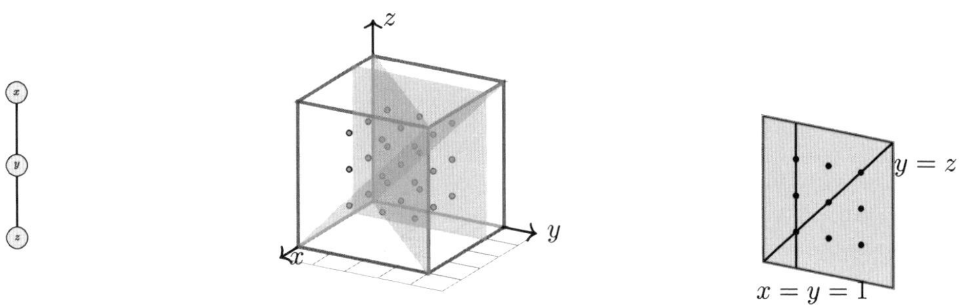

Fig. 7. P_3 with vertex x colored by 1, the corresponding induced inside-out polytope $(P, \mathcal{H})$, and the *induced* inside-out polytope $(\overline{P}, \overline{\mathcal{H}})$.

5. Lattice Polytopes Meet Statistical Mechanics

In this section, we will explicitly determine chromatic polynomials for graphs of the form $G \times P_n$ and give asymptotic results about graphs of the form $G \times C_n$. It is important to note that one should always use as much symmetry as possible, even when only some graph automorphisms are known.

Example 5.1 (Example 4.2 continued, [EK18, Ex 5.22]). Let again $G = P_3$. Recall that

$$L = \begin{pmatrix} k^2 - 3k + 3 & k^3 - 6k^2 + 13k - 10 \\ k^2 - 4k + 5 & k^3 - 6k^2 + 14k - 13 \end{pmatrix}.$$

Theorem 2.5 states that the chromatic polynomial of $P_3 \times P_n$ is

$$\chi(k) = (w_1, w_2) \begin{pmatrix} k^2 - 3k + 3 & k^3 - 6k^2 + 13k - 10 \\ k^2 - 4k + 5 & k^3 - 6k^2 + 14k - 13 \end{pmatrix}^{n-1} \begin{pmatrix} 1 \\ 1 \end{pmatrix},$$

where $w_1 = k(k-1)$ and $w_2 = k(k-1)(k-2)$ are the sizes of the orbit o_1 and o_2, respectively. Proposition 4.1 implies that the biggest eigenvalue $\lambda_{\max}$ of L can now be bounded by

$$k^3 - 5k^2 + 10k - 8 \le \lambda_{\max} \le k^3 - 5k^2 + 10k - 7.$$

This example is too small to witness the difference between o and $\tilde{o}$, since there was no graph symmetry identifying orbits $\tilde{o}$ and $\widetilde{o'}$. However, the next example shows that one should always use as much symmetry as possible.

Example 5.2 (Example 4.1 continued [EK18, Ex 5.16]). Let $G = C_5$ be the graph for which we want to calculate the chromatic polynomial of $G \times P_n$. We label the edges of C_5 by $12, 23, 34, 45$ and 51. We first quotient by permutations of colors. The 11 partitions of the vertices into independent sets are (as computer generated in no particular order):

14 2 3 5	14 2 35	14 25 3	1 24 3 5	1 24 35	1 2 3 4 5
1 2 35 4	1 25 3 4	13 24 5	13 2 4 5	13 25 4	

Since every partition corresponds to an orbit $\tilde{o}$ as defined in Definition 4.1, we get the 11×11-matrix in Figure 8.

$$\begin{vmatrix}
k^4-10k^3+41k^2-84k+73 & k^3-7k^2+19k-20 & k^3-7k^2+19k-20 & k^4-11k^3+49k^2-106k+94 & k^3-8k^2+24k-26 & k^5-15k^4+94k^3-313k^2+559k-428 & k^4-11k^3+49k^2-105k+90 & k^4-11k^3+49k^2-105k+90 & k^3-8k^2+24k-27 & k^4-11k^3+49k^2-106k+94 & k^3-8k^2+24k-26 \\
k^4-10k^3+40k^2-77k+60 & k^3-6k^2+14k-13 & k^3-7k^2+18k-17 & k^4-11k^3+48k^2-98k+78 & k^3-7k^2+18k-17 & k^5-15k^4+93k^3-301k^2+510k-360 & k^4-10k^3+40k^2-77k+60 & k^4-11k^3+48k^2-98k+78 & k^3-8k^2+23k-23 & k^4-11k^3+48k^2-99k+81 & k^3-8k^2+23k-23 \\
k^4-10k^3+40k^2-77k+60 & k^3-7k^2+18k-17 & k^3-6k^2+14k-13 & k^4-11k^3+48k^2-99k+81 & k^3-8k^2+23k-23 & k^5-15k^4+93k^3-301k^2+510k-360 & k^4-11k^3+48k^2-98k+78 & k^4-10k^3+40k^2-77k+60 & k^3-8k^2+23k-23 & k^4-11k^3+48k^2-98k+78 & k^3-7k^2+18k-17 \\
k^4-11k^3+49k^2-106k+94 & k^3-8k^2+24k-26 & k^3-8k^2+24k-27 & k^4-10k^3+41k^2-84k+73 & k^3-7k^2+19k-20 & k^5-15k^4+94k^3-313k^2+559k-428 & k^4-11k^3+49k^2-105k+90 & k^4-11k^3+49k^2-106k+94 & k^3-7k^2+19k-20 & k^4-11k^3+49k^2-105k+90 & k^3-8k^2+24k-26 \\
k^4-11k^3+48k^2-98k+78 & k^3-7k^2+18k-17 & k^3-8k^2+23k-23 & k^4-10k^3+40k^2-77k+60 & k^3-6k^2+14k-13 & k^5-15k^4+93k^3-301k^2+510k-360 & k^4-10k^3+40k^2-77k+60 & k^4-11k^3+48k^2-99k+81 & k^3-7k^2+18k-17 & k^4-11k^3+48k^2-98k+78 & k^3-8k^2+23k-23 \\
k^4-11k^3+50k^2-113k+107 & k^3-8k^2+25k-30 & k^3-8k^2+25k-30 & k^4-11k^3+50k^2-113k+107 & k^3-8k^2+25k-30 & k^5-15k^4+95k^3-325k^2+609k-501 & k^4-11k^3+50k^2-113k+107 & k^4-11k^3+50k^2-113k+107 & k^3-8k^2+25k-30 & k^4-11k^3+50k^2-113k+107 & k^3-8k^2+25k-30 \\
k^4-11k^3+49k^2-105k+90 & k^3-7k^2+19k-20 & k^3-8k^2+24k-26 & k^4-11k^3+49k^2-105k+90 & k^3-7k^2+19k-20 & k^5-15k^4+94k^3-313k^2+559k-428 & k^4-10k^3+41k^2-84k+73 & k^4-11k^3+49k^2-106k+94 & k^3-8k^2+24k-26 & k^4-11k^3+49k^2-106k+94 & k^3-8k^2+24k-27 \\
k^4-11k^3+49k^2-105k+90 & k^3-8k^2+24k-26 & k^3-7k^2+19k-20 & k^4-11k^3+49k^2-106k+94 & k^3-8k^2+24k-27 & k^5-15k^4+94k^3-313k^2+559k-428 & k^4-11k^3+49k^2-106k+94 & k^4-10k^3+41k^2-84k+73 & k^3-8k^2+24k-26 & k^4-11k^3+49k^2-105k+90 & k^3-7k^2+19k-20 \\
k^4-11k^3+48k^2-99k+81 & k^3-8k^2+23k-23 & k^3-8k^2+23k-23 & k^4-10k^3+40k^2-77k+60 & k^3-7k^2+18k-17 & k^5-15k^4+93k^3-301k^2+510k-360 & k^4-11k^3+48k^2-98k+78 & k^4-11k^3+48k^2-98k+78 & k^3-6k^2+14k-13 & k^4-10k^3+40k^2-77k+60 & k^3-7k^2+18k-17 \\
k^4-11k^3+49k^2-106k+94 & k^3-8k^2+24k-27 & k^3-8k^2+24k-26 & k^4-11k^3+49k^2-105k+90 & k^3-8k^2+24k-26 & k^5-15k^4+94k^3-313k^2+559k-428 & k^4-11k^3+49k^2-106k+94 & k^4-11k^3+49k^2-105k+90 & k^3-7k^2+19k-20 & k^4-10k^3+41k^2-84k+73 & k^3-7k^2+19k-20 \\
k^4-11k^3+48k^2-98k+78 & k^3-8k^2+23k-23 & k^3-7k^2+18k-17 & k^4-11k^3+48k^2-98k+78 & k^3-8k^2+23k-23 & k^5-15k^4+93k^3-301k^2+510k-360 & k^4-11k^3+48k^2-99k+81 & k^4-10k^3+40k^2-77k+60 & k^3-7k^2+18k-17 & k^4-10k^3+40k^2-77k+60 & k^3-6k^2+14k-13
\end{vmatrix}$$

Fig. 8. L matrix where we only quotient out by permutations of colors.

The automorphism group of C_5 is the dihedral group generated by (12345) and $(1)(25)(34)$. The 11 partitions of the vertex set end up in 3 orbits after quotienting by the dihedral group. The classes are represented by:

$$1\ 2\ 3\ 4\ 5 \,|\, 1\ 24\ 35 \,|\, 1\ 2\ 4\ 35$$

Adding up entries from columns — indexed by orbits $\tilde{o}_i$ that get mapped to the same orbit o — of the 11×11-matrix gives an even more compactified version, the 3×3 matrix L,

$$\begin{pmatrix} k^5 - 15k^4 + 95k^3 - 325k^2 + 609k - 501 & 5k^3 - 40k^2 + 125k - 150 & 5k^4 - 55k^3 + 250k^2 - 565k + 535 \\ k^5 - 15k^4 + 93k^3 - 301k^2 + 510k - 360 & 5k^3 - 36k^2 + 96k - 93 & 5k^4 - 53k^3 + 224k^2 - 449k + 357 \\ k^5 - 15k^4 + 94k^3 - 313k^2 + 559k - 428 & 5k^3 - 38k^2 + 110k - 119 & 5k^4 - 54k^3 + 237k^2 - 506k + 441 \end{pmatrix}.$$

Again, the biggest eigenvalue $\lambda_{\max}$ is bounded below by the smallest row sum and above by the biggest row sum.

References

[AH77a] K. Appel and W. Haken, *Every planar map is four colorable. I. Discharging*, Illinois J. Math. **21** (1977), no. 3, 429–490.

[AH77b] ———, *Supplement to: "Every planar map is four colorable. I. Discharging" (Illinois J. Math. 21 (1977), no. 3, 429–490) by Appel and Haken; "II. Reducibility" (ibid. 21 (1977), no. 3, 491–567) by Appel, Haken and J. Koch*, Illinois J. Math. **21** (1977), no. 3, 1–251. (microfiche supplement).

[AHK77] K. Appel, W. Haken, and J. Koch, *Every planar map is four colorable. II. Reducibility*, Illinois J. Math. **21** (1977), no. 3, 491–567.

[Bat94] Victor V. Batyrev, *Dual polyhedra and mirror symmetry for Calabi-Yau hypersurfaces in toric varieties*, J. Algebraic Geom. **3** (1994), no. 3, 493–535.

[BBK+15] Matthias Beck, Benjamin Braun, Matthias Köppe, Carla D. Savage, and Zafeirakis Zafeirakopoulos, *s-lecture hall partitions, self-reciprocal polynomials, and Gorenstein cones*, Ramanujan J. **36** (2015), no. 1-2, 123–147.

[BEMPS10] Laura Beaudin, Joanna Ellis-Monaghan, Greta Pangborn, and Robert Shrock, *A little statistical mechanics for the graph theorist*, Discrete Math. **310** (2010), no. 13-14, 2037–2053.

[BH93] Winfried Bruns and Jürgen Herzog, *Cohen-Macaulay rings*, Cambridge Studies in Advanced Mathematics, vol. 39, Cambridge University Press, Cambridge, 1993.

[BK14] Matthias Beck and Florian Kohl, *Rademacher-Carlitz polynomials*, Acta Arith. **163** (2014), no. 4, 379–393.

[BP03] Matthias Beck and Dennis Pixton, *The Ehrhart polynomial of the Birkhoff polytope*, Discrete Comput. Geom. **30** (2003), no. 4, 623–637.

[BR15] Matthias Beck and Sinai Robins, *Computing the continuous discretely*, second ed., Undergraduate Texts in Mathematics, Springer, New York, 2015.

[BZ06] Matthias Beck and Thomas Zaslavsky, *Inside-out polytopes*, Adv. Math. **205** (2006), no. 1, 134–162.

[Ciu98] Mihai Ciucu, *An improved upper bound for the 3-dimensional dimer problem*, Duke Math. J. **94** (1998), no. 1, 1–11.

[CLS11] David A. Cox, John B. Little, and Henry K. Schenck, *Toric varieties*, Graduate Studies in Mathematics, vol. 124, American Mathematical Society, Providence, RI, 2011.

[Cox15] David A. Cox, *Mirror symmetry and polar duality of polytopes*, Symmetry **7** (2015), no. 3, 1633–1645.

[Ehr62] Eugène Ehrhart, *Sur les polyèdres rationnels homothétiques à n dimensions*, C. R. Acad. Sci. Paris **254** (1962), 616–618.

[EK18] Alexander Engström and Florian Kohl, *Transfer-matrix methods meet Ehrhart theory*, Adv. Math. **330** (2018), 1–37.

[HM07] Agnes M. Herzberg and M. Ram Murty, *Sudoku squares and chromatic polynomials*, Notices Amer. Math. Soc. **54** (2007), no. 6, 708–717.

[Huh12] June Huh, *Milnor numbers of projective hypersurfaces and the chromatic polynomial of graphs*, J. Amer. Math. Soc. **25** (2012), no. 3, 907–927.

[JS14] Katharina Jochemko and Raman Sanyal, *Arithmetic of marked order polytopes, monotone triangle reciprocity, and partial colorings*, SIAM J. Discrete Math. **28** (2014), no. 3, 1540–1558.

[Koh18] Florian Kohl, *Lattice Polytopes — Applications and Properties*, PhD thesis, Freie Universität Berlin (2018), x + 127 pages.

[Mac71] Ian G. Macdonald, *Polynomials associated with finite cell-complexes*, J. London Math. Soc. (2) **4** (1971), 181–192.

[Mey00] Carl Meyer, *Matrix analysis and applied linear algebra*, Society for Industrial and Applied Mathematics (SIAM), Philadelphia, PA, 2000.

[Pom93] James E. Pommersheim, *Toric varieties, lattice points and Dedekind sums*, Math. Ann. **295** (1993), no. 1, 1–24.

[Rea68] Ronald C. Read, *An introduction to chromatic polynomials*, J. Combinatorial Theory **4** (1968), 52–71.

[Sta96] Richard P. Stanley, *Combinatorics and commutative algebra*, second ed., Progress in Mathematics, vol. 41, Birkhäuser Boston, Inc., Boston, MA, 1996.

[Sta12] ———, *Enumerative combinatorics. Volume 1*, second ed., Cambridge Studies in Advanced Mathematics, vol. 49, Cambridge University Press, Cambridge, 2012.

[Stu96] Bernd Sturmfels, *Gröbner bases and convex polytopes*, University Lecture Series, vol. 8, American Mathematical Society, Providence, RI, 1996.

[Tut54] W. T. Tutte, *A contribution to the theory of chromatic polynomials*, Canadian J. Math. **6** (1954), 80–91.

[Whi32] Hassler Whitney, *A logical expansion in mathematics*, Bull. Amer. Math. Soc. **38** (1932), no. 8, 572–579.

A brief survey about moment polytopes of subvarieties of products of Grassmanians

Laura Escobar

Department of Mathematics and Statistics, Washington University in St. Louis St. Louis, MO 63120, USA

E-mail: laurae@wustl.edu

The Grassmannian has a natural action by an algebraic torus. In this survey we describe the diagonal action of this torus on subvarieties of products of Grassmannians. From this action we describe how to construct an associated moment polytope. The main varieties considered are Schubert varieties, Richardson varieties, and their desingularizations. The moment polytopes of these varieties include the permutahedron and associahedron. We also look at the Barbasch-Evens-Magyar varieties, which are desingularizations of symmetric orbit closures in the flag manifold.

Keywords: Moment polytope, Grassmannian, flag manifold, Schubert variety, Bott-Samelson variety, Richardson variety, brick variety, Barbasch-Evens-Magyar variety, permutahedron, associahedron.

1. Introduction

The algebraic torus $\mathsf{T} := (\mathbb{C}^*)^n$ acts on a product of Grassmannians $X = \mathsf{Gr}_{i_1,n} \times \cdots \times \mathsf{Gr}_{i_m,n}$ diagonally. Embed X into a projective space and let Y be a T-invariant subvariety of X. The moment polytope of Y with respect to this action and embedding is a polytope whose combinatorial data encodes the T-action. For example, the 1-skeleton of this polytope can be used to compute the equivariant cohomology of Y, see [19] for an introduction. The main goal of this expository paper is to describe how to construct moment polytopes for subvarieties of X. We then apply our construction to Schubert, Richardson, Bott-Samelson, brick, and certain Barbasch-Evens-Magyar varieties. Along this survey we will encounter some classical polytopes like the permutahedron and associahedron.

2. Moment polytopes

Let Z be a projective variety with an action of an algebraic torus $\mathsf{T} := (\mathbb{C}^*)^n$. Let $z \in Z$ be an element such that $\overline{\mathsf{T} \cdot z}$ is the largest toric variety inside Z with respect to this action. An embedding of Z into a projective space induces an embedding of $\overline{\mathsf{T} \cdot z}$ so there is a polytope associated to this toric variety We call this polytope a *moment polytope* of Z and denote it by $\Phi(Z)$. Concretely, the polytope $\Phi(Z)$ is the convex hull of the weights by which T acts on z. These weights are the T-*weights* of Z. For readers interested in how moment polytopes arise in the context of symplectic geometry, see [6].

The *Grassmannian* $\mathsf{Gr}_{i,n}$ consists of the i-dimensional vector spaces in $\mathbb{C}^n$. The algebraic torus $\mathsf{T} := (\mathbb{C}^*)^n$ acts on the Grassmannian $\mathsf{Gr}_{i,n}$ as follows. Given an i-dimensional vector space V and $(t_1, \dots, t_n) \in \mathsf{T}$ we let M be a matrix whose column span is V and T be the diagonal matrix with entries $(t_1, \dots, t_n)$. The action is

$$(t_1, \dots, t_n) \cdot V := \text{column span}(TM).$$

Throughout this paper we use the *Plücker embedding.* We remark that we can obtain other polytopes by applying the Veronese embedding afterwards, see [13] for details. Let $V \in \mathsf{Gr}_{i,n}$ and M be any matrix with column span V. We will use the following notation

$$[n] := \{1, 2, \dots, n\}.$$

Given $C \subset [n]$ with $|C| = i$, the *Plücker coordinate* $p_C(V)$ is the minor of M associated to the rows indexed by C. The Plücker embedding $\mathsf{Gr}_{i,n} \to \mathbb{P}^{\binom{n}{i}-1}$ maps V to

$$(p_C(V) \mid C \subset [n], |C| = i) \in \mathbb{P}^{\binom{n}{i}-1}.$$

To compute the moment polytope of $\mathsf{Gr}_{i,n}$, we now set V to be any vector space such that all its Plücker coordinates are nonzero. The T-weights of $\mathsf{Gr}_{i,n}$ are the vectors $(a_1, \dots, a_n)$ such that

$$(t_1, \dots, t_n) \cdot p_C(V) = \prod_{j=1}^{n} t_j^{a_j} p_C(V).$$

Let $t_C := \prod_{j \in C} t_j$. Since

$$p_C(V) = t_C p_C(V)$$

then the T-weights are

$$e_C := \sum_{j \in C} e_j,$$

where $e_1, \dots, e_n$ denotes the standard basis of $\mathbb{C}^n$. The following proposition follows.

Proposition 2.1. *The moment polytope $\Phi(\mathsf{Gr}_{i,n})$ is the hypersimplex*

$$\Delta_{i,k} = \text{conv}\{e_C \mid C \subset [n], |C| = i\}.$$

Now consider the product of Grassmannians $X = \mathsf{Gr}_{i_1,n} \times \cdots \times \mathsf{Gr}_{i_m,n}$ for some sequence $(i_1, \dots, i_m)$. Let T act on X diagonally. Concretely, for $(V_1, \dots, V_m) \in X$ the action is

$$(t_1, \dots, t_n) \cdot (V_1, \dots, V_m) := (\text{column span}(TM_1), \dots, \text{column span}(TM_m)), \quad (1)$$

where each M_i is a matrix whose column span is V_i and T is the diagonal matrix with entries $(t_1, \dots, t_n)$. Again, by considering a point $x \in X$ such that $\overline{\mathsf{T} \cdot x}$ is the largest toric variety in X with respect to T and an embedding of X into a projective space we obtain a moment polytope $\Phi(X)$.

We first embed X into a product of projective spaces using various Plücker embeddings:

$$\mathsf{Gr}_{i_1,n} \times \cdots \times \mathsf{Gr}_{i_m,n} \hookrightarrow \mathbb{P}^{\binom{n}{i_1}-1} \times \cdots \times \mathbb{P}^{\binom{n}{i_m}-1}.$$

We then apply the Segre embedding multiple times so that

$$\mathbb{P}^{\binom{n}{i_1}-1} \times \cdots \times \mathbb{P}^{\binom{n}{i_m}-1} \hookrightarrow \mathbb{P}^{\binom{n}{i_1}\cdots\binom{n}{i_m}-1}.$$

The *Segre embedding* is given by the map

$$\mathbb{P}^{r-1} \times \mathbb{P}^{s-1} \hookrightarrow \mathbb{P}^{rs-1}$$
$$([x_0,\ldots,x_{r-1}],[y_0,\ldots,y_{s-1}]) \mapsto [x_iy_j \mid i=0,\ldots,r-1, \quad j=0,\ldots,s-1].$$

Given a T-action on $\mathbb{P}^{r-1}\times\mathbb{P}^{s-1}$, the Segre embedding induces a T-action on $\mathbb{P}^{rs-1}$. Given $(a_1,\ldots,a_n)$ the T-weight on the i-th coordinate of the T-action on $\mathbb{P}^{r-1}$ and $(b_1,\ldots,b_n)$ the T-weight on the j-th coordinate of the T-action on $\mathbb{P}^{s-1}$. Then the T-weight of the (i,j)-th coordinate of the T-action on $\mathbb{P}^{rs-1}$ is $(a_1+b_1,\ldots,a_n+b_n)$. Therefore under this embedding the T-weights of X are the sums of the T-weights of each of the $\mathsf{Gr}_{i_j,n}$. We conclude the following proposition.

Proposition 2.2. *For* $X = \mathsf{Gr}_{i_1,n} \times \cdots \times \mathsf{Gr}_{i_m,n}$,

$$\Phi(X) = \operatorname{conv}\left\{\sum_{j=1}^{m} e_{C_j} \mid C_j \subset [n], |C_j| = i_j\right\}.$$

Let Y be a subvariety of $X = \mathsf{Gr}_{i_1,n} \times \cdots \times \mathsf{Gr}_{i_m,n}$ which is invariant under the T-action, i.e. $\mathsf{T}\cdot Y \subset Y$. The moment polytope $\Phi(Y)$ for this action and the embedding above is the convex hull of the T-weights of Y. Since $Y \subset X$, then the weights are a subset of the weights for X. To find these weights it is useful to relate the T-weights with the T-fixed points.

The T-fixed points of $\mathsf{Gr}_{i,n}$ are the i-dimensional vector spaces spanned by subsets of $\{e_1,\ldots,e_n\}$. Given $C \subset [n]$ with $|C| = i$, the Plücker coordinates corresponding to $V = \operatorname{span}\{e_j \mid j \in C\}$ are all zero except $p_C(V) = 1$. There is only one weight when T acts on this point and it is e_C. This gives a correspondence between the T-weights and the T-fixed points of $\mathsf{Gr}_{i,n}$. This is also true for X and Y. The T-fixed points of Y are the $(x_1,\ldots,x_m) \in Y$ such that each x_j is a T-fixed points of $\mathsf{Gr}_{i_j,n}$. In other words the T-fixed points of Y are the

$$(\operatorname{span}\{e_j \mid j \in C_1\},\ldots,\operatorname{span}\{e_j \mid j \in C_m\}) \in Y \tag{2}$$

such that $C_j \subset [n]$ and $|C_j| = i_j$ for each $j = 1,\ldots,m$. The T-weight corresponding to the T-fixed point in the equation (2) is $\sum_{j=1}^m e_{C_j}$. We summarize this discussion in the following theorem.

Theorem 2.1. *The moment polytope of a* T*-invariant subvariety* $Y \subset \mathsf{Gr}_{i_1,n} \times \cdots \times \mathsf{Gr}_{i_m,n}$ *is*

$$\Phi(Y) = \operatorname{conv}\left\{ \sum_{j=1}^{m} e_{C_j} \mid (\operatorname{span}\{e_j \mid j \in C_1\}, \ldots, \operatorname{span}\{e_j \mid j \in C_m\}) \in Y \right\}.$$

In the following sections we describe the moment polytopes for some important subvarieties of $\mathsf{Gr}_{i_1,n} \times \cdots \times \mathsf{Gr}_{i_m,n}$.

3. Flag manifold

The *flag manifold* F_n is the subset of $\mathsf{Gr}_{1,n} \times \cdots \times \mathsf{Gr}_{n,n}$ consisting of the sequences $(V_1, \ldots, V_n)$ such that $V_i \subset V_j$ for all $i < j$. Such a sequence is called a *flag.*

The action of the torus on $\mathsf{Gr}_{i,n}$ extends diagonally to F_n, see the equation (1). The T-fixed points of F_n are the flags $(V_1, \ldots, V_n)$ such that each V_i is spanned by a subset of $\{e_1, \ldots, e_n\}$ of cardinality i. Throughout this paper we use one-line notation for permutations:

$$w = [w(1)w(2)\ldots w(n)] \in \mathfrak{S}_n.$$

The T-fixed points are indexed by $\mathfrak{S}_n$, the permutations of $[n]$. Namely, $w \in \mathfrak{S}_n$ determines the flag

$$F(w) := (\operatorname{span}\{e_{w(1)}\}, \operatorname{span}\{e_{w(1)}, e_{w(2)}\}, \ldots, \operatorname{span}\{e_{w(1)}, \ldots, e_{w(n)}\}).$$

The T-weight associated to this point is

$$\begin{aligned} \sum_{i=1}^{n}(n-i+1)e_{w(i)} &= \sum_{i=1}^{n} w_0(i)e_{w(i)} \\ &= \sum_{j=1}^{n} w_0 w^{-1}(j)e_j \\ &= (w_0 w^{-1}(1), \ldots, w_0 w^{-1}(n)), \end{aligned}$$

where $w_0 \in \mathfrak{S}_n$ is the longest permutation, i.e. $w_0(i) := n-i+1$. The *permutahedron* in $\mathbb{R}^n$ is the convex hull of the vectors $(w(1), \ldots, w(n))$ for all $w \in \mathfrak{S}_n$. Since composition by w_0 is an involution then the T-weights of F_n are $(w(1), \ldots, w(n))$ for all $w \in \mathfrak{S}_n$. Combining these observations with Theorem 2.1 we have the following proposition.

Proposition 3.1. *The moment polytope of the flag manifold is the permutahedron, i.e.*

$$\Phi(\mathsf{F}_n) = \operatorname{conv}\{(w(1), \ldots, w(n)) \mid w \in \mathfrak{S}_n\}.$$

Example 3.1. Consider the flag manifold F_3. The T-fixed point for $w = [231]$ is

$$F(w) = (\operatorname{span}\{e_2\}, \operatorname{span}\{e_2, e_3\}, \operatorname{span}\{e_2, e_3, e_1\})$$

and the corresponding T-weight is

$$e_2 + e_{\{2,3\}} + e_{\{1,2,3\}} = (1, 3, 2) = (w_0 w^{-1}(3), w_0 w^{-1}(2), w_0 w^{-1}(1)).$$

The moment polytope of F_3 is depicted in Fig. 1.

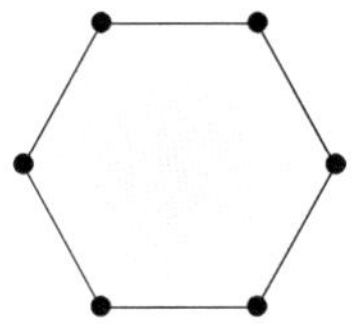

Fig. 1. The moment polytope $\Phi(\mathsf{F}_3)$.

3.1. *Schubert and Bott-Samelson varieties*

We now study the moment polytopes for Schubert varieties inside the flag manifold and Bott-Samelson varieties. Schubert varieties consist of flags $(V_1, \ldots, V_n)$ that satisfy some predetermined lower bounds of the dimension of $V_i \cap \operatorname{span}\{e_1, \ldots, e_j\}$ for all i, j. These varieties have important applications in combinatorics, algebraic geometry, and representation theory. Most Schubert varieties are singular. Bott-Samelson varieties provide desingularizations of Schubert varieties.

The set B of invertible $n \times n$ upper triangular matrices acts on the Grassmannian $\mathsf{Gr}_{i,n}$ as follows. Given an i-dimensional vector space V let M be a matrix whose column span is V. The matrix $B \in \mathsf{B}$ acts on V by

$$B \cdot V := \text{column span}(BM). \tag{3}$$

For example,

$$\mathsf{B} \cdot \operatorname{span}\{e_i\} = \{\operatorname{span}\{\alpha_1 e_1 + \cdots + \alpha_{i-1} e_{i-1} + e_i\} \mid \alpha_j \in \mathbb{R}\}.$$

The B-action extends diagonally to F_n. Concretely, for $(V_1, \ldots, V_n) \in \mathsf{F}_n$ and $B \in \mathsf{B}$

$$B \cdot (V_1, \ldots, V_n) := (\text{column span}(BM_1), \ldots, \text{column span}(BM_n)),$$

where each M_i is a matrix whose column span is V_i.

The *Schubert variety* X_w for $w \in \mathfrak{S}_n$ is the B-orbit closure of $F(w)$:

$$\mathsf{X}_w := \overline{\mathsf{B} \cdot F(w)}.$$

Since B acts diagonally on F_n then $(V_1, \ldots, V_n) \in \mathsf{X}_w$ if and only if $V_i \in \overline{\mathsf{B} \cdot \operatorname{span}\{e_{w(1)}, \ldots, e_{w(i)}\}}$ for each i. Notice that for $1 \leq j_1 < \ldots < j_i \leq n$

$$\overline{\mathsf{B} \cdot \operatorname{span}\{e_{j_1}, \ldots, e_{j_i}\}} = \{\operatorname{span}\{\alpha_1 e_1 + \alpha_2 e_2 + \cdots + \alpha_{j_i} e_{j_i}\} \mid \alpha_j \in \mathbb{R}\}. \tag{4}$$

Example 3.2. The flag corresponding to $w = [2431]$ is

$$F(w) = (\operatorname{span}\{e_2\}, \operatorname{span}\{e_2, e_4\}, \operatorname{span}\{e_2, e_4, e_3\}, \operatorname{span}\{e_2, e_4, e_3, e_1\}).$$

By applying the equation (4) to span$\{e_{w(1)},\ldots,e_{w(i)}\}$ for each i we conclude that $(V_1,V_2,V_3,\mathbb{C}^4)\in \mathsf{X}_w$ if and only if there exist $\alpha_{ij}\in\mathbb{R}$ such that

$$\begin{aligned}V_1 &= \operatorname{span}\{\alpha_{11}e_1+\alpha_{12}e_2\},\\ V_2 &= V_1\oplus \operatorname{span}\{\alpha_{21}e_1+\alpha_{22}e_2+\alpha_{23}e_3+\alpha_{24}e_4\}, \text{ and}\\ V_3 &= V_2\oplus \operatorname{span}\{\alpha_{31}e_1+\alpha_{32}e_2+\alpha_{33}e_3\}.\end{aligned}$$

The geometry of Schubert varieties is encoded in the combinatorics of $\mathfrak{S}_n$. Let s_i denote the transposition interchanging i and $i+1$. Every permutation w can be factored as a composition of s_i's. A *word* for w is a sequence $\underline{w}=(q_1,\ldots,q_m)$ of numbers in $[n-1]$ such that $w=s_{q_1}\cdots s_{q_m}$. The word is *reduced* whenever m is as small as possible. In such case the dimension of X_w equals m. Furthermore, $\mathsf{X}_v\subset\mathsf{X}_w$ if and only if the reduced word $\underline{w}$ contains a subsequence which is a reduced word for v. This defines a partial order $\leq$ of $\mathfrak{S}_n$ called the *Bruhat order.* The T-fixed points of X_w are the T-fixed points in F_n that lie in X_w. Therefore, they are the flags $F(v)$ such that $v\leq w$. We conclude that

$$\Phi(\mathsf{X}_w)=\operatorname{conv}\{(w_0v^{-1}(n),\ldots,w_0v^{-1}(1))\mid v\leq w\}.$$

Consider the following subfamily of the Bruhat interval polytopes of Tsukerman-Williams [18]:

$$\mathsf{Q}_{u,w_0}:=\operatorname{conv}\{(v(n),\ldots,v(1))\mid u\leq v\}.$$

From properties of the Bruhat order, see [3], we have that

$$\{w_0v^{-1}\mid v\leq w\}=\{v'\mid w_0w^{-1}\leq v'\}.$$

Therefore, $\Phi(\mathsf{X}_w)=\mathsf{Q}_{w_0w^{-1},w_0}$, i.e. the polytopes $\Phi(\mathsf{X}_w)$ are a subfamily of Bruhat interval polytopes.

Example 3.3. A reduced word for the permutation $w=[231]$ is $\underline{w}=(1,2)$. The subsequences that give reduced permutations are $\underline{[123]}=\varnothing$, $\underline{[132]}=(2)$, $\underline{[213]}=(1)$, and $\underline{w}$. Therefore

$$\Phi(\mathsf{X}_w)=\operatorname{conv}\{(3,2,1),(3,1,2),(2,3,1),(1,3,2)\}.$$

Figure 2 shows the moment polytope of X_w.

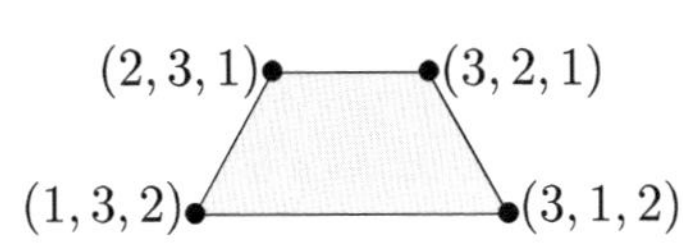

Fig. 2. The moment polytope $\Phi(\mathsf{X}_{[231]})$.

Most Schubert varieties are singular. Demazure [8] and Hansen [12] independently constructed smooth varieties which provide desingularizations of Schubert varieties. Demazure named them Bott-Samelson varieties due to a similar construction present in work by Bott-Samelson [4].

We now give a definition of these varieties which is a rephrasing of the description by Magyar [14]. Let $\underline{q} = (q_1, \dots, q_m)$ be a sequence of numbers in $[n-1]$. We construct a diagram $\mathcal{B}_{\underline{q}}$ called the *brick diagram* of $\underline{q}$. Start by drawing lines $\ell_i \subset \mathbb{R}^2$ for $i = 1, \dots, n$ where ℓ_i is defined by $y = i$. For $j = 1, \dots, m$ draw the line segment from (j, q_j) to $(j, q_j + 1)$, i.e. we connect the lines ℓ_{q_j} and ℓ_{q_j+1} with a line segment. The resulting diagram is $\mathcal{B}_{\underline{q}}$. Figure 3 shows the brick diagram for $\underline{q} = (1, 3, 2, 3)$.

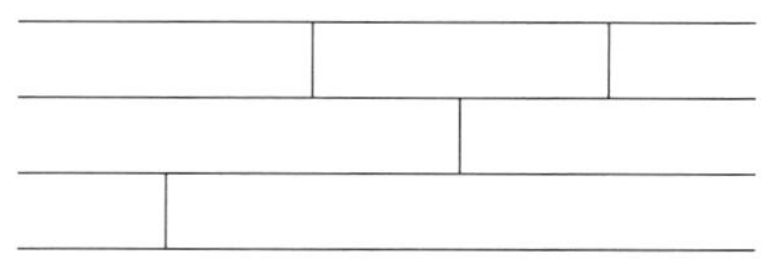

Fig. 3. The brick diagram $\mathcal{B}_{(1,3,2,3)}$.

Let $\Gamma_{\underline{q}}$ denote the set of the connected components of the complement of $\mathcal{B}_{\underline{q}}$. We label each region $\gamma \in \Gamma_{\underline{q}}$ with a vector space V_γ such that:

(i) If γ is bounded below by ℓ_k then $V_\gamma \in \mathsf{Gr}_{k,n}$.
(ii) If γ is not bounded on the left then $V_\gamma = \operatorname{span}\{e_1, \dots, e_k\}$.
(iii) If region γ is adjacent to and below the region γ' then $V_\gamma \subset V_{\gamma'}$.

The *Bott-Samelson variety* associated to $\underline{q}$ is

$$\mathsf{Z}_{\underline{q}} \subset \mathsf{Gr}_{q_1,n} \times \mathsf{Gr}_{q_2,n} \times \cdots \times \mathsf{Gr}_{q_m,n}$$

consisting of all labelings $(V_\gamma \mid \gamma \in \Gamma_{\underline{q}})$ satisfying (i)-(iii).

Example 3.4. Consider the labeling of $\Gamma_{(1,3,2,3)}$ shown in Fig. 4. Throughout this paper we omit the labels of the bottom and top connected components which are 0 and $\mathbb{C}^n$, respectively. For this labeling to give a point in $\mathsf{Z}_{(1,3,2,3)}$ we require that $V_1 \subset \operatorname{span}\{e_1, e_2\} \cap V_2$, $V_2 \subset V_3 \cap V_4$, and $V_3 \supset \operatorname{span}\{e_1, e_2\}$.

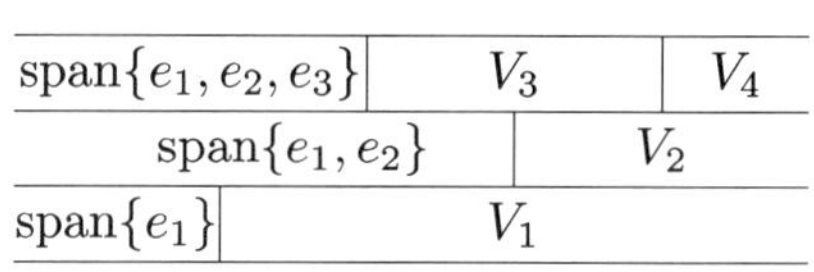

Fig. 4. A labeling in $\mathsf{Z}_{(1,3,2,3)}$.

The variety $\mathsf{Z}_{\underline{q}}$ is T-invariant. The T-fixed points of $\mathsf{Z}_{\underline{q}}$ are the labelings $(V_\gamma \mid \gamma \in \Gamma_{\underline{q}})$ such that each $V_\gamma = \operatorname{span}\{e_j \mid j \in C_\gamma\}$ for some $C_\gamma \subset [n]$ with $\dim V_\gamma = |C_\gamma|$.

To simplify notation, for the T-fixed points we label the regions $\gamma \in \Gamma_{\underline{q}}$ by the subsets C_γ. One can adapt conditions (i)-(iii) to these subsets. Therefore, by applying Theorem 2.1 we have that the moment polytope of $\mathsf{Z}_{\underline{q}}$ is

$$\Phi(\mathsf{Z}_{\underline{q}}) = \mathrm{conv}\left\{ \sum_{\gamma\in\Gamma_{\underline{q}}} e_{C_\gamma} \;\middle|\; \begin{array}{l} |C_\gamma| = k \text{ for } \gamma \text{ bounded below by } \ell_k \\ C_\gamma = [k] \text{ for } \gamma \text{ not bounded on the left} \\ C_\gamma \subset C_{\gamma'} \text{ if } \gamma \text{ is adjacent to and below } \gamma' \end{array} \right\}. \tag{5}$$

Notice that $\Phi(\mathsf{Z}_{\underline{q}}) \subset \mathbb{R}^n$ and it lies on the hyperplane

$$x_1 + \cdots + x_n = q_1 + \cdots + q_m + \frac{n(n+1)}{2}.$$

Example 3.5. In this example we describe the polytope $\Phi(\mathsf{Z}_{(1,2,1)})$ and show that not all the T-fixed points determine vertices. Consider C_1, C_2, and C_3 given in Fig. 5. We can uniquely determine a T-fixed point by choosing whether $C_1 = 1$

12		C_2	
1	C_1		C_3

Fig. 5. The brick diagram and labeling for T-fixed points of $\mathsf{Z}_{(1,3,2,3)}$.

or not, $C_2 = 12$ or not, and $C_3 = C_1$ or not, c.f. Proposition 3.2. Therefore $\mathsf{Z}_{(1,2,1)}$ has 8 T-fixed points. Table 1 shows these points together with their corresponding weights. Notice that indeed $x_1+x_2+x_3 = 1+2+1+\frac{3\cdot 4}{2}$. The polytope $\Phi(\mathsf{Z}_{(1,2,1)})$

Table 1. The T-fixed points for $\mathsf{Z}_{(1,2,1)}$ and their corresponding weights.

T-fixed point (C_1, C_2, C_3)	Weight in $\mathbb{R}^3$ $(3,2,1) + e_{C_1} + e_{C_2} + e_{C_3}$
$(1,12,1)$	$(6,3,1)$
$(1,12,2)$[a]	$(5,4,1)$
$(2,12,1)$[a]	$(5,4,1)$
$(2,12,2)$	$(4,5,1)$
$(1,13,1)$	$(6,2,2)$
$(1,13,3)$	$(5,2,3)$
$(2,23,2)$	$(3,5,2)$
$(2,23,3)$	$(3,4,3)$

Note: [a] By changing the embedding into $\mathbb{P}^N$ the weights for these fixed points can be different.

is shown in Fig. 6. It is combinatorially equivalent to the permutahedron in Fig. 1.

We now give an alternate description of the T-fixed points of $\mathsf{Z}_{\underline{q}}$. A *subword* of $\underline{q} = (q_1, \ldots, q_m)$ is a sequence $\underline{p} = (p_1, \ldots, p_m)$ where some of the entries of $\underline{q}$

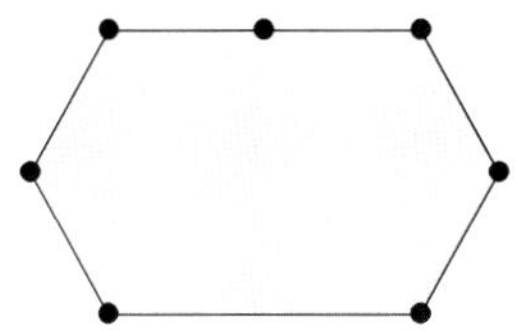

Fig. 6. The moment polytope $\Phi(\mathsf{Z}_{(1,2,1)})$.

are replaced by $-$. Given a subword $\underline{p}$ of $\underline{q}$ construct the labeling $\mathcal{C}_{\underline{p}}$ as follows. Let $\gamma_j \in \Gamma_{\underline{q}}$ be the region bounded on the left by the segment (j, q_j) and $\gamma_{j'}$ be the region bounded on the right by this line segment. If $p_j = -$ set $C_{\gamma_j} = C_{\gamma_{j'}}$. Otherwise, C_{γ_j} is determined by requiring that $C_{\gamma_j} \neq C_{\gamma_{j'}}$. This is because C_{γ_j} and $C_{\gamma_{j'}}$ have the same dimension, contain a vector space of one dimension less, and are contained in a vector space of one dimension more.

Proposition 3.2 ([8], [12]). *The correspondence between subwords of $\underline{q}$ and* T*-fixed points of $\mathsf{Z}_{\underline{q}}$ described above is a bijection. Therefore $\mathsf{Z}_{\underline{q}}$ has 2^m* T*-fixed points.*

Given a labeling determining a point in $\mathsf{Z}_{\underline{q}}$ the labelings of the regions not bounded on the right form a flag. Therefore we have the map

$$\beta : \mathsf{Z}_{\underline{q}} \longrightarrow \mathsf{F}_n$$
$$(V_\gamma \mid \gamma \in \Gamma_{\underline{q}}) \mapsto (V_\gamma \mid \gamma \text{ is not bounded on the right}).$$

Actually the image of this map is a Schubert variety X_w; we will concretely describe the relation between $\underline{q}$ and w in the equation (6). For now, we restrict to the case when $\underline{q}$ is a reduced word. For a reduced word $\underline{w}$ of w the map $\beta : \mathsf{Z}_{\underline{w}} \longrightarrow \mathsf{X}_w$ is the desingularizations of Demazure [8] and Hansen [12]. Since this map is T-equivariant then the T-fixed points of X_w are the images of the T-fixed points of $\mathsf{Z}_{\underline{w}}$. However, many T-fixed points of $\mathsf{Z}_{\underline{w}}$ can map to the same T-fixed point of X_w.

Example 3.6. The map $\beta : \mathsf{Z}_{(1,2,1)} \longrightarrow \mathsf{X}_{[321]}$ is described in Fig. 7. Note that the

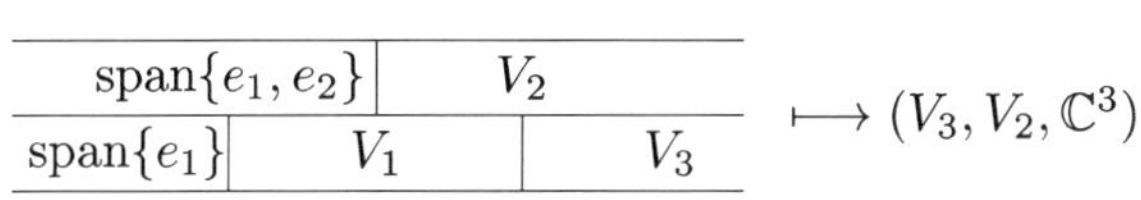

Fig. 7. The map $\beta : \mathsf{Z}_{(1,2,1)} \longrightarrow \mathsf{X}_{[321]}$.

T-fixed points of $\mathsf{Z}_{(1,2,1)}$ corresponding to the subwords $(-,-,-)$ and $(1,-,1)$ both map to the T-fixed point $F([123]) \in \mathsf{X}_{[321]}$.

3.2. *Richardson and brick varieties*

We now study the moment polytopes for Richardson varieties and brick varieties. A Richardson variety is a transversal intersections of two Schubert varieties. They are fundamental in Schubert calculus, which studies enumerative problems of linear subspaces. Most Richardson varieties are singular and brick varieties provide desingularizations for them.

The set B_- of invertible $n \times n$ lower triangular matrices acts on $\mathsf{Gr}_{i,n}$ by the equation (3), i.e. by the same formula as B. The *Richardson variety* X_w^v is

$$\mathsf{X}_w^v := \overline{\mathsf{B} \cdot F(w)} \cap \overline{\mathsf{B}_- \cdot F(v)}.$$

This variety is nonempty whenever $v \leq w$. Note that you recover Schubert varieties by setting $v = id$. The T-fixed points of X_w^v are the T-fixed points in F_n that lie in X_w^v. Therefore, they are the flags $F(u)$ such that $v \leq u \leq w$. We conclude that

$$\Phi(\mathsf{X}_w^v) = \operatorname{conv}\{(w_0u^{-1}(n), \dots, w_0u^{-1}(1)) \mid v \leq u \leq w\}.$$

The *Bruhat interval polytopes* of Tsukerman-Williams [18] are

$$\mathsf{Q}_{v,w} := \operatorname{conv}\{(v(n), \dots, v(1)) \mid v \leq u \leq w\}.$$

Now, from properties of the Bruhat order, see [3], we have that

$$\{w_0u^{-1} \mid v \leq u \leq w\} = \{u' \mid w_0w^{-1} \leq u' \leq w_0v^{-1}\}.$$

Therefore, $\Phi(\mathsf{X}_w^v) = \mathsf{Q}_{w_0w^{-1}, w_0v^{-1}}$, i.e. the polytopes $\Phi(\mathsf{X}_w^v)$ are a subfamily of Bruhat interval polytopes.

Some Richardson varieties are singular. Brion [5] constructed desingularizations for Richardson varieties as fiber products of Bott-Samelson varieties. Balan [1] showed that these desingularizations are general fibers of the Bott-Samelson map $\beta : \mathsf{Z}_{\underline{q}} \to \mathsf{F}_n$. The author studied the general fibers of β in [9] and named them brick varieties. We now define these varieties.

Given a sequence $\underline{q}$ there is a unique permutation w such that

$$w = \max\{s_{p_1} \cdots s_{p_m} \mid (p_1, \dots, p_m) \text{ is a subword of } \underline{q}\},$$

where $s_- := id$ and the maximum is taken with respect to the Bruhat order. The *Demazure product* $\delta(\underline{q})$ of $\underline{q}$ is defined to be this permutation. For example if $\underline{w}$ is a reduced word for w then $\delta(\underline{w}) = w$. For any $\underline{q}$ we can compute the Demazure product in a greedy way. To compute the Demazure product of $\underline{q} = (q_1, \dots, q_m)$ we start with the word $\underline{w_1} := (q_1)$. At the beginning of step $i > 1$ we have a reduced word $\underline{w_{i-1}} = (p_1, \dots, p_k)$. If $(p_1, \dots, p_k, q_i)$ is a reduced word then we define $\underline{w_i} := (p_1, \dots, p_k, q_i)$, otherwise $\underline{w_i} := (p_1, \dots, p_k)$. Table 2 computes the Demazure product of $\underline{q} = (3, 1, 3, 2, 1, 2)$.

The image of the Bott-Samelson map is

$$\beta(\mathsf{Z}_{\underline{q}}) = \mathsf{X}_{\delta(\underline{q})}. \tag{6}$$

We define the *brick variety* $\mathsf{Br}_{\underline{q}}$ to be the fiber of $F(\delta(\underline{q}))$ under β, i.e.

$$\mathsf{Br}_{\underline{q}} := \beta^{-1}(F(\delta(\underline{q})).$$

Table 2. Steps to compute $\delta(3,1,3,2,1,2) = s_3s_1s_2s_1$.

Step i	$\underline{w_i}$
1	$\underline{w_1} = (3)$
2	$\underline{w_2} = (3,1)$
3	$\underline{w_3} = (3,1)$
4	$\underline{w_4} = (3,1,2)$
5	$\underline{w_5} = (3,1,2,1)$
6	$\underline{w_6} = (3,1,2,1)$

The dimension of $\mathsf{Br}_{\underline{q}}$ is the difference between the length of $\underline{q}$ and the length of a reduced word for $\delta(\underline{q})$, see [9] for details. Therefore $\mathsf{Br}_{\underline{w}}$ is a point when $\underline{w}$ is a reduced word of w. Because of this we will consider $\underline{q}$ that are not reduced words.

Since $\mathsf{Br}_{\underline{q}}$ is a fiber in $\mathsf{Z}_{\underline{q}}$ we can use the labelings of the previous section to describe the points in $\mathsf{Br}_{\underline{q}}$. To ease notation, let $w = \delta(\underline{q})$. We label each region $\gamma \in \Gamma_{\underline{q}}$ with a vector space V_γ such that:

(i) If γ is bounded below by ℓ_k then $V_\gamma \in \mathsf{Gr}_{k,n}$.
(ii) If γ is not bounded on the left then $V_\gamma = \text{span}\{e_1, \dots, e_k\}$.
(iii) If region γ is adjacent to and below the region γ' then $V_\gamma \subset V_{\gamma'}$.
(iv) If γ is not bounded on the right then $V_\gamma = \text{span}\{e_{w(1)}, \dots, e_{w(k)}\}$.

The brick variety

$$\mathsf{Br}_{\underline{q}} \subset \mathsf{Gr}_{q_1,n} \times \mathsf{Gr}_{q_2,n} \times \cdots \times \mathsf{Gr}_{q_m,n}$$

associated to $\underline{q}$ consists of all labelings of the regions satisfying (i)-(iv).

Example 3.7. For $\underline{q} = (1,2,1,2,1)$ the Demazure product is

$$\delta(\underline{q}) = s_1s_2s_1 = s_2s_1s_2 = [321].$$

The labeling shown in Fig. 8 gives a point in $\mathsf{Br}_{(1,2,1,2,1)}$. Therefore, we must choose vector spaces $(V_1, V_2, V_3) \in \mathsf{Gr}_{1,3} \times \mathsf{Gr}_{2,3} \times \mathsf{Gr}_{3,3}$ such that $V_1 \subset \text{span}\{e_1, e_2\} \cap V_2$ and $V_3 \subset \text{span}\{e_2, e_3\} \cap V_2$.

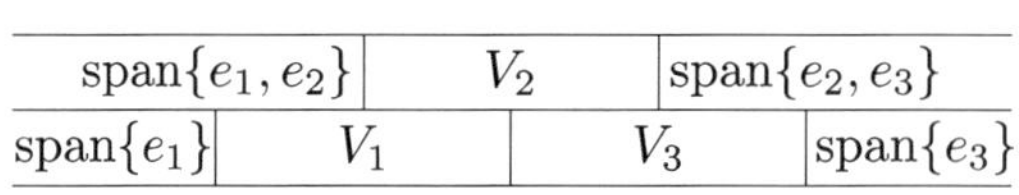

Fig. 8. A labeling in $\mathsf{Br}_{(1,2,1,2,1)}$.

The brick variety is a T-invariant subvariety of a product of Grassmannians. The T-fixed points of $\mathsf{Br}_{\underline{q}}$ are the T-fixed points of $\mathsf{Z}_{\underline{q}}$ that lie in $\mathsf{Br}_{\underline{q}}$. From conditions (i)-(iv) together with the equation (5) we have that

$$\Phi(\mathsf{Br}_{\underline{q}}) = \operatorname{conv}\left\{ \sum_{\gamma \in \Gamma_{\underline{q}}} e_{C_\gamma} \;\middle|\; \begin{array}{l} |C_\gamma| = k \text{ for } \gamma \text{ bounded below by } \ell_k \\ C_\gamma = [k] \text{ for } \gamma \text{ unbounded on the left} \\ C_\gamma \subset C_{\gamma'} \text{ if } \gamma \text{ is adjacent to and below } \gamma' \\ C_\gamma = \{w(1), \ldots, w(k)\} \text{ for } \gamma \text{ unbounded on the right} \end{array} \right\},$$

where $w = \delta(\underline{q})$.

Example 3.8. We now describe the polytope $\Phi(\mathsf{Br}_{(1,2,1,2,1)})$. The brick diagram together with the set-labelings are shown in Fig. 9. Table 3 shows the 5 T-fixed

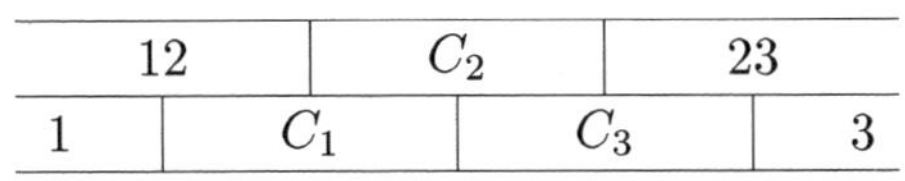

Fig. 9. The brick diagram and labeling for the T-fixed points of $\mathsf{Br}_{(1,2,1,2,1)}$.

points together with their corresponding subwords and points in $\mathbb{R}^3$. The polytope

Table 3. The T-fixed points for $\mathsf{Br}_{(1,2,1,2,1)}$ and their corresponding weights.

Subword	T-fixed point (C_1, C_2, C_3)	Weight in $\mathbb{R}^3$ $(3,3,3) + e_{C_1} + e_{C_2} + e_{C_3}$
(1, 2, 1, -, -)	(2, 23, 3)	(3, 5, 5)
(-, 2, 1, 2, -)	(1, 13, 3)	(5, 3, 5)
(-, -, 1, 2, 1)	(1, 12, 2)	(5, 5, 3)
(1, -, -, 2, 1)	(2, 12, 2)	(4, 6, 3)
(1, 2, -, -, 1)	(2, 23, 2)	(3, 6, 4)

$\Phi(\mathsf{Z}_{(1,2,1)})$ is shown in Fig. 10.

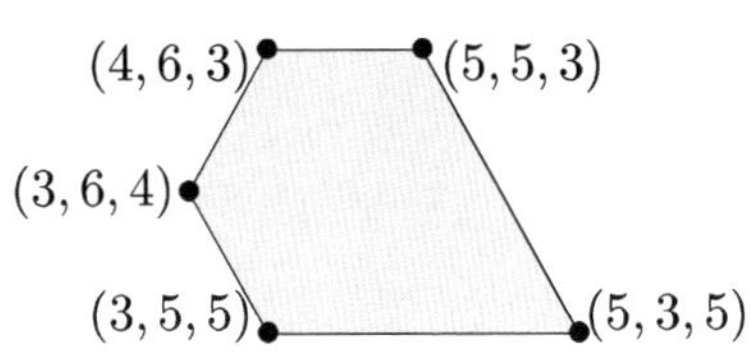

Fig. 10. The moment polytope $\Phi(\mathsf{Br}_{(1,2,1,2,1)})$.

An *associahedron* $\mathcal{A}_n \subset \mathbb{R}^n$ is a polytope whose faces encode the subdivisions of a fixed $(n+2)$-gon. The vertices correspond to the triangulations of this $(n+2)$-gon using diagonals, which is the *Catalan number* $C_n = \frac{1}{n+1}\binom{2n}{n}$. The edges correspond to subdivisions obtained from a triangulation by removing exactly one diagonal. In general, the faces of $\mathcal{A}_n$ correspond to subdivisions obtained from removing diagonals from a triangulation.

Stasheff [17] showed that this process yields an abstract cell complex, see Fig. 11 for an example. There are many different realizations of the associahedron as a polytope, see the survey [7] for details. The following theorem gives a way to obtain the associahedron as the moment polytope of a brick variety.

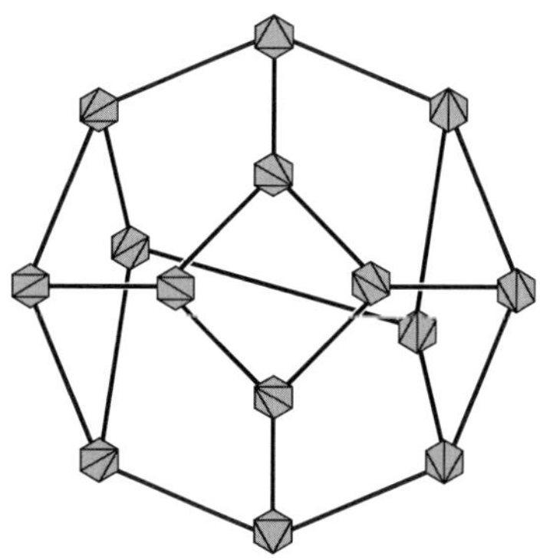

Fig. 11. The cell complex consisting of triangulations of a regular hexagon.

Theorem 3.1 ([9]). *For*

$$\underline{q} = (1, 2, \ldots, n, 1, 2, \ldots, n, 1, 2, \ldots, n-1, \ldots, 1),$$

the polytope $\Phi(\mathsf{Br}_{\underline{q}})$ *is an associahedron* $\mathcal{A}_{n-1}$.

In [9, 16] the reader can find other $\underline{q}$ giving associahedra.

The polytope $\Phi(\mathsf{Br}_{(1,2,1,2,1)})$ is an associahedron for the triangulations of a pentagon. The associahedron $\Phi(\mathsf{Br}_{\underline{q}})$ for $\underline{q} = (1, 2, 3, 1, 2, 3, 1, 2, 1)$ is depicted in Fig. 12.

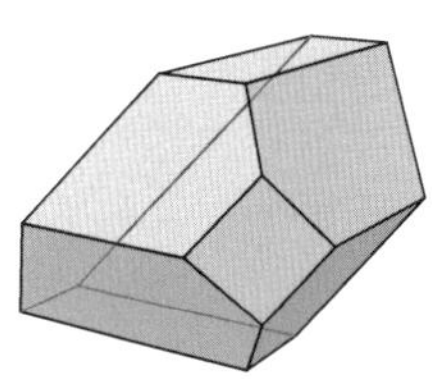

Fig. 12. The associahedron $\Phi(\mathsf{Br}_{(1,2,3,1,2,3,1,2,1)})$.

Building up on work by Pilaud-Pocchiola [15], Pilaud-Santos [16] showed that the associahedron can be constructed as a brick polytope. Theorem 3.1 follows since brick polytopes are moment polytopes of brick varieties, see [9, Theorem 21].

Gaussent [11] studied other fibers of the Bott-Samelson map. It would be interesting to look at their moment polytopes.

4. Barbasch-Evens-Magyar varieties

We have seen that given a sequence $\underline{q}$ we can construct subvariety of a product of Grassmannians consisting of labelings $(V_\gamma \mid \gamma \in \Gamma_{\underline{q}})$ by imposing that

(i) if γ is bounded below by ℓ_k then $V_\gamma \in \mathsf{Gr}_{k,n}$,
(ii) if region γ is adjacent to and below the region γ' then $V_\gamma \subset V_{\gamma'}$,

and some conditions on the unbounded regions in $\Gamma_{\underline{q}}$ such that the resulting variety is T-invariant. In [10] the authors obtain the desingularizations by Barbasch-Evens [2] of symmetric orbit closures by imposing conditions on the regions of $\Gamma_{\underline{q}}$ unbounded on the left. They also study the moment polytopes of these desingularizations. We now explore some of these varieties and polytopes. See [10] for more general results and details about the connection of these varieties to symmetric orbit closures.

Fix $a \leq n$ and $\underline{q} = (q_1, \ldots, q_m)$ a sequence of numbers in $[n-1]$. We label each region $\gamma \in \Gamma_{\underline{q}}$ with a vector space V_γ such that:

(i) If γ is bounded below by ℓ_k then $V_\gamma \in \mathsf{Gr}_{k,n}$.
(ii) If region γ is adjacent to and below the region γ' then $V_\gamma \subset V_{\gamma'}$.
(iii) If γ is the region bounded below by ℓ_a and unbounded on the left then $V_\gamma = \mathrm{span}\{e_1, \ldots, e_a\}$.

The *Barbasch-Evens-Magyar* variety

$$\mathsf{BEM}^a_{\underline{q}} \subset (\mathsf{Gr}_{1,n} \times \mathsf{Gr}_{2,n} \times \cdots \times \mathsf{Gr}_{n-1,n}) \times (\mathsf{Gr}_{q_1,n} \times \mathsf{Gr}_{q_2,n} \times \cdots \times \mathsf{Gr}_{q_m,n})$$

associated to $\underline{q}$ consists of all labelings of the regions satisfying (i)-(iii).

Example 4.1. Let $\underline{q} = (1,3,2,3)$. A point in $\mathsf{BEM}^2_{\underline{q}}$ is determined by the $V_1, \ldots, V_6$ in Fig. 13 satisfying conditions (i)-(iii) above. For example, $V_2 \subset \mathrm{span}\{e_1, e_2\} \cap V_3$.

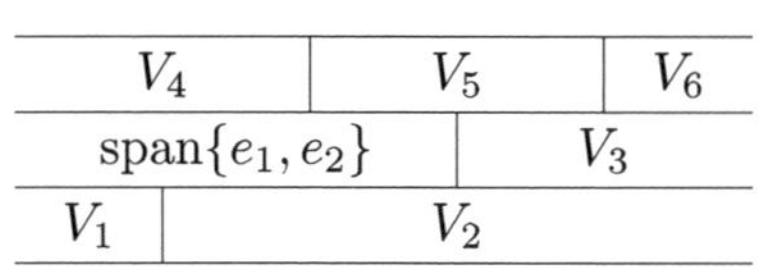

Fig. 13. A labeling in $\mathsf{BEM}_{(1,3,2,3)}$.

The variety $\mathsf{BEM}_{\underline{q}}$ is T-invariant. Therefore the T-fixed points of $\mathsf{BEM}_{\underline{q}}$ are the T-fixed points of the product of the Grassmannians that lie in $\mathsf{BEM}_{\underline{q}}$. We then have that the moment polytope of $\mathsf{Z}_{\underline{q}}$ is

$$\Phi(\mathsf{BEM}^a_{\underline{q}}) = \mathrm{conv}\left\{ \sum_{\gamma\in\Gamma_{\underline{q}}} e_{C_\gamma} \;\middle|\; \begin{array}{c} |C_\gamma| = k \text{ for } \gamma \text{ bounded below by } \ell_k \\ C_\gamma \subset C_{\gamma'} \text{ if } \gamma \text{ is adjacent to and below } \gamma' \\ C_\gamma = [a] \text{ for } \gamma \text{ bounded below by } \ell_a \\ \text{and unbounded on the left} \end{array} \right\}. \tag{7}$$

Example 4.2. Let $\underline{q} = (1,2)$ and $a = 1$. The brick diagram together with the set-labelings are shown in Fig. 14. Table 4 shows the 8 T-fixed points together

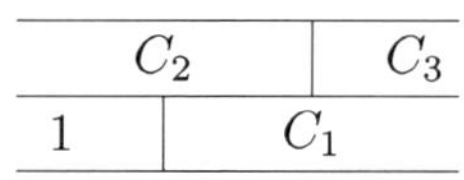

Fig. 14. The brick diagram and labeling for the T-fixed points of $\mathsf{BEM}^1_{(1,2)}$.

with their corresponding subwords and points in $\mathbb{R}^3$. The last 4 T-fixed points

Table 4. The T-fixed points for $\mathsf{BEM}^1_{(1,2)}$ and their corresponding weights.

T-fixed point (C_1, C_2, C_3)	Weight in $\mathbb{R}^3$ $(2,1,1) + e_{C_1} + e_{C_2} + e_{C_3}$
$(1, 12, 12)$	$(5, 3, 1)$
$(1, 12, 13)$	$(5, 2, 2)$
$(2, 12, 12)$	$(4, 4, 1)$
$(2, 12, 23)$	$(3, 4, 2)$
$(1, 13, 13)$	$(5, 1, 3)$
$(1, 13, 12)$	$(5, 2, 2)$
$(3, 13, 13)$	$(4, 1, 4)$
$(3, 13, 23)$	$(3, 2, 4)$

can be obtained from the first 4 by exchanging 2 with 3. This implies that the corresponding weights are obtained by exchanging the second and third coordinates. If we restrict the polytope to the convex hull of the top 4 points in Table 4 we obtain $\Phi(\mathsf{Z}_{(1,2)})$. Therefore, we can obtain $\mathsf{BEM}^1_{(1,2)}$ by first reflecting $\Phi(\mathsf{Z}_{(1,2)})$ across the hyperplane $x_2 = x_3$ and then taking the convex hull of the points in $\Phi(\mathsf{Z}_{(1,2)})$ and the reflected polytope. This is stated for general $\mathsf{BEM}^a_{\underline{q}}$ in Theorem 4.1. The polytope $\Phi(\mathsf{BEM}^1_{(1,2)})$ is shown in Fig. 15. This polytope is combinatorially equivalent to $\Phi(\mathsf{Z}_{(1,2,1)})$ from Example 3.5. However this is not the case in general; e.g. see [10, Example 6.1].

Theorem 4.1 ([10]). *The moment polytope* $\Phi(\mathsf{BEM}^a_{\underline{q}})$ *is the convex hull of the points in the reflections of* $\Phi(\mathsf{Z}_{\underline{q}})$ *using the transformations of the form* $(x_1, \ldots, x_n) \mapsto (x_{w(1)}, \ldots, x_{w(n)})$ *for any permutation* w *such that* $[w(1) \ldots w(a)]$

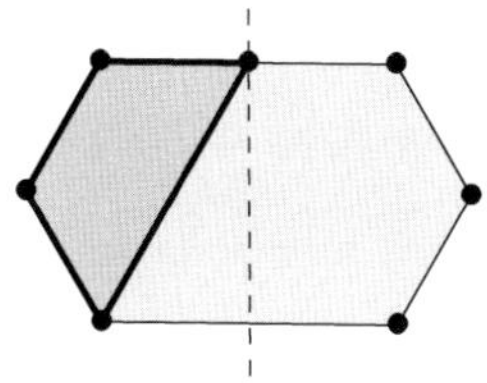

Fig. 15. The darker polytope is the moment polytope $\Phi(\mathsf{Z}_{(1,2)})$. To obtain the moment polytope $\Phi(\mathsf{BEM}^1_{(1,2)})$ we reflect $\Phi(\mathsf{Z}_{(1,2)})$ across the dashed line which is the intersection of the hyperplanes $x_2 = x_3$ and $x_1 + x_2 + x_3 = 9$.

is a permutation of $[a]$. *Furthermore, the dimension of* $\Phi(\mathsf{BEM}^a_{\underline{q}})$ *is* $n-1$ *if* a *appears in* $\underline{q}$ *and* $n-2$ *if it doesn't.*

References

[1] M. Balan, Standard monomial theory for desingularized Richardson varieties in the flag variety $\mathrm{GL}(n)/B$, *Transform. Groups* **18**, 329 (2013).

[2] D. Barbasch and S. Evens, K-orbits on Grassmannians and a PRV conjecture for real groups, *J. Algebra* **167**, 258 (1994).

[3] A. Björner and F. Brenti. *Combinatorics of Coxeter groups*, volume 231 of *Graduate Texts in Mathematics.* Springer, New York, 2005.

[4] R. Bott and H. Samelson, Applications of the theory of Morse to symmetric spaces, *Amer. J. Math.* **80**, 964 (1958).

[5] M. Brion, Lectures on the geometry of flag varieties, in *Topics in cohomological studies of algebraic varieties*, Trends Math. (Birkhäuser, Basel, 2005) pp. 33–85.

[6] A. Cannas da Silva, Symplectic toric manifolds, in *Symplectic geometry of integrable Hamiltonian systems (Barcelona, 2001)*, Adv. Courses Math. CRM Barcelona (Birkhäuser, Basel, 2003) pp. 85–173.

[7] C. Ceballos, F. Santos and G. M. Ziegler, Many non-equivalent realizations of the associahedron, *Combinatorica* **35**, 513 (2015).

[8] M. Demazure, Désingularisation des variétés de Schubert généralisées, *Ann. Sci. École Norm. Sup. (4)* **7**, 53 (1974), Collection of articles dedicated to Henri Cartan on the occasion of his 70th birthday, I.

[9] L. Escobar, Brick manifolds and toric varieties of brick polytopes, *Electron. J. Combin.* **23**, Paper 2.25, 18 (2016).

[10] L. Escobar, B. Wyser and A. Yong, K-orbit closures and Barbasch-Evens-Magyar varieties (2017), arXiv:1708.06663.

[11] S. Gaussent, The fibre of the Bott-Samelson resolution, *Indag. Math. (N.S.)* **12**, 453 (2001).

[12] H. C. Hansen, On cycles in flag manifolds, *Math. Scand.* **33**, 269 (1973).

[13] A. Knutson, The symplectic and algebraic geometry of Horn's problem, *Linear Algebra Appl.* **319**, 61 (2000), Special Issue: Workshop on Geometric and Combinatorial Methods in the Hermitian Sum Spectral Problem (Coimbra, 1999).

[14] P. Magyar, Schubert polynomials and Bott-Samelson varieties, *Comment. Math. Helv.* **73**, 603 (1998).

[15] V. Pilaud and M. Pocchiola, Multitriangulations, pseudotriangulations and primitive sorting networks, *Discrete Comput. Geom.* **48**, 142 (2012).

[16] V. Pilaud and F. Santos, The brick polytope of a sorting network, *European J. Combin.* **33**, 632 (2012).
[17] J. D. Stasheff, Homotopy associativity of H-spaces. I, II, *Trans. Amer. Math. Soc. 108 (1963), 275-292; ibid.* **108**, 293 (1963).
[18] E. Tsukerman and L. Williams, Bruhat interval polytopes, *Adv. Math.* **285**, 766 (2015).
[19] J. S. Tymoczko, An introduction to equivariant cohomology and homology, following Goresky, Kottwitz, and MacPherson, in *Snowbird lectures in algebraic geometry*, Contemp. Math. Vol. 388 (Amer. Math. Soc., Providence, RI, 2005) pp. 169–188.

Cubical Dehn–Sommerville equations and self-reciprocal cubical complexes

Magda Hlavacek

Mathematics Department, UC Berkeley
Berkeley, CA 94704, USA
E-mail: mhlava@math.berkeley.edu

The classical Dehn–Sommerville equations, relating the face numbers of simplicial polytopes, have an analogue for cubical polytopes. These relations can be generalized to apply to simplicial and cubical Eulerian complexes. We outline a proof of the simplicial version given in M. Beck and R. Sanyal's preprint *Combinatorial Reciprocity Theorems* that uses the idea of self-reciprocal complex of lattice polytopes, and describe how to adapt these proof ideas to the cubical case.

Keywords: Dehn-Sommerville equations, cubical complexes, simplicial complexes, Eulerian complexes, self-reciprocity, Ehrhart polynomial.

1. Introduction

In their original form, the Dehn–Sommerville equations give a set of linear relations that the face vector of a simplicial polytope must satisfy. These equations first appeared in 1905, when Max Dehn proved in [5] that they hold in dimensions 4 and 5, and conjectured that they hold in all dimensions. In 1927, Duncan Sommerville proved Dehn's conjecture in [12]. Later, it became apparent that these relations are involved in answering the following face enumeration question: Which vectors $(f_{-1}, f_0, \ldots, f_{d-1})$ are the face vector of a simplicial polytope P in $\mathbb{R}^d$? The g-theorem, conjectured in 1971 by McMullen [11] and proved by Stanley, Billera-Lee in 1980 [13][4], gives three conditions that completely characterizes the f-vectors of simplicial polytopes. One of these conditions is that $(f_{-1}, f_0, \ldots, f_{d-1})$ satisfy the Dehn-Sommerville relations.

In addition to realizing their importance in face enumeration questions, people have also generalized these relations to apply to broader collections of objects. For example, in [2], Bayers and Billera generalize the Dehn-Sommerville equations to apply to completely balanced spheres and Eulerian poset complexes. In this article, we focus our discussion on two specific subcases of the latter: Eulerian simplicial complexes and Eulerian cubical complexes. For the statement of the simplicial case, see Definition 3.1, and for the cubical case, see Definition 3.2.

There are many proofs in the literature of the simplicial case. One of the more well-known proofs is a combinatorial proof that relies on double-counting techniques. In this article, we summarize a more geometric proof method, appearing in [3], that uses Ehrhart theory of the standard geometric realization of the simplicial complex. We then discuss how to adapt this line of reasoning to apply to the cubical case.

In Sections 2 through 4, we give necesary background information, and give statements of the Dehn-Sommerville equations in various generalities. In Section 5.1, we introduce the Ehrhart polynomial of a simplicial complex, and use the self-reciprocity of these polynomials to outline a proof in [3] of the Dehn–Sommerville equations. This proof relies on a geometric realization of simplicial complexes, which does not always exist for cubical complexes, discussed in Section 4. In Section 5.2, we adjust some definitions and show that we can still apply these ideas to prove that the Dehn–Sommerville equations hold for cubical Eulerian comples. In Section 6, we discuss some remaining questions brought up by this proof approach.

2. Dehn-Sommerville Equations for Simplicial and Cubical Polytopes

In this section, we give necessary definitions and background, culminating in the statement of the Dehn–Sommerville equations for cubical and simplicial polytopes.

Definition 2.1. A *polytope* is the convex hull of finitely many points in $\mathbb{R}^d$. A *lattice polytope* is the convex hull of finitely many points in $\mathbb{Z}^d$. A *face* of a polytope P in $\mathbb{R}^d$ is determined by a linear functional $w \in \mathbb{R}^d$, and is the set of the points for which w is minimized:

$$\text{face}_w(P) = \{x \in P : wx \leq wy \text{ for all } y \in P\}.$$

Definition 2.2. A *d-simplex* is the convex hull of $d+1$ affinely independent points. For example, a 2-simplex is a triangle, a 3-simplex is a tetrahedron.
A polytope is *simplicial* if all of its facets (faces of codimension–1) are simplices.

Definition 2.3. Let P be a d-dimensional polytope. Then we define the *f-vector* of P as:

$$f(P) = (f_{-1}, f_0, \dots, f_{d-1})$$

where $f_{-1} = 1$ by convention, and for $0 \leq i \leq d-1$, f_i is the number of i-dimensional faces of P.

It is often more convenient use a specific transformation of the f-vector, called the h-vector. For example, the Dehn-Sommerville equations can be expressed much more simply in terms of the h-vector.

Definition 2.4. Let P be a simplicial polytope. The h-vector of P is

$$h(P) = (h_0, \dots, h_d)$$

where the h_i are defined implicitly by

$$\sum_{i=0}^{d} f_{i-1} z^i (1-z)^{d-i} = \sum_{i=0}^{d} h_k z^i.$$

The Dehn–Sommerville equations relate the entries of the f-vector (and the h-vector) of a simplicial polytope. As mentioned earlier, hey were first conjectured and proven for low dimensions by Dehn in 1905, [5]. Sommerville proved that these relations hold for all d in 1927, [12].

Theorem 2.1 (Dehn-Sommerville). *Let P be a simplical polytope with dimension d, and let f_i be the number of $i-$dimensional faces of P. Then, for each $1 \leq j \leq d$*

$$f_{j-1} = \sum_{k=j}^{d} (-1)^{d-k} \binom{k}{j} f_{k-1}.$$

This is equivalent to the following symmetry relation on the h-vector:

$$h_i = h_{d-i}$$

for all $0 \leq i \leq d$.

The definition of a simplicial polytope is natural because every face of a simplex is a lower dimensional simplex. Thus, one may wonder if we can apply the same treatment to other collections of polytopes closed under taking faces, such as cubes.

Definition 2.5. The standard d-cube is the polytope described as follows:

$$\square^d = \{(x_1, \ldots, x_d) : 0 \leq x_i \leq 1 \text{ for all } 1 \leq i \leq d\}.$$

A polytope is *cubical* if all of its facets are combinatorially equivalent to a standard d-cube for some d.

In [7], Grünbaum gives the following analogue of the Dehn-Sommerville equations for cubical polytopes.

Theorem 2.2. *Let P be a cubical polytope of dimension d. Then, for each $1 \leq j \leq d$*

$$f_{j-1} = \sum_{k=j}^{d} (-1)^{d-k} 2^{k-j} \binom{k-1}{j-1} f_{k-1}.$$

3. Dehn-Sommerville Equations for Simplicial and Cubical Eulerian Complexes

The Dehn-Sommerville equations can be naturally extended to apply to a wider class of objects.

Definition 3.1. An *(abstract) simplicial complex* is a nonempty collection Γ of subsets of a finite set V such that

- if $\sigma \in \Gamma$ and $\tau \subset \sigma$, then $\tau \in \Gamma$.

The σ are called *faces* of Γ, and we use the following notion of dimension:

$$\dim(\sigma) := |\sigma| - 1$$

We call a simplicial complex *d-dimensional* if d is the maximum dimension of its faces. We view Γ as a poset ordered by inclusion.

The f-vector $f(\Gamma)$ and h-vector $h(\Gamma)$ are defined in a similar way as for simplicial polytopes:

Definition 3.2. Let Γ be a d-dimensional simplicial complex. Then the f-vector of Γ is

$$f(\Gamma) = (f_{-1}, f_0, \ldots, f_d),$$

where $f_{-1} = -1$ by convention, and for $0 \leq i \leq d$, f_i is the number of faces σ of Γ such that $\dim(\sigma) = i$.
The h-vector of Γ is

$$h(\Gamma) = (h_0, \ldots, h_d, h_{d+1})$$

where the h_i are defined implicitly by

$$\sum_{i=1}^{d+1} f_{i-1}(\Gamma) z^i (1-z)^{d+1-i} = \sum_{i=0}^{d+1} h_i(\Gamma) z^i$$

Its natural to ask whether the Dehn–Sommerville equations extend to all simplicial complexes. The answer to this is no. However, Dehn–Sommerville relations do hold for certain simplicial complexes:

Definition 3.3. A graded poset Π with $\hat{0}, \hat{1}$ is *Eulerian* if

$$\mu_\Pi(x, y) = (-1)^{\ell(x,y)}$$

where $\ell(x, y)$is the length of a maximal chain in the interval $[x, y]$.
We call a simplicial complex Γ *Eulerian* if $\Gamma \cup \hat{1}$ is an Eulerian poset, where $\Gamma \cup \hat{1}$ is Γ with a maximum element $\hat{1}$ appended.

The boundary complex of a polytope, for example, is Eulerian. We can now state the Dehn-Sommerville equations for Eulerian simplicial complexes, as appears in [3]. This result is a special case of some broader results in [2].

Theorem 3.1. *Let Γ be a d-dimensional Eulerian simplicial complex. Then for* $0 \leq i \leq d+1$,

$$h_i = h_{d-i+1}$$

which is equivalent to

$$f_{j-1} = \sum_{k=j}^{d+1} (-1)^{d+1-k} \binom{k}{j} f_{k-1}$$

for $0 \leq j \leq d+1$.

Later on we will outline a proof appearing in [3].

Just as we can generalize the idea of a simplicial polytope to abstract simplicial complexes, we can also generalize the idea of a cubical polytope to abstract cubical complexes. We use definitions and notation from [9].

Definition 3.4. An *(abstract) cubical complex* is a nonempty collection $\mathcal{C}$ of subsets of a finite set V such that:

- For every $\sigma \in \mathcal{C}$, the elements of σ can be represented as the vertices of a d-dimensional cube for some d, where faces contained in σ are identified with the vertex sets of the faces of the cube.
- If $\sigma, \tau \in \mathcal{C}$, then $\sigma \cap \tau \in \mathcal{C}$.

Again, the σ are called *faces* of $\mathcal{C}$, and we use the following notion of dimension:

$$2^{\dim(\sigma)} = |\sigma|.$$

We call a cubical compled *d-dimensional* if d is the maximum dimension of its faces.

Similarly as for the simplicial case, we can see that the face lattice of a cubical polytope is an example of a an abstract cubical complex. Accordingly, the cubical version of the Dehn-Sommerville equations can be generalized similarly. The following appears in Adin's 1995 paper defining a cubical analogue to the simplicial h-vector, [1].

Theorem 3.2. *Let* Γ *be a d-dimensional Eulerian cubical complex. Then for* $1 \leq j \leq d+1$,

$$f_{j-1} = \sum_{k=j}^{d+1} (-1)^{d-k+1} 2^{k-j} \binom{k-1}{j-1} f_{k-1}.$$

Notice that in this statement of the theorem, we do not yet present an equivalent formulation in terms of an h-vector. Several definitions of a cubical h-vector exist, each preserving a nice property of the simplicial h-vector. In Section 6, we will discuss a cubical h-vector defined by Adin in [1] that can be used to express the cubical Dehn-Sommerville relations.

4. A Note on Geometric Realizeability

One difference between simplicial and cubical complexes arises when we try to find a geometric realization of them in terms of lattice polytopes.

Definition 4.1. A *lattice polytope* is the convex hull of finitely many points in $\mathbb{Z}^d \subset \mathbb{R}^d$.

In the following discussion of geometric realizations of abstract complexes, we again use definitions and notation consistent with [9]. In order to define what we mean by a geometric realization, we need the idea of a complex of polytopes.

Definition 4.2. A *complex of (lattice) polytopes* is a collection $\mathcal{K}$ of (lattice) polytopes such that:

- If F is a face of G and $G \in \mathcal{K}$ then $F \in \mathcal{K}$.
- If $F, G \in \mathcal{K}$ then $F \cap G$ is a face of both F and G (and is thus in $\mathcal{K}$).

The *dimension* of $\mathcal{K}$ is the maximum dimension of elements of $\mathcal{K}$.

In general terms, a geometric realization of an abstract complex is a complex of polytopes that has the same face structure.

Definition 4.3. Let Γ be an abstract complex with base set V. A map $\phi \to \mathbb{R}^n$ is a *geometric realization* of Γ if:

- For each nonempty $\sigma \in \Gamma$, the restricted map $\phi|_\sigma$ is an injective map from σ to $\mathbb{R}^d$ such that $\phi|_\sigma(\sigma)$ is the vertex set of a convex polytope, and the faces contained in σ are inverse images under $\phi|_\sigma$ of the faces of $\operatorname{conv}(\phi|_\sigma(\sigma))$.
- For all $\sigma, \tau \in \mathcal{K}$,

$$\operatorname{conv}(\phi(\sigma)) \cap \operatorname{conv}(\phi(\tau)) = \operatorname{conv}(\phi(\sigma \cap \tau)).$$

A simplicial complex always has a geometric realization. In fact, there is a standard way of constructing such a realization.

Definition 4.4. Let Γ be an abstract simplicial complex with base set V, and let $\{e_v : v \in V\}$ be the standard basis of $\mathbb{R}^V$. Let $\sigma \in V$. Then we define the standard simplex associated to σ as:

$$\Delta[\sigma] = \operatorname{conv}\{e_v : v \in \sigma\}.$$

We call the following the *canonical realization* of Γ:

$$\mathcal{R}[\Gamma] = \{\Delta[\sigma] : \sigma \in \Gamma\}.$$

Definition 4.5. A d-simplex $\Delta = \operatorname{conv}\{\mathbf{v_0}, \mathbf{v_1}, \dots, \mathbf{v_d}\}$ is *unimodular* if and only if the vectors $\mathbf{v_1} - \mathbf{v_0}, \dots \mathbf{v_d} - \mathbf{v_0}$ form a $\mathbb{Z}$-basis of $\mathbb{Z}^d$.

Proposition 4.1. *$\mathcal{R}[\Gamma]$ is a complex of unimodular simplices, and its face lattice is isomorphic to Γ.*

This canonical realization is a key ingredient to the proof of Theorem 3.2 outlined in [3]. However, it turns out that not only is there no standard realization of an abstract cubical complex; some cubical complexes have no geometric realization at all. In [8], Hetyei remarks that the cubical complex obtained from the faces of three squares glued so that they form a mobius strip cannot be realized geometrically. See Figure 1 for this example.

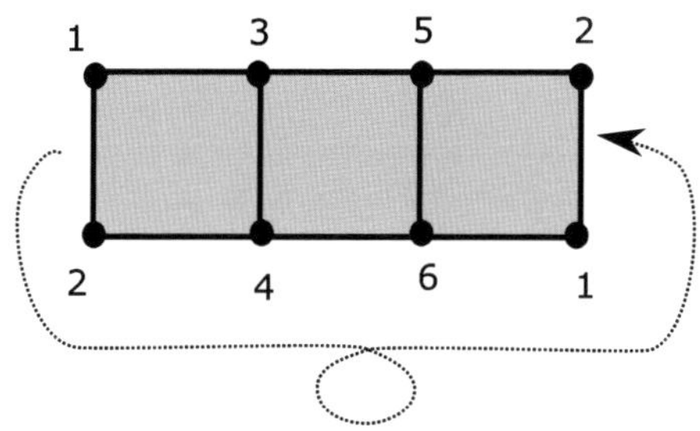

Fig. 1. The poset formed by the faces of these squared, glued in the specified way, was shown by Hetyei to be a cubical complex with no geometric realization.

However, Hetyei defines the following weaker form of a geometric realization for a cubical complex in [9], [8]:

Definition 4.6. Let $\mathcal{C}$ be a cubical complex over a set V. A weak geometric realization of $\mathcal{C}$ is a collection of maps $\Phi = \{\phi_\sigma\}$ indexed by the nonempty faces σ of ϕ, such that

- the map ϕ_σ is an injective map from σ to $\mathbb{R}^d$ such that $\phi_\sigma(\sigma)$ is the vertex set of a convex polytope, and the faces contained in σ are inverse images under ϕ_σ of the faces of $\text{conv}(\phi_\sigma(\sigma))$.

The *standard weak geometric realization* of a cubical complex maps each face σ to a standard d-cube.

In Section 5.2 we show that this weak geometric realization is sufficient to apply the proof techniques in Section 5.6 of [3] to the cubical case.

5. Cubical Dehn–Sommerville and Self-reciprocal Complexes

5.1. *The Simplicial Case*

In this section, we give an overview of section 5.6 of [3], resulting in an outline of a proof of 3.1 We first recall the Ehrhart function, introduced by Eugene Ehrhart in [6].

Definition 5.1. Let P be a lattice polytope in $\mathbb{R}^d$. Then the *Ehrhart function* associated to P counts the number of lattice points in the nth dilate of P:

$$i_P(n) = \left|nP \cap \mathbb{Z}^d\right|.$$

The Ehrhart function of a polytope agrees with a polynomial, and admits a nice reciprocity theorem, see [6], [10], which we will use later in our discussion of self-reciprocal complexes.

Theorem 5.1 (Ehrhart-Macdonald). *Let P be a lattice polytope in $\mathbb{R}^d$. Then for $n \in \mathbb{Z}^+$,*

$$(-1)^{\dim(P)} i_P(-n) = i_{P^\circ}(n)$$

where P° denotes the relative interior of P.

We can also consider the *Ehrhart series* of a lattice polytope:

$$\mathrm{E}_P(z) := \sum_{n\geq 0} i_P(n) z^n.$$

Since $i_P(n)$ is a polynomial of degree d, we can express $\mathrm{E}_P(z)$ in the following form:

$$\mathrm{E}_P(z) := \sum_{n\geq 0} i_P(n) z^n = \frac{h_0^* + h_1^* z + \cdots + h_{d+1}^* z^{d+1}}{(1-z)^{d+1}}.$$

We call $(h_0^*, \ldots h_{d+1}^*)$ the h^**−vector* of P.

We now address complexes of lattice polytopes: Let $\mathcal{K}$ be a complex of lattice polytopes in $\mathbb{R}^d$. We can extend the definition of the Ehrhart polynomial:

$$ehr_{\mathcal{K}}(n) = \left|n\mathcal{K} \cap \mathbb{Z}^d\right|.$$

Ehr($\mathcal{K}$) always agrees with a polynomial of degree dim($\mathcal{K}$), and thus we can define the Ehrhart series and the h^*-vector in a similar way as for polytopes:

$$Ehr_{\mathcal{K}}(z) := 1 + \sum_{n\geq 1} ehr_{\mathcal{K}}(n) z^n = \frac{h_0^* + h_1^* z + \cdots + h_{d+1}^* z^{d+1}}{(1-z)^{d+1}}.$$

We call $(h_0^*, \ldots h_{d+1}^*)$ the h^**-vector* of $\mathcal{K}$.

We are now finally able to define self-reciprocity.

Definition 5.2. $\mathcal{K}$ is *self-reciprocal* if for all $n > 0$

$$(-1)^{\dim(\mathcal{K})} \mathrm{ehr}_{\mathcal{K}}(-n) = \mathrm{ehr}_{\mathcal{K}}(n).$$

In some cases, self-reciprocity is equivalent to a symmetry in the h^*-vector:

Theorem 5.2. *Let $\mathcal{K}$ be a d-dimensional complex of lattice polytopes, such that $\chi(\mathcal{K}) = 1 - (-1)^{d+1}$, where $\chi(\mathcal{K})$ is the Euler charactistic of $\mathcal{K}$. Then $\mathcal{K}$ is self-reciprocal if and only if*

$$h^*_{d+1-i}(\mathcal{K}) = h^*_i(\mathcal{K}).$$

One example of a self-reciprocal complex is the boundary complex of a lattice polytope. This example can be generalized further:

Proposition 5.1. *Eulerian complexes of lattice polytopes are self reciprocal.*

In all these cases, $\chi(\mathcal{K}) = 1 - (-1)^{d+1}$ so the h^*-vector is symmetric.

We are now ready to sketch the proof of Theorem 3.1 outlined in [3]. Suppose we have some Eulerian simplicial complex Γ. We consider its canonical realization as a complex of unimodular simplices, $\mathcal{R}[\Gamma]$. By Proposition 5.1 and Theorem 5.2, this is self-reciprocal and thus has a symmetric h^*-vector. One can verify that since all faces of $\mathcal{R}[\Gamma]$ are unimodular, the h^*-vector and h-vector are the same. Since the h^*-vector is symmetric, so is the h-vector, so Theorem 3.1 must hold.

5.2. *The Cubical Case*

In the previous section, we gave a definition for the Ehrhart polynomial of a complex of lattice polytope. In this section, we extend that definition to cover abstract cubical complexes. Since these are not geometric objects, it may seem strange to give them an Ehrhart polynomial; however, we use their standard weak geometric realizations to define the following:

Definition 5.3. Let $\mathcal{C}$ be a cubical complex with a standard weak geometric realization of $\mathbf{\Phi} = \{\phi_\sigma\}$. Then we define the Ehrhart polynomial of $\mathcal{C}$ as

$$i_{\mathcal{C}}(n) = \sum_{\sigma \in \mathcal{C}} -\mu_{\hat{\mathcal{C}}}(\sigma, \hat{1}) i_{\phi_\sigma(\sigma)}(n).$$

Note that an equivalent expression for this is

$$\mathrm{ehr}_{\mathcal{C}}(n) = \sum_{\sigma \in \mathcal{C}} \mathrm{ehr}_{\phi_\sigma(\sigma)^\circ}(n).$$

For convenience, we will use i_σ and $\dim(\sigma)$ to refer to $i_{\phi_\sigma(\sigma)}$ and $\dim \phi_\sigma(\sigma)$ respectively.

Figure 2 shows one way of thinking about the Ehrhart polynomial of our example of a cubical complex with no geometric realization.

Since $i_{\mathcal{C}}(n)$ is the sum of expressions that agree with a polynomial, it must agree with a polynomial itself.

We define self-reciprocity the same way as earlier:

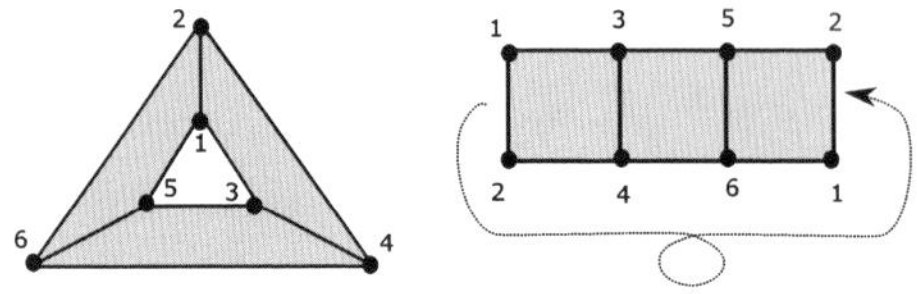

Fig. 2. These two cubical complexes have the same Ehrhart polynomial as given in Definition 5.3.

Definition 5.4. A d-dimensional cubical complex $\mathcal{C}$ is *self-reciprocal* if its Ehrhart polynomial satisfies

$$(-1)^d i_{\mathcal{C}}(-n) = i_{\mathcal{C}}(n).$$

We now show that an Eulerian cubical complex is self-reciprocal.

Theorem 5.3. *Let $\mathcal{C}$ be an Eulerian cubical complex. Then $\mathcal{C}$ is self-reciprocal.*

Proof. Let σ be a face of $\mathcal{C}$. We first note that $\ell(\sigma, \hat{1}) = \dim(\mathcal{C}) - \dim(\sigma) + 1$, Thus, $-\mu(\sigma, \hat{1}) = \dim \mathcal{C} - \dim(\sigma)$. Substituting this into our definition for $i_{\mathcal{C}}(n)$,

$$i_{\mathcal{C}}(n) = \sum_{\sigma \in \mathcal{C}} -\mu_{\hat{\mathcal{C}}}(\sigma, \hat{1}) i_\sigma(n) = \sum_{\sigma \in \mathcal{C}} (-1)^{\dim(\mathcal{C}) - \dim(\sigma)} i_\sigma(n).$$

We now use Theorem 5.1 to compute $(-1)^{\dim(\mathcal{C})} i_{\mathcal{C}}(-n)$:

$$(-1)^{\dim(\mathcal{C})} i_{\mathcal{C}}(-n) = \sum_{\sigma \in \mathcal{C}} (-1)^{\dim(\sigma)} i_\sigma(n) = \sum_{\sigma \in \mathcal{C}} i_{\sigma^\circ}(n) = i_{\mathcal{C}}(n).$$

□

We now have the machinery necessary to show that Theorem 3.2 holds. Let $\mathcal{C}$ be an Eulerian cubical complex of dimension d. Then, using the fact that $ehr_{(\square^d)^\circ} = (n-1)^r$, we can write down the Ehrhart polynomial as:

$$\mathrm{ehr}_{\mathcal{C}}(n) = \sum_{\sigma \in \mathcal{C}} \mathrm{ehr}_{\phi(\sigma)^\circ}(z) = \sum_{\sigma \in \mathcal{C}} (n-1)^{\dim(\sigma)} = \Sigma_{i=0}^d f_i(\mathcal{C})(n-1)^i. \tag{5.2.1}$$

We can now use the fact that $\mathcal{C}$ is Eulerian, and thus self-reciprocal. Since

$$(-1)^d i_{\mathcal{C}}(-n) = i_{\mathcal{C}}(n) \tag{5.2.2}$$

we substitute our expression for the Ehrhart polynomials given in (5.2.1) into (5.2.2) and obtain

$$(-1)^d \sum_{i=0}^{d} f_i(-(n+1))^i = (-1)^{d-i} \sum_{i=0}^{d} f_i(n+1)^i = \sum_{i=0}^{d} f_i(n-1)^i \tag{5.2.3}$$

We now do a shift, replacing $(n+1)$ and $(n-1)$ with $(n+2)$ and n:

$$(-1)^{d-i} \sum_{i=0}^{d} f_i(n+2)^i = \sum_{i=0}^{d} f_i(n)^i. \tag{5.3.4}$$

We expand the left hand side:

$$(-1)^{d-i} \sum_{i=0}^{d} f_i(n+2)^i = (-1)^{d-i} \sum_{i=0}^{d} f_i \sum_{j=0}^{i} \binom{i}{j} 2^{i-j} n^j = (-1)^{d-1} \sum_{j=0}^{d} \left(\sum_{i=j}^{d} \binom{i}{j} f_i 2^{i-j} \right) n^j \tag{5.3.5}$$

Putting things together, we get

$$(-1)^{d-1} \sum_{j=0}^{d} \left(\sum_{i=j}^{d} \binom{i}{j} f_i 2^{i-j} \right) n^j = \sum_{j=0}^{d} f_j n^j. \tag{5.3.6}$$

Setting the coefficients of n^{j-1} in each side of (5.3.6) equal to each other, we get

$$f_{j-1} = (-1)^{d-j+1} \sum_{i=j}^{d+1} \binom{i-1}{j-1} f_{i-1} 2^{i-j}, \tag{5.3.7}$$

which gives us the desired equations for Theorem 3.2.

6. Future Work

In this article, we have given a brief overview of simplicial and cubical Dehn-Sommerville equations and, following the treatment in [3], discussed how to use the idea of a self-reciprocal complex to prove the Dehn-Sommerville equations for Eulerian simplicial and cubical complexes. In our discussion of the Eulerian case, we used the Ehrhart series and the h-vector to prove that the Dehn-Sommerville equations hold, since the Ehrhart series of the open standard simplex is easy to work with. On the other hand, in our discussion of the cubical case, we used Ehrhart polynomials, since the Ehrhart polynomial of a unit cube is easy to work with.

However, the approach of using the Ehrhart series in the cubical case is interesting because it may give a way to express the Dehn-Sommerville relations as a statement of symmetry of some analogue of the simplicial h-vector. Several people have defined cubical h-vectors, each mimicing different nice properties of the simplicial h-vector. In particular, Adin shows that the symmetry of his long cubical

h-vector, defined in [1], is equivalent to the Dehn-Sommerville relations for Eulerian cubical complexes. Using the weak geometric realization of a cubical complex $\mathcal{C}$, we can define the h^*-vector of an abstract cubical complex as follows. Since $ehr_{\mathcal{C}}(n)$ agrees with a polynomial, we can define the Ehrhart series and the h^*-vector of $\mathcal{C}$ in a similar way as for polytopes and for simplicial complexes,

$$Ehr_{\mathcal{C}}(z) := 1 + \sum_{n\geq 1} ehr_{\mathcal{C}}(n)z^n = \frac{h_0^* + h_1^* z + \cdots + h_{d+1}^* z^{d+1}}{(1-z)^{d+1}}.$$

We call $(h_0^*, \ldots h_{d+1}^*)$ the *h^*-vector* of $\mathcal{C}$. From computing small examples, it seems like the h^*-vector of a cubical complex $\mathcal{C}$ is different from Adin's cubical h-vector, and it may be interesting to further the relation between these two vectors.

It could also be interesting to explore how to adapt other proofs of the simplicial Dehn-Sommerville relations to the cubical case. In Section 4.9 of [3], Beck and Sanyal define a chain partition. They give an expression for a (Π, σ)-chain-partition of a finite poset Π with $\hat{0}$ and $\hat{1}$ equipped with order preserving map $\phi : \Pi \to \mathbb{Z}_{>0}$

Definition 6.1. A (Π, ϕ)-*chain partition* of n is of the form

$$n = \phi(c_1) + \cdots + \phi(c_m)$$

for some multichain $\hat{0} \prec c_1 \preceq \cdots \preceq c_m \prec \hat{1}$.

We let $cp_{\Pi,\phi}(n)$ be the number of (Π, ϕ)-chain partitions of n.

They compute $cp_{\Gamma,\phi}$, where $\phi(\sigma) = |\sigma|$ for a simplicial complex Γ of dimension d. They then show that if Γ is Eulerian, then

$$CP_{\Gamma,\phi}(z) = \sum_{n\geq 0} cp_{\Gamma,\phi}(n)z^n = \frac{h(z)}{(1-q)^{d+1}},$$

where $h(z)$ is a polynomial with symmetric coefficients. They then show that these coefficents are indeed the h-vector of Γ. This leads us to ask whether there is a nice choice of order preserving map ϕ such that one can use the idea of $(\mathcal{C}, \phi)$-chain partitions to prove Theorem 3.2.

Acknowledgements

The author would like to thank Matthias Beck for support, guidance, and many insightful conversations. She would also like to thank Sebastian Manecke for his insights and suggestions about how to define an Ehrhart polynomial for an abstract cubical complex, and an anonymous referee for very helpful comments.

References

[1] Ron M. Adin. A new cubical h-vector. In *Proceedings of the 6th Conference on Formal Power Series and Algebraic Combinatorics (New Brunswick, NJ, 1994)*, volume 157, pages 3–14, 1996.

[2] Margaret M. Bayer and Louis J. Billera. Generalized Dehn-Sommerville relations for polytopes, spheres and Eulerian partially ordered sets. *Invent. Math.*, 79(1):143–157, 1985.

[3] Matthias Beck and Raman Sanyal. Combinatorial reciprocity theorems: An invitation to enumerative geometric combinatorics. *Graduate Studies in Mathematics, American Mathematical Society, to appear*, 2018.

[4] Louis J. Billera and Carl W. Lee. A proof of the sufficiency of McMullen's conditions for f-vectors of simplicial convex polytopes. *J. Combin. Theory Ser. A*, 31(3):237–255, 1981.

[5] M. Dehn. Die Eulersche Formel im Zusammenhang mit dem Inhalt in der Nicht-Euklidischen Geometrie. *Math. Ann.*, 61(4):561–586, 1906.

[6] Eugène Ehrhart. Sur les polyèdres rationnels homothétiques à n dimensions. *C. R. Acad. Sci. Paris*, 254:616–618, 1962.

[7] Branko Grünbaum. *Convex polytopes*, volume 221 of *Graduate Texts in Mathematics*. Springer-Verlag, New York, second edition, 2003. Prepared and with a preface by Volker Kaibel, Victor Klee and Günter M. Ziegler.

[8] G. Hetyei. On the Stanley ring of a cubical complex. *Discrete Comput. Geom.*, 14(3):305–330, 1995.

[9] Gabor Hetyei. *Simplicial and cubical complexes: Analogies and differences.* ProQuest LLC, Ann Arbor, MI, 1994. Thesis (Ph.D.)–Massachusetts Institute of Technology.

[10] I. G. Macdonald. Polynomials associated with finite cell-complexes. *J. London Math. Soc.*, 4:181–192, 1971.

[11] P. McMullen. The numbers of faces of simplicial polytopes. *Israel J. Math.*, 9:559–570, 1971.

[12] Duncan MY Sommerville. The relations connecting the angle-sums and volume of a polytope in space of n dimensions. *Proc. R. Soc. Lond. A*, 115(770):103–119, 1927.

[13] Richard P. Stanley. The number of faces of a simplicial convex polytope. *Adv. in Math.*, 35(3):236–238, 1980.

Hollow lattice polytopes: Latest advances in classification and relations with the width

Oscar Iglesias-Valiño

Departamento de Matemáticas, Estadísticay Computación, Universidad de Cantabria
Santander, 39005, Spain
E-mail: oscar.iglesias@unican.es
http://personales.unican.es/iglesiasvo/

Volume and width are two of the most studied ways to measure a body. With the concept of width, natural questions arise on lattice polytopes.

A particular question someone can ask is the maximum width that a lattice polytope can have for arbitrary dimension or which is the finiteness threshold of a hollow lattice polytope. These questions are the main topic of this survey.

Focusing in these concepts have helped to bring new upper bounds in the volume of hollow polytopes with lower bounded width, which lead into new advances in their classification. Some examples are maximal 3-hollow polytopes or empty 4-simplices.

In this survey, we recap all background knowledge regarding this concept and get together all new results that have been published during the recent years. These advances in lattice polytopes may lead to new results in big questions of convex geometry as it is finding new examples or bounds in relation with the Flatness Theorem for convex bodies and polytopes.

Keywords: Width; lattice polytope; hollow; empty; threshold; flatness.

1. Introduction

Through this survey we will mostly talk about lattice polytopes. By a polytope we mean the convex hull of finitely many points in $\mathbb{R}^d$, when we say lattice polytope we mean that the vertices belong to the d-dimensional lattice $\mathbb{Z}^d$.

For a lattice polytope P we will say that P is *hollow* if it has no lattice points in its relative interior and we will say it is *empty* if the only lattice points that belong to P are its vertices. We will denote the *normalized volume* of P, $\mathrm{Vol}(P) := d!\mathrm{vol}(P)$ and we will use it during all survey.

As width is the main topic in this survey we start with the definition of the concept of width for an arbitrary convex body:

Definition 1.1. The width of a convex body $K \in \mathbb{R}^d$ respect to an affine functional $f : \mathbb{R}^d \to \mathbb{R}^d$ is defined as the

$$\mathrm{w}_f(K) := \max_{x,y \in K} |(f(x) - f(y)|$$

The *(lattice) width* of a lattice polytope P is the minimum value of the widths among all non-constant integer functionals and we will denote it as $\mathrm{w}(P)$. Lattice width can also be seen as the minimum lattice distance of two hyperplanes that can enclose a lattice polytope.

One of the main results in hollow bodies is the following Flatness Theorem, whose study was motivated by integer programming [14, 16] among others [5, 21]:

Theorem 1.1 (Flatness theorem). *There exists a function $f : \mathbb{N} \to \mathbb{R}$ such that for any hollow convex body $K \in \mathbb{R}^d$ and with respect to the lattice $\mathbb{Z}^d$,* $\mathrm{w}(K) \leq f(d)$.

With the appearance of the flatness theorem stated above, the interest in getting this "flatness constant" as small as possible increases (this constant only depends on the dimension).

In Section 2, we discuss upper and lower bounds for width in general dimensions and some particular cases of lattice polytopes where the maximum and minimum width are known or known to be tight.

In Section 3, we talk about the non-existence of upper bound for the volume of hollow polytopes. For lattice polytopes with interior points, Hensley [10] proved an upper bound for the volume that force to separate the study of hollow polytopes and lattice polytopes with interior points.

In Section 4 we will check the concept of finiteness threshold width introduced by Blanco et al. in [8] and look at how this relates with the maximum width of certain class of polytopes like empty or hollow lattice polytopes.

Even more, Section 4 present recent results in classification for certain classes of lattice polytopes. Some of these results have been found using important facts for projections of hollow polytopes.

Section 5 talks about the difficulty of computing the width of a polytope. As explained in section 3, it is possible to find volume upper bounds when we restrict the width of a hollow polytope to be greater than some value, nevertheless, we explain some of the reasons why an exhaustive enumeration of hollow polytopes seems not reasonable to that theoretical upper bound.

Section 6 deals with some open questions like determining the values of better upper bounds for the maximum width of hollow polytopes, find values of finiteness threshold width or some other questions related to the width of hollow lattice polytopes.

The objective of this survey is clarify the current situation on hollow lattice polytope classification and make emphasis in the relation of the latest results in the topic with the understanding of the width of a lattice polytope. Some results of classification for empty simplices [13] are introduced to the reader in this survey, but we mostly talk about recent works that have been developed in last years and might give room to some improvement for several open questions stated in the last section of this survey.

2. Flatness and width bound on hollow polytopes

In this Section we discuss about the maximum and minimum width that a hollow lattice polytope can have. More generally, the maximum width that a hollow convex body can have is referred to as the flatness constant as defined in Theorem 1.1. This term was introduced by Kannan and Lovász [14], but several people before had studied this problem [7]. In [14], they give a list of works in the developing lower flatness constant value, the first version with an exponential upper bound is due to Khintchine [16].

In the same paper, Kannan and Lovász give an upper bound for the width of hollow convex bodies in $O(d^2)$, but still they conjecture this bound is not the best possible and expect that the right upper bound may be in $O(d)$.

Banazczyk et al. [5] got a result of $O(d^{3/2})$ for the flatness constant. This result is still stated to be the best known bound in some works [8] but a better bound can be derived by Rudelson work in 1998 [21], who proved constants in $O(d^{4/3}(1+\log d))$. In some recent works, they do not mention Rudelson et al. upper bound. This fact may have been forgotten as they do not mention width in that work.

In [5] there are also statements about this upper bounds if we constrain our body. For empty simplices, Banazczyk et al. give an upper bound in $O(d \log d)$. They give bounds for more general classes of polytopes in which the number of vertices or faces is polynomial in the dimension (Remark 2.7 in [5]).

With these results a linear upper bound seems reasonable, but there is still a huge gap between the know upper bounds and the understanding of the flatness constant in terms of lattice polytopes examples.

In terms of lower bounds for the maximal width of hollow lattice polytopes the best known is given by the *dth*-dilation of an unimodular simplex in dimension d.

If we restrict to empty simplices, J.-M. Kantor showed with a non-constructive proof that the lattice width of a d-dimensional empty simplex can grow linearly in d:

Theorem 2.1 (Kantor, 1999 [15]). *For any β slightly less than $1/e$, there exists for sufficiently large d a sequence of empty simplices of growing dimension k and width $\mathrm{w} > d\beta$.*

Sebő [23], constructed explicit examples of simplices such that have width $w \geq d-2$, $\forall d \geq 3$ with the following example:

Let $k \in \mathbb{N}$ and let us call the simplex $S_d(k) := \operatorname{conv}(s_0, s_1, \cdots, s_d)$, with the vertices satisfying the following conditions:

Let $s_{i,j}$ denote the j-th coordinate of s_i, that is, $s_{i,j} = 0$ if $j \notin \{i, i+1\}$ mod d, $s_{i,i} = 1$, $s_{i,i+1} = k$, $(i,j \in \{1, \cdots, d\})$. In this example the width of $S_d(k)$ is k unless $k = 1$ and n is even.

More over, Bárány point out that for even d the facet of $S_d(k)$ opposite to the origin s_0 is an empty $(d-1)$-simplex of width $d-2$.

In the case of empty d-simplices, their maximum width is bounded by:

$$2\lfloor d/2 \rfloor - 1 \leq w(d) \leq O(d \log d)$$

In dimension 4 Haase and Ziegler [9] showed families of empty simplices with low volume.

$$\sigma(v) = \text{conv}(e_1, e_2, e_3, e_4, v), \qquad v = (2, 2, 3, D - 6)$$

for every $D \geq 8$ with $gcd(D, 6) = 1$, $\sigma(v)$ is empty and has width 2. Here, D is the normalized volume of the simplex $\sigma(v)$. In fact, this family is one of the families described in terms of algebraic geometry by Mori et al. [18] listing it as a family of *terminal quotient singularities.*

In [9], Haase and Ziegler also give an empty 4-simplex of width 4: $\sigma(6, 14, 17, 65)$ with $\sigma(v)$ as define above. They conjecture that this is the only one. This was proven by Iglesias-Valiño and Santos [12] and keeps the linear upper bound for the width in hollow polytopes still reasonable.

3. Volume of hollow polytopes

Hollow polytopes may seem the easiest example of lattice polytopes as they have no interior points but when trying to look at "how big a lattice polytope can be?", hollow polytopes seem more complicated.

There are easy examples of families of hollow simplices with volume as big as you want. Reeve tetrahedron, denoted here by Δ_h, are examples of this fact [20]:

$$\Delta_h := \text{conv}((0, 0, 0), (1, 0, 0), (0, 1, 0), (1, 1, h))$$

h is the volume of the simplex Δ_h and no matter the value of h, the Reeve tetrahedron will always be empty, and so hollow.

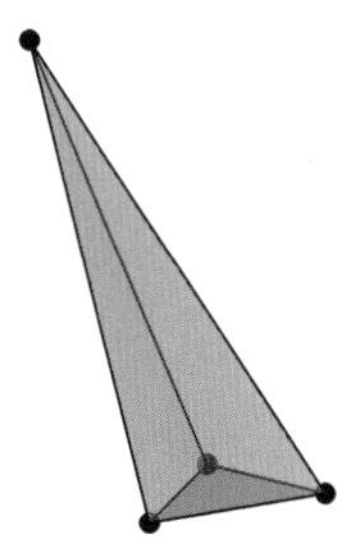

Fig. 1. Reeve tetrahedra with volume $h = 3$.

Attending to this fact, finding families of a big volume for lattice polytopes with interior points might look feasible but the following theorem by Hensley proves this fact to be wrong:

Theorem 3.1 (Hensley, 1967 [10]). *For fixed d and $k > 0$, there is a bound on the volume of lattice d-polytopes with k interior lattice points.*

Even more, there is a bound in the number of lattice polytopes that you can find with a certain number of lattice points:

Theorem 3.2 (Lagarias and Ziegler, 1991 [17]). *There is only finitely many lattice d-polytopes with k interior lattice points for every d and k.*

These strong theorems make a huge difference between the family of polytopes with interior points and hollow polytopes.

Hensley gives a proof of the volume upper bound but this bound for lattice polytopes with interior points may not be "nice". Best bounds are found to be double exponential [1] and occurring in simplices as Hensley conjectured.

For classification purposes, having an upper bound in the volume for some families of hollow polytopes is important, and work involved in this area has been produced in last years. Fixing a minimal width for hollow polytopes, their volume gets bounded as we will see below.

Averkov and Wagner [3] found a relation between the area and the lattice width of arbitrary lattice-free convex sets in the plane:

Theorem 3.3. *For a lattice-free convex body in* $\mathbb{R}^2$ *with* $\mathrm{w} := \mathrm{w}(K)$ *and* A *the area of* K*:*

$$\begin{aligned}
&A \leq \infty && \textit{for} && 0 < \mathrm{w}(K) \leq 1\\
&A \leq \tfrac{\mathrm{w}^2}{2(\mathrm{w}-1)} && \textit{for} && 1 < \mathrm{w}(K) \leq 2\\
&A \leq \tfrac{3\,\mathrm{w}^2}{3\,\mathrm{w}+1-\sqrt{1+6\,\mathrm{w}-3\,\mathrm{w}^2}} && \textit{for} && 2 < \mathrm{w}(K) \leq 1+2/\sqrt{3}\\
&A \geq \tfrac{3}{8}\,\mathrm{w}^2 && \textit{for} && 0 < \mathrm{w}(K) \leq 1+2/\sqrt{3}
\end{aligned}$$

Going in further dimension, Averkov et al. get an upper bound for the volume of hollow lattice polytopes in dimension 3 and width greater or equal to 3 [2]. This result have been improved lately for any convex hollow 3-body of width larger than 2.155 by Iglesias-Valiño and Santos:

Theorem 3.4 (Iglesias Valiño and Santos, 2018 [12]). *Let* K *be a hollow convex 3-body of lattice width* w*, with* $w > 1 + 2/\sqrt{3} = 2.155$*. Then,* $\mathrm{vol}(K)$ *is bounded above by*

$$\begin{aligned}
&\frac{8w^3}{(w-1)^3}, && \text{if } w \geq \frac{2}{\sqrt{3}}(\sqrt{5}-1)+1 \approx 2.427, \text{ and}\\
&\frac{3w^3}{4(w-(1+2/\sqrt{3}))}, && \text{if } w \leq 2.427.
\end{aligned}$$

Here the lower bound in the width is in fact, the flatness constant for convex bodies in dimension 2 ([3, 11]).

The proof of this upper bound for the volume uses the same idea of Proposition 2 in Section 4 of [2] that relies in finding relations between the width bound and the succesive minima and covering minima [12].

4. Width threshold and classification of hollow polytopes

In this section we recall the definition of what is called the width threshold defined by Blanco et al. in [8] and check the relations of this number with the maximum width of hollow and empty lattice polytopes.

In this section, when we talk about classification of hollow polytopes, we say that P and Q are *unimodular equivalent* if there exist an affine integer isomorphism, $f : \mathbb{R}^d \to \mathbb{R}^d$ that fix the lattice and send P in Q, i.e., $f(\mathbb{Z}^d) = \mathbb{Z}^d$ and $f(P) = Q$ with determinant 1.

Definition 4.1. We define *finiteness threshold width* to be the minimum number $w^\infty(d)$, such that for every $w \in \mathbb{Z}$ with $w > w^\infty(d)$ there is only finitely many lattice polytopes of width $> w$ modulo unimodular equivalence. This constant depends only on the dimension d.

For each d, there is a $w^\infty(d)$ such that for every $n \in \mathbb{N}$ all but finitely many d-polytopes with n lattice points have width $\leq w^\infty(d)$ [8].

It is interesting to know which are the values known for the width threshold for hollow and empty polytopes. In Table 1 we can see values set up by new advances in last years.

Relations between maximum width or width threshold between empty and hollow polytopes are shown in the following theorem by Blanco et al.:

Theorem 4.1 (Theorem 1.5 in [8]).

(1) $w^\infty(d) \leq w^\infty(d+1)$ *for all* d.
(2) $w^\infty(d) \leq w_H(d-1) < \infty$.
(3) $w_E(d-1) \leq w^\infty(d)$ *for* $d \geq 3$.
(4) $w_H(d-2) \leq w^\infty(d)$.

The constants $w_E(d)$ and $w_H(d)$ are the maximum width achieved by an empty or a hollow d-polytope, respectively.

Inequality (1) follow from constructive arguments that show that it is not possible to lower the width while increasing the dimension of a polytope.

The proof of inequality (2) relies in the following lemma:

Lemma 4.1. *All but finitely many polytopes in dimension d with size n are hollow.*

n is the number of lattice points in the polytope, $n := |P \cap \mathbb{Z}^d|$, this parameter is called size of the polytope.

This lemma together with the finiteness of hollow d-polytopes not having a projection to $d-1$ [19] imply the inequality $w^\infty(d) \leq w_H(d-1)$, and in particular, inequality (2) implies the finiteness of width threshold for any dimension d because the maximum width of a hollow polytope is bounded by flatness theorem introduced in section 2.

In table 1, we can see the values of these threshold width for low dimesion, and it is easy to realise that the inequalities $w^\infty(d) \leq w_H(d-1)$ and $w_H(d-2) \leq w^\infty(d)$ are not tight for $d \geq 3$.

Table 1. Information known about maximum width of different classes of hollow polytopes and finiteness width threshold.

Dimension	$w_E(d-1)$	$w_H(d-2)$	$w^\infty(d)$	$w_H(d-1)$
$d=1$	-	-	0	-
$d=2$	1	-	0	1
$d=3$	1	1	1	2
$d=4$	1 [22]	2	2 [8]	3
$d=5$	≥ 4 [9]	3 [2, 4]	≥ 4	≥ 4

Note: This table is taken from [8].

The value of the finiteness threshold width is unclear for higher dimension but in the same paper where they introduce the concept, Blanco et al. proof the value for dimension 4.

Theorem 4.2 (Theorem 1.2 in [8]). $w^\infty(4) = 2$.

It is uncertain which is the value for dimension 5, so some new information about the finiteness threshold width will be interesting.

4.1. *New results in classification of hollow polytopes*

Information about projecting hollow polytopes and some advances trying to get the value of the flatness constant and finiteness threshold width have lead into new results that allow to reach classification for hollow polytopes:

Theorem 4.3 (Treutlin, 2008 [24]). *Let $P \in \mathbb{Z}^3$ be a lattice polytope without interior lattice points, then either:*

(1) P projects to a Δ_1, the unimodular simplex in dimension 1.
(2) P can be projected to the double unimodular simplex $2\Delta_2$.
(3) P is in a list of finitely many hollow 3-polytopes.

In this theorem, Treutlin gives the clue to a new approach to classify hollow polytopes and he ask if this theorem can be generalized for arbitrary dimension.

Looking at this theorem, we can realise that the polytope $2\Delta_2$ appears here because it is the only hollow polytope in dimension 2 with width 2, and so, all hollow 3-polytopes with a lattice projection to this simplex have width 2.

After the result of Treutlin, Nill and Ziegler [19] prove the conjectured result in general dimension:

Theorem 4.4 (Nill and Ziegler, 2011 [19]). *Any hollow lattice d-polytope P admits a lattice projection onto a hollow lattice $d-1$-polytope, except if P belongs to one of finitely many exceptions.*

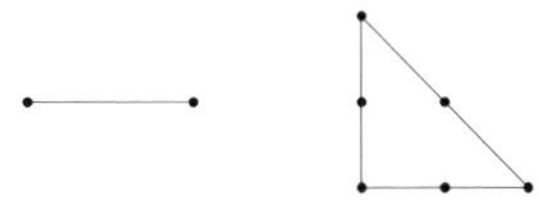

Fig. 2. $\Delta_1 := \text{conv}(0,(0,1))$ and $2\Delta_2 := \text{conv}(0,(2,0),(0,2))$.

As a corollary of this theorem, there is only two possibilities for hollow d-polytopes:

- They project onto a hollow lattice polytope of dimension $d' < d$ that do not project onto hollow polytopes of dimension lower than d', or
- they are contained in a hollow maximal d-polytope. And the number of these maximal hollow d-polytopes is finite.

According to Nill and Ziegler theorem, a classification of maximal hollow d-polytopes is theoretically possible for every dimension as they construct an upper bound for the volume of hollow polytopes in the proof of theorem 4.4. Even though, find a reasonable upper bound for the volume or the number of vertices of these maximal hollow polytopes is a challenging task and definitely of some interest.

Averkov et al. have pursued this objective for dimension 3 and they found the 12 maximal hollow polytopes of width greater than 1 that contain all hollow 3-polytopes that do not project to the unit segment or to $2\Delta_2$. See references [2, 4] for explicit description of these maximal hollow 3-polytopes.

In last years, Iglesias-Valiño and Santos [12, 13] also use this approach in order to get a complete classification of empty 4-simplices. Empty 3-simplices have been classified in 1964 by White [25] who gave a description for every empty simplex realising that all of them have width 1.

In the case of empty 4-simplices, Iglesias-Valiño and Santos give the number of the finitely many exceptions that do not project to a hollow polytope of dimension $d \leq 3$ and they characterize which families of empty 4-simplices project to Δ_1 and $2\Delta_2$.

There are also some families that project to 3-dimensional hollow polytopes. Some of these families were described before in a work by Mori-Morrison-Morrison where they refer to empty 4-simplices in terms of *stable quintuples and terminal quotient singularities* [18].

In these works, it can be clearly seen that there are exceptional empty 4-simplices of width greater than two do not project to dimension less than 4, and there is also some exceptions of width 2 that they do not project to dimension less than 4. Except for these "sporadic simplices", every other 4-simplex can be added to a family of simplices that project to either:

- The unit segment (equivalently, they have width one). These are one family with 3 parameters.
- $2\Delta_2$. There are 2 different families with 2 parameters.

- 52 families of hollow triangular bipyramids that depend on 1 parameter.

5. A computation-proof problem

In this section, we discuss the complexity and difficulty of computing the width and problems that arise in the classification of hollow d-polytopes as said in section 4. In particular, the calculation of width for one particular family of simplices and classification of empty simplices in dimension 4 are addressed throughout the section.

The existence of hollow polytopes of arbitrarily large volume and the need of bounding the width by below in order to get a nice upper bound for the volume of certain hollow polytope is treated in this section. Trying to enumerate all hollow polytopes up to these bounds is computationally hard and some techniques are required to get better upper bounds and algorithms.

Even more, Sebő [23] states that is difficult to compute the width of empty simplices of the following form:

$$\sigma(v) := \operatorname{conv}(e_1, e_2, \cdots, e_n, v), \qquad \text{with} \qquad v \in \mathbb{Z}^n.$$

This way to express an empty simplex may seem quite particular, but for $d \leq 4$ all empty d-simplices have at least one unimodular facet so all of them are equivalent to one of this form.

Theorem 5.1 (Sebő, 1999 [23]). *It is NP-complete decide if the width of an empty simplex of the form $\sigma(v)$ is ≤ 1.*

However, if the dimension is fixed there are algorithms that find the width of these particular empty simplices in polynomial time. As an example [9] propose a solution based in a integer linear program:

Theorem 5.2 (Lemma 11 in [9]). *Let W be an upper bound for the width of a simplex of the form $\sigma(v)$. Then $\mathrm{w}(\sigma(v))$ is the optimal value of the following program:*

$$\begin{array}{c}\textit{minimize } w \textit{ subject to:}\\ \begin{array}{rcll} w_0 \leq & l_i & \leq w_0 + w & \textit{for} \quad 1 \leq i \leq d, \\ w_0 \leq & \sum_{i=1}^{d} l_i v_i & \leq w_0 + w, & \\ & \sum_{i=1}^{d} l_i W^{i-1} & \geq 1, & \end{array}\end{array}$$

with l_i being the linear functional that attains the width and w and w_0 integer values for the width. But different programs can be used as in [23].

By looking at the complexity of "the easiest example" of hollow polytopes and realising that is computationally intense, it does not seem reasonable trying to give a complete classification on certain class of hollow polytopes just with an exhaustive enumeration.

As an example of the difficulty of this task, Iglesias-Valiño and Santos in [12] use an enumeration approach that spent more than a year of CPU-time in order to reach a complete enumeration of empty 4-simplices with volume lower than 7600.

There are some other results that will difficult the enumeration of empty simplices in dimension greater than 4.

- The quotient group of $\mathbb{Z}^d$ by the sublattice generated by the vertices of an empty d-simplex is cyclic when $d \leq 4$ [6], but this is not true in bigger dimension.
- Every empty 4-simplex has an unimodular facet. In dimension 5, there are examples of empty simplices without unimodular facets (Theorem 8 [9]).

These 2 facts have been used in the classification work in [12] to speed up and make the enumeration easier. The impossibility to use them in dimension 5 make things even more difficult.

Looking at this example, it does not seem feasible reach a complete enumeration with an exhaustive enumeration of more complicated lattice polytopes. Even though, trying to get a classification of hollow 4-simplices, empty 4-polytopes or empty 5-simplices would be a possible next step to continue developing the classifications results in the line of Section 4.

6. Open questions and future work

We would like to finish this survey with a list of open questions and future directions of work related with the topic discussed in the sections above.

(1) Is there a hollow lattice d-polytope of width larger than d for some d? Is there such a hollow d-simplex?

(2) It would be quite interesting to define the exact value for the finiteness threshold width in every dimension d. Answering this question it would give $w_H(d)$ values as inequalities in Theorem 4.1 show.
Some small dimension values of the finiteness width threshold are known, $w^\infty(3) = 1$, $w^\infty(4) = 2$ and we do not know the exact value in dimension 5, but it is known that it is bounded from below $w^\infty(5) \geq 4$ because of the existence of an empty 4-simplex of width 4. One possible question before attaining the exact value would be to ask if this width threshold is strictly increasing with $d \geq 3$.

(3) The question 1.12 in [8] is an interesting one. Blanco et al. ask if the finiteness threshold width of an arbitrary dimension with $d > 4$ is always a value that does not depend on the number of lattice points of a polytope.

$$w^\infty(d) = w^\infty(d, d+1) \qquad \text{for all } d > 4$$

in which $w^\infty(d, n)$ is the finiteness threshold with in dimension d for polytopes with size n.
This would imply that it is just sufficient to find the value of the width threshold only for empty lattice simplices.

(4) In this survey, some approaches to classify hollow polytopes for low dimension have been discussed. As discussed in Section 5 getting classification as a result

of enumeration does not seem reachable if no new methods are found as the latest classification results are computationally expensive [12].
It is still worth to ask if there is feasible to attain a classification of hollow 4-polytopes. At least it would be nice to know the exact value of $w_H(4)$. Even more, enumerating the finitely many maximal hollow 4-polytopes would be a great achievement. Perhaps, trying to get a complete classification of empty 4-polytopes it would be a interesting middle step.

(5) There are some applications interested in classification of some families of hollow polytopes. Characterizing Ehrhart polynomials and h^*-vector of polytopes is a frequent topic in which a lot of people is interested in. Iglesias-Valiño and Santos results give a complete description of the volume of the facets of empty 4-simplices [13]. This information is equivalent to give h^*-vectors of the form $(1, 0, h_2^*, h_3^*, 0)$ which may give some information for the people studying h^*-vectors.

Acknowledgments

Thank to the organizers of Summer Workshop on Lattice Polytopes at Osaka University during the summer of 2018. Specially Professor T. Hibi for the funding and A. Tsuchiya for the organization of travel, accomodation and the whole event. The author also wants to thank Francicso Santos for suggestions and valuable comments. The author is supported by grants MTM2017-83750-P, MTM2014-54207-P; BES-2015-073128 of the Spanish Ministry of Science (AEI/FEDER, UE)

References

[1] G. Averkov, J. Krümpelmann and B. Nill, Lattice Simplices with a Fixed Positive Number of Interior Lattice Points: A Nearly Optimal Volume Bound. *International Mathematics Research Notices*, rny130 (2018). DOI: `https://academic.oup.com/imrn/advance-article-abstract/doi/10.1093/imrn/rny130/5036933`

[2] G. Averkov, J. Krümpelmann and S. Weltge, Notions of maximality for integral lattice-free polyhedra: the case of dimension three. *Math. Oper. Res.*. Vol 42-4 (2017). DOI: `https://doi.org/10.1287/moor.2016.0836`

[3] G. Averkov and C. Wagner, Inequalities for the lattice width of lattice-free convex sets in the plane. *Beitr. Algebra Geom.* (2012) 53:1–23. `http://dx.doi.org/10.1007/s13366-011-0028-8`

[4] G. Averkov, C. Wagner and R. Weismantel, Maximal lattice-free polyhedra: finiteness and an explicit description in dimension three, *Math. Oper. Res.* **36** (2011), no. 4, 721–742. `http://www.jstor.org/stable/41412333`

[5] W. Banaszczyk, A. E. Litvak, A. Pajor, and S. J. Szarek, The flatness theorem for nonsymmetric convex bodies via the local theory of Banach spaces. *Math. Oper. Res.* 24 (1999), no. 3, 728–750. `https://doi.org/10.1287/moor.24.3.728`

[6] M. Barile, D. Bernardi, A. Borisov and J.-M. Kantor, On empty lattice simplices in dimension 4. *Proc. Am. Math. Soc.* **139** (2011), no. 12, 4247–4253. `https://www.jstor.org/stable/41291992`

[7] A. Barvinok, Lattice points and lattice polytopes in Handbook of Discrete and Computational Geometry, Third Edition, edited by Jacob E. Goodman, Joseph O'Rourke, and Csaba D. Tóth. *CRC Press*, November 2017, pp 447–474. ISBN 9781498711395

[8] M. Blanco, C. Haase, J. Hoffman, F. Santos, The finiteness threshold width of lattice polytopes, preprint 2016. `https://arxiv.org/abs/1607.00798`

[9] C. Haase and G. M. Ziegler, On the maximal width of empty lattice simplices. *Eur. J. Combin.* 21 (2000), 111–119. `https://doi.org/10.1006/eujc.1999.0325`

[10] D. Hensley. Lattice vertex polytopes with interior lattice points. *Pacific J. Math.* 105:1 (1983), 183–191. `https://projecteuclid.org/euclid.pjm/1102723501`

[11] C. A. J. Hurkens, Blowing up convex sets in the plane. *Linear Algebra Appl.* 134, 121–128 (MR 91i:52009) (1990) `https://doi.org/10.1016/0024-3795(90)90010-A`

[12] O. Iglesias Valiño and F. Santos, Classification of empty lattice 4-simplices of width larger than two. *Accepted for publication in Tran. Amer. Math. Soc.* DOI:10.1090/tran/7531. `https://arxiv.org/pdf/1704.07299.pdf`

[13] O. Iglesias Valiño and F. Santos, The complete classification of empty lattice 4-simplices In preparation 2018.

[14] R. Kannan, L. Lovász, Covering minima and lattice-point-free convex bodies. *Annals of Mathematics*, 128 (1988), 577–602. `https://www.jstor.org/stable/1971436`

[15] J.-M. Kantor. On the width of lattice-free simplices. *Compositio Math.*, 118:235–241, 1999. `https://doi.org/10.1023/A:1001164317215`

[16] A. Khintchine. A quantitative formulation of Kronecker's theory of aproximation. *Izv. Akad. Nauk. SSSR, Ser. Mat.* 12 pp 113–122, 1948. `http://mi.mathnet.ru/eng/izv3022`

[17] J. C. Lagarias and G. M. Ziegler. Bounds for lattice polytopes containing a fixed number of interior points in a sublattice. *Canadian J. Math.*, 43:5 (1991), 1022–1035. `https://cms.math.ca/10.4153/CJM-1991-058-4`

[18] S. Mori, D. R. Morrison and I. and Morrison, On four-dimensional terminal quotient singularities. *Math. Comput.* 51 (1988), no. 184, 769–786. `http://www.jstor.org/stable/2008778`

[19] B. Nill and G. M. Ziegler, Projecting lattice polytopes without interior lattice points. *Math. Oper. Res.*, 36:3 (2011), 462–467. `https://doi.org/10.1287/moor.1110.0503`

[20] J. E. Reeve, On the Volume of lattice polyhedra. *Proc. Lond. Math. Soc.* (3) 7 (1957), 378–395. `https://doi.org/10.1112/plms/s3-7.1.378`

[21] M. Rudelson. Distances between non-symmetric convex bodies and the MM*-estimate. *Positivity* (2000) 4: 161. `https://doi.org/10.1023/A:1009842406728`

[22] H. E. Scarf, Integral polyhedra in three space. *Math. Oper. Res.* 10 (1985), 403–438. `https://doi.org/10.1287/moor.10.3.403`

[23] A. Sebő, An introduction to empty lattice simplices, in *Integer Programming and Combinatorial Optimization (Graz 1999)* `https://doi.org/10.1007/3-540-48777-8_30`

[24] J. Treutlein, 3-dimensional lattice polytopes without interior lattice points. `https://arxiv.org/abs/0809.1787` (2008).

[25] G. K. White. Lattice tetrahedra. *Canadian J. Math.* 16 (1964), 389–396. `https://doi.org/10.4153/CJM-1964-040-2`

The Lecture Hall cone as a toric deformation

Lukas Katthän

Institut für Mathematik, Goethe-Universität Frankfurt
Frankfurt am Main, Germany
E-mail: katthaen@math.uni-frankfurt.de

The Lecture Hall cone is a simplicial cone whose lattice points naturally correspond to Lecture Hall partitions. The celebrated Lecture Hall Theorem of Bousquet-Mélou and Eriksson states that a particular specialization of its multivariate Ehrhart series factors in a very nice and unexpected way. Over the years, several proofs of this result have been found, but it is still not considered to be well-understood from a geometric perspective.

In this note we propose two conjectures which aim at clarifying this result. Our main conjecture is that the Ehrhart ring of the Lecture Hall cone is actually an initial subalgebra A_n of a certain subalgebra of a polynomial ring, which is itself isomorphic to a polynomial ring. As passing to initial subalgebras does not affect the Hilbert function, this explains the observed factorization. We give a recursive definition of certain Laurent polynomials, which generate the algebra A_n. Our second conjecture is that these Laurent polynomials are in fact polynomials.

We computationally verified that both conjectures hold for Lecture Hall partitions of length at most 12.

Keywords: Lecture Hall partition, Lecture Hall Theorem, SAGBI basis, toric deformation.

1. Introduction

A *Lecture Hall partition*[a] is a finite sequence $\lambda = (\lambda_1, \lambda_2, \ldots, \lambda_n) \in \mathbb{Z}^n$ satisfying

$$\frac{\lambda_1}{n} \geq \frac{\lambda_2}{n-1} \geq \cdots \geq \frac{\lambda_n}{1} \geq 0.$$

The set $\mathbf{L}_n$ of Lecture Hall partitions can be viewed as the set of lattice points in the *Lecture Hall cone*, which is the cone over the simplex with vertices

$$\begin{pmatrix} 1 \\ 0 \\ 0 \\ \vdots \\ 0 \\ 0 \end{pmatrix}, \begin{pmatrix} n \\ n-1 \\ 0 \\ \vdots \\ 0 \\ 0 \end{pmatrix}, \begin{pmatrix} n \\ n-1 \\ n-2 \\ \vdots \\ 0 \\ 0 \end{pmatrix}, \ldots, \begin{pmatrix} n \\ n-1 \\ n-2 \\ \vdots \\ 2 \\ 0 \end{pmatrix}, \begin{pmatrix} n \\ n-1 \\ n-2 \\ \vdots \\ 2 \\ 1 \end{pmatrix}.$$

A Hilbert Basis for $\mathbf{L}_n$ was given in [1, Theorem 5.3], see Remark 4.1 below. Lecture Hall partitions were introduced by Bousquet-Mélou and Eriksson in [2]. Their original motivation for considering Lecture Hall partitions is their *Lecture Hall Theorem.*

[a]Our indexing of the entries of λ is reversed with respect to the usual convention.

Theorem 1.1 (Lecture Hall Theorem, [2, Theorem 1.1]). *The number of Lecture Hall Partitions in* $\mathbf{L}_n$ *which sum to* $N \in \mathbb{N}$ *is equal to the number of partitions of* N *into odd parts less than* $2n$.

The generating function version of the Lecture Hall Theorem is the identity

$$\sum_{\lambda \in \mathbf{L}_n} q^{|\lambda|} = \prod_{i=1}^{n} \frac{1}{1 - q^{2i-1}}, \tag{1}$$

where $|\lambda| := \sum_i \lambda_i$. In this note we consider a bivariate refinement of (1) which also due to Bousquet-Mélou and Eriksson. In order to state it we define

$$|\lambda|_o := \lambda_1 + \lambda_3 + \lambda_5 + \cdots$$
$$|\lambda|_e := \lambda_2 + \lambda_4 + \lambda_6 + \cdots .$$

Theorem 1.2 (Lecture Hall Theorem, bivariate version [2, Eq. (2)]). *It holds that*

$$\sum_{\lambda \in \mathbf{L}_n} q_1^{|\lambda|_o} q_2^{|\lambda|_e} = \prod_{i=1}^{n} \frac{1}{1 - q_1^i q_2^{i-1}}. \tag{2}$$

Clearly, Theorem 1.1 follows from Theorem 1.2 by setting $q_1 = q_2$. There are numerous extensions and generalizations of Theorem 1.1 and Theorem 1.2 in the literature. We refer the reader to the survey article by Savage [7] for a wealth of references. However, even though there are a number of proofs of Theorem 1.1, the result is still not considered to be well understood:

> [...], Theorem 1.2 is hardly understood at all. This is in spite of the fact that by now there are many proofs, including those of Bousquet-Mélou and Eriksson [8–10], Andrews [1], Yee [55,56], Andrews, Paule, Riese, and Strehl [3], Eriksen [31], and Bradford et al. [11]. We have also contributed to the collection of proofs with co-authors Corteel [25], Corteel and Lee [20], Andrews and Corteel [2], Bright [15], and, most recently, Corteel and Lovejoy [23].

C.D. Savage, in [7]

In this note, we propose a new approach to the Lecture Hall Theorem. Roughly speaking, we interpret the left-hand side of (2) as the Hilbert series of the Ehrhart ring of $\mathbf{L}_n$ with respect to a particular grading, and the right-hand side of (2) as the Hilbert series of a polynomial ring. Then, we conjecture that Ehrhart ring of $\mathbf{L}_n$ can be obtained as a "deformation" of a polynomial ring.

To make this idea precise, we recall some background on toric deformations. Let $\Bbbk$ be a field, $S := \Bbbk[x_1, \ldots, x_m]$ the polynomial ring over it and $\prec$ a term order on S. For a finitely generated graded sub-$\Bbbk$-algebra $A \subseteq S$ we consider the *initial subalgebra* $\mathrm{In}_\prec(A) := \Bbbk[\mathrm{In}_\prec(f) \colon f \in A] \subset S$ of A, where $\mathrm{In}_\prec(f)$ denotes the leading term of f. Initial subalgebras have been studied in the context of *SAGBI bases* (also known as *canonical bases*), and we give some background on their theory in Section 4. We offer the following conjecture:

Conjecture 1.1. *Fix $n \in \mathbb{N}$. There exist a $\mathbb{Z}^2$-graded polynomial ring $\Bbbk[x_1, \ldots, x_m]$ for some $m \in \mathbb{N}$, a graded subalgebra $A_n \subseteq \Bbbk[x_1, \ldots, x_m]$ and a term order $\prec$ on $\Bbbk[x_1, \ldots, x_m]$, such that*

(a) A_n is isomorphic to a polynomial ring in n variables of degrees $(1,0), (2,1), \ldots, (n, n-1)$, and
(b) $\mathrm{In}_\prec(A_n)$ is isomorphic to the Ehrhart ring of $\mathbf{L}_n$.

Note that the Hilbert series of A_n in the conjecture equals

$$\prod_{i=1}^{n} \frac{1}{1 - q_1^i q_2^{i-1}}.$$

Moreover, the Hilbert series of A_n and $\mathrm{In}_\prec(A)$ coincide [4, Proposition 2.4], therefore Conjecture 1.1 implies the Lecture Hall Theorem.

We are going to give a much more precise version of Conjecture 1.1 below as Conjecture 4.1. For this, we are first going to define the *Lecture Hall polynomials* in Section 3. These are Laurent polynomials which we conjecture to be polynomials (Conjecture 3.1) and which generate our candidate for the algebra A_n in Conjecture 1.1. We also have a conjecture for a minimal SAGBI basis of A_n, and in Theorem 4.1 we give some computational evidence.

If Conjecture 4.1 is true, then it gives rise to a bijection φ_n from the set of subsets of $[n-1]$ to the Hilbert basis of $\mathbf{L}_n$. In Section 5 we give some properties of this map, and in Section 6 we list its values for $n \leq 8$. Finding a description of this map for all n might be a problem of independent interest.

Acknowledgments

The author thanks Matthias Beck for bringing the Lecture Hall Theorem to the authors attention. Moreover, I would like thank Victor Reiner, Benjamin Braun and Volmar Welker for several helpful discussions.

Research that led to this paper was supported by the National Science Foundation under Grant No. DMS-1440140 while the author was in residence at the Mathematical Sciences Research Institute in Berkeley, California, during the Fall 2017 semester on *Geometric and Topological Combinatorics*.

2. Notation and conventions

For $n \in \mathbb{N}$ we set $[n] := \{1, \ldots, n\}$. For a finite set S we write 2^S for its power set. For an $(n \times n)$-matrix M and two sets $S, T \subseteq [n]$ of the same cardinality we write $\Delta_T^S(M)$ for the submatrix of M using the rows and columns with indices in S and T, respectively.

Most of our discussion is independent on the group field $\Bbbk$. Therefore, we chose to work over the rationals $\mathbb{Q}$ for concreteness.

3. The Lecture Hall polynomials

In this section, we are going to define the Lecture Hall polynomals, which generate our candidate for the algebra A_n of Conjecture 1.1.

Let $S_n := \mathbb{Q}[y_1, y_2, \ldots, y_n]$ be a polynomial ring in n variables. It is convenient to set $S_\infty := \bigcup_n S_n$, using the natural inclusion $S_{n-1} \hookrightarrow S_n$. For a sequence of polynomials $\mathbf{P} := P_1, P_2, \ldots$ in S_∞ we define an infinite matrix $M(\mathbf{P})$ by setting

$$M(\mathbf{P})_{i,j} := \begin{cases} -P_{j-i+1} & \text{if } j \geq i \\ 0 & \text{otherwise.} \end{cases}$$

Explicitly, $M(\mathbf{P})$ looks as follows:

$$M(\mathbf{P}) = \begin{pmatrix} -P_1 & -P_2 & -P_3 & -P_4 & \ldots \\ 0 & -P_1 & -P_2 & -P_3 & \ldots \\ 0 & 0 & -P_1 & -P_2 & \ldots \\ 0 & 0 & 0 & -P_1 & \ldots \\ \vdots & \vdots & \vdots & \vdots & \ddots \end{pmatrix}$$

For $i \in \mathbb{N}$ let

$$\mathcal{E}_i(\mathbf{P}) := -\det \Delta^{[\lceil i/2 \rceil]}_{\{\lfloor i/2 \rfloor, \ldots, i\}}(M(\mathbf{P})),$$

i.e., the negative of the minor of $M(\mathbf{P})$ which uses the $\lceil \frac{i}{2} \rceil$ many top rows and the columns with indices $\lfloor \frac{i}{2} \rfloor, \ldots, i$. One might think of $\mathcal{E}_i(\mathbf{P})$ as the negative of the determinant the maximal top-aligned square submatrix of $M(\mathbf{P})$, whose top right corner is $-P_i$ and which does not contain any of the zeros of $M(\mathbf{P})$.

Definition 3.1. We define the *Lecture Hall sequence* to be the sequence $\mathbf{P}_{\mathrm{LH}} := \ell_1, \ell_2, \ldots$ of rational functions in $\mathrm{Quot}(S_\infty)$ which are defined by requiring that

$$\mathcal{E}_i(\mathbf{P}_{\mathrm{LH}}) = y_1^i y_2^{i-1} \cdots y_{i-1}^2 y_i \tag{LHS$_i$}$$

for all $i \geq 1$. We call the elements of $\mathbf{P}_{\mathrm{LH}}$ *Lecture Hall polynomials* and denote them with ℓ_1, ℓ_2, and so on.

We talk about "Lecture Hall polynomials" instead of "Lecture Hall rational functions", because they are Laurent polynomials (Proposition 3.1) and we conjecture them to be actual polynomials.

Example 3.1. Here we compute the first Lecture Hall polynomials. The equations (LHS$_i$) for $i = 1, \ldots, 4$ are:

$$\begin{aligned} y_1 &= \mathcal{E}_1(\mathbf{P}_{\mathrm{LH}}) = -\det(-\ell_1) = \ell_1 \\ y_1^2 y_2 &= \mathcal{E}_2(\mathbf{P}_{\mathrm{LH}}) = -\det(-\ell_2) = \ell_2 \\ y_1^3 y_2^2 y_3 &= \mathcal{E}_3(\mathbf{P}_{\mathrm{LH}}) = -\det \begin{pmatrix} -\ell_2 & -\ell_3 \\ -\ell_1 & -\ell_2 \end{pmatrix} = \ell_1 \ell_3 - \ell_2^2 \\ y_1^4 y_2^3 y_3^2 y_4 &= \mathcal{E}_4(\mathbf{P}_{\mathrm{LH}}) = -\det \begin{pmatrix} -\ell_3 & -\ell_4 \\ -\ell_2 & -\ell_3 \end{pmatrix} = \ell_2 \ell_4 - \ell_3^2 \end{aligned}$$

We read off the first two equations that $\ell_1 = y_1$ and $\ell_2 = y_1^2 y_2$. Using this we can solve the third one for ℓ_3 and obtain that $\ell_3 = y_1^3 y_2^2 + y_1^2 y_2^2 y_3 = y_1^2 y_2^2(y_1 + y_3)$. This in turn allows us to solve the fourth equation for ℓ_4, which yields that

$$\ell_4 = y_1^2 y_2^3 (y_1 + y_3)^2 + y_4^2 y_3^2 y_2^2 y_1.$$

Proposition 3.1.

(a) The Lecture Hall sequence is well-defined.
(b) Each ℓ_i is a Laurent polynomial and has coefficients in $\mathbb{Z}$.
(c) For each $i \geq 0$, the i-th Lecture Hall polynomial ℓ_i depends only on the variables $y_1, \ldots, y_i$, and it is non-constant as a function of y_i.

Proof. Let $\mathbf{P} := P_1, P_2, \ldots$ be a sequence of polynomials in S_∞. We note that $\mathcal{E}_i(\mathbf{P})$ depends only on $P_1, \ldots, P_i$, and it is linear in P_i. Therefore, (LHS$_i$) implies that each P_i is a rational function in the P_j for $j < i$, and we can solve the equations (LHS$_i$) in an iterative way for each i. Thus, $\mathbf{P}_{\mathrm{LH}}$ is well-defined.

For the second item, note that for $i \geq 3$, the coefficient of P_i in (LHS$_i$) equals the determinant of $\Delta^{\{2,\ldots,\lceil i/2 \rceil\}}_{\{\lfloor i/2 \rfloor,\ldots,i-1\}}(M(\mathbf{P}_{\mathrm{LH}}))$ up to a sign. Since the entries of $M(\mathbf{P}_{\mathrm{LH}})$ are constant along diagonals, the latter equals $\mathcal{E}_{i-2}(\mathbf{P}_{\mathrm{LH}})$. By construction, this minor is a monomial, and thus solving for ℓ_i yields a Laurent polynomial with coefficients in $\mathbb{Z}$.

For the last item, note that it follows from our discussion that ℓ_i is determined by (LHS$_1$), (LHS$_2$), …, (LHS$_i$). But those equations involve only the variables $y_1, \ldots, y_i$ and thus ℓ_i depends only on them. Further, the right-hand side of (LHS$_i$) is non-constant as a function of y_i, and thus the same holds for the left-hand side. But the left-hand side is a polynomial in the $\ell_1, \ldots, \ell_i$, and all ℓ_j for $j < i$ do not depend on y_i. Thus ℓ_i cannot be constant as a function of y_i. □

Based on computational evidence, we offer the following conjecture:

Conjecture 3.1. *Each Lecture Hall polynomial ℓ_i is a polynomial.*

We verified this conjecture for $i \leq 12$. A list of the ℓ_i for $i \leq 8$ is given below in Section 6. In view of our application to the Lecture Hall cone, We define a $\mathbb{Z}^2$-grading on S_∞ by setting

$$\deg y_i := \begin{cases} (0,1) & \text{if } i \text{ is even,} \\ (1,0) & \text{if } i \text{ is odd.} \end{cases}$$

Proposition 3.2. *Each Lecture Hall polynomial ℓ_i is homogeneous of degree $(i, i-1)$ with respect to the given $\mathbb{Z}^2$-grading on S_∞.*

Proof. This is a tedious but straightforward computation. We omit the details. □

The Propositions 3.1 and 3.2 together imply the following corollary:

Corollary 3.1. *The algebra* $A_n := \mathbb{Q}[\ell_1, \dots, \ell_n] \subset \mathrm{Quot}(S_\infty)$ *generated by the first* n *Lecture Hall polynomials is isomorphic to a* $\mathbb{Z}^2$*-graded polynomial ring. Its Hilbert series equals*

$$\prod_{i=1}^{n} \frac{1}{1 - q_1^i q_2^{i-1}}$$

Proof. Part (c) of Proposition 3.1 implies that all ℓ_i are algebraically independent, and thus A_n is a polynomial ring. The claim for the Hilbert series is then immediate from Proposition 3.2. □

Since each ℓ_i depends only on $y_1, \dots, y_i$ and is a Laurent polynomial, the algebra A_n is actually a subalgebra of $\mathbb{Q}[y_1^{\pm}, \dots, y_n^{\pm}]$. Moreover, if Conjecture 3.1 holds, then we even have that $A_n \subseteq \mathbb{Q}[y_1, \dots, y_n]$.

4. Realizing the Ehrhart ring as an initial subalgebra

Let us return to a general discussion of initial subalgebras. Let $A \subseteq S := \mathbb{Q}[x_1, \dots, x_m]$ be a subalgebra. A finite collection of polynomials $p_1, \dots, p_r \in A$ is called a SAGBI basis (Subalgebra Analogue to Gröbner Basis for Ideals) if $\mathrm{In}_\prec(A)$ is generated by $\mathrm{In}_\prec(p_1), \dots, \mathrm{In}_\prec(p_r)$ as $\mathbb{Q}$-algebra. SAGBI bases were introduced by Robbiano and Sweedler [6], and independently by Kapur and Madeler [5]. Their theory is in many ways similar to the theory of Gröbner bases, with the important difference that not every finitely-generated subalgebra admits a finite SAGBI basis. We refer the reader to Chapter 11 of [8] and to the short survey by Bravo [3] for more information about these concepts.

We now introduce our candidate for a SAGBI basis for the algebra generated by the Lecture Hall polynomials.

Definition 4.1. For a finite set $S \subseteq \mathbb{N}, S \neq \emptyset$ let

$$\ell_S := -\det \Delta_{S+1}^{[\#S]}(M(\mathbf{P}_{\mathrm{LH}})),$$

where $S + 1 := \{s + 1 \colon s \in S\}$. In other words, ℓ_S is the negative of the minor of $M(\mathbf{P}_{\mathrm{LH}})$ using $\#S$ many top rows and the columns in $S + 1$. In addition, we set $\ell_\emptyset := \ell_1$.

Note that $\ell_i = \ell_{\{i-1\}}$ for $i \in \mathbb{N}$, and that $\ell_{\{\lfloor i/2 \rfloor, \dots, i-1\}} = \mathcal{E}_i(\mathbf{P}_{\mathrm{LH}}) = y_1^i y_2^{i-1} \cdots y_{i-1}^2 y_i$. Now we can state the precise version of Conjecture 1.1:

Conjecture 4.1. *Let* $\prec$ *be the degree-lexicographic term order on* $S_n := \mathbb{Q}[y_1, \dots, y_n]$ *with* $y_1 \succ y_2 \succ \dots \succ y_n$. *Assume that Conjecture 3.1 holds, i.e., that the* ℓ_i *are polynomials. Further, let* $A_n = \mathbb{Q}[\ell_1, \dots, \ell_n]$. *Then:*

(i) *The initial subalgebra* $\mathrm{In}_\prec(A_n)$ *equals the Ehrhart ring of* $\mathbf{L}_n$.

(ii) *For each $S \subseteq [n-1]$, the leading term of ℓ_S is a monomial whose exponent vector is a Lecture Hall partition, and the Lecture Hall partitions arising in this way form the Hilbert basis of $\mathbf{L}_n$. In particular, the set $\{\ell_S : S \subseteq [n-1]\}$ is a SAGBI basis for A_n.*

Remark 4.1.

(a) Part (i) of Conjecture 4.1 implies the Lecture Hall Theorem.

(b) Our conjectured SAGBI basis has 2^{n-1} elements, and this is also the cardinality of the Hilbert basis H_n of $\mathbf{L}_n$. Indeed, by [1, Theorem 5.3], the Hilbert basis H_n consists of the vectors of the form

$$(a_1, a_2, \ldots, a_{i-1}, a_i, 0, \ldots, 0) \tag{3}$$

with $a_1, \ldots a_i \in \mathbb{Z}$, $a_2 > a_3 > \cdots > a_i > 0$ and $a_1 - a_2 + 1$. Omitting the first coordinate yields a bijection from H_n to $2^{[n-1]}$, and thus $\#H_n = 2^{n-1}$.

(c) Part (ii) of Conjecture 4.1 alone implies that the Ehrhart ring of $\mathbf{L}_n$ is contained in $\mathrm{In}_\prec(A_n)$. Hence part (ii), Corollary 3.1 and Lecture Hall Theorem together imply part (i). Moreover, since the cardinality of H_n and of $\{\ell_S : S \subseteq [n-1]\}$ are the same, Conjecture 4.1 follows once one can show the following:

   (1) For each $S \subseteq [n-1]$, the exponent vector of the leading term of ℓ_S is of the form (3), and
   (2) for any two distinct sets $S, S' \subseteq [n-1]$, $S \neq S'$, the exponent vectors of the leading terms of ℓ_S and $\ell_{S'}$ are different.

   On the other hand, it seems desirable to find a proof of Conjecture 4.1 which does not rely on the Lecture Hall Theorem.

Let us verify the conjecture in two small cases.

Example 4.1. The cases $n = 1$ and $n = 2$ are rather trivial, so we consider the case $n = 3$. We have that $\mathrm{In}_\prec(\ell_\emptyset) = \mathrm{In}_\prec(\ell_1) = y_1$, $\mathrm{In}_\prec(\ell_{\{1\}}) = \mathrm{In}_\prec(\ell_2) = y_1^2 y_2$ and $\mathrm{In}_\prec(\ell_{\{2\}}) = \mathrm{In}_\prec(\ell_3) = y_1^3 y_2^2$. Moreover, by definition we have that $\ell_{1,2} = -\mathcal{E}_3(\mathbf{P}_{\mathrm{LH}}) = y_1^3 y_2^2 y_3$. Hence the criterion of part (c) of Remark 4.1 is satisfied.

Example 4.2. Next we consider the case $n = 4$. In addition to the computations above, we have that $\mathrm{In}_\prec(\ell_{\{3\}}) = \mathrm{In}_\prec(\ell_4) = y_1^4 y_2^3$ and $\ell_{\{2,3\}} = -\mathcal{E}_4(\mathbf{P}_{\mathrm{LH}}) = y_1^4 y_2^3 y_3^2 y_4$. For the remaining two polynomials we compute that

$$\begin{aligned}
\ell_{\{1,3\}} &= -\det\begin{pmatrix} -\ell_2 & -\ell_4 \\ -\ell_1 & -\ell_3 \end{pmatrix} = \ell_1\ell_4 - \ell_2\ell_3 \\
&= y_1^3 y_2^3 (y_1 + y_3)^2 + y_1^3 y_2^2 y_3^2 y_4 - y_1^4 y_2^3 (y_1 + y_3) \\
&= y_1^4 y_2^3 y_3 + y_1^3 y_2^3 y_3^2 + y_1^3 y_2^2 y_3^2 y_4,
\end{aligned}$$

and thus $\mathrm{In}_\prec(\ell_{\{1,3\}}) = y_1^4 y_2^3 y_3$. Moreover,

$$\begin{aligned}\ell_{\{1,2,3\}} &= -\det\begin{pmatrix}-\ell_2 & -\ell_3 & -\ell_4\\ -\ell_1 & -\ell_2 & -\ell_3\\ 0 & -\ell_1 & -\ell_2\end{pmatrix} = \ell_2\det\begin{pmatrix}-\ell_2 & -\ell_3\\ -\ell_1 & -\ell_2\end{pmatrix} - \ell_1\det\begin{pmatrix}-\ell_3 & -\ell_4\\ -\ell_1 & -\ell_2\end{pmatrix}\\ &= -\ell_2 y_1^3 y_2^2 y_3 + \ell_1(y_1^4 y_2^3 y_3 + y_1^3 y_2^3 y_3^2 + y_1^3 y_2^2 y_3^2 y_4)\\ &= y_1^4 y_2^3 y_3^2 + y_1^4 y_2^2 y_3^2 y_4\end{aligned}$$

which implies that $\mathrm{In}_\prec(\ell_{\{1,2,3\}}) = y_1^4 y_2^3 y_3^2$. In conclusion, the criterion of part (c) of Remark 4.1 is satisfied.

As in these examples, the items in part (c) of Remark 4.1 can be verified computationally for small n. Using `Maple`, we found the following computational evidence for Conjecture 4.1:

Theorem 4.1. *Conjecture 4.1 holds for $n \leq 12$.*

Remark 4.2. In all examples we computed so far, the leading term of ℓ_S has coefficient 1. This is the reason for our choice of signs in the definition of the Lecture Hall polynomials.

Remark 4.3. Conjecture 4.1 implies a particular description of the toric ideal associated to $\mathbf{L}_n$. Consider the polynomial ring $R_n := \mathbb{Q}[x_S : S \subseteq [n-1]]$ and let $I_n \subseteq R_n$ be the kernel of the natural map $R_n \to A_n, x_S \mapsto \ell_S$. Since the ℓ_i are algebraically independent for $i \geq 1$, it is not difficult to see that I_n is generated by the defining relations of the ℓ_S, i.e., by equations of the form $x_S + \det \Delta_{S+1}^{[\#S]}(M(x_\emptyset, x_{\{1\}}, \dots))$.

If Conjecture 4.1 is true, then by Theorem 11.4 of [8] the defining ideal of $\mathrm{In}_\prec(A_n)$, (i.e., the toric ideal of $\mathbf{P}_{\mathrm{LH}}$) is an initial ideal of I_n with respect to a particular non-generic weight order. Therefore one can compute the toric ideal by computing a Gröbner basis of I_n with respect to that term order.

We computed this Gröbner basis using `Macaulay2` for $n \leq 7$, and, as expected from Theorem 4.1, its initial forms generate the toric ideal. We would like to point out that I_7 has $2^{7-1} - 7 = 53$ generators, while the toric ideal and the Gröbner basis have 1351 generators each. Therefore, these considerations might be helpful in understanding the toric ideals of $\mathbf{L}_n$.

5. A certain bijection on finite sets

For a subset $S \subseteq [n-1]$ let $\varphi_n(S)$ be the exponent vector of $\mathrm{In}_\prec(\ell_S)$. If Conjecture 4.1 is true, then $\varphi_n(S)$ a bijection between $2^{[n-1]}$ and the Hilbert basis H_n of $\mathbf{L}_n$. As mentioned above in Remark 4.1, H_n can itself be identified with the power set of $[n-1]$, so φ_n can also be considered as a permutation of this set.

Problem 5.1. *Find a combinatorial description of φ_n.*

A table of the values φ_n for $n \leq 8$ is provided in Section 6. We caution the reader that because Conjecture 4.1 is just a conjecture, a priory the map φ_n might fail to be injective or might fail to take values in H_n for large n, even though we do not expect this to happen. We close this note with a few properties of φ_n.

Proposition 5.1. *Assume that Conjecture 4.1 is true. Then the following holds for $n \in \mathbb{N}$:*

(a) For each $i \leq n$, it holds that $\varphi_n(2^{[i-1]}) = H_i$.
(b) For $S \subseteq [n-1], S \neq \emptyset$, the alternating sum of the entries of $\varphi_n(S)$ equals $\#S$.
(c) For $S \subseteq [n-1], S \neq \emptyset$, the sum of the entries of $\varphi_n(S)$ equals $2\sum_{s \in S} s + 2\#S - \#S^2$.

Proof. It follows from the construction that the restriction of φ_n to $2^{[i-1]}$ equals φ_i, and thus Conjecture 4.1 implies the first claim.

The second and third claim follow from considering the $\mathbb{Z}^2$-grading on S_n. Let $r := \#S$ and $S = \{s_1, \ldots, s_r\}$ with $s_1 < s_2 < \cdots < s_R$. Expanding the minor from the definition of ℓ_S, one sees that one of the summands is $\prod_{i=1}^r \ell_{s_i+2-i}$, and hence $\deg \ell_S = \deg \prod_{i=1}^r \ell_{s_i+2-i}$. The sum and the difference between the two components of the degree correspond to the sum and the alternating sum of the exponent vector of $\mathrm{In}_\prec(\ell_S)$. From this the claimed formulas follow from a straightforward computation. □

6. Some computational data

Here is a list of the Lecture Hall polynomials ℓ_i for $i = 1, \ldots, 8$.

$$
\begin{aligned}
\ell_1 &= y_1 \\
\ell_2 &= y_1^2 y_2 \\
\ell_3 &= y_1^2 y_2^2 (y_1 + y_3) \\
\ell_4 &= y_1^2 y_2^3 (y_1 + y_3)^2 + y_1^2 y_2^2 y_3^2 y_4 \\
\ell_5 &= y_1^2 y_2^4 (y_1 + y_3)^3 + 2 y_1^2 y_2^3 y_3^2 y_4 (y_1 + y_3) + y_1^2 y_2^2 y_3^2 y_4^2 (y_3 - y_5) \\
\ell_6 &= y_1^2 y_2^5 (y_1 + y_3)^4 + 3 y_1^2 y_2^4 y_3^2 y_4 (y_1 + y_3)^2 + y_1^2 y_2^2 y_3^2 y_4^3 (y_3 - y_5)^2 \\
&\quad + y_1^2 y_2^2 y_3^2 y_4^2 (2 y_1 y_2 y_3 + 3 y_2 y_3^2 - 2 y_1 y_2 y_5 - 2 y_2 y_3 y_5 - y_5^2 y_6) \\
\ell_7 &= y_1^2 y_2^6 (y_1 + y_3)^5 + 4 y_1^2 y_2^5 y_3^2 y_4 (y_1 + y_3)^3 + y_1^2 y_2^2 y_3^2 y_4^4 (y_3 - y_5)^3 \\
&\quad + y_1^2 y_2^2 y_3^2 y_4^2 (3 y_1^2 y_2^2 y_3 - 3 y_1^2 y_2^2 y_5 + 9 y_1 y_2^2 y_3^2 - 6 y_1 y_2^2 y_3 y_5 + 2 y_1 y_2 y_3^2 y_4 - 4 y_1 y_2 y_3 y_4 y_5 \\
&\quad + 2 y_1 y_2 y_4 y_5^2 - 2 y_1 y_2 y_5^2 y_6 + 6 y_2^2 y_3^3 - 3 y_2^2 y_3^2 y_5 + 4 y_2 y_3^3 y_4 - 6 y_2 y_3^2 y_4 y_5 + 2 y_2 y_3 y_4 y_5^2 \\
&\quad - 2 y_2 y_3 y_5^2 y_6 - 2 y_3 y_4 y_5^2 y_6 + 2 y_4 y_5^3 y_6 - y_5^3 y_6^2 + y_5^2 y_6^2 y_7) c \\
\ell_8 &= y_1^2 y_2^7 (y_1 + y_3)^6 + 5 y_1^2 y_2^6 y_3^2 y_4 (y_1 + y_3)^4 + y_1^2 y_2^2 y_3^2 y_4^5 (y_3 - y_5)^4 \\
&\quad + y_1^2 y_2^2 y_3^2 y_4^2 (4 y_1^3 y_2^3 y_3 - 4 y_1^3 y_2^3 y_5 + 18 y_1^2 y_2^3 y_3^2 - 12 y_1^2 y_2^3 y_3 y_5 + 3 y_1^2 y_2^2 y_3^2 y_4 - 6 y_1^2 y_2^2 y_3 y_4 y_5 \\
&\quad + 3 y_1^2 y_2^2 y_4 y_5^2 - 3 y_1^2 y_2^2 y_5^2 y_6 + 24 y_1 y_2^3 y_3^3 - 12 y_1 y_2^3 y_3^2 y_5 + 12 y_1 y_2^2 y_3^3 y_4 - 18 y_1 y_2^2 y_3^2 y_4 y_5 \\
&\quad + \text{[see next page]}
\end{aligned}
$$

$$+ 6y_1y_2^2y_3y_4y_5^2 - 6y_1y_2^2y_3y_5^2y_6 + 2y_1y_2y_3^3y_4^2 - 6y_1y_2y_3^2y_4^2y_5 + 6y_1y_2y_3y_4^2y_5^2 - 4y_1y_2y_3y_4y_5^2y_6$$
$$- 2y_1y_2y_4^2y_5^3 + 4y_1y_2y_4y_5^3y_6 - 2y_1y_2y_5^3y_6^2 + 2y_1y_2y_5^2y_6^2y_7 + 10y_2^3y_3^4 - 4y_2^3y_3^3y_5 + 10y_2^2y_3^4y_4$$
$$- 12y_2^2y_3^3y_4y_5 + 3y_2^2y_3^2y_4y_5^2 - 3y_2^2y_3^2y_5^2y_6 + 5y_2y_3^4y_4^2 - 12y_2y_3^3y_4^2y_5 + 9y_2y_3^2y_4^2y_5^2 - 6y_2y_3^2y_4y_5^2y_6$$
$$- 2y_2y_3y_4^2y_5^3 + 4y_2y_3y_4y_5^3y_6 - 2y_2y_3y_5^3y_6^2 + 2y_2y_3y_5^2y_6^2y_7 - 3y_3^2y_4^2y_5^2y_6 + 6y_3y_4^2y_5^3y_6 - 2y_3y_4y_5^3y_6^2$$
$$+ 2y_3y_4y_5^2y_6^2y_7 - 3y_4^2y_5^4y_6 + 3y_4y_5^4y_6^2 - 2y_4y_5^3y_6^2y_7 - y_5^4y_6^3 + 2y_5^3y_6^3y_7 - y_5^2y_6^3y_7^2 + y_5^2y_6^2y_7^2y_8)$$

Finally, we give a table of the values of φ_n for $n \leq 8$. Trailing zeros are omitted.

S	$\varphi_n(S)$
$\emptyset$	(1)
$\{1\}$	$(2,1)$
$\{2\}$	$(3,2)$
$\{1,2\}$	$(3,2,1)$
$\{3\}$	$(4,3)$
$\{1,3\}$	$(4,3,1)$
$\{2,3\}$	$(4,3,2,1)$
$\{1,2,3\}$	$(4,3,2)$
$\{4\}$	$(5,4)$
$\{1,4\}$	$(5,4,1)$
$\{2,4\}$	$(5,4,2,1)$
$\{1,2,4\}$	$(5,4,2)$
$\{3,4\}$	$(5,4,3,2)$
$\{1,3,4\}$	$(5,4,3,1)$
$\{2,3,4\}$	$(5,4,3,2,1)$
$\{1,2,3,4\}$	$(5,4,3)$
$\{5\}$	$(6,5)$
$\{1,5\}$	$(6,5,1)$
$\{2,5\}$	$(6,5,2,1)$
$\{1,2,5\}$	$(6,5,2)$
$\{3,5\}$	$(6,5,3,2)$
$\{1,3,5\}$	$(6,5,3,1)$
$\{2,3,5\}$	$(6,5,3,2,1)$
$\{1,2,3,5\}$	$(6,5,3)$
$\{4,5\}$	$(6,5,4,3)$
$\{1,4,5\}$	$(6,5,4,2)$
$\{2,4,5\}$	$(6,5,4,3,1)$
$\{1,2,4,5\}$	$(6,5,4,1)$
$\{3,4,5\}$	$(6,5,4,3,2,1)$
$\{1,3,4,5\}$	$(6,5,4,2,1)$
$\{2,3,4,5\}$	$(6,5,4,3,2)$
$\{1,2,3,4,5\}$	$(6,5,4)$

S	$\varphi_n(S)$
$\{6\}$	$(7,6)$
$\{1,6\}$	$(7,6,1)$
$\{2,6\}$	$(7,6,2,1)$
$\{1,2,6\}$	$(7,6,2)$
$\{3,6\}$	$(7,6,3,2)$
$\{1,3,6\}$	$(7,6,3,1)$
$\{2,3,6\}$	$(7,6,3,2,1)$
$\{1,2,3,6\}$	$(7,6,3)$
$\{4,6\}$	$(7,6,4,3)$
$\{1,4,6\}$	$(7,6,4,2)$
$\{2,4,6\}$	$(7,6,4,3,1)$
$\{1,2,4,6\}$	$(7,6,4,1)$
$\{3,4,6\}$	$(7,6,4,3,2,1)$
$\{1,3,4,6\}$	$(7,6,4,2,1)$
$\{2,3,4,6\}$	$(7,6,4,3,2)$
$\{1,2,3,4,6\}$	$(7,6,4)$
$\{5,6\}$	$(7,6,5,4)$
$\{1,5,6\}$	$(7,6,5,3)$
$\{2,5,6\}$	$(7,6,5,4,1)$
$\{1,2,5,6\}$	$(7,6,5,2)$
$\{3,5,6\}$	$(7,6,5,4,2,1)$
$\{1,3,5,6\}$	$(7,6,5,3,1)$
$\{2,3,5,6\}$	$(7,6,5,4,2)$
$\{1,2,3,5,6\}$	$(7,6,5,1)$
$\{4,5,6\}$	$(7,6,5,4,3,2)$
$\{1,4,5,6\}$	$(7,6,5,3,2,1)$
$\{2,4,5,6\}$	$(7,6,5,4,3,1)$
$\{1,2,4,5,6\}$	$(7,6,5,2,1)$
$\{3,4,5,6\}$	$(7,6,5,4,3,2,1)$
$\{1,3,4,5,6\}$	$(7,6,5,3,2)$
$\{2,3,4,5,6\}$	$(7,6,5,4,3)$
$\{1,2,3,4,5,6\}$	$(7,6,5)$

S	$\varphi_n(S)$
$\{7\}$	$(8,7)$
$\{1,7\}$	$(8,7,1)$
$\{2,7\}$	$(8,7,2,1)$
$\{1,2,7\}$	$(8,7,2)$
$\{3,7\}$	$(8,7,3,2)$
$\{1,3,7\}$	$(8,7,3,1)$
$\{2,3,7\}$	$(8,7,3,2,1)$
$\{1,2,3,7\}$	$(8,7,3)$
$\{4,7\}$	$(8,7,4,3)$
$\{1,4,7\}$	$(8,7,4,2)$
$\{2,4,7\}$	$(8,7,4,3,1)$
$\{1,2,4,7\}$	$(8,7,4,1)$
$\{3,4,7\}$	$(8,7,4,3,2,1)$
$\{1,3,4,7\}$	$(8,7,4,2,1)$
$\{2,3,4,7\}$	$(8,7,4,3,2)$
$\{1,2,3,4,7\}$	$(8,7,4)$
$\{5,7\}$	$(8,7,5,4)$
$\{1,5,7\}$	$(8,7,5,3)$
$\{2,5,7\}$	$(8,7,5,4,1)$
$\{1,2,5,7\}$	$(8,7,5,2)$
$\{3,5,7\}$	$(8,7,5,4,2,1)$
$\{1,3,5,7\}$	$(8,7,5,3,1)$
$\{2,3,5,7\}$	$(8,7,5,4,2)$
$\{1,2,3,5,7\}$	$(8,7,5,1)$
$\{4,5,7\}$	$(8,7,5,4,3,2)$
$\{1,4,5,7\}$	$(8,7,5,3,2,1)$
$\{2,4,5,7\}$	$(8,7,5,4,3,1)$
$\{1,2,4,5,7\}$	$(8,7,5,2,1)$
$\{3,4,5,7\}$	$(8,7,5,4,3,2,1)$
$\{1,3,4,5,7\}$	$(8,7,5,3,2)$
$\{2,3,4,5,7\}$	$(8,7,5,4,3)$
$\{1,2,3,4,5,7\}$	$(8,7,5)$

S	$\varphi_n(S)$
$\{6,7\}$	$(8,7,6,5)$
$\{1,6,7\}$	$(8,7,6,4)$
$\{2,6,7\}$	$(8,7,6,5,1)$
$\{1,2,6,7\}$	$(8,7,6,3)$
$\{3,6,7\}$	$(8,7,6,5,2,1)$
$\{1,3,6,7\}$	$(8,7,6,4,1)$
$\{2,3,6,7\}$	$(8,7,6,5,2)$
$\{1,2,3,6,7\}$	$(8,7,6,2)$
$\{4,6,7\}$	$(8,7,6,5,3,2)$
$\{1,4,6,7\}$	$(8,7,6,4,2,1)$
$\{2,4,6,7\}$	$(8,7,6,5,3,1)$
$\{1,2,4,6,7\}$	$(8,7,6,3,1)$
$\{3,4,6,7\}$	$(8,7,6,5,3,2,1)$
$\{1,3,4,6,7\}$	$(8,7,6,4,2)$
$\{2,3,4,6,7\}$	$(8,7,6,5,3)$
$\{1,2,3,4,6,7\}$	$(8,7,6,1)$
$\{5,6,7\}$	$(8,7,6,5,4,3)$
$\{1,5,6,7\}$	$(8,7,6,4,3,2)$
$\{2,5,6,7\}$	$(8,7,6,5,4,2)$
$\{1,2,5,6,7\}$	$(8,7,6,3,2,1)$
$\{3,5,6,7\}$	$(8,7,6,5,4,3,1)$
$\{1,3,5,6,7\}$	$(8,7,6,4,3,1)$
$\{2,3,5,6,7\}$	$(8,7,6,5,4,1)$
$\{1,2,3,5,6,7\}$	$(8,7,6,2,1)$
$\{4,5,6,7\}$	$(8,7,6,5,4,3,2,1)$
$\{1,4,5,6,7\}$	$(8,7,6,4,3,2,1)$
$\{2,4,5,6,7\}$	$(8,7,6,5,4,2,1)$
$\{1,2,4,5,6,7\}$	$(8,7,6,3,2)$
$\{3,4,5,6,7\}$	$(8,7,6,5,4,3,2)$
$\{1,3,4,5,6,7\}$	$(8,7,6,4,3)$
$\{2,3,4,5,6,7\}$	$(8,7,6,5,4)$
$\{1,2,3,4,5,6,7\}$	$(8,7,6)$

References

[1] Matthias Beck, Benjamin Braun, Matthias Köppe, Carla D. Savage, and Zafeirakis Zafeirakopoulos, *Generating functions and triangulations for lecture hall cones*, SIAM J. Discrete Math. **30** (2016), no. 3, 1470–1479.

[2] Mireille Bousquet-Mélou and Kimmo Eriksson, *Lecture hall partitions*, Ramanujan J. **1** (1997), no. 1, 101–111.

[3] Ana Bravo, *Some facts about canonical subalgebra bases*, Trends in commutative al-

gebra, Math. Sci. Res. Inst. Publ., vol. 51, Cambridge Univ. Press, Cambridge, 2004, pp. 247–254.

[4] Aldo Conca, Jürgen Herzog, and Giuseppe Valla, *Sagbi bases with applications to blow-up algebras*, J. Reine Angew. Math. **474** (1996), 113–138.

[5] Deepak Kapur and Klaus Madlener, *A completion procedure for computing a canonical basis for a k-subalgebra*, Computers and mathematics (Cambridge, MA, 1989), Springer, New York, 1989, pp. 1–11.

[6] Lorenzo Robbiano and Moss Sweedler, *Subalgebra bases*, Commutative algebra (Salvador, 1988), Lecture Notes in Math., vol. 1430, Springer, Berlin, 1990, pp. 61–87.

[7] Carla D. Savage, *The mathematics of lecture hall partitions*, J. Combin. Theory Ser. A **144** (2016), 443–475.

[8] Bernd Sturmfels, *Gröbner bases and convex polytopes*, University Lecture Series, vol. 8, American Mathematical Society, Providence, RI, 1996.

A short survey on Tesler matrices and Tesler polytopes

Yonggyu Lee
University of California, Davis
Davis, California 95616, USA
E-mail: ygulee@ucdavis.edu

This article is a small collection of theorems and conjectures related to Tesler matrices and Tesler polytopes that includes Morales' conjecture on Ehrhart positivity of the Tesler polytope of hook sum $(1, \dots, 1)$ and the relation between Tesler matrices and the Hilbert series of the quotient ring of diagonal covariants.

Keywords: Tesler polytopes, Tesler matrices.

1. Introduction

Tesler matrices have been drawing attention due to its relation to the Macdonald polynomial, parking functions and diagonal harmonics. Recently, Mészáros, Morales and Rhoades defined the Tesler polytope, which has as its integer points as Tesler matrices and showed that the Tesler polytopes have nice properties and are unimodular equivalent to flow polytopes [10]. In this article, we will collect some known properties of Tesler polytopes. In Section 3, we discuss how to approach the conjecture by Morales (will be stated in Section 3). Also, we will take a look at some known connections between Tesler matrices and diagonal harmonics.

2. Background

This section provides some basic knowledge in polyhedra theory, Tesler polytopes and the study of Ehrhart positivity.

Definition 2.1. For $\alpha = (\alpha_1, \dots, \alpha_n) \in \mathbb{Z}^n_{\geq 0}$, we say that an $n \times n$ matrix with integer entries $A = (a_{i,j})$ is a Tesler matrix of hook sum α if

- A is an upper triangular matrix with non-negative entries.
- (Hook sum condition) For all $1 \leq k \leq n$ we have

$$hs_k(A) = (a_{k,k} + a_{k,k+1} + \cdots + a_{k,n}) - (a_{1,k} + a_{2,k} + \cdots + a_{k-1,k}) = \alpha_k.$$

Also, let $T(\alpha)$ be the set of all Tesler matrices of hook sum α.

Definition 2.2 ([10]). Let $\alpha = (\alpha_1, \dots, \alpha_n) \in \mathbb{Z}^n_{\geq 0}$ and $\mathbb{U}(n)$ be the set of $n \times n$ upper triangular matrices. Then the Tesler polytope of hook sum α is

$$\mathrm{Tes}_n(\alpha) = \{A = (a_{i,j}) \in \mathbb{U}(n) \mid hs_k(A) = \alpha_k \textit{ for } 1 \leq k \leq n \textit{ and } x_{i,j} \geq 0 \textit{ for every } 1 \leq i \leq j \leq n\}.$$

By the definition, it is easy to see that the set of lattice points of $\text{Tes}_n(\alpha)$ is exactly the set of the Tesler matrices with hook sum α. It turns out that the Tesler polytopes have very nice properties.

Proposition 2.3 ([10]). *For any $\alpha \in \mathbb{Z}^n_{>0}$, $\text{Tes}_n(\alpha)$ has the following properties :*

(1) $\text{Tes}_n(\alpha)$ is a simple polytope.
(2) $\text{Tes}_n(\alpha)$ is combinatorially equivalent to $\Delta_1 \times \cdots \times \Delta_{n-1}$.
(3) The set of all vertices of $\text{Tes}_n(\alpha)$ is
$\{A \mid A$ is permutational Tesler matrix of hook sum $\alpha\}$.
Where the permutation Tesler matrices are the $n\times n$ Tesler matrices with exactly one nonzero entry in each row.
(4) Two vertices v and w of $\text{Tes}_n(\alpha)$ are adjacent iff $s(v)$ can be obtained from $s(w)$ by moving 1 in a row to a different column in the same row. Where $s(A) = (s_{i,j})$,

$$s_{i,j} = \begin{cases} 1, & \text{if } a_{i,j} \neq 0, \\ 0, & \text{if } a_{i,j} = 0, \end{cases}$$

for any $n \times n$ matrix $A = (a_{i,j})$.

By third and fourth part of the above proposition, it is not very hard to show that $\text{Tes}_n(\alpha)$ is totally unimodular in the subspace defined by the hook sum conditions. Also, the Tesler polytopes can be thought of as flow polytopes.

Definition 2.4. For any $\alpha = (\alpha_1, \ldots, \alpha_n) \in \mathbb{Z}^n$, the flow polytope $\text{Flow}_n(\alpha)$ of complete graph K_{n+1} with net flow α_i on vertex i for $i = 1, 2, \ldots, n$ and $-\sum_{i=1}^n \alpha_i$ on $n+1$ is the set of functions $f : E \to \mathbb{R}_{\geq 0}$ where E is the edge set of K_{n+1} such that $\sum_{j>s} f(k,j) - \sum_{i<k} f(i,k) = \alpha_k$.

For $\alpha = (\alpha_1, \ldots, \alpha_n) \in \mathbb{Z}^n_{\geq 0}$, *the type A_n Kostant partition function $K_{A_n}(\alpha)$* is the number of ways to write $\alpha' := (\alpha_1, \ldots, \alpha_n, -\sum_{i=1}^n \alpha_i)$ as an $\mathbb{N}$ combination of the type A_n roots $e_i - e_j$, $1 \leq i < j \leq n+1$. It can be shown that $K_{A_n}(\alpha)$ is exactly the number of integer points in $\text{Flow}_n(\alpha)$.

Theorem 2.5 ([10]). *For $\alpha \in \mathbb{Z}_{\geq 0}$, $\text{Flow}(\alpha)$ is unimodular equivalent to $\text{Tes}_n(\alpha)$. (i.e, there exists an unimodular transformation u such that $u(\text{Flow}(\alpha)) = \text{Tes}_n(\alpha)$)*

Proof. Let $\Phi : \text{Tes}_n(\alpha) \to \text{Flow}_n(\alpha)$ defined by $\Phi : A = (a_{i,j}) \mapsto f_A$, where

$$f_A(i,j) = \begin{cases} a_{i,j} & \text{if } j \neq n+1, \\ a_{i,i} & \text{if } j = n+1. \end{cases}$$

The map Φ is a linear transformation that simply permutes the coordinates of $\text{Tes}_n(\alpha)$. Therefore the determinant of Φ is ± 1. □

Definition 2.6. Let V be a vector space. A subset $\Lambda \subset V$ is called a *lattice*, if

- Λ is an additive subgroup of V.
- For any bounded set $B \subset V$, $B \cap \Lambda$ is finite.
- Λ spans V.

Definition 2.7. A subset $P \subset V$ is called a *lattice polytope* if it is a convex hull of finitely many points in Λ.

Let $E_P(t)$ be a function that maps $t \in \mathbb{Z}_{\geq 0}$ to the number of lattice points in the t-th dilate of a d-dimensional lattice polytope $P \subset V$ (i.e, $E_P(t) = |tP \cap \Lambda|$). Ehrhart proved that this function is a polynomial in t of degree d and $E_P(t)$ is called the *Ehrhart polynomial* of P [5]. A condition that all the coefficients of $E_P(t)$ being positive is called the *Ehrhart positivity* of P.

By Theorem 2.5, the Ehrhart polynomial for $\mathrm{Tes}_n(\alpha)$ can be calculated using the following formula which was first given by Lidskii in [3] for complete graphs K_{n+1} and then generalized by Baldoni and Vergne in [14] to arbitrary graphs G. Most recently, Mészáros and Morales in [9] recover Baldoni-Vergne's result by using polytopal subdivisions on flow polytopes.

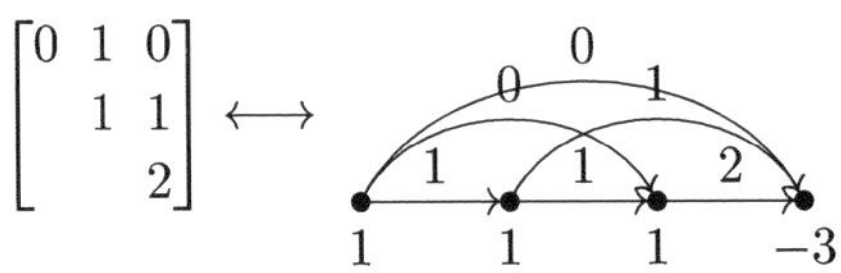

Fig. 1. The numbers below the vertices indicate the net flows on corresponding vertices and the numbers indicate the weight on corresponding edges.

Theorem 2.8 (Lidskii [3], Baldoni-Vergne [14], Mészáros-Morales [9]). *Suppose G is a connected graph on the vertex set $[n+1]$, with m edges directed $i \to j$ if $i < j$, and with at least one outgoing edge at vertex i for $i = 1, ..., n$. Let out_k (and in_k, respectively) denote the outdegree (and the indegree, respectively) of vertex k in G minus 1. Then for any $\alpha = (\alpha_1, \dots, \alpha_n) \in \mathbb{Z}^n_{\geq 0}$, the Ehrhart polynomial of* the flow polytope on a graph G of flow α $\mathrm{Flow}_G(a)$ *is given by*

$$i(\mathrm{Flow}_G(\alpha), t) = \sum_{j} \prod_{k=1}^{n} \binom{\alpha_k t + \mathrm{out}_k}{j_k} \cdot K_G(j_1 - \mathrm{out}_1, j_2 - \mathrm{out}_2, \dots, j_n - \mathrm{out}_n, 0)$$

$$= \sum_{j} \prod_{k=1}^{n} \left(\!\!\binom{\alpha_k t - \mathrm{in}_k}{j_k}\!\!\right) \cdot K_G(j_1 - \mathrm{out}_1, j_2 - \mathrm{out}_2, \dots, j_n - \mathrm{out}_n, 0)$$

where each summation is over all weak compositions $j = (j_1, \dots, j_n)$ of $m - n$ that are $\geq (\mathrm{out}_1, \dots, \mathrm{out}_n)$ in dominance order.

From the above formula, when the indegree of any vertices are either 0 or 1, we easily get the positivity result. For the Tesler polytope case, we take G to be the complete graph on $n+1$ vertices with the direction $i \to j$ whenever $i < j$ and

$\alpha \in \mathbb{Z}^n_{\geq 0}$. Thus, with $\text{out}_k = k-1$, $\text{in}_k = k-2$ and $\left(\!\binom{\alpha_k t - \text{in}_k}{j_k}\!\right) = \binom{\alpha_k t - \text{in}_k + j_k - 1}{j_k}$ we get,

$$i(\text{Flow}_n(\alpha), t) = \sum_j \prod_{k=1}^n \binom{\alpha_k t + k - 1}{j_k} \cdot K_G(j_1 - \text{out}_1, j_2 - \text{out}_2, \ldots, j_n - \text{out}_n, 0)$$
$$= \sum_j \prod_{k=1}^n \binom{\alpha_k t - k + j_k + 1}{j_k} \cdot K_G(j_1 - \text{out}_1, j_2 - \text{out}_2, \ldots, j_n - \text{out}_n, 0)$$

But even with the above formula, it is still unclear if all the coefficient are positive or not. One possible way to approach the problem of Ehrhart positivity is "alpha positivity" which arises from McMullen's formula.

Theorem 2.9 ([13, McMullen's formula]). *Let P be a d-dimensional polytope. Then there exists a funtion α such that the following is true :*

$$|P \cap \mathbb{Z}^n| = \sum_{F \,:\, \textit{face of } P} \alpha(F, P)\, \text{Vol}_{dimF}(F).$$

Where $\alpha(F, P)$ only depends on the normal cone of P at F and Vol_d *is the relative volume.*

One important fact of $\alpha(F, P)$ is that the value is invariant under dilation of polytope. So, if we plug in tP (the t-th dilate of a polytope P) to the above formula, then we obtain

$$|tP \cap \mathbb{Z}^n| = E_P(t) = \sum_{F \,:\, \text{face of } P} \alpha(F, P)\, \text{Vol}_{dimF}(F) t^{dimF}.$$

which leads to the following very useful corollary.

Corollary 2.1. *Let P be a polytope and $E_P(t) = e_d t^d + \cdots + e_1 t + 1$ be it's Ehrhart polynomial. Then,*

$$e_i = \sum_{F \,:\, i\textit{-dimension face of } P} \alpha(F, P)\, \text{Vol}_i(F).$$

By the above formula, we see that if all the alpha values $\alpha(F, P)$ are positive for any face F of P, we get the positivity of all the coefficients e_i's of the Ehrhart polynomial of P. This condition of $\alpha(F, P) > 0$ for any face F of P is called *alpha positivity* of P. But it is well-known that the function α is not unique, so the alpha positivity of a polytope depends on the choice of the function alpha. There are three different constructions of alpha's by Berline-Vergne [12], Pommersheim-Thomas [8] and Ring-Schürmann [11]. In this article we will refer to Berline-Vergne's alpha as *BV-alpha* and Ring-Schürmann's as *RS-mu*. Even though the alpha positivity is a stronger statement than the Ehrhart positivity, the fact that it can be explored without even knowing the Ehrhart polynomial makes it a very powerful tool. Also, alpha positivity is not necessarily harder than Ehrhart positivity but it is rather a diffrent approach. Another great benefit of this method is the reduction theorem.

Definition 2.10. Let P be a polytope in a vector space V.

- The *feasible cone* of P at F is :

$$\text{fcone}(F, P) := \{u \in V \mid x + \delta u \in P \text{ for sufficiently small } \delta\},$$

  where x is any relative interior point of F.
  The *pointed feasible cone of* P at F is

$$\text{fcone}^p(F, P) = \text{fcone}(F, P)/\operatorname{lin}(F),$$

  where $\operatorname{lin}(F)$ is the vector subspace of V obtained by translating the affine hull of F.
- Given a polytope P in a subspace V of $\mathbb{R}^n$. Let V^* be the dual space of V in $\mathbb{R}^n$. Then the *normal cone* of P at F with respect to V is

$$\text{ncone}_V(F, P) := \{u \in V^* \mid \langle u, p_1\rangle \geq \langle u, p_2\rangle \ \forall p_1 \in F, \forall p_2 \in P\}.$$

- The *normal fan* $\sum_V(P)$ of P with respect to V is the collection of all normal fans of P.

If a normal fan $\sum_V Q$ of a polytope Q is a refinement of $\sum_V P$ of a polytope P, then intuitively this means that Q has almost the same face structure as P but some faces of P are shrunk to points by moving the facet defining planes around (parallel movement). For certain alpha's that are the solutions for the McMullen's formula, this kind of facet movement does not affect the positivity.

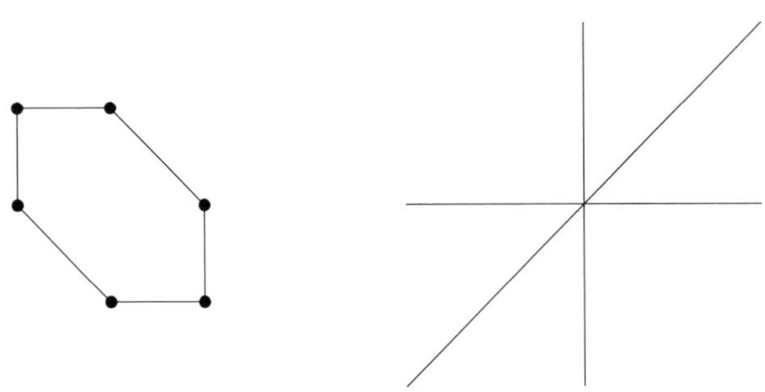

Fig. 2. A hexagon (left) and it's normal fan (right).

Theorem 2.11 (Reduction Theorem [6]). *Suppose Ψ is a function on indicator functions of rational cones C in a vector space V such that*

- *Ψ is a linear transformation on the algebra of rational cones in V.*
- *If a cone C contains a line, then $\Psi([C]) = 0$.*

Let P and Q be two polytopes in V such that $\sum_V Q$ is a refinement of $\sum_V P$. Then for any fixed k, if we set $\alpha(F,P) = \Psi([\text{ncone}(F,P)/\text{lin}(F)])$ and $\alpha(F,P) > 0$ for every k-dimensional face F of P, then $\alpha(G,Q) > 0$ for every k-dimensional face G of Q.

By this theorem, whenever we prove the alpha positivity (with a function alpha satisfying the conditions of the above theorem) of a polytope P, we obtain the alpha positivity of any polytopes that has normal fan which is a refinement of $\sum_V P$. Currently, BV-alpha is known to satisfy the conditions of the reduction theorem.

BV-alpha and RS-mu in general gives different values, even for a very simple 2 dimensional triangle. But in some cases, the values coincide for an unknown reason.

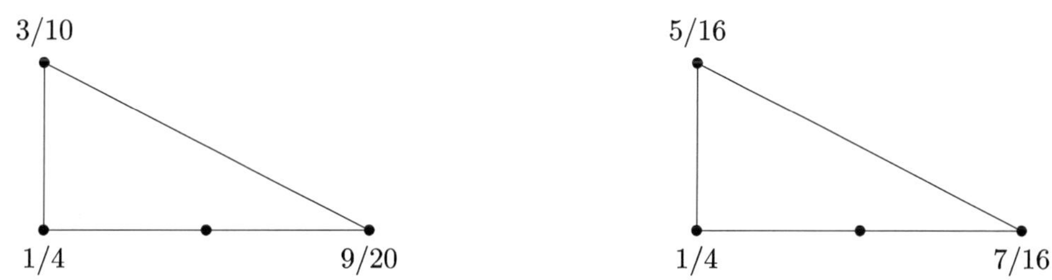

Fig. 3. The left figure shows the BV-alpha values on the vertices of the triangle and the right figure shows the RS-mu values on the vertices of the same triangle.

For example, Castillo and Liu showed in [6] that any function which satisfies the McMullen's formula that is invariant under the change of coordinates agree with BV-alpha on any faces of any polytopes which normal fan is the Braid fan. And for an unknown reason, BV-alpha and RS-mu agrees on the faces of projected $\text{Tes}_3(1,1,1)$ (will be defined in Example 3.1).

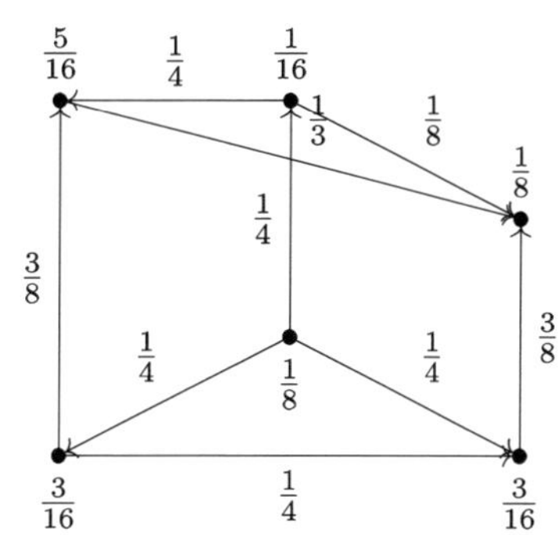

Fig. 4. The numbers indicate the BV-alpha or RS-mu values on the vertices and edges of projected Tesler one.

3. Alpha value calculation for the Tesler polytopes

Conjecture 3.1 (Morales [13]). *For each positive integer n, $\mathrm{Tes}_n(1,0,0,\ldots,0)$ and $\mathrm{Tes}_n(1,\ldots,1)$ are both Ehrhart positive.*

In this section, to approach the conjecture of the Ehrhart positivity of $\mathrm{Tes}_n(1,\ldots,1)$, we will discuss how to efficiently calculate the BV-alpha values on the faces of the projected Tesler polytopes.

3.1. *Projected Tesler polytope*

Definition 3.1. Let *hook-sum space* to be the subspace of $\mathbb{R}^{\binom{n}{2}}$ defined by the k-th hook sum conditions for $1 \le k \le n$,

$$hs_k(A) = (a_{k,k} + a_{k,k+1} + \cdots + a_{k,n}) - (a_{1,k} + a_{2,k} + \cdots + a_{k-1,k}) = 1.$$

We consider a family of maps which will induce unimodular transformation from the hook-sum space. The reason for considering these maps is that the Tesler polytopes are not full dimensional which often cause some complications when we want to apply theorems or formulas in polyhedral theory, in particular, BV-alpha calculation. We can instead calculate the alpha values on the image of a Tesler polytope under the unimodular transformations which is now a full dimensional polytope. Since the map is unimodular, the Tesler polytope and the image have the same Ehrhart polynomial. Thus, showing the alpha positivity for the image will imply the Ehrhart positivity of the Tesler polytope. Notice that despite the Ehrhart polynomial being the same, the alpha values on individual faces may differ and it is possible for the image polytope to be alpha positive, but the preimage is not or vice versa.

Theorem 3.2. *Let $\mathbb{U}(n)$ be the set of $n \times n$ upper triangular matrices with non-negative entries and $(x_{i,j}) \in \mathbb{U}(n)$. Then for any $j = (j_1, j_2, \ldots, j_{n-1}) \in [n] \times [n-1] \times [n-1] \times \cdots \times [2]$,*
A map $\phi_j : \mathbb{U}(n) \longrightarrow \mathbb{U}(n)$ defined by

$$\phi_j((x_{i,j})) = (y_{i,j}) \quad \textit{where} \quad y_{i,j} = \begin{cases} x_{i,j} & \textit{if } j \neq j_i \\ 0 & \textit{if } j = j_i \end{cases}$$

Defines a unimodular transformation between the hook-sum space and $\mathbb{R}^{\binom{n}{2}}$.

Example 3.1. The most natural way to obtain such map would be just by ignoring the diagonal. $\phi_d :=$

$$\begin{bmatrix} d_1 & x_{1,1} & x_{1,2} & \cdots & x_{1,n-1} \\ & d_2 & x_{2,2} & \cdots & x_{2,n-1} \\ & & \ddots & \cdots & \vdots \\ & & & d_{n-1} & x_{n-1,n-1} \\ & & & & d_n \end{bmatrix} \longrightarrow \begin{bmatrix} x_{1,1} & x_{1,2} & \cdots & x_{1,n-1} \\ & x_{2,2} & \cdots & x_{2,n-1} \\ & & \ddots & \vdots \\ & & & x_{n-1,n-1} \end{bmatrix}$$

We will call $\phi_d(\text{Tes}_n(\alpha))$ *the projected Tesler polytope of hook sum* α (or projected $\text{Tes}_n(\alpha)$) and $\phi_d(\text{Tes}_n(1,\ldots,1))$ as *the projected Tesler one*. In most of the cases, n will be fixed, so this should not cause any confusions.

3.2. *The BV-alpha*

The BV-alpha values of the projected Tesler one in Figure 4 were calculated via nice formula of BV-alpha. However, in this case, we were lucky that the polytope was unimodular, because otherwise, we have to decompose the pointed feasible cones into unimodular cones which will drastically complicate the calculation. The Tesler polytope being unimodular is the main reason for the author to try BV-alpha before other alpha constructions.

Theorem 3.3 (Berline-Vergne [12]). *There exists a function Φ on indicator functions of rational cones C in a vector space V with the following properties:*

(1) Φ is a linear transformation on the vector space over $\mathbb{Q}$ spanned by the indicator functions $[C]$ of all rational cones $C \subset V$.
(2) If a cone C contains a line, then $\Phi([C]) = 0$.
(3) It is invariant under orthogonal unimodular transformation, thus, is symmetric about coordinates, that is, invariant under rearranging coordinates with signs.
(4) Setting

$$\alpha(F,P) := \Phi([\text{fcone}^p(F,P)]),$$

gives a solution to McMullen's formula.

BV-alpha has a nice formula for the codimension 2 and 3 faces which were given in [2], Example 19.3. For the lower dimensional faces, we have an algorithm to find a formula, but it gets way more complicated to explicitly calculate the values.

Lemma 3.1 ([2]). *Let $C = \text{fcone}(F,P)/\text{lin}(F) = \text{Cone}(u_1,u_2)$ where u_1 and u_2 form a basis for the orthogonal projection of $\mathbb{Z}^n$ to $\mathbb{R}^n / \text{lin}(F)$ and α be the BV-alpha. Then,*

$$\alpha(F,P) = \frac{1}{4} + \frac{1}{12}\left(\frac{\langle u_1,u_2\rangle}{\langle u_1,u_1\rangle} + \frac{\langle u_1,u_2\rangle}{\langle u_2,u_2\rangle}\right).$$

Lemma 3.2. *[2] Let $C = \text{fcone}(F,P)/\text{lin}(F) = \text{Cone}(u_1,u_2,u_3)$. Where u_1 and u_2 form a basis for the orthogonal projection of $\mathbb{Z}^n$ to $\mathbb{R}^n / \text{lin}(F)$ and α be the BV-alpha. Then,*

$$\alpha(F,P) = \tfrac{1}{8} + \tfrac{1}{24}\left(\tfrac{\langle u_1,u_2\rangle}{\langle u_1,u_1\rangle} + \tfrac{\langle u_1,u_2\rangle}{\langle u_2,u_2\rangle} + \tfrac{\langle u_1,u_3\rangle}{\langle u_1,u_1\rangle} + \tfrac{\langle u_1,u_3\rangle}{\langle u_3,u_3\rangle} + \tfrac{\langle u_2,u_3\rangle}{\langle u_2,u_2\rangle} + \tfrac{\langle u_2,u_3\rangle}{\langle u_3,u_3\rangle}\right).$$

Remark 3.4. The formula for codimension 4 faces has more than 1000 terms...

4. Diagonal Hilbert series and Tesler matrices

Here we give the definition of diagonal Hilbert series along with Haglund's theorem and some important special cases as remarks.

Definition 4.1. We define the *double algebra*

$$DS := \mathbb{C}[x_1, \ldots, x_n, y_1, \ldots, y_n],$$

the *polarized power sums*

$$p_{k,l} := \sum_{i=1}^{n} x_i^k y_i^l \text{ for } k + l > 0,$$

and the *diagonal covariant algebra*

$$DR := DS/(p_{k,l} = \sum_{i=1}^{n} x_i^k y_i^l \ : \ k + l > 0),$$

where $(p_{k,l} = \sum_{i=1}^{n} x_i^k y_i^l \ : \ k+l > 0)$ is the ideal generated by polarized power sums (note that DR has a natural bigrading structure from the degrees with respect to x and y).

Finally, *diagonal Hilbert series* is the Hilbert series of DR and denoted by $\mathrm{Hilb}(n; q, t)$. Therefore,

$$\mathrm{Hilb}(n; q, t) := \sum_{ij} \dim(DR_{ij}) q^i t^j.$$

The following is the Haglund's formula which expresses the diagonal Hilbert series as a weighted sum over the Tesler matrices.

Theorem 4.2 ([7]). *For a given* $\mathbf{a} \in \mathbb{Z}^n_{\geq 0}$ *and a Tesler matrix* $A = (a_{i,j})$, *we define*

$$\mathrm{weight}(A) = (-(1-q)(1-t))^{\mathrm{Pos}(A)-n} \prod_{a_{i,j}>0} [a_{i,j}]_{q,t}.$$

where $\mathrm{Pos}(A)$=# *of positive entries of* A *and* $[k]_{q,t} = \frac{q^k - t^k}{q-t}$. *Let*

$$P_{\mathbf{a}}(q, t) := \sum_{A \text{ : Tesler matrix of hook sum } \mathbf{a}} \mathrm{weight}(A),$$

Then $\mathrm{Hilb}(n; q, t) = P_{(1,\ldots,1)}(q, t)$.

Remark 4.3. When $\mathbf{a} \in \mathbb{Z}^n_{>0}$, $\mathrm{Pos}(A) \geq n$. So $P_{\mathbf{a}}(q, t)$ becomes a polynomial. The above identity proves that $P_{(1,\ldots,1)}(q, t)$ is (q, t)-*positive*, that is, all the coefficients as a polynomial of q and t are positive.

Remark 4.4. When $t = 1$, $P_{(1,\ldots,1)}(q, 1) = \sum_{A \in \Pi_n(1,\ldots,1)} \prod_{a_{i,j}} [a_{i,j}]_q$ where

$$\Pi_n(1, \ldots, 1) := \{A \in \mathrm{Tes}_n(1, \ldots, 1) : \text{A has exactly one nonzero entry for each row}\}$$

and is called the *set of permutational Tesler matrices* (by Prop 2.3, it is the set of the vertices of $\mathrm{Tes}_n(1, \ldots, 1)$).

However, the proof of theorem 4.2 is an algebraic proof which leaves the combinatorial nature of the weighted sum unknown.

Open problem 1. Find a combinatorial proof of the (q,t)-positivity of $P_{\mathbf{a}}(q,t)$.

Also, there is a refinement of the above open problem. To state it, we need to associate the Dyck path to a Tesler matrix. Let us say an $n \times n$ matrix A *factors* at $1 \leq k \leq n-1$ if A is block diagonal with an initial block of size $k \times k$.

Proposition 4.1 ([4]). *For any $\alpha \in \mathbb{Z}^n_{\geq}$, and Tesler matrix $A \in T(\alpha)$, we have the equality*

$$\sum_{s=1}^{k} a_{s,s} + \sum_{i=1}^{k} \sum_{j=k+1}^{n} a_{i,j}\chi(i<j) = \sum_{s=1}^{k} u_s \qquad (\text{for all } 1 \leq k \leq n).$$

where $\chi(\cdot) = \begin{cases} 1 & \text{when } \cdot \text{ is true} \\ 0 & \text{when } \cdot \text{ is false} \end{cases}.$
Thus, we have the inequality

$$\sum_{s=1}^{k} a_{s,s} \leq \sum_{s=1}^{k} u_s. \tag{1}$$

with equality for $k=n$. Moreover, we see that A factors at k if and only if

$$\sum_{s=1}^{k} a_{s,s} = \sum_{s=1}^{k} u_s.$$

The inequality (1) allows us to associate a Dyck path $Dyck(A)$ to a given Tesler matrix A in the following way done in [4]:
Given $A = (a_{i,j}) \in T(1,\ldots,1)$, starting at $(0,0)$ of the $n \times n$ lattice square, and for each diagonal element $a_{i,i}$, we go successively NORTH one unit step and then EAST $a_{i,i}$ unit steps. The inequality in (1) (when all $u_i = 1$) assures that the resulting path remains weakly above the diagonal (i,i). Moreover, since each time equality holds in (1), the factorization criterion applies, and there is a direct connection between the Dyck path and the block diagonal structure of the corresponding matrix.

Conjecture 4.1 ([4]). *For each composition $p \models n$ the sum*

$$Q_p(q,t) = \sum_{A \in \mathrm{Tes}_n(1,\ldots,1)} weight(A)\chi(p(Dyck(A)) = p)$$

is (q,t)-positive.

Acknowledgements

The author thanks Takayuki Hibi for the financial support (The Birth of Modern Trends in Commutative Algebra and Convex Polytope with Statistical and Computational Strategies) provided in "2018 Summer Workshop in Lattice Polytopes". Also, the author thanks Fu Liu, Brendon Rhoades and Federico Castillo for helpful comments.

References

[1] A. H. Morales, *Erhart polynomials of examples of flow polytopes*, https://sites.google.com/site/flowpolytopes/ehrhart.
[2] A. Barvinok. *Integer points in polyhedra.* Zurich Lectures in Advanced Mathematics. European Mathematical Society (EMS), Zürich, 2008.
[3] B. V. Lidskii, *The Kostant function of the system of roots An*, Funktsional. Anal. i Prilozhen. 18 (1984), no. 1, 76–77. MR 739099
[4] D. Armstrong, A. Garsia, J. Haglund, B. Rhoades, and B. Sagan. *Combinatorics of Tesler matrices in the theory of parking functions and diagonal harmonics.* J. Comb. 3 (2012), 451–494.
[5] E. Ehrhart (1962), *Sur les polyédres rationnels homothétiques à n dimensions*, C. R. Acad. Sci. Paris. 254, 616–618.
[6] F. Castillo and F. Liu, *Berline-Vergne valuation and generalized permutohedra*, Discrete and Computational Geometry, to appear.
[7] J. Haglund, A polynomial expression for the Hilbert series of the quotient ring of diagonal coinvariants, Adv. Math. 227 (2011) 2092-2106]
[8] J. Pommersheim and H. Thomas, *Cycles representing the Todd class of a toric variety*, J. Amer. Math. Soc. 17 (2004), no. 4, 983–994. MR 2083474
[9] K. Mészáros and A. H. Morales, *Volumes and Ehrhart polynomials of flow polytopes*, arXiv:1710.00701.
[10] K. Mészáros, A. H. Morales, B. Rhoades, The polytope of Tesler matrices.
[11] M. H. Ring and A. Schürmann, *Local formulas for Ehrhart coefficients from lattice tiles*, arXiv:1709.10390.
[12] N. Berline and M. Vergne, *Local Euler-Maclaurin formula for polytopes*, Mosc. Math. J. 7 (2007), no. 3, 355–386, 573. MR 2343137
[13] P. McMullen, *Valuations and dissections*, Handbook of convex geometry, Vol. A, B, NorthHolland, Amsterdam, 1993, pp. 933–988. MR 1243000 (95f:52018)
[14] W. Baldoni and M. Vergne, *Kostant partitions functions and flow polytopes*, Transform. Groups 13 (2008), no. 3-4, 447–469. MR 2452600

Eberhard-type theorems with two kinds of polygons

Sebastian Manecke

FB 12 - Institut für Mathematik
Goethe-Universität Frankfurt, Robert-Mayer-Str. 10
D-60325 Frankfurt am Main, Germany
E-mail: manecke@math.uni-frankfurt.de

Eberhard-type theorems are statements about the realizability of a polytope (or more general polyhedral maps) given the valency of its vertices and sizes of its polygonal faces up to a linear degree of freedom. We present new theorems of Eberhard-type where we allow adding two kinds of polygons and one type of vertices. We also hint towards a full classification of these types of results.

Keywords: Eberhard-type theorems; polyhedral maps; 3-polytopes; topological graph theory.

1. Overview

The classical Eberhard theorems are two results on the constructability of r-valent 3-polytopes for a given sequence $(p_k)_{k\geq 3}$, where $p_k \in \mathbb{N} = \{0, 1, 2, \dots\}$ describes the number of occurrences of each k-gon. For us, a **sequence** a will always be a function $\mathbb{N} \setminus \{0, 1, 2\} \to \mathbb{N}$ with finite support.

The original formulations were (see [4]):

Theorem 1.1 (Eberhard's theorem for 3-valent 3-polytopes). *Let* $(p_3, p_4, p_5, p_7, \dots, p_m)$ *be a sequence of natural numbers for which*

$$\sum_{\substack{3\leq k\leq m\\ k\neq 6}} (6-k)\cdot p_k = 12, \tag{1}$$

holds. Then there exists a number p_6 *and a* 3*-valent* 3*-polytope which has exactly* p_k k*-gons for each* $3 \leq k \leq m$.

Theorem 1.2 (Eberhard's theorem for 4-valent 3-polytopes). *Let* $(p_3, p_5, \dots, p_m)$ *be a sequence of natural numbers for which*

$$\sum_{\substack{3\leq k\leq m\\ k\neq 4}} (4-k)\cdot p_k = 8, \tag{2}$$

holds. Then there exists a number p_4 *and a* 4*-valent* 3*-polytope which has exactly* p_k k*-gons for each* $3 \leq k \leq m$.

If p_k is the number of k-gons of a 3-valent (resp. 4-valent) polytope, then (1) (resp. (2)) holds as an immediate consequence of Euler's relation and double counting of the number of edges. Thus these theorems answer the question under which condition a sequence p that suffices the natural combinatorial conditions has a realization as a polytope.

In the 1970's Barnette, Ernest, Grünbaum, Jendrol', Jucovič, Trenkler and Zaks [1, 3, 6, 7, 9] extended these results to polyhedral maps, which are graph embeddings on a closed topological 2-manifold (or **surface**) generalizing the combinatorics of 3-polytopes. In the two setups described by the classical Eberhard theorems there is now a complete characterization for which sequences $(p_k)_{k\geq 3}$ and $(v_k)_{k\geq 3}$ and surfaces S there exists a polyhedral map on S with p_k k-gons and v_k k-valent vertices when choosing the value of p_6 and v_3, resp. p_4 and v_4, appropriately.

We want to call the sequences $p = (p_k)_{k\geq 3}$ and $v = (v_k)_{k\geq 3}$ which count the number of k-gons and k-valent vertices of a polyhedral map M the **p-vector** and the **v-vector** of M and call the pair (p, v) to be realizable on a surface S, if there exists a polyhedral map M on S with p-vector p and v-vector v.

An easy construction shows that we can in fact find infinitely many p_6 and v_3, resp. p_4 and v_4, such that (p, v) is realizable as a polyhedral map on a fixed S. By using Euler's relation and an easy double counting argument one cannot increase p_6 and v_3, resp. p_4, v_4 independently from each other and thus one can deduce that there is a linear relation between these numbers. We want to propose the generalized Eberhard problem:

Question 1.1. Let $p = (p_k)_{k\geq 3}$, $v = (v_k)_{k\geq 3}$, $q = (q_k)_{k\geq 3}$ and $w = (w_k)_{k\geq 3}$ be sequences and S be a surface. Does there exist infinitely many $c, d \in \mathbb{N}$ and a polyhedral map on S with p-vector $p + c \cdot q$ and v-vector $v + d \cdot w$?

Theorem 1.1 and its generalization to polyhedral maps answer this question for $q = (0, 0, 0, 1, 0, \dots)$ and $w = (1, 0, \dots)$, whereas Theorem 1.2 and its generalizations answer this question for $q = (0, 1, 0 \dots)$, $w = (0, 1, 0, \dots)$. It is easy to check, that the only missing possibility for q and w with exactly one non-zero entry, where such a statement can be true, is $q = (1, 0, \dots)$, $w = (0, 0, 0, 1, 0, \dots)$. This case is in fact just the dual of Theorem 1.1, and therefore all cases with exactly one non-zero entry in both q and w have been classified.

Question 1.1 for q, w with more than one non-zero entry was first considered by DeVos et al. [2], who gave an answer in the case of $v = (1, 0, \dots)$, $q = (0, 0, 1, 0, 1, 0, \dots)$ and $w = (1, 0, \dots)$ for any surface. We will also consider similar theorems of this type in this article. To state them more easily, let us introduce the following notation: Define $[i]$ to be the sequence a with $a_i = 1$ and $a_j = 0$ for $i \neq j$. We then set $[a_{k_1} \times k_1, a_{k_2} \times k_2, \dots, a_{k_n} \times k_n] := \sum_{i=1}^{n} a_{k_i}[k_i]$, where only entries a_{k_i} not equal to zero occur. If $a_{k_i} = 1$, we will just write k_i instead of $1 \times k_i$.

In his master thesis [8], the author gave a complete answer to Question 1.1 in the case that q has precisely two non-zero and coprime entries and w has one non-zero entry. We will state the full theorem in Sec. 2 and give the ideas for the constructions used in the proofs in Sec. 3. The last Section, Sec. 4, will show how these constructions yield one case of the full statement in [8], that is to say, the following two theorems:

Theorem 1.3. *Let p and v be a pair of admissible sequences for an orientable closed 2-manifold S and $k \in \mathbb{N}$. Then there exists infinitely many $c, d \in \mathbb{N}$ for which there exists a polyhedral map on S with p-vector $p + c \cdot [(3k+1) \times 3, 3k+5]$ and v-vector $v + d \cdot [4]$.*

Theorem 1.4. *Let p and v be a pair of admissible sequences for an orientable closed 2-manifold S and $k \in \mathbb{N}$. Then there exists infinitely many $c, d \in \mathbb{N}$ for which there exist a polyhedral map on S with p-vector $p + c \cdot [(3k+3) \times 3, 3k+7]$ and v-vector $v + d \cdot [4]$.*

2. Polyhedral maps and generalized Eberhard problems

We will review basic notions from (topological) graph theory. A **simple graph** G is a finite undirected graph without loops and multi-edges. If G' is a subgraph of G this is denoted by $G' \subseteq G$. We want to write $u_1 - \cdots - u_k$ for paths and $u_1 - \cdots - u_k - u_1$ for cycles. The **valence** of a vertex is the number of incident edges.

All of our graphs are considered to be embedded into a closed (topological) 2-manifold, which we call **surfaces** for brevity. We assume our 2-manifolds to be oriented in this article. An embedding of a simple graph with vertices V, edges E and faces F is called a **map**, provided that G is simple, every vertex $v \in V$ has valence at least 3 and every $f \in F$ is a closed 2-cell (i.e. homeomorphic to a disk). The faces of a map incident to k edges (or equivalently, k vertices) will be called **k-gonal** faces or simply **k-gons**. A map on a closed 2-manifold is called **polyhedral**, if for every two faces $f, f', f \neq f'$ there is either no vertex, a single vertex or a single edge incident to both f and f'. In these cases the two faces are said to **meet properly**.

In Section 3 we will weaken the definition of a map to some extent to allow for 2-valent vertices. This does not warrant a whole new definition, so we state it here for completeness and to avoid confusion.

An important property of polyhedral maps is that each edge contains exactly two vertices and is contained in exactly two faces. From this fact one can see that the concept of polyhedral maps dualizes perfectly, i.e. if an embedding is a polyhedral map, then the dual of the embedding is again a polyhedral map.

Example 2.1. We can view every 3-polytope as a map on a surface, where the graph of the map is the graph of the 3-polytope and the embedding is held by radial projection onto $\mathbb{S}^2$. In this context each face of the 3-polytope corresponds to one of the map. It is easy to see that this map is polyhedral, which gives rise to the property's name. Also note, that the dual map corresponding to a 3-polytope is the corresponding map of the dual polytope.

To further strengthen the link between 3-polytopes and polyhedral maps on the sphere $\mathbb{S}^2$, we mention the following two results:

Proposition 2.1. *Every graph G of a polyhedral map M is 3-**connected**, i.e. G has at least 3 vertices and the deletion of any 2 vertices leaves the graph connected.*

Theorem 2.1 (Steinitz's theorem). *A graph is the edge graph of a 3-polytope if and only if it is planar and 3-connected.*

These two theorems combined yield that every polyhedral map on the sphere $\mathbb{S}^2$ comes from a 3-polytope and vice versa.

We will now turn our focus to Eberhard theorems for polyhedral maps on surfaces. The **p-vector** of a map M on a 2-manifold is the sequence $(p_3, \ldots, p_m)$, where p_k denotes the number of faces with exactly k vertices. Similarly the **v-vector** of M is the sequence $(v_3, \ldots, v_n)$ where each v_k is the number of vertices of M with valence k.

A pair of sequences (p, v) is said to be **realizable as a polyhedral map** on the closed 2-manifold S (or short: realizable on S), if there exists such a map having p as its p-vector and v as its v-vector. In this language we can state the following two generalizations of Theorems 1.1 and 1.2, which will be central in our constructions:

Theorem 2.2 (Jendrol', Jucovič [6], 1977). *Each pair of sequences $p = (p_3, \ldots, p_m)$ and $v = (v_3, \ldots, v_n)$ is realizable on a closed orientable 2-manifold with Euler characteristic χ for some $p_6 \in \mathbb{N}$, $v_3 \in \mathbb{N}$ if and only if*

$$\sum_{k=3}^{m}(6-k)p_k + 2\sum_{k=4}^{n}(3-k)v_k = 6\chi,$$

$$\sum_{\substack{k=3\\2\nmid k}}^{m} p_k \neq 0 \quad \text{or} \quad \sum_{\substack{k=4\\3\nmid k}}^{n} v_k \neq 1 \qquad \text{if } \chi = 2,$$

$$p \neq [5,7] \quad \text{or} \quad v \neq [v_3 \times 3] \qquad \text{if } \chi = 0.$$

Theorem 2.3 (Barnette, Grünbaum, Jendrol', Jucovič, Zaks [1, 3, 7, 9]). *Each pair of sequences $p = (p_3, \ldots, p_m)$ and $v = (v_3, \ldots, v_n)$ is realizable on a closed orientable 2-manifold with Euler characteristic χ for some $p_4, v_4 \in \mathbb{N}$ if and only if*

$$\sum_{k=3}^{m}(4-k)p_k + \sum_{k=3}^{n}(4-k)v_k = 4\chi,$$

$$\sum_{k=3}^{m} kp_k \equiv 0 \qquad (\text{mod } 2),$$

$$p \neq [3,5] \quad \text{or} \quad v \neq [v_4 \times 4] \qquad \text{if } \chi = 0,$$

$$p \neq [p_4 \times 4] \quad \text{or} \quad v \neq [3,5] \qquad \text{if } \chi = 0.$$

Note that special cases arise in both theorems if the surface is a torus, i.e. if $\chi = 0$. Izmestiev et al. [5] gave a simple argument using holonomy groups to explain why these special cases occur.

The rest of this section is devoted to find an easy characterization for when we cannot hope Question 1.1 to have a positive answer. Let M be a polyhedral map on a surface S with vertices V, edges E, and faces F, p-vector $p = (p_3, \ldots, p_n)$ and v-vector $v = (v_3, \ldots, v_m)$. Let $\chi = \chi(S)$ be the Euler characteristic of S and $e := |E|$. The two basic combinatorial results here are double-counting

$$2e = \sum_{k=3}^{m} k \cdot p_k = \sum_{k=3}^{n} k \cdot v_k, \tag{3}$$

and the Euler-Poincaré relation

$$|V| - |E| + |F| = \chi. \tag{4}$$

One can easily deduce from these relations, that the following two equivalent conditions are necessary for two sequences p, v being the p- and v-vector of a polyhedral map:

Proposition 2.2. *Let p, v be the p- and v-vector of a polyhedral map on a surface S. Then* (3) *is true for some $e \in \mathbb{N}$ and the following equivalent conditions hold:*

$$\sum_{k=3}^{m}(6-k)p_k + 2\sum_{k=4}^{n}(3-k)v_k = 6\chi(S), \quad \textit{and}$$
$$\sum_{k=3}^{m}(4-k)p_k + \sum_{k=3}^{n}(4-k)v_k = 4\chi(S).$$

Equivalent here means, that together with $\sum_{k\geq 3} p_k = \sum_{k\geq 3} v_k$ each equation can be deduced from the other. If p and v satisfy these equations, we will call the pair (p, v) **admissible** (on S). We remark that we gain precisely the conditions of Theorems 2.2 and 2.3.

In light of Question 1.1 and using the same arguments, it is not difficult to see, that we can always assume p and v to be admissible. Also important to note is, that from the same arguments we can derive similar conditions on q and w which have to be fulfilled in order for Question 1.1 to be answered in the positive. We will not go into the details here and simply state the cases resulting from these restrictions.

We will restrict our setting to $q = [q_s \times s, q_l \times l]$ having only two non-negative entries and $w = [w_r \times r]$ having one. Let us further assume that $\gcd(q_s, q_l) = 1$. These conditions are quite natural, as any obstruction on finding a Eberhard-type theorem for some q will also give an obstruction for $c \cdot q$, $c \in \mathbb{N}$. As stated above, not all values s, l, and r can be obtained in this setting, only the following cases can occur:

$$\begin{aligned}
(s,r)=(3,3):\quad & q=[q_3\times 3, q_l\times l], && w=[3], && q_3=\tfrac{l-6}{\gcd(l,3)}, && q_l=\tfrac{3}{\gcd(l,3)}\\
(s,r)=(4,3):\quad & q=[q_4\times 4, q_l\times l], && w=[3], && q_4=\tfrac{l-6}{\gcd(l,2)}, && q_l=\tfrac{2}{\gcd(l,2)}\\
(s,r)=(5,3):\quad & q=[q_5\times 5, q_l\times l], && w=[3], && q_5=l-6, && q_l=1\\
(s,r)=(3,4):\quad & q=[q_3\times 3, q_l\times l], && w=[4], && q_3=l-4, && q_l=1\\
(s,r)=(3,5):\quad & q=[q_3\times 3, q_l\times l], && w=[5], && q_3=3l-10, && q_l=1
\end{aligned}$$

We will answer the fourth case in this article. The full result by the author is the following:

Theorem 2.4 (Manecke [8], 2016). *Let $q = [q_s \times s, q_l \times l]$, $w = [r]$ as before. Then there exist inifitely many $c, d \in \mathbb{N}$ and a polyhedral map M on a surface S for all admissible sequences (p, v) if and only if* $\gcd(s, l) = 1$ *and if $s = r = 3$, then $l < 11$.*

We close the section by noting that by duality this result gives also a full classification for w having two non-zero entries and q having only one.

3. Construction

Let $r \in \mathbb{N}$ be the valence of those vertices we are free to add to a polyhedral map. All constructions later in this article will utilize the concept of replacing each face of a polyhedral map with a larger patch. It can be quite challenging to see whether the resulting structures fit together. This section introduces the necessary formalism for these kinds of constructions. All statements are presented without proof, all proofs can be found in [8]. Note that throughout this section we allow 2-valent vertices in special maps we call patches.

Definition 3.1 (Patch). A map $\mathcal{P}$ on the Euclidean plane with possibly 2-valent vertices on the unbounded **outer face** is called a **patch**. The vertices and edges of the outer face form the **boundary** $\partial\mathcal{P}$ of the patch. A patch is an **r-patch**, if each of its vertices except the ones on the boundary is r-valent, while for the valence $\deg(v)$ of a vertex v on the outer face $2 \leq \deg(v) \leq r$ holds. The **p-vector of a patch** is the sequence $(p_3, p_4, \ldots)$, where p_k denotes the number of k-gonal inner faces of the patch.

We say that two r-patches $\mathcal{P}_1$ and $\mathcal{P}_2$ fit together along a path $v_1 - \cdots - v_n$ on $\partial\mathcal{P}_1$ and $u_1 - \cdots - u_n$ on $\partial\mathcal{P}_2$, if, after gluing them together such that $v_i = u_{n+1-i}$ the resulting patch is still an r-patch. Define $w(v) := \deg(v) - 1$. Then the condition for fitting together is just $w(v_i) + w(u_{n+1-i}) = r$ for all $i \in \{2, \ldots, n-1\}$ and $w(v_i) + w(u_{n+1-i}) \leq r$, $w(v_{n+1-i}) + w(u_i) \leq r$. We say a tuple $(w_1, \ldots, w_n)$ is **self-fitting**, if $w_i + w_{n+1-i} = r$ for all $i \in \{1, \ldots, n\}$.

Essential for our constructions will be w-expansions. We will use them, when we replace all k-gons in a polyhedral map with larger structures.

Definition 3.2. Let $w = (w_1, \ldots, w_n) \in \mathbb{N}^n$. A **$w$-expansion** of an r-patch $\mathcal{P}$ with boundary $\partial\mathcal{P} = v_1 - v_2 - \cdots - \cdots - v_m - v_1$ is an r-patch $\mathcal{P}'$ with boundary

$$\partial\mathcal{P}' = v'_1 - \underbrace{v'^{(1)}_1 - \cdots - v'^{(1)}_n}\; - v'_2 - \underbrace{v'^{(2)}_1 - \cdots - v'^{(2)}_n} - \quad \cdots \\ - v'_m - \underbrace{v'^{(m)}_1 - \cdots - v'^{(m)}_n} - v'_1,$$

such that $w(v_i) = w(v'_i)$ and $w(v'^{(i)}_j) = w_j$ for all $i \in \{1, \ldots, m\}$, $j \in \{1, \ldots, n\}$. We call the vertices v'_i **corner vertices** and the vertices $v'^{(i)}_j$ **side vertices**. Furthermore, a patch is called **w-k-gonal**, if it is the w-expansion of the patch consisting of only a k-gon, i.e. if $w(v'_i) = 1$ for $i \in \{1, \ldots, k\}$.

Using this notation we describe the following construction scheme:

Algorithm 3.1.
Input: *A map on a surface S with p-vector p, v-vector v, underlying graph $G = (V, E)$ and faces F.*

A self-fitting tuple $w = (w_1, \ldots, w_n)$.

For each k-gonal face $f \in F$ a w-k-gonal r-patch $\mathcal{P}(f)$ with p-vector $p^{(f)}$.

Output: *A map on S with v-vector $v + d\cdot[r]$ for some $d \in \mathbb{N}$ and p-vector $\sum_{f\in F} p^{(f)}$.*

Description: *Divide each edge $e \in E$ in the embedding of G in S by n vertices and draw into each face f the dedicated r-patch $\mathcal{P}(f)$ such that the corner vertices of $\mathcal{P}(f)$ coincide with the original vertices V and the side vertices are the new vertices added by the subdivision, see Fig. 1. Here we use the fact, that our surfaces are oriented and assume that all patches are glued with the same orientation. These patches form a combined graph, which is embedded by construction into S (there is a homeomorphism between each subdivided face $f \in F$ and the corresponding patch $\mathcal{P}(f)$). It is straightforward to see, that this gives a map with the desired properties.*

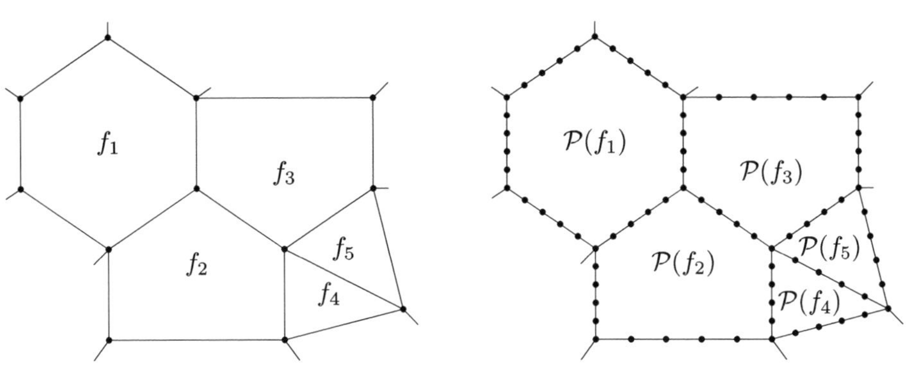

Fig. 1.

As previously stated, these definitions are used to formalize the construction step "replace each face with a patch". Up until now, there is no requirement explicitly stated on the interior of the patch. If we expect the result of such a construction to be a polyhedral map, further conditions have to be met. Additionally, when using Algorithm 3.1 we have the problem of assigning a patch for each face of the map. While we might need only one type of patch for a k-gon for each $k \geq 3$, we could still have to deal with a huge amount of values of k. We now want to define a construction scheme for patches for arbitrary k which additionally allow to create polyhedral maps, even from non-polyhedral ones.

Definition 3.3. Let $\mathcal{P}$ be an r-patch with boundary

$$\partial\mathcal{P} = i_0 - i_1 - \cdots - i_{m-1} - (i_m = o_0) - \cdots - o_s - \cdots - o_{n-1} - (o_n = i'_m) \\ - i'_{m-1} - \cdots - i'_1 - i'_0 - i_0$$

(i_m and o_0 denote the same vertex, the same holds for o_n and i'_m), $1 \leq s < n$, $m > 0$, such that:

- $w(i_0) + w(i'_0) = r - 1$,
- $\mathcal{P}$ fits to itself along $i_1 - \cdots - i_{m-1}$ and $i'_{m-1} - \cdots - i'_1$,
- $w(o_s) = 1$, and
- $(w(o_{s+1}), \ldots, w(o_{n-1}), w(o_n) + w(o_0), w(o_1), \ldots, w(o_{s-1}))$ is a self-fitting tuple.

Such a patch will be called **expansion patch with outer tuple** $(w(o_{s+1}), \ldots, w(o_{n-1}), w(o_n) + w(o_0), w(o_1), \ldots, w(o_{s-1}))$.

Example 3.1. We want to review the last definition with two examples. A hexagon can be interpreted as an expansion 3-patch $\mathcal{H}$ with outer tuple $(w_{\mathcal{H}}(o_3) + w_{\mathcal{H}}(o_0), w_{\mathcal{H}}(o_1)) = (2, 1)$, with vertices labeled according to Definition 3.3 in Fig. 2. Similarly two quadrangles which share a common edge build an expansion 4-patch $\mathcal{Q}_2$ with outer tuple $(w_{\mathcal{Q}_2}(o_3) + w_{\mathcal{Q}_2}(o_0), w_{\mathcal{Q}_2}(o_1)) = (2, 2)$, as seen in the same figure.

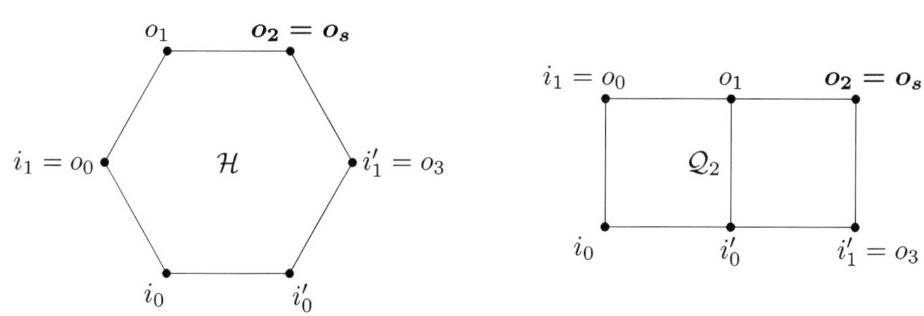

Fig. 2. Two expansion patches

We want to pull apart this definition a bit to give a geometric intuition. The first thing to note is that by definition we are able to glue two copies of an expansion patch along the paths $i_0 - \cdots - i_m$ and $i'_m - \cdots - i'_0$. When doing this the new patch has a boundary path $o_{s+1} - \cdots - o_{n-1} - (o_n = o_0) - o_1 - \cdots - o_{s-1}$, which we require to be self-fitting. Therefore we can glue two of those patches along this boundary to get an even larger patch (see Fig. 3).

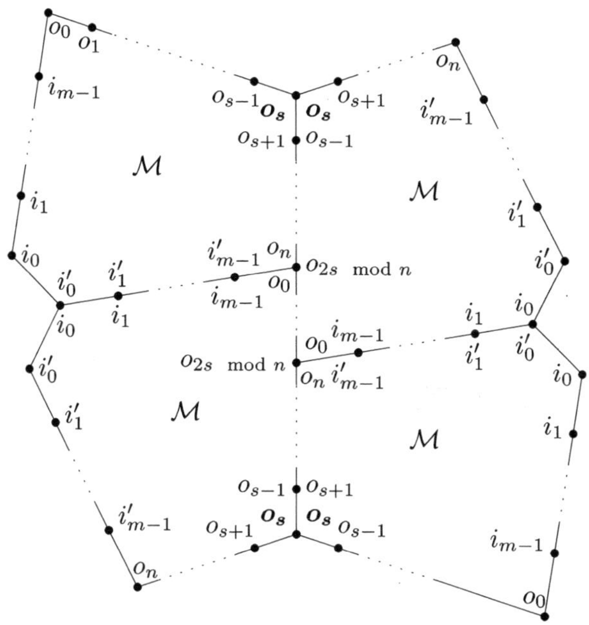

Fig. 3. An edge patch

For an expansion patch $\mathcal{M}$, we want to call the patch obtained by gluing four copies of $\mathcal{M}$ as stated the **edge patch** of $\mathcal{M}$. An expansion patch will be said to have the **polyhedral property** if every two inner faces in the corresponding edge patch meet properly.

Example 3.2. The examples in Example 3.1 do in fact have the polyhedral property, which can be verified by looking at the edge patch in Fig. 4.

Expansion patches will be our basic building block for all our constructive proofs. We can use them to obtain larger o-k-gonal patches for any $k \geq 3$:

Algorithm 3.2.

Input: *An expansion r-patch $\mathcal{M}$ with outer tuple o and p-vector p.*

Output: *For every $k \geq 3$ an o-k-gonal r-patch $\mathcal{M}(k)$ with p-vector $[k] + k \cdot p$.*

If $\mathcal{M}$ has the polyhedral property, then all inner faces of $\mathcal{M}(k)$ meet properly.

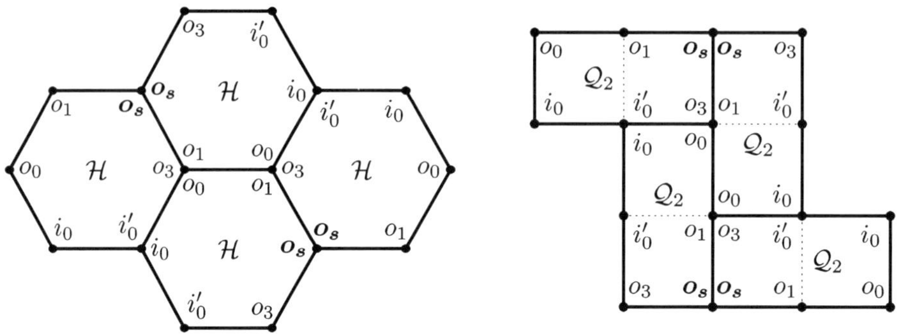

Fig. 4. Two edge patches

Description: *We construct $\mathcal{M}(k)$ from k copies of $\mathcal{M}$ and a single k-gon. Let*

$$\partial\mathcal{P} = i_0 - i_1 - \cdots - (i_m = o_0) - o_1 - \cdots - o_s - \cdots - o_{n-1} - (o_n = i'_m) - \dots \\ - i'_1 - i'_0 - i_0$$

be the boundary of $\mathcal{M}$ as in Definition 3.3. We now form a larger patch by gluing the edge $\{i_0, i'_0\}$ of each of the k copies of $\mathcal{M}$ to an edge of the k-gon and also gluing the vertex associated to i_l, $1 \leq l \leq m$ from one copy to the vertex associated to i'_l from the adjacent copy. Graphically speaking, we form a ring of k patches of the form $\mathcal{M}$ around the k-gon. The p-vector of $\mathcal{M}(k)$ is therefore $[k] + k \cdot p$. We leave out the proof that the inner faces of $\mathcal{M}(k)$ meet properly in the case of $\mathcal{M}$ being polyhedral.

With these constructions at hand we can now finally design a scheme to create a polyhedral map from a non-polyhedral one.

Proposition 3.1. *Given a map M on an orientable closed 2-manifold S and an expansion r-patch $\mathcal{M}$ with outer tuple o, Algorithm 3.1 returns a polyhedral map on S when we take $\mathcal{P}(f) = \mathcal{M}(k)$ for each k-gonal face f of M.*

Example 3.3. Using the expansion patches $\mathcal{H}$ and $\mathcal{Q}_2$ from Examples 3.1 and 3.2, we can construct from a polyhedral map a new one with arbitrarily many hexagons (or quadrangles) added, while inserting only 3-valent (or 4-valent) vertices. Given a map M on a closed oriented 2-manifold, we can simply use Theorem 3.1 repeatedly on M with either $\mathcal{H}$ or $\mathcal{Q}_2$ to get the desired result. The theorem inserts at least a single hexagon or quadrangle during each step (which is quite an understatement, the number of polygons added is by far larger), so repeating this step eventually leads to a map that has more than a specified amount of hexagons or quadrangles.

Putting all these constructions together, we can formulate a proof strategy for Question 1.1 in the case of $q = [q_s \times s, q_l \times l]$, $w = [r]$. We state it for $r = 4$ only, but the ideas carry over to $r = 3$, too.

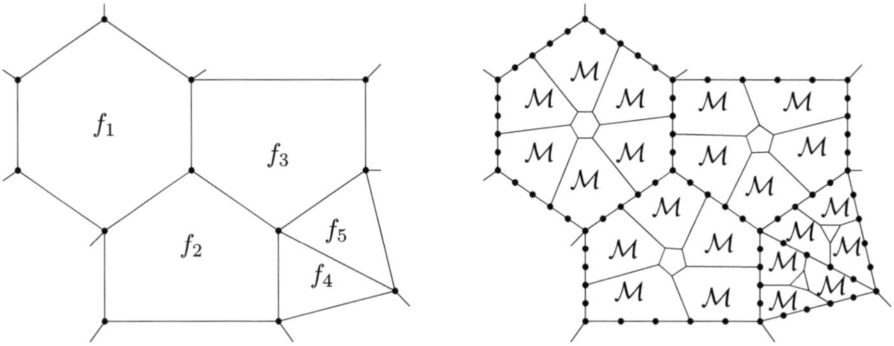

Fig. 5. Algorithm 3.1 for $\mathcal{P}(f) = \mathcal{M}(k)$

Proposition 3.2. *Let* $r = 4$, $p = (p_3, \ldots, p_m)$ *and* $v = (v_3, \ldots v_n)$ *be an admissible pair of sequences. Let* $w = [4]$ *and* $q = [q_s \times s, q_l \times l]$, *where* $s = 3$, $l > 4$, $q_s = l.- 4$ *and* $q_l = 1$. *Assume there exist*

- *an expansion* r*-patch* $\mathcal{P}_N$ *with outer tuple* o *consisting of* s*-gons and* l*-gons,*
- *an* o*-4-gonal* r*-patch* $\mathcal{P}_F$ *consisting of* s*-gons and* l*-gons, and*
- *an expansion* r*-patch* $\mathcal{P}_P$ *with the polyhedral property consisting of* s*-gons and* l*-gons.*

Then there exists a polyhedral map on S *with* p*-vector* $p + c \cdot q$ *and* v*-vector* $v + d \cdot w$ *for infinitely many* $c, d \in \mathbb{N}$.

Idea of the proof. Use Theorem 2.2 or Theorem 2.3 as a starting map and apply Proposition 3.2 for the given patches. □

Remark 3.1. We will use Theorem 3.2 heavily in the next section. Therefore, we want to stress what is needed to check to see if the prerequisites of Theorem 3.2 are fulfilled. As stated, we need three r-patches, $\mathcal{P}_N$, $\mathcal{P}_F$ and $\mathcal{P}_P$, which are called in this manner for the rest of the article. The list of properties is:

- $\mathcal{P}_N$, $\mathcal{P}_F$ and $\mathcal{P}_P$ consist of only s-gons and l-gons and all inner vertices have valence r.
- $\mathcal{P}_N$ and $\mathcal{P}_P$ are expansion patches:
  - i_0 and i'_0 are in sum incident to $r - 1$ faces,
  - i_k and i'_k, are in sum incident to r faces, $1 \leq k < m$,
  - starting at the vertex o_s and going in both directions for each pair of vertices $o_{s+k \bmod n}$ and $o_{s-k \bmod n}$ the identity $w_{\mathcal{P}_X}(o_{s+k \bmod n}) + w_{\mathcal{P}_X}(o_{s-k \bmod n}) = r$ holds, where we "identify" o_n with o_0. For ease of comparison we also state the outer tuple o for $\mathcal{P}_N$.
- $\mathcal{P}_F$ is o-4-gonal, i.e. if starting at some vertex and looking at the number of inner

faces incident to this vertex we see the pattern

$$1, o_1, \dots o_n, 1, o_1, \dots o_n, 1, o_1, \dots o_n, 1, o_1, \dots o_n,$$

where o is the outer tuple of $\mathcal{P}_N$.

- $\mathcal{P}_P$ has the polyhedral property. For this we provide the corresponding edge patch to make the verification easier.

4. 4-valent Eberhard-type theorems with triangles

In this section we want to prove 4-valent Eberhard-type theorems with triangles, i.e. for $q = [q_3 \times 3, q_l \times l]$, $w = [w_4 \times 4]$, $l > 4$, $\gcd(q_3, q_l) = 1$.

For all the proofs, we want to use Theorem 3.2, therefore we need to have a construction scheme for patches with arbitrarily large l-gons. These we get by the next three constructions:

Algorithm 4.1. *When we want to use this construction in this section we label an edge (the specified edge) with a square and point with arrows to a k_1-gon and a k_2-gon.*

Input: *A 4-patch with p-vector p and a specified edge with exactly one vertex incident to some k_1-gon and the other vertex incident to some k_2-gon. We require that in the cyclic order around both end points, starting at the specified edge, the k_1-gon and the k_2-gon have the same position.*

Output: *A new 4-patch with p-vector $p - [k_1, k_2] + [2k \times 3] + [k_1 + k, k_2 + k]$ for all $k \in \mathbb{N}$.*

If every two faces of the 4-patch meet properly, then this is carried over to the new patch.

Description: *Using the replacement of the single edge as seen in Fig. 6 results in a new 4-patch with p-vector $p - [k_1, k_2] + [2 \times 3] + [k_1 + 1, k_2 + 1]$. The line on the left labeled with a square is the specified edge and the line on the right labeled with a square is a new edge that we can use to repeat the construction. Every time we use this construction we add two new triangles while increasing the number of vertices of the left and right polygon by one; doing this k times gives the desired 4-patch. That all faces meet properly follows by induction as this property is preserved in each step.*

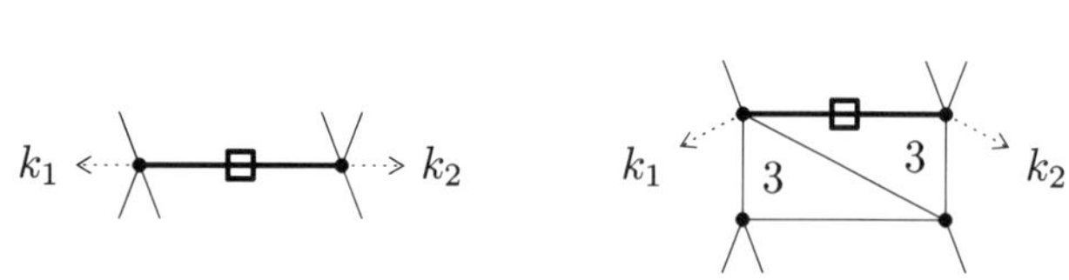

Fig. 6.

Algorithm 4.2. *When we want to use this construction in this section we label an edge (the specified edge) with a diamond and point with arrows to a k_1-gon and a k_2-gon.*
Input: *A 4-patch with p-vector p and a specified edge which is the common edge of a k_1-gon and a k_2-gon.*
Output: *A 4-patch with p-vector $p - [k_1, k_2] + [(6k) \times 3] + [k_1 + 3k, k_2 + 3k]$ for all $k \in \mathbb{N}$.*
Description: *Using the replacement of the single edge as seen in Fig. 7 results in a new 4-patch with p-vector $p - [k_1, k_2] + [(6k) \times 3] + [k_1 + 3, k_2 + 3]$. The line on the left labeled with a diamond is the specified edge and the line on the right labeled with a diamond is a new edge which we can use to repeat the construction. Every time we use this construction we add six triangles while increasing the number of vertices of the left and right polygon by three; doing this k times gives the desired 4-patch.*

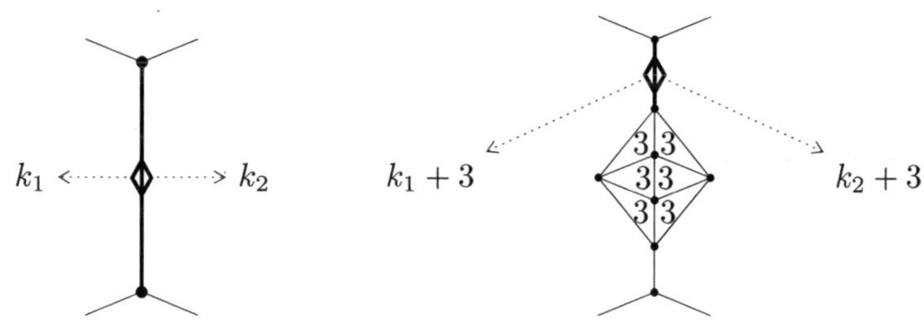

Fig. 7.

Algorithm 4.3. *When we want to use this construction in this section we encircle a vertex (the specified vertex) and point with arrows to a k_1-gon and a k_2-gon.*
Input: *A 4-patch with p-vector p and a specified vertex which is adjacent to both k_1-gon and a k_2-gon which do not share an edge containing this vertex.*
Output: *A new 4-patch with p-vector $p - [k_1, k_2] + [(6k) \times 3] + [k_1 + 3k, k_2 + 3k]$ for all $k \in \mathbb{N}$.*
Description: *Using the replacement of the vertex as seen in Figure 8 results in a new 4-patch with p-vector $p - [k_1, k_2] + [(6k) \times 3] + [k_1 + 3, k_2 + 3]$. The encircled vertex on the left is the specified vertex and the encircled vertex on the right is a new vertex which we can use to repeat the construction. Every time we use this construction we add six triangles while increasing the number of vertices of the left and right polygon by three; doing this k times gives the desired 4-patch.*

With our whole machinery at work, we can now state the proofs of our main theorems easily:

Proof of Theorem 1.3. An expansion 4-patch $\mathcal{P}_N$ with outer tuple $o = (1, 2, 1, 3, 2, 3)$ is shown in Fig. 9(a) and a corresponding o-4-gonal 4-patch $\mathcal{P}_F$ is

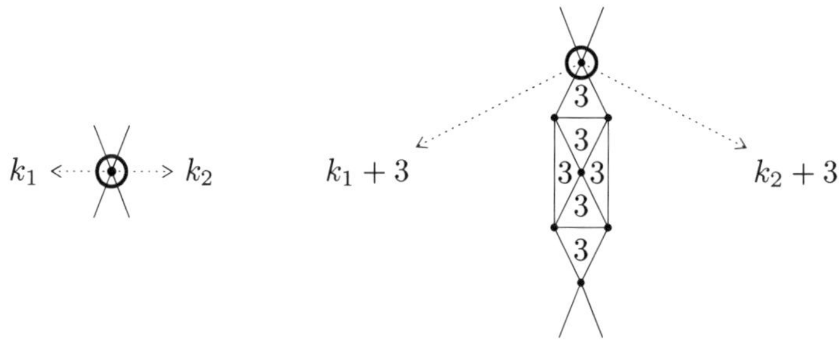

Fig. 8.

shown in Fig. 9(d), both consisting of triangles and pentagons. By using Algorithm 4.1, Algorithm 4.2 and Algorithm 4.3 as indicated we get 4-patches consisting of only triangles and $(3k+5)$-gons, $k \in \mathbb{N}$. We can see in Fig. 9(b) that $\mathcal{P}_N$ has the polyhedral property, thus we can apply Theorem 3.2 with $\mathcal{P}_P := \mathcal{P}_N$. □

Proof of Theorem 1.4. An expansion 4-patch $\mathcal{P}_N$ with outer tuple $o = (2,2,3,2,1,3,2,1,2,2)$ is shown in Fig. 9(a) and a corresponding o-4-gonal 4-patch $\mathcal{P}_F$ is shown in Fig. 9(c), both of which consist of only triangles and heptagons. By using Algorithm 4.1, Algorithm 4.2 and Algorithm 4.3 as indicated we get 4-patches consisting of only triangles and $(3k+7)$-gons, $k \in \mathbb{N}$. We can reuse the 4-patch $\mathcal{P}_P$ with the polyhedral property in Fig. 9(a) from the last theorem and after application of Algorithm 4.1 it likewise contains triangles and $(3k+7)$-gons. Thus we can apply Theorem 3.2. □

References

[1] D. W. Barnette, E. Jucovič and M. Trenkler, Toroidal Maps With Prescribed Types Of Vertices And Faces, *Mathematika* **20**, 82 (1971).

[2] M. DeVos, A. Georgakopoulos, B. Mohar and R. Šámal, An Eberhard-like theorem for pentagons and heptagons, *Discrete & Computational Geometry* **44**, 931 (2010).

[3] B. Grünbaum, Planar maps with prescribed types of vertices and faces, *Mathematika* **16**, 28 (1969).

[4] B. Grünbaum *et al.*, *Convex Polytopes* (Springer, 2003).

[5] I. Izmestiev, R. B. Kusner, G. Rote, B. Springborn and J. M. Sullivan, There is no triangulation of the torus with vertex degrees 5, 6,..., 6, 7 and related results: Geometric proofs for combinatorial theorems, *Geometriae Dedicata* **166**, 15 (2013).

[6] S. Jendrol' and E. Jucovič, Generalization of a theorem by V. Eberhard, *Mathematica Slovaca* **27**, 383 (1977).

[7] E. Jucovič and M. Trenkler, A theorem on the structure of cell–decompositions of orientable 2–manifolds, *Mathematika* **20**, 63 (1973).

[8] S. Manecke, *New Eberhard-like theorems*, Master's thesis, Technische Universität Dresden, (2016).

[9] J. Zaks, The analogue of Eberhard's theorem for 4-valent graphs on the torus, *Israel Journal of Mathematics* **9**, 299 (1971).

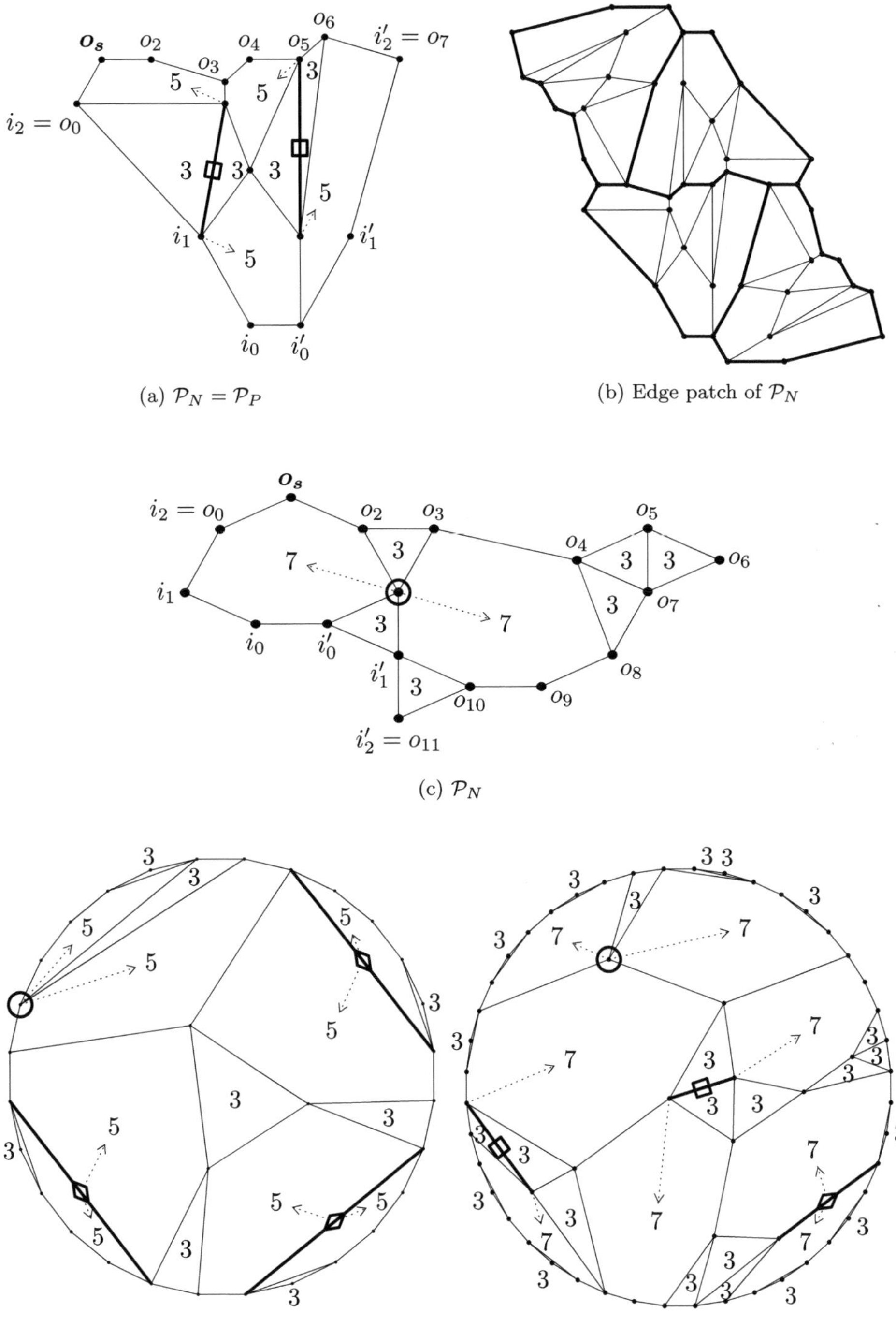

(a) $\mathcal{P}_N = \mathcal{P}_P$

(b) Edge patch of $\mathcal{P}_N$

(c) $\mathcal{P}_N$

(d) $\mathcal{P}_F$

(e) $\mathcal{P}_F$

Fig. 9. $\mathcal{P}_N = \mathcal{P}_P$, the edge patch of $\mathcal{P}_N$ and $\mathcal{P}_F$

Complete intersection Calabi–Yau threefolds in Hibi toric varieties and their smoothing

Makoto Miura

Korea Institute for Advanced Study
85 Hoegiro, Dongdaemun-gu, Seoul, 130-722, Republic of Korea
E-mail: miura@kias.re.kr

In this article, we summarize a combinatorial description of complete intersection Calabi–Yau threefolds in Hibi toric varieties. Such Calabi–Yau threefolds have at worst conifold singularities, and are often smoothable to non-singular Calabi–Yau threefolds. We focus on such non-singular Calabi–Yau threefolds of Picard number one, and illustrate the calculation of topological invariants, using new motivating examples.

Keywords: Calabi–Yau threefolds; conifold transitions; mirror symmetry; order polytopes; Hibi toric varieties.

1. Introduction

A *Hibi toric variety* is defined as a projective toric variety $\mathbb{P}_{\Delta(P)}$ associated with an order polytope

$$\Delta(P) = \{(x_u)_{u\in P} \mid 0 \leq x_u \leq x_v \leq 1 \text{ for } u \prec v \in P\} \subset \mathbb{R}^P, \tag{1}$$

for a finite poset $P = (P, \prec)$. For example, all products of projective spaces are Hibi toric varieties; hence at least 2590 topologically distinct non-singular Calabi–Yau threefolds are obtained as complete intersections [12]. In general, complete intersection Calabi–Yau threefolds in Hibi toric varieties have finite number of nodes, and are often smoothable to non-singular Calabi–Yau threefolds by flat deformations. Complete intersections in Grassmannians (or more generally in minuscule Schubert varieties) give basic examples of such smoothing [5, 17].

The purpose of this article is to provide a brief summary on combinatorial descriptions of complete intersection Calabi–Yau threefolds in Hibi toric varieties and their smoothing. Based on [7], we describe the smoothability in terms of posets (Proposition 3.2), and survey the calculation of topological invariants for resulting non-singular simply-connected Calabi–Yau threefolds (Subsection 4.2), by focusing on the case of Picard number one for simplicity. In addition to the summary, we show the simply-connectedness as a corollary of the result on small resolutions for Hibi toric varieties (Proposition 2.1). To illustrate the calculation, we introduce several new examples of such non-singular Calabi–Yau threefolds of Picard number one (Subsection 4.3, Table 1).

A Calabi–Yau threefold is a complex projective threefold X with at worst canonical singularities satisfying $\omega_X \simeq \mathcal{O}_X$ and $H^1(X, \mathcal{O}_X) = 0$. There are a huge number of such threefolds, even non-singular. Mirror symmetry is a conjectural duality between a non-singular Calabi–Yau threefold X and another non-singular Calabi–Yau threefold X^*, called a *mirror manifold* for X. Various non-trivial relations between X and X^* are expected. For example, Hodge numbers satisfy

$$h^{i,j}(X) = h^{3-j,i}(X^*) \quad \text{for all } i \text{ and } j. \tag{2}$$

One of the big mysteries of mirror symmetry is whether every non-singular Calabi–Yau threefold X has a mirror manifold X^* or not. Note an obvious exception in the case with $h^{2,1}(X) = 0$, and that the mirror manifold X^* is not unique in general, even as topological manifolds. There is an excellent class of non-singular Calabi–Yau threefolds such that the above question has an affirmative answer; for crepant resolutions of complete intersection Calabi–Yau threefolds in Gorenstein toric Fano varieties, we have mirror manifolds in the same class, called the *Batyrev–Borisov mirrors* [3, 8]. In order to expand this class, the conjectural mirror construction via conifold transitions seems to be a promising direction.

Let X_0 be a Calabi–Yau threefold with finitely many nodes. Suppose that X_0 admits a smoothing $X \rightsquigarrow X_0$ by a flat deformation, and a small resolution $Y \to X_0$. The composite operation connecting two non-singular Calabi–Yau threefolds X and Y is called a *conifold transition*:

$$X \rightsquigarrow X_0 \leftarrow Y. \tag{3}$$

There is a natural closed immersion of the Kuranishi space $\mathrm{Def}(Y)$ to $\mathrm{Def}(X_0)$ [19, Proposition 2.3], and hence, it makes sense to put them together into some giant moduli space. There is a question, commonly referred to as (a version of) *Reid's fantasy*, which asks whether all simply-connected non-singular Calabi–Yau threefolds fit together into a single irreducible family via conifold transitions [20]. Suppose that X and Y have torsion-free homology for a conifold transition (3). *Morrison's conjecture* in [18] says that the mirror manifolds are also connected via a conifold transition of the opposite direction:

$$Y^* \rightsquigarrow Y_0^* \leftarrow X^*. \tag{4}$$

Together with the spirit of Reid's fantasy, one may expect a mirror construction for a large number of non-singular Calabi–Yau threefolds from the Batyrev–Borisov mirror pairs.

We still do not know the existence of a mirror manifold X^*, even for the smoothing $X \rightsquigarrow X_0$ of a complete intersection Calabi–Yau threefold X_0 in a Hibi toric variety. Nevertheless, we can discuss the mirror symmetry by calculating periods and Picard–Fuchs operators for the conjectural mirror family, as we see in Remark 4.1 for example.

2. Hibi toric varieties

2.1. *Examples*

Let us begin with simple examples of Hibi toric varieties. For the empty poset, we set the Hibi toric variety $\mathbb{P}_{\Delta(\varnothing)}$ to be a point. For a singleton $u := \{u\}$ (by abuse of notation), the order polytope is a line segment $\Delta(u) = [0,1]$, and hence, the Hibi toric variety $\mathbb{P}_{\Delta(u)}$ is a projective line $\mathbb{P}^1$.

Let P be a finite poset consisting of $n := |P|$ elements. If P is a *chain*, i.e., a totally ordered set, the order polytope $\Delta(P)$ is a regular simplex defined by the inequalities $0 \leq x_1 \leq \cdots \leq x_n \leq 1$, so that the Hibi toric variety $\mathbb{P}_{\Delta(P)}$ is a projective space $\mathbb{P}^n$. It is equally clear the case that P is an *anti-chain*, i.e., the poset in which every pair of elements is incomparable. In this case, the order polytope $\Delta(P)$ is a unit hypercube $[0,1]^n$, and the Hibi toric variety $\mathbb{P}_{\Delta(P)}$ is the product of n copies of $\mathbb{P}^1$.

Example 2.1. A first non-trivial example is a poset $P = \{u, v, w\}$ with the partial order defined by $u \succ w$ and $v \succ w$. The defining inequalities of the order polytope $\Delta(P)$ is shown in the left of Figure 1, also depicted symbolically in the middle. It becomes a pyramid in $\mathbb{R}^P \simeq \mathbb{R}^3$ as shown in the right of Figure 1. Therefore, the associated Hibi toric variety $\mathbb{P}_{\Delta(P)}$ is a projective cone over $\mathbb{P}^1 \times \mathbb{P}^1$ with a general apex in $\mathbb{P}^3$.

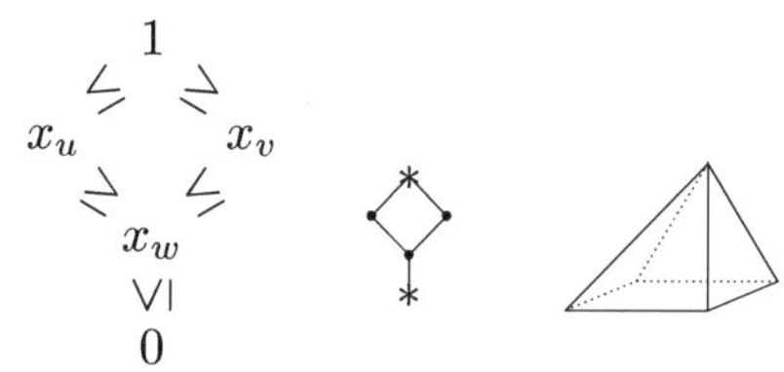

Fig. 1. An example of order polytopes

A *disjoint union* $P = P_1 + P_2$ of finite posets P_1 and P_2 is a disjoint union as sets equipped with the partial order $\prec$ satisfying (i) $u \in P_1$, $v \in P_1$ and $u \prec v \in P_1$ imply $u \prec v \in P$, (ii) $u \in P_2$, $v \in P_2$ and $u \prec v \in P_2$ imply $u \prec v \in P$, and (iii) $u \in P_1$ and $v \in P_2$ imply $u \not\prec v \in P$ (i.e., u and v are incomparable in P). The corresponding Hibi toric variety is projectively equivalent to the product of two Hibi toric varieties,

$$\mathbb{P}_{\Delta(P_1+P_2)} \simeq \mathbb{P}_{\Delta(P_1)} \times \mathbb{P}_{\Delta(P_2)}. \tag{5}$$

A *ordinal sum* $P = P_1 \oplus P_2$ of P_1 and P_2 is a disjoint union as sets equipped with the partial order $\prec$ satisfying the same (i) and (ii) as the disjoint union $P_1 + P_2$ above, and (iii)$'$ $u \in P_1$ and $v \in P_2$ imply $u \prec v \in P$. Note that the operation $\oplus$ is not commutative though it is associative. The corresponding Hibi toric variety is a special hyperplane section of a projective join of two Hibi toric varieties with general positions,

$$\mathbb{P}_{\Delta(P_1 \oplus u \oplus P_2)} \simeq \mathrm{Join}\left(\mathbb{P}_{\Delta(P_1)}, \mathbb{P}_{\Delta(P_2)}\right). \tag{6}$$

These operations generalize the examples, a chain $P = \bigoplus_{i=1}^n u_i$, an anti-chain $P = \sum_{i=1}^n u_i$, and $P = w \oplus (u+v) = \varnothing \oplus w \oplus (u+v)$ in Example 2.1.

The posets built up by disjoint unions and ordinal sums from singletons are sometimes called series-parallel posets. One of the simplest examples that are not series-parallel is the poset with the *Hasse diagram*:

 (7)

Recall that, in a Hasse diagram for a poset P, a vertex represents an element of P and an oriented edge represents a *covering relation* $u \lessdot v$ on P, that is,

$$u \prec v \text{ and there is no } w \in P \text{ such that } u \prec w \prec v. \tag{8}$$

For example, the Hasse diagram (7) represents the poset $P = \{a, b, c, d\}$ with $a \prec b \succ c \prec d$. The associated Hibi toric variety $\mathbb{P}_{\Delta(P)}$ is a limit of a toric degeneration of a general linear section fourfold in a Grassmannian $G(2, 5)$.

2.2. *Invariant subvarieties and singularities*

Invariant subvarieties of Hibi toric varieties are again (projectively equivalent to) lower dimensional Hibi toric varieties. We follow the description of invariant subvarieties by Wagner [22].

Let P be a finite poset. The associated bounded poset is defined as

$$\hat{P} := \hat{0} \oplus P \oplus \hat{1}, \tag{9}$$

where $\hat{0}$ and $\hat{1}$ are singletons. By definition, the elements $\hat{0}$ and $\hat{1}$ are the unique minimal and the maximal elements in $\hat{P}$, respectively. Note that the Hasse diagram of $\hat{P}$ can be regarded as the graph describing the defining inequalities of order polytope $\Delta(P)$, as we see in the middle of Figure 1. We use this identification between inequalities with edges, and variables with vertices for the Hasse diagram of $\hat{P}$. Furthermore, by abuse of notation, we write the same symbol P as the Hasse diagram of P. For example, we say that P is *connected* if the Hasse diagram of P is connected, and P is a *cycle* if the Hasse diagram of P is a cycle as an unoriented graph, and so on.

Definition 2.1. Let $\hat{P}$ be a bounded poset. A surjective order-preserving map

$$\varphi \colon \hat{P} \to \hat{P}' \tag{10}$$

with $\varphi(\hat{0}) = \hat{0}$ and $\varphi(\hat{1}) = \hat{1}$ is called a *contraction* of $\hat{P}$ if every fiber is connected and there exists a covering relation $u \lessdot v \in \hat{P}$ for all $\bar{u} \lessdot \bar{v} \in \hat{P}'$ such that $\bar{u} = \varphi(u)$ and $\bar{v} = \varphi(v)$.

There is a one-to-one correspondence between faces of order polytope $\Delta(P)$ and contractions of the associated bounded poset $\hat{P}$. More precisely, a face θ_φ corresponding to a contraction $\varphi \colon \hat{P} \to \hat{P}'$ is unimodular equivalent to the order polytope $\Delta(P')$. In other words, the associated invariant subvariety of a Hibi toric variety $\mathbb{P}_{\Delta(P)}$ is projectively equivalent to the Hibi toric variety $\mathbb{P}_{\Delta(P')}$, as mentioned at the beginning of this subsection. In particular, we have one-to-one correspondences between facets of $\Delta(P)$ and edges of $\hat{P}$, and vertices of $\Delta(P)$ and *order ideals* of P. Here an order ideal is defined as a subset $\tau \subset P$ satisfying

$$u \in \tau, v \in P \text{ and } u \succ v \text{ imply } v \in \tau. \tag{11}$$

Let us write the set of edges of $\hat{P}$ as $E = \text{Edges}(\hat{P})$, and the set of order ideals of P as $J(P)$.

We illustrate the correspondences by using the poset P in Example 2.1. We have five facets corresponding to E, and five vertices corresponding to $J(P)$. The defining equalities of a face θ_φ can be obtained by making all variables in a fiber $\varphi^{-1}(\bar{u})$ equal.

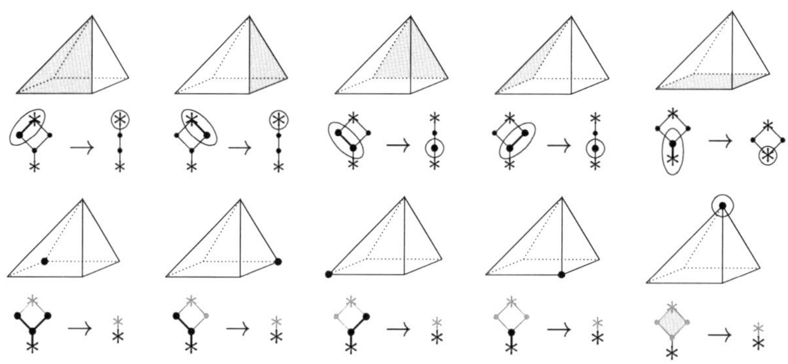

Fig. 2. One-to-one correspondence between faces and contractions

In Figure 2, four facets of the order polytope $\Delta(P)$ are meeting at the same vertex circled. Hence the corresponding point should be singular in $\mathbb{P}_{\Delta(P)}$. In general, a singular locus comes from a contraction replacing more inequalities to equalities than codimension. A subposet $C \subset \hat{P}$ is said to be *convex* if it satisfies

$$u \in C, v \in C \text{ and } u \prec w \prec v \text{ imply } w \in C. \tag{12}$$

Furthermore, let us call $C \subset \hat{P}$ a *minimal convex cycle* if C is (i) a full subposet not containing both $\hat{0}$ and $\hat{1}$, and (ii) a convex cycle such that all convex full subposets $C' \subset C$ are trees.

Theorem 2.1 ([22, Corollary 2.4]). *Let P be a finite poset. An irreducible singular locus of $\mathbb{P}_{\Delta(P)}$ corresponds to a minimal convex cycle $C \subset \hat{P}$. For the corresponding contraction, all the fibers are singletons except one fiber C.*

Remark 2.1. Now it is worth noting the homogeneous coordinate rings of Hibi toric varieties. Since all the lattice points in $\Delta(P)$ are vertices, the homogeneous coordinate ring of $\mathbb{P}_{\Delta(P)}$ is a *Hibi algebra*,

$$A_{J(P)} = \mathbb{C}\left[J(P)\right]/I_{J(P)}, \tag{13}$$

where $\mathbb{C}[J(P)]$ is the polynomial $\mathbb{C}$-algebra in variables p_τ for $\tau \in J(P)$, and $I_{J(P)}$ is the ideal coming from linear relations of vertices of $\Delta(P)$. In fact, the ideal $I_{J(P)}$ has quadratic generators,

$$p_\alpha p_\beta - p_{\alpha\vee\beta} p_{\alpha\wedge\beta} \text{ for all } \alpha \not\sim \beta \in J(P). \tag{14}$$

Note here that $J(P)$ is a *lattice*, i.e., a poset with the least upper bound $\alpha \vee \beta$ and the greatest lower bound $\alpha \wedge \beta$ for each pair of elements. In fact, $J(P)$ with the partial order given by set inclusions is equipped with $\alpha \vee \beta = \alpha \cup \beta$ and $\alpha \wedge \beta = \alpha \cap \beta$. Furthermore, $J(P)$ becomes a *distributive lattice*, i.e., the lattice with distributive laws,

$$\alpha \wedge (\beta \vee \gamma) = (\alpha \wedge \beta) \vee (\alpha \wedge \gamma) \quad \text{for all } \alpha, \beta, \gamma \in J(P). \tag{15}$$

One may start from finite distributive lattices instead of finite posets, which gives another description of Hibi toric varieties in literature.

Example 2.2. Under the notation in Remark 2.1, the Hibi toric variety $\mathbb{P}_{\Delta(P)}$ in Example 2.1 is embedded as a quadric threefold in $\mathbb{P}^4$ defined by

$$p_{\diagdown} p_{\diagup} - p_{\vee} p_{\bullet} = 0. \tag{16}$$

2.3. *Divisors*

Let P be a finite poset. For each edge $e \in E = \mathrm{Edges}(\hat{P})$, we have the corresponding invariant prime divisor, denoted by D_e. We write $D_{E'} = \sum_{e \in E'} D_e$ for each subset $E' \subset E$. The linear equivalences in the divisor class group $\mathrm{Cl}(\mathbb{P}_{\Delta(P)})$ are generated by the following relations:

$$\sum_{s(e)=u} D_e \simeq \sum_{t(e')=u} D_{e'} \quad \text{for all } u \in P, \tag{17}$$

where we write $t(e) \lessdot s(e)$ for each $e = (s(e), t(e)) \in E \subset \hat{P} \times \hat{P}$.

Let us describe the Picard group of $\mathbb{P}_{\Delta(P)}$. First, suppose P is connected. For an order ideal $\tau \subset P$ and a subset $E' \subset E$, a set of edges

$$E'(\tau) = \{e \in E' \mid t(e) \in \hat{0} \oplus \tau \text{ and } s(e) \notin \hat{0} \oplus \tau\} \tag{18}$$

defines a Weil divisor $D_{E'(\tau)}$ on $\mathbb{P}_{\Delta(P)}$. We have $D_{E(\tau)} \simeq D_{E(P)}$ for all $\tau \in J(P)$. The divisor $D_{E(P)}$ is, in fact, the Cartier divisor corresponding to the lattice polytope $\Delta(P)$ itself. One can show that the Picard group is isomorphic to $\mathbb{Z}$ generated by the associated very ample invertible sheaf $\mathcal{O}(1) = \mathcal{O}(D_{E(P)})$.

More generally, suppose $P = \sum_{j=1}^{\rho} P_j$ with ρ connected components $P_1, \ldots, P_\rho$. Note that there is a natural decomposition as sets

$$E = \bigsqcup_{j=1}^{\rho} E_j, \tag{19}$$

where $E_j := \mathrm{Edges}(\hat{P}_j) \subset E$ for each j. The Picard group $\mathrm{Pic}(\mathbb{P}_{\Delta(P)})$ becomes a free abelian group of rank ρ generated by

$$\mathcal{L}_j := \mathcal{O}\left(D_{E_j(P_j)}\right) \tag{20}$$

for each connected component $P_j \subset P$. We have

$$\mathcal{O}(1) = \mathcal{O}(D_{E(P)}) = \bigotimes_{j=1}^{\rho} \mathcal{L}_j. \tag{21}$$

Next, we note a formula for the self-intersection number,

$$\left(D^n_{E(P)}\right) = n!\,\mathrm{Vol}\,\Delta(P) = c_{J(P)}, \tag{22}$$

where $c_{J(P)}$ denotes the number of maximal chains on $J(P)$. It follows from a formula for the Hilbert–Poincaré series of Hibi algebra $A_{J(P)}$ obtained by [15, Corollary of Lemma 5].

Lastly, let us suppose P is *pure*. Recall that a finite poset P is called pure if every maximal chain on P has the same length. We define a height $h(u)$ of $u \in \hat{P}$ as the length of the longest chain bounded above by u in $\hat{P}$, and write $h_P = h(\hat{1})$. Thus an anti-canonical divisor $-K_{\mathbb{P}_{\Delta(P)}} = D_E$ is written as

$$-K_{\mathbb{P}_{\Delta(P)}} = \sum_{k=1}^{h_P} D_{E(\tau_k)} \simeq h_P D_{E(P)}, \tag{23}$$

where $\tau_k := \{u \in P \mid h(u) < k\} \in J(P)$ for $k = 1, \dots, h_P$. Together with (21), it turns out that the Hibi toric variety $\mathbb{P}_{\Delta(P)}$ for a pure poset P is a Gorenstein Fano variety with $\omega^\vee = \mathcal{O}(-K_{\mathbb{P}_{\Delta(P)}}) \simeq \mathcal{O}(h_P)$. Moreover, one can show that it has at worst terminal singularities, by looking at the normal fan Σ of $\Delta(P)$ (see [13, Lemma 1.4]).

2.4. *Small resolution*

Let P be a finite pure poset. The associated Hibi toric variety $\mathbb{P}_{\Delta(P)}$ is a Gorenstein terminal Fano variety with $\omega^\vee \simeq \mathcal{O}(h_P)$. If $\mathbb{P}_{\Delta(P)}$ is $\mathbb{Q}$-factorial in addition, it turns out to be non-singular, and even more, a product of projective spaces by [13, Corollary 2.4]. Even if it is not $\mathbb{Q}$-factorial, we have the following property indicating the mildness of singularities of $\mathbb{P}_{\Delta(P)}$.

Proposition 2.1. *For a finite pure poset P, any toric crepant $\mathbb{Q}$-factorialization of the Hibi toric variety $\mathbb{P}_{\Delta(P)}$ is a small resolution.*

Proof. Let P be a finite poset, $N = \mathbb{Z}P$ and $M = \mathbb{Z}^P$ the free abelian groups of rank $n = |P|$ dual to each other, and $N_\mathbb{R} = \mathbb{R}P$ and $M_\mathbb{R} = \mathbb{R}^P$ the real scalar extensions, respectively.

First, let us see a description of the normal fan Σ in $N_\mathbb{R}$ for the order polytope $\Delta(P) \subset M_\mathbb{R}$. By definition, a one-dimensional cone in Σ is generated by the normal vector of a facet of $\Delta(P)$. Hence it corresponds to an edge of $\hat{P}$. Let $\delta(e) \in N$ denote such primitive vector associated with $e \in E$, The map δ is extended to be the composite linear map $\delta = \mathrm{pr} \circ \partial \colon \mathbb{Z}E \to N$ of

$$\partial \colon \mathbb{Z}E \to \mathbb{Z}\hat{P} = N \oplus \mathbb{Z}\hat{0} \oplus \mathbb{Z}\hat{1}, \quad e \mapsto \partial(e) := t(e) - s(e) \tag{24}$$

and a projection $\mathrm{pr} \colon N \oplus \mathbb{Z}\hat{0} \oplus \mathbb{Z}\hat{1} \to N$. By using the same symbol δ as the real extension, each maximal dimensional cone in Σ associated with an order ideal $\tau \in J(P)$ is written as

$$\sigma_\tau := \mathrm{Cone}\, \delta\left(E \setminus E(\tau)\right) \subset N_{\mathbb{R}}, \tag{25}$$

where $E(\tau)$ is defined by (18). On the other hand, $\mathrm{Conv}\, \delta(E)$ is a Gorenstein terminal Fano polytope by [13, Lemma 1.3–1.5]. Namely, for any $\tau \in J(P)$, all the primitive generators of the cone σ_τ, i.e., the elements in $\delta(E \setminus E(\tau))$, lie on an affine hyperplane with integral distance one from the origin, and it holds

$$\left(\mathrm{Conv}\, \delta(E \setminus E(\tau))\right) \cap N = \delta\left(E \setminus E(\tau)\right). \tag{26}$$

Suppose P is pure, and let $X_{\widehat{\Sigma}} \to \mathbb{P}_{\Delta(P)}$ be a toric crepant $\mathbb{Q}$-factorialization. In other words, $\widehat{\Sigma}$ is a maximal simplicial refinement of the normal fan Σ of $\Delta(P)$ such that $X_{\widehat{\Sigma}}$ denotes the corresponding $\mathbb{Q}$-factorial toric variety. Since $\mathbb{P}_{\Delta(P)}$ has at worst terminal singularities, the crepant birational morphism $X_{\widehat{\Sigma}} \to \mathbb{P}_{\Delta(P)}$ is a small modification by definition. Hence it is sufficient to show that $X_{\widehat{\Sigma}}$ is non-singular.

Fix a maximal dimensional cone σ in $\widehat{\Sigma}$. Since (25) and (26), there exist an order ideal $\tau \subset P$ and a subset $B \subset E \setminus E(\tau)$ consisting of $n+1$ elements such that $\sigma = \mathrm{Cone}\, \delta(B) \subset \sigma_\tau$. As in the example shown in Figure 3, the subgraph $(\hat{P}, E \setminus E(\tau))$ of the Hasse diagram of $\hat{P}$ defining σ_τ consists of two connected graphs, and the subgraph $(\hat{P}, B)$ defining σ consists of two connected tree graphs. In fact, if $(\hat{P}, B)$ contains a cycle, σ cannot have maximal dimension. Therefore, we have a unique unoriented path in $(\hat{P}, B)$ from any $u \in P$ to $\hat{0}$ or $\hat{1}$, which attains a value $\pm u \in \mathbb{Z}\delta(B)$ by summing up and mapping by δ. Hence $\delta(B)$ forms a $\mathbb{Z}$-basis of $N = \mathbb{Z}P$. Since σ is arbitrary, it follows that $X_{\widehat{\Sigma}}$ is non-singular.

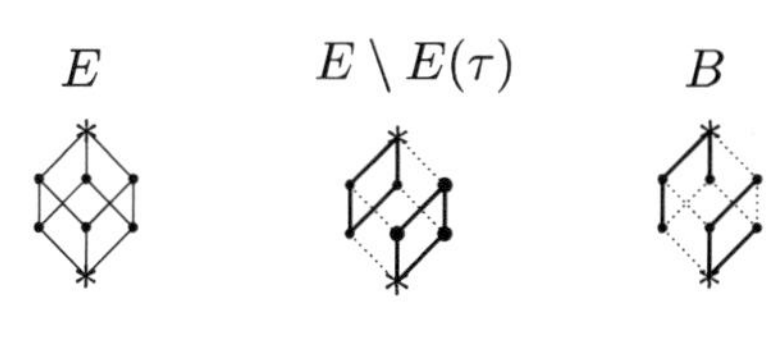

Fig. 3. An example of subgraphs corresponding to σ_τ and σ

□

3. CICY threefolds in Hibi toric varieties

3.1. *Examples*

We describe Calabi–Yau threefolds obtained as a complete intersection of general sections of invertible sheaves in Hibi toric varieties. We call such Calabi–Yau threefolds *complete intersection Calabi–Yau (CICY)* threefolds in Hibi toric varieties.

Let P be a finite poset, and X_0 a CICY threefold in $\mathbb{P}_{\Delta(P)}$. From the adjunction formula, $\mathbb{P}_{\Delta(P)}$ has at worst Gorenstein singularities. In other words, all connected components of P are pure. If P is a disjoint union of several pure connected posets, we have a number of Calabi–Yau threefolds as complete intersections of nef divisors in $\mathbb{P}_{\Delta(P)}$, e.g., complete intersection Calabi–Yau threefolds in products of projective spaces. However, we assume in the sequel that P is pure connected for simplicity. Under this assumption, X_0 is merely a complete intersection of very ample divisors in $\mathbb{P}_{\Delta(P)}$. Let $(d_1, \ldots, d_r) \subset \mathbb{P}_{\Delta(P)}$ denote a complete intersection variety of very ample divisors defined by general sections of $\mathcal{O}(d_1), \ldots, \mathcal{O}(d_r)$, respectively. Then $X_0 = (d_1, \ldots, d_r) \subset \mathbb{P}_{\Delta(P)}$ is a CICY threefold if and only if

$$\sum_{j=1}^{r} d_j = h_P \quad \text{and} \quad |P| - r = 3. \tag{27}$$

Example 3.1. The poset in (7) gives an example of hypersurface Calabi–Yau threefolds in Hibi toric varieties. Thus $X_0 = (d_1 = 3) \subset \mathbb{P}_{\Delta(P)}$ in this case.

Example 3.2. As an example to illustrate calculations, we introduce a finite pure connected poset $P = \mathcal{P}_1$:

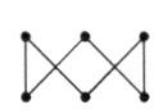

(28)

We have $|\mathcal{P}_1| = 6$ and $h_{\mathcal{P}_1} = 3$, and hence, the associated Hibi toric variety $\mathbb{P}_{\Delta(\mathcal{P}_1)}$ is a six-dimensional Gorenstein terminal Fano variety with $\omega^\vee \simeq \mathcal{O}(3)$. We have a linear section Calabi–Yau threefold $X_0 = (1^3) \subset \mathbb{P}_{\Delta(\mathcal{P}_1)}$.

The first part of $J(\mathcal{P}_1)$ corresponds to order ideals in the left of Figure 4. By continuing while focusing on set inclusions, we obtain the lattice $J(\mathcal{P}_1)$ as in the middle of Figure 4, consisting of $|J(\mathcal{P}_1)| = 18$ elements. Moreover, we have $c_{J(\mathcal{P}_1)} = 48$, the number of maximal chains on $J(\mathcal{P}_1)$, by counting as in the right in Figure 4.

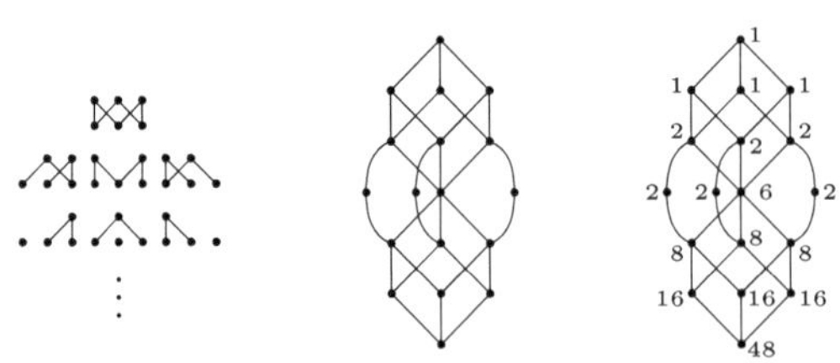

Fig. 4. On the Hasse diagram of $J(\mathcal{P}_1)$

3.2. *Stringy Hodge numbers*

Let P be a pure connected poset and $X_0 = (d_1, \dots, d_r) \subset \mathbb{P}_{\Delta(P)}$. We have a small resolution $Y \to X_0$, by taking the strict transform of X_0 for a small toric resolution $X_{\widehat{\Sigma}} \to \mathbb{P}_{\Delta(P)}$ for example. In this case, the stringy Hodge numbers of X_0 are nothing but usual Hodge numbers of Y. From [4, Proposition 8.6], the following combinatorial formulas hold.

Proposition 3.1.

$$h^{1,1}_{\mathrm{st}}(X_0) = h^{1,1}(Y) = |E| - |P|, \tag{29}$$

$$\begin{aligned} h^{1,2}_{\mathrm{st}}(X_0) = h^{1,2}(Y) = &\sum_{i\in[r]} \left[\sum_{J\subset[r]} (-1)^{|J|} l\left((d_i - d_J)\Delta(P)\right) \right] \\ &- \sum_{J\subset[r]} (-1)^{r-|J|} \left[\sum_{e\in E} l^*(d_J\theta_e) \right] - |P|, \end{aligned} \tag{30}$$

where $l(\theta)$ and $l^*(\theta)$ denote the number of lattice points in a face $\theta \subset M_{\mathbb{R}}$ and in the interior of θ, respectively; $[r] = \{1, \dots, r\}$, $d_J = \sum_{j\in J} d_j$ and θ_e is the facet of $\Delta(P)$ corresponding to an edge $e \in E$.

Note that a nonzero contribution in the first term of (30) comes only from the range of $d_i - d_J \geq 0$, and in the second term it comes only from the range $d_J = h_P - 1$ or h_P. In particular, if (i) $d_j = 1$ for all j, and (ii) P has no ordinal summand of singleton, i.e., $P \neq P_1 \oplus u \oplus P_2$ for any P_1 and P_2, we have

$$h^{1,2}_{\mathrm{st}}(X_0) = h^{1,2}(Y) = h_P\left(|J(P)| - h_P\right) - \sum_{e\in E} l^*\left(h_P\theta_e\right) - |P|. \tag{31}$$

Example 3.3. For the example $P = \mathcal{P}_1$ and a complete intersection $X_0 = (1^3)$, we obtain $h^{1,1}_{\mathrm{st}}(X_0) = 12 - 6 = 6$ from (29). Since the both conditions (i) and (ii) are satisfied, the simplified formula (31) holds in this case. By the symmetry $\mathfrak{S}_3 \times \mathfrak{S}_3$ of $\hat{\mathcal{P}}_1$ as a poset and the order duality, it suffices to see the following two types of facets:

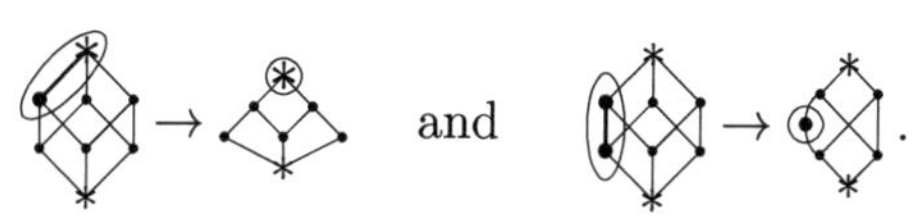 (32)

There are six facets for each type, and clearly $l^*(3\theta) = 1$ (resp. 0) for the former (resp. the latter) type. Therefore $h^{1,2}_{\mathrm{st}}(X_0) = 3(18 - 3) - 6 - 6 = 33$.

3.3. *Numbers of nodes*

Recall that three-dimensional Gorenstein terminal toric singularities are at worst nodes (i.e, ordinary double points), since they are presented by three-dimensional cones over a unit triangle or a unit square. Together with the Bertini-type theorem for toroidal singularities, the singularities of X_0 are also at worst nodes. We count the number of nodes $\mathrm{dp}(X_0)$ on X_0 in the following.

Each node on X_0 lies on one singular locus of codimension three of $\mathbb{P}_{\Delta(P)}$, corresponding to a minimal convex cycle $C \subset \hat{P}$ with four elements. There are four types of such minimal convex cycles:

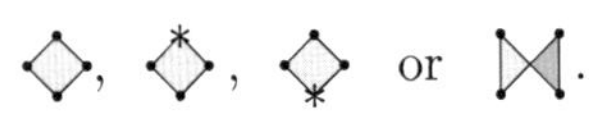
(33)

Let $\Lambda_4(\hat{P})$ denote the set of such minimal convex cycles consisting of four elements on $\hat{P}$. For each $C \in \Lambda_4(\hat{P})$, there is the contraction $\hat{P} \to \hat{P}_C$ such that all the fibers are singleton except one fiber C. Of course it holds $|P_C| = |P| - 3$ for all $C \in \Lambda_4(\hat{P})$. Hence C defines a singular locus of codimension three and of degree $\deg \Delta(P_C) = c_{J(P_C)}$ from (22). Therefore, the number of nodes $\mathrm{dp}(X_0)$ is calculated by a formula

$$\mathrm{dp}(X_0) = \prod_{j=1}^{r} d_j \sum_{C \in \Lambda_4(\hat{P})} c_{J(P_C)}. \tag{34}$$

Example 3.4. There are six minimal convex cycles on $\hat{\mathcal{P}}_1$, each of which consists of four elements. By symmetry, they are all equivalent to

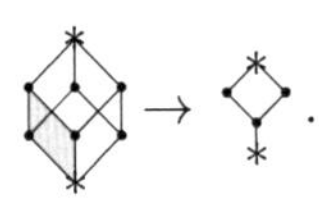
(35)

Since the locus is a quadric threefold in Example 2.2, we obtain $\mathrm{dp}(X_0) = 6 \cdot 2 = 12$.

3.4. *Smoothability*

For smoothability, we follow the argument in the case of hypersurfaces in toric varieties by [7]. Let $\{p_1, \ldots, p_{\mathrm{dp}}\}$ be the set of nodes on X_0, where $\mathrm{dp} = \mathrm{dp}(X_0)$, and $f : Y \to X_0$ be a small resolution. The exceptional lines $L_i := f^{-1}(p_i) \simeq \mathbb{P}^1$ for $i = 1, \ldots, \mathrm{dp}$ form a linear subspace of $H_2(Y, \mathbb{C})$. By [19, Theorem 2.5], the Calabi–Yau threefold X_0 is smoothable by a flat deformation if and only if the homology classes $[L_i] \in H_2(Y, \mathbb{C})$ satisfy a relation,

$$\sum_{i=1}^{\mathrm{dp}} \alpha_i [L_i] = 0, \tag{36}$$

where $\alpha_i \neq 0$ for all $i = 1, \ldots, \mathrm{dp}$.

Note that one can identify

$$H_2(Y, \mathbb{Q}) \simeq \left\{ (\lambda_e)_{e \in E} \;\middle|\; \sum_{e \in E} \lambda_e \delta(e) = 0 \right\} \subset \mathbb{Q}^E. \tag{37}$$

Under this identification, the homology class $[L_i]$ coincides with a relation,

$$\rho_C : \delta(e_p) + \delta(e_q) - \delta(e_r) - \delta(e_s) = 0 \tag{38}$$

up to signs, where the corresponding node p_i lies on a singular locus associated with a minimal convex cycle $C \in \Lambda_4(\hat{P})$, and the cycle C with an orientation passes through the four edges; e_p, e_q in the forward direction and e_r, e_s in the opposite direction.

Proposition 3.2. *Let P be a pure connected poset, X_0 a CICY threefold of degree $(d_1, \ldots, d_r)$ in the Hibi toric variety $\mathbb{P}_{\Delta(P)}$. If $\prod_{j=1}^r d_j > 1$, X_0 is smoothable. If $\prod_{j=1}^r d_j = 1$, X_0 is smoothable if and only if for any $C \in \Lambda_4(\hat{P})$ such that P_C is a chain the element ρ_C is a linear combination of the remaining elements $\rho_{C'}$ with $C' \in \Lambda_4(\hat{P})$, $C \neq C'$.*

Example 3.5. For the example $P = \mathcal{P}_1$, $X_0 = (1^3)$ is smoothable since P_C is not a chain for all $C \in \Lambda_4(\hat{\mathcal{P}}_1)$ as we see in (35), although $\prod_{j=1}^3 d_j = 1$.

Example 3.6. There are cases that X_0 is not smoothable. For example, in (39) we present the two cycles on the depicted $\hat{P}$ satisfying the condition that P_C is a chain,

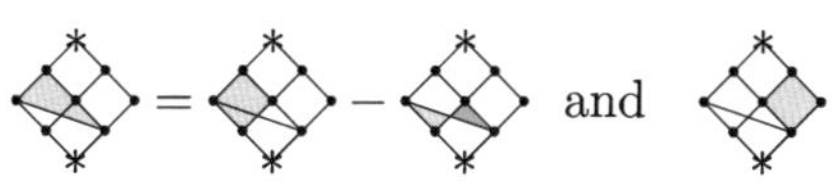
(39)

The former cycle is a linear combination of remaining cycles as expressed by abuse of notation. However, the latter cycle is linearly independent to other cycles. Therefore, $X_0 = (1^4)$ is not smoothable by Proposition 3.2.

4. Smoothing of CICY threefolds in Hibi toric varieties

4.1. *Simply-connectedness*

Proposition 4.1 (Corollary of Proposition 2.1). *Let P be a pure poset, X_0 a CICY threefold in $\mathbb{P}_{\Delta(P)}$, and $Y \to X_0$ a small resolution. Then Y is simply-connected. If a smoothing $X \rightsquigarrow X_0$ exists, X is also simply-connected.*

Proof. By Proposition 2.1, we have a small resolution $X_{\widehat{\Sigma}} \to X_\Sigma = \mathbb{P}_{\Delta(P)}$. Since $X_{\widehat{\Sigma}}$ is a compact toric variety, it is simply-connected (see for example [10, Theorem 9.1]). Let $\Sigma^{(1)}$ denote the subfan of Σ consisting of cones of dimension less than or equal to one. Note $\widehat{\Sigma}^{(1)} = \Sigma^{(1)}$. The quasi-projective toric variety $X_{\Sigma^{(1)}}$ is simply-connected as well. In fact, the difference between $X_{\widehat{\Sigma}}$ and $X_{\Sigma^{(1)}}$ in subsets of real codimension four does not effect fundamental groups. By the Lefschetz theorem for non-singular quasi-projective manifolds [11, 14], a complete intersection $X_{\Sigma^{(1)}} \cap X_0 = X_{\widehat{\Sigma}^{(1)}} \cap Y$ is also simply-connected. Similarly as above, the difference between $X_{\widehat{\Sigma}^{(1)}} \cap Y$ and Y does not effect fundamental groups. Therefore Y is also simply-connected. The latter statement follows from the fact that a conifold transition does not change fundamental groups. □

Let X be a smoothing of a CICY threefold in Hibi toric variety. We do not know whether homology groups of X can have torsion or not. Suppose that X has torsion-free homology and $h^{1,1}(X) = 1$. In this case, by Wall's theorem [23, Theorem 5], the diffeomorphism class of X is determined only by the three topological invariants,

$$\deg X,\ c_2(X)\cdot H, \quad \text{and} \quad \chi(X), \tag{40}$$

where H is the hyperplane class, $c_2(X)$ is the second Chern class and $\chi(X) = 2\left(h^{1,1}(X) - h^{2,1}(X)\right)$ is the topological Euler number of X. We summarize the calculation of these topological invariants in the next subsection.

4.2. *Topological invariants*

For a conifold transition $X \rightsquigarrow X_0 \leftarrow Y$, Hodge numbers satisfy

$$h^{1,1}(X) = h^{1,1}(Y) - \mathrm{rk}, \tag{41}$$

$$h^{1,2}(X) = h^{1,2}(Y) + \mathrm{dp} - \mathrm{rk}, \tag{42}$$

where $\mathrm{rk} = \mathrm{rk}(X_0)$ is the dimension of linear subspace of $H_2(Y, \mathbb{C})$ spanned by classes of exceptional lines $[L_i]$ for $i = 1, \ldots \mathrm{dp}$, i.e.,

$$\mathrm{rk}(X_0) = \mathrm{rank}\left(\rho_C \,\middle|\, C \in \Lambda_4(\hat{P})\right) \tag{43}$$

and $\mathrm{dp} = \mathrm{dp}(X_0)$ is the number of nodes on X_0, which we compute by (34). In particular, from $h^{1,1}(Y) = |E| - |P| = b_1(\hat{P}) + 1$, we have $h^{1,1}(X) = 1$ if and only if all minimal cycles on $\hat{P}$ are generated by cycles in $\Lambda_4(\hat{P})$.

Assume $h^{1,1}(X) = 1$. From (22) and the invariance by a flat deformation, it holds

$$\deg X = \deg X_0 = c_{J(P)} \prod_{j=1}^{r} d_j. \tag{44}$$

Since also the invariance $\chi(X, \mathcal{O}_X(1)) = \chi(X_0, \mathcal{O}_{X_0}(1))$ and a standard cohomology calculation for complete intersection varieties in $\mathbb{P}_{\Delta(P)}$, we obtain

$$\chi(X, \mathcal{O}_X(1)) = \dim H^0(X_0, \mathcal{O}_{X_0}(1)) \tag{45}$$

$$= \dim H^0(\mathbb{P}_{\Delta(P)}, \mathcal{O}_{\mathbb{P}_{\Delta(P)}}(1)) - r_1 \tag{46}$$

$$= |J(P)| - r_1, \tag{47}$$

where $r_1 = \#\left\{j \mid d_j = 1\right\}$. Therefore the Hirzebruch–Riemann–Roch theorem gives

$$c_2(X)\cdot H = 12\chi(X, \mathcal{O}_X(1)) - 2\deg X \tag{48}$$

$$= 12\left(|J(P)| - r_1\right) - 2c_{J(P)} \prod_{j=1}^{r} d_j. \tag{49}$$

Example 4.1. For $P = \mathcal{P}_1$, $h^{1,1}(X) = 1$ holds since all minimal cycles are in $\Lambda_4(\hat{\mathcal{P}}_1)$. Hence $\mathrm{rk}(X_0) = 5$ by (41). From $h^{1,2}(Y) = 33$ and $\mathrm{dp}(X_0) = 12$, it holds $h^{1,2}(X) = 40$ and $\chi(X) = 2(h^{1,1}(X) - h^{1,2}(X)) = -78$ by (42). We also obtain $\deg(X) = 48$ and $c_2(X) \cdot H = 12(18-3) - 2 \cdot 48 = 84$ by (44) and (49), respectively.

4.3. *Examples*

Let P be a pure poset and X_0 a CICY threefold in $\mathbb{P}_{\Delta(P)}$. We write $X = X_P$ for a smoothing $X \rightsquigarrow X_0$ if it exists. In spite of the large number of smoothable CICY threefolds in Hibi toric varieties, there are few examples of X_P with $h^{1,1}(X_P) = 1$. From [17, Proposition 3.1], we have twelve such threefolds as complete intersections in minuscule Schubert varieties up to deformation equivalence; five in projective spaces, five in other Grassmannians $G(k,n)$, one in an orthogonal Grassmannian $OG(5,10)$, and one in a singular Schubert variety of the Cayley plane $\mathbb{OP}^2$. The latter two threefolds are also regarded as complete intersections of some homogeneous vector bundles on Grassmannians, i.e., No. 4 and No. 7 in [16, Table 1], respectively. Apart from these twelve examples, we introduce six more examples in Table 1. The topological invariants are computed by formulas in the previous subsection.

Table 1. Examples of X of Picard number one.

posets $\mathcal{P}_i$	$\mathcal{P}_1$	$\mathcal{P}_2$	$\mathcal{P}_3$	$\mathcal{P}_4$	$\mathcal{P}_5$	$\mathcal{P}_6$
Hasse diagrams	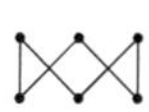	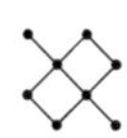	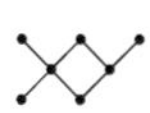			
degrees	(1^3)	(1^5)	(1^4)	(1^3)	$(1,2)$	(1^5)
$\deg X_{\mathcal{P}_i}$	48	29	42	61	32	25
$c_2(X_{\mathcal{P}_i}) \cdot H$	84	74	84	94	80	70
$\chi(X_{\mathcal{P}_i})$	−78	−100	−96	−86	−116	−100

Note that $X_{\mathcal{P}_4}$, $X_{\mathcal{P}_5}$ and $X_{\mathcal{P}_6}$ are deformation equivalent to complete intersections of some homogeneous vector bundles on Grassmannians, i.e., No. 23, No. 20 and No. 5 in [16, Table 1], respectively (from a private communication with Daisuke Inoue and Atsushi Ito for $X_{\mathcal{P}_4}$). Furthermore, $X_{\mathcal{P}_5}$ and $X_{\mathcal{P}_6}$ are also regarded as a complete intersection $(1^2, 2)$ in a Lagrangian Grassmannian $LG(3,6)$, and a complete intersection of two Grassmannians $G(2,5)$ in $\mathbb{P}^9$, respectively.

Remark 4.1. Let us discuss mirror symmetry for examples in Table 1. For each conifold transition $X \rightsquigarrow X_0 \leftarrow Y$, we expect another conifold transition $Y^* \rightsquigarrow Y_0^* \leftarrow X^*$ in the mirror side, if X and Y have torsion-free homology. We have a Batyrev–Borisov mirror Y^* for $Y \subset \mathbb{P}_{\widehat{\Sigma}}$, and the degeneration $Y^* \rightsquigarrow Y_0^*$ corresponding to the small resolution $Y \to X_0$ based on the argument by [2]. However, we do not know whether Y_0^* has the same number of nodes as X_0 and admits a small resolution $X^* \to Y_0^*$ or not. In spite of that, periods and the Picard–Fuchs operator vanishing the periods for the conjectural mirror family are computable in advance. The resulting operators in the case of $\mathcal{P}_2$, $\mathcal{P}_3$, $\mathcal{P}_4$, $\mathcal{P}_5$ and $\mathcal{P}_6$ coincide with already known operators, #195, #28, #124, #42, and #101 in [1, 21], respectively.

The operator for $\mathcal{P}_1$ seems unknown, thus we write it here. A formula for the fundamental period $\omega_0(z) = \sum_{m=0}^{\infty} A_m z^m$ is given in [17, Eq. (5,2)], where

$$A_m = \sum_{s,t,u,v,w} \binom{s}{u}\binom{v}{s}\binom{t}{s}\binom{t}{v}\binom{w}{t}\binom{m}{t}\binom{m}{w}\binom{v-t+w}{u-s+v}\binom{m}{v-t+w} \tag{50}$$

is read from the (slightly modified) dual graph of $\hat{\mathcal{P}}_1$, by associating binomial coefficients with oriented edges and linear relations with pairs of dashed edges:

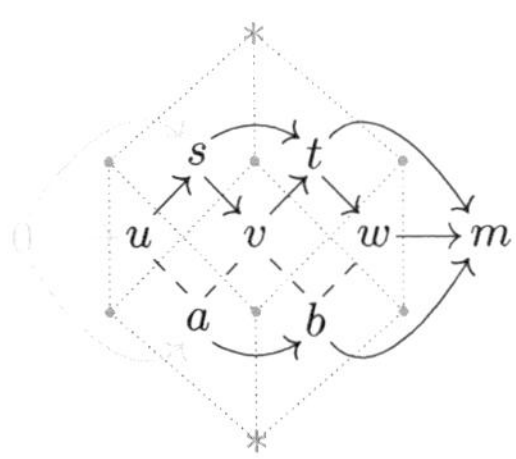

$$\begin{cases} a = u - s + v, \\ b = v - t + w. \end{cases} \tag{51}$$

With the aid of numerical method, we obtain the following Picard–Fuchs operator for a conjectural mirror family for $X_{\mathcal{P}_1}$,

$$\begin{aligned} \mathcal{D} = \theta^4 &- 2z(33\theta^4 + 58\theta^3 + 48\theta^2 + 19\theta + 3) \\ &+ 4z^2(174\theta^4 + 448\theta^3 + 527\theta^2 + 314\theta + 75) \\ &- 8z^3(332\theta^4 + 1096\theta^3 + 1507\theta^2 + 953\theta + 228) \\ &+ 96z^4(\theta+1)^2(6\theta+5)(6\theta+7), \end{aligned} \tag{52}$$

where $\theta = z\partial_z$ and $\mathcal{D}\omega_0(z) = 0$. We observe that the operator generates integral BPS numbers for genus 0 and genus 1 with small degrees, by standard methods for the computation [6, 9].

Acknowledgments

This article is an extended version of the author's talk at Algebraic and Geometric Combinatorics on Lattice Polytopes 2018. He would like to thank the organizers for the kind invitation. He is also grateful to Tomoyuki Hisamoto, Atsushi Ito and Taro Sano for helpful discussions.

References

[1] G. Almkvist, C. van Enckevort, D. van Straten, and W. Zudilin, *Tables of Calabi–Yau equations*, [arXiv:math.AG/0507430].

[2] V. V. Batyrev, *Toric degenerations of Fano varieties and constructing mirror manifolds*, in The Fano Conference (Univ. Torino, Turin, 2004), pp. 109–122.

[3] V. V. Batyrev, *Dual polyhedra and mirror symmetry for Calabi–Yau hypersurfaces in toric varieties*, J. Algebraic Geom. 3, 493–535 (1994).

[4] V. V. Batyrev and L. A. Borisov, *On Calabi-Yau complete intersections in toric varieties*, in Higher-dimensional complex varieties (Trento, 1994) (de Gruyter, Berlin, 1996), pp. 39–65.

[5] V. V. Batyrev, I. Ciocan-Fontanine, B. Kim, and D. van Straten, *Conifold transitions and mirror symmetry for Calabi-Yau complete intersections in Grassmannians*, Nuclear Phys. B **514**, 640–666 (1998).

[6] M. Bershadsky, S. Cecotti, H. Ooguri, and C. Vafa, *Holomorphic anomalies in topological field theories*, Nuclear Phys. B **405**, 279–304 (1993).

[7] V. Batyrev and M. Kreuzer, *Constructing new Calabi-Yau 3-folds and their mirrors via conifold transitions*, Adv. Theor. Math. Phys. **14**, 879–898 (2010).

[8] L. Borisov, *Towards the Mirror Symmetry for Calabi-Yau Complete intersections in Gorenstein Toric Fano Varieties*, [arxiv:alg-geom/9310001].

[9] P. Candelas, X. C. de la Ossa, P. S. Green, and L. Parkes, *A pair of Calabi–Yau manifolds as an exactly soluble superconformal theory*, Nuclear Phys. B **359**, 21–74 (1991)

[10] V. I. Danilov, *The geometry of toric varieties*, Uspekhi Mat. Nauk **33**, 85–134, 247 (1978).

[11] M. Goresky and R. MacPherson, *Stratified Morse theory*, Vol. 14, Ergebnisse der Mathematik und ihrer Grenzgebiete (3) [Results in Mathematics and Related Areas (3)] (Springer-Verlag, Berlin, 1988), pp. xiv+272.

[12] A. He and P. Candelas, *On the number of complete intersection Calabi–Yau manifolds*, Comm. Math. Phys. **135**, 193–199 (1990)

[13] T. Hibi and A. Higashitani, *Smooth Fano polytopes arising from finite partially ordered sets*, Discrete Comput. Geom. **45**, 449–461 (2011).

[14] H. A. Hamm and L. D. Trang, *Lefschetz theorems on quasiprojective varieties*, Bull. Soc. Math. France 113, 123–142 (1985).

[15] T. Hibi and K. Watanabe, *Study of three-dimensional algebras with straightening laws which are Gorenstein domains. I*, Hiroshima Math. J. **15**, 27–54 (1985).

[16] D. Inoue, A. Ito, and M. Miura, *Complete intersection Calabi–Yau manifolds with respect to homogeneous vector bundles on Grassmannians*, [arXiv:math.AG/1607.07821].

[17] M. Miura, *Minuscule Schubert varieties and mirror symmetry*, SIGMA Symmetry Integrability Geom. Methods Appl. **13**, Paper No. 067, 25 (2017).

[18] D. R. Morrison, *Through the looking glass*, in Mirror symmetry, III (Montreal, PQ, 1995), Vol. 10, AMS/IP Stud. Adv. Math. (Amer. Math. Soc., Providence, RI, 1999), pp. 263–277.

[19] Y. Namikawa, *Stratified local moduli of Calabi-Yau threefolds*, Topology **41**, 1219–1237 (2002).

[20] M. Reid, *The moduli space of 3-folds with* $K = 0$ *may nevertheless be irreducible*, Math. Ann. **278**, 329–334 (1987).

[21] D. van Straten, Calabi–Yau Database, Version 2.0 (`www2.mathematik.uni-mainz.de/CYequations/db/`), Version 3.0 (`cydb.mathematik.uni-mainz.de`).

[22] D. G. Wagner, *Singularities of toric varieties associated with finite distributive lattices*, J. Algebraic Combin. **5**, 149–165 (1996).

[23] C. T. C. Wall, *Classification problems in differential topology. V. On certain 6-manifolds*, Invent. Math. 1 (1966), 355–374; corrigendum, ibid **2**, 306 (1966).

Technically, squares are polytopes

Louis Ng

Department of Mathematics, San Francisco State University
San Francisco, CA 94132, USA
E-mail: louisng@mail.sfsu.edu
www.sfsu.edu

This article is a survey on the combinatorics of semimagic/pandiagonal magic squares and explains their connection to polytopes. On the side of combinatorics, we count magic squares with fixed strict upperbounds or fixed sum. As for polytopes, we show the relationship between magic squares and various polytopes such as the Birkhoff polytopes.

Keywords: Magic squares, semimagic, pandiagonal, Birkhoff polytope, hypersimplex, inside-out polytope, Ehrhart reciprocity.

1. Introduction

Let us embark on a magical journey, a combinatorial journey about magic squares and polytopes.

Traditionally, magic squares are $n \times n$ array using integers $1, 2, \dots, n^2$ where the row sums, the column sums, and the two main diagonal sums are all equal. There are 880 magic squares of size 4×4 and 275305224 magic squares of size 5×5 (we consider two squares the same if they are identical after rotations and reflections) (OEIS A006052). The number of 6×6 magic squares is not known.

As mathematicians, we like to add and remove various conditions and see what happens, so we will do just that. First, some definitions. A magic square is **semimagic** if the diagonal sums are ignored, while a **pandiagonal** magic square cares about all diagonals, including the wrap-around ones. If we are to denote the cells of the magic squares with a_{ij} like matrices, then all the diagonal sums (including wrap-around ones) can be represented as

$$\sum_{i+j \equiv k \pmod n} a_{ij} \qquad \text{or} \qquad \sum_{i-j \equiv k \pmod n} a_{ij}$$

for $k \in \{1, 2, \dots, n\}$.

We can also change the rules around which entries we are allowed to use. **Weak** magic squares has no clause on duplicate entries, whereas **strong** magic squares requires the integers used to not repeat. In between strong and weak, **Latin** magic squares only require entries in rows and in columns to not repeat. Since the numbers we use are not bounded, we get either no magic squares or infinitely many magic squares depending on which type we are counting. To make it more interesting, we can bound the integers used or the magic sum. Let the **cubical count** be the number of magic squares using nonnegative integers less than some upperbound t, and let the **affine count** be the number of magic squares using nonnegative integers with some fixed magic sum s. For example,

9	2	7
4	6	8
5	10	3

is a strong magic square, though not pandiagonal since $9 + 2 + 7 \neq 2 + 8 + 5$. It is counted when we do the cubical count with $t \geq 11$, or when we do the affine count with $s = 18$.

2. Applications

Before we set sail for magic squares and polytopes, let us explore the applications of magic squares.

To most, magic squares are purely for fun, but people have found uses for magic squares outside of recreational mathematics. Let us start with art. On his famous engraving *Melencolia I*, German Renaissance artist Dürer Albrecht included various mathematical object, amongst which is a 4×4 magic square at the upper right corner [14]. This square was specifically crafted (pun intended) for the engraving, in that the two bottom center cells read 1514, the year of the engraving. The magic square in the engraving is one of ten squares featured in Macau's stamps in 2014 [9]. Other squares include the Franklin square by Benjamin Franklin and the Luo Shu square.

Fun and art aside, magic squares can also be used in encryption [12, 16].

3. Background

As promised, this journey is also about polytopes, so we will first define terms related to polytopes.

A **hyperplane** $H_{\vec{a},b}$ is defined as

$$H_{\vec{a},b} := \{\vec{x} = (x_1, x_2, \ldots, x_d) \in \mathbb{R}^d : \vec{a} \cdot \vec{x} = b\},$$

for some $\vec{a} = (a_1, a_2, \ldots, a_d) \in \mathbb{R}^d \setminus (0, 0, \ldots, 0)$ and $b \in \mathbb{R}$. A hyperplane splits $\mathbb{R}^d$ into two **open halfspaces**

$$H^{>}_{\vec{a},b} := \{\vec{x} = (x_1, x_2, \ldots, x_d) \in \mathbb{R}^d : \vec{a} \cdot \vec{x} > b\} \text{ and}$$
$$H^{<}_{\vec{a},b} := \{\vec{x} = (x_1, x_2, \ldots, x_d) \in \mathbb{R}^d : \vec{a} \cdot \vec{x} < b\}.$$

$H_{\vec{a},b}$ is the boundary of the open halfspaces $H^{>}_{\vec{a},b}$ and $H^{<}_{\vec{a},b}$. Taking the union of open halfspaces and their boundaries, we get the **closed halfspaces**

$$H^{\geq}_{\vec{a},b} := \{\vec{x} = (x_1, x_2, \ldots, x_d) \in \mathbb{R}^d : \vec{a} \cdot \vec{x} \geq b\} \text{ and}$$
$$H^{\leq}_{\vec{a},b} := \{\vec{x} = (x_1, x_2, \ldots, x_d) \in \mathbb{R}^d : \vec{a} \cdot \vec{x} \leq b\}.$$

A **convex**[a] **polytope** P is the bounded intersection of finitely many closed half-spaces. In other words,

$$P = \bigcap_{i=1}^{n_1} H_{\vec{a_i},b_i} \cap \bigcap_{i=1}^{n_2} H^{\geq}_{\vec{c_i},d_i} \cap \bigcap_{i=1}^{n_3} H^{\leq}_{\vec{e_i},f_i}.$$

The **affine hull** aff(P) of P is $\bigcap\{H_{\vec{a},b} : P \subseteq H_{\vec{a},b}\}$. The **dimension** of a polytope is the dimension of its affine hull. The **relative interior of a polytope** P° is the interior of P relative to aff(P). The set P° has the representation

$$P^\circ = \bigcap_{i=1}^{n_1} H_{\vec{a_i},b_i} \cap \bigcap_{i=1}^{n_2} H^{>}_{\vec{c_i},d_i} \cap \bigcap_{i=1}^{n_3} H^{<}_{\vec{e_i},f_i}.$$

The dimension of the relative interior of a polytope is the same as the dimension of the polytope.

We can also represent a polytope as the convex hull of a finite set of points [6], where the **convex hull** of $\vec{p_1}, \vec{p_2}, \ldots, \vec{p_n} \in \mathbb{R}^d$ is

$$\{k_1\vec{p_1} + k_2\vec{p_2} + \cdots + k_n\vec{p_n} : k_j \geq 0 \text{ and } k_1 + k_2 + \cdots + k_n = 1\}.$$

The set of **vertices** of a polytope is the smallest set of points whose convex hull is said polytope. The vertices of the relative interior of a polytope are the vertices of the polytope.

One combinatorial aspect of polytopes is the count of **lattice points** (points with integer coordinates). Given set S, let $L(S,n)$ be the number of lattice points in the nth dilate of S. In other words $L(S,n) = |nS \cap \mathbb{Z}^d|$. For polytope P, the associated generating function is the **Ehrhart series**

$$\mathrm{Ehr}_P(z) := 1 + \sum_{n\geq 1} L(P,n)z^n.$$

For P°, the interior of a polytope, the Ehrhart series is

$$\mathrm{Ehr}_{P^\circ}(z) := \sum_{n\geq 1} L(P^\circ,n)z^n.$$

In general, the **generation function** of a sequence a_n is $\sum_{n\geq 0} a_n z^n$.

4. Semimagic Squares

4.1. *Weak Semimagic Squares*

We begin our adventure with weak $n\times n$ semimagic squares, since they have the most relaxed conditions. A polytope with points in $\mathbb{R}^{n^2}$ whose points are matrices with row sums and column sums 1 is known as the **Birkhoff polytope** B_n. Therefore, the affine count of weak semimagic squares with magic sum s is the same as the number of lattice points in the s-th dilate

$$\{s\vec{x} : \vec{x} \in B_n\}$$

[a] All polytopes in this article will be convex, so we will just call them polytopes

of the Birkhoff polytope in the matching dimension. Another way to see the connection between weak semimagic squares and the Birkhoff polytopes is with the vertices of B_n, which are the permutation matrices. These matrices have exactly one 1 in each row/column, and thus spans weak semimagic squares. Even though the method for computing the affine count is known, the computation scales exponentially with the dimension [5]. To date, $L_{B_n}(s) = |sB_n \cap \mathbb{Z}^{n^2}|$ is known for $n \leq 9$. The list is available at [4].

4.2. *Weak Semimagic Rectangles*

Let us shift our attention to magic rectangles. Of course, if the row sums and the column sums of an $m \times n$ rectangular matrix are both s, then ms = sum of all cells = ns, which is only true when $m = n$. Ergo, when talking about semimagic rectangles, we allow the row sums to be different from the column sums. Specifically, row sum r and column sum c are related by $r = \frac{nc}{m}$.

Just as weak semimagic squares are associated with the Birkhoff polytope, weak $2 \times 2n$ semimagic rectangles are associated with the $(2n, n)$-hypersimplex [15]. A (d, k)-**hypersimplex** [6] is a polytope of the form

$$\{(x_1, \ldots, x_d) : 0 \leq x_j \leq 1 \text{ and } x_1 + \cdots + x_d = k\}.$$

This association is because $2 \times n$ semimagic squares are determined by one row and the column sum. The bounds of 0 and 1 for each x_j ensures that all entries are nonnegative.

The vertices of a (d, k)-hypersimplex are d-dimension points with k ones and $d - k$ zeroes for their coordinates. On the other hand, the polytope associated with weak $2 \times (2n + 1)$ semimagic rectangles is not a hypersimplex. Rather, the vertices have n ones, n zeros, and a single $\frac{1}{2}$.

Let $[z^k]f(z)$ denotes the coefficient of z^k of the polynomial/series $f(z)$. The affine counts (base on the column sum c) of weak $2 \times n$ semimagic rectangles are given by [15]

$$\left[z^{\frac{nc}{2}}\right] \left(\frac{1 - z^{c+1}}{1 - z}\right)^n.$$

Table 1 to Table 6 show the number of weak 2×2 to 2×7 magic rectangles.

Table 1. Number of weak 2×2 magic rectangles with column sum c.

c	0	1	2	3	4	5	6	7	8	9	10	11
affine count	1	2	3	4	5	6	7	8	9	10	11	12

4.3. *Latin Semimagic Squares*

Matthias Beck and Thomas Zaslavsky computed the cubical and affine count for Latin 3×3 semimagic squares [8]. Here is a summary of the techniques used. 3×3

Table 2. Number of weak 2×3 magic rectangles with column sum c.

c	0	2	4	6	8	10	12	14	16	18
affine count	1	7	19	37	61	91	127	169	217	271

Table 3. Number of weak 2×4 magic rectangles with column sum c.

c	0	1	2	3	4	5	6	7	8	9
affine count	1	6	19	44	85	146	231	344	489	670

Table 4. Number of weak 2×5 magic rectangles with column sum s.

c	0	2	4	6	8	10	12	14	16	18
affine count	1	51	381	1451	3951	8801	17151	30381	50101	78151

semimagic squares can all be reduced to the form

0	β	$2\alpha+\beta$ $+\gamma$
$\alpha+\beta$	$\alpha+\beta$ $+\gamma-\delta$	δ
$\alpha+\beta$ $+\gamma$	$\alpha+\delta$	$\beta-\delta$

,

by subtracting the smallest integer used from every entries to obtain a zero, move it to the corner through swapping rows/columns, and again rearranging rows/columns or reflect across the $i = j$ diagonal such that $a_{12} < a_{13}$, $a_{21} \leq a_{31}$, and $a_{12} < a_{21}$. Therefore, we can count the number of squares of the above form first and adjust the count later. To ensure that all entries except for the top-left are positive, we have the inequalities

$$
\begin{aligned}
&0 \leq \alpha, \\
&0 < \beta, \gamma, \text{ and} \\
&0 \leq \delta \leq \beta.
\end{aligned}
$$

Table 5. Number of weak 2×6 magic rectangles with column sum c.

c	0	1	2	3	4	5	6	7	8	9
affine count	1	20	141	580	1751	4332	9331	18152	32661	55252

Table 6. Number of weak 2×7 magic rectangles with column sum c.

c	0	2	4	6	8	10	12	14
affine count	1	393	8135	60691	273127	908755	2473325	5832765

These inequalities along with the upperbound from the cubical count or the sum equations from the affine count form a half-open polytope in $\mathbb{R}^4$ for each counting method. The main reason to use this over the Birkhoff polytope is that we need only seven inequalities instead of eighteen to take care of the Latin condition. The inequalities are

$$\delta \neq \frac{\beta-\alpha}{2}, \frac{\beta}{2}, \frac{\beta+\gamma}{2}, \frac{\alpha+\beta+\gamma}{2}, \beta-\alpha, \alpha+\gamma, \gamma.$$

Each equation is a hyperplane in $\mathbb{R}^4$.

Let $\mathcal{H}$ be a collection of hyperplanes and P be a polytope. The pair $(P, \mathcal{H})$ is an **inside-out polytope** [7]. The Ehrhart reciprocity theorem [13] states that given a rational polytope[b] $P \in \mathbb{R}^d$, the lattice point counting functions $L_P(n)$ and $L_{P^\circ}(n)$ are polynomials in n. Furthermore,

$$L_P(-n) = (-1)^{\dim P} L_{P^\circ}(n).$$

If we consider their Ehrhart series $\mathrm{Ehr}_P(z) := 1 + \sum_{n\geq 1} L_P(n) z^n$ and $\mathrm{Ehr}_{P^\circ}(z) := \sum_{n\geq 1} L_{P^\circ}(n) z^n$, they are related by

$$\mathrm{Ehr}_P\left(\frac{1}{z}\right) = (-1)^{1+\dim P}\mathrm{Ehr}_{P^\circ}(z). \tag{1}$$

Just as rational polytopes, rational inside-out polytopes have a nice reciprocity formula, which follows from Equation (1) and the inclusion-exclusion principle [7];

$$-\mathrm{Ehr}_{P^\circ \setminus \cup\mathcal{H}}\left(\frac{1}{z}\right) = \sum_{u\in\mathcal{L}} |\mu(u)| \mathrm{Ehr}_{\bar{u}}(z),$$

where

- $\mathrm{Ehr}_{P^\circ\setminus\cup\mathcal{H}}(z) := \sum_{n\geq 1} |n(P^\circ \setminus \cup\mathcal{H}) \cap \mathbb{H}^d| z^n$,
- the partially ordered set $\mathcal{L} = \{\bigcap_{i\in S\subseteq[n]}(H_i \cap P^\circ) : H_i \in \mathcal{H}\}$ is ordered by $A \prec B$ if and only if $B \subset A$,
- the Möbius function $\mu(A) = \begin{cases} 1 & \text{if } P^\circ = A \\ -\sum_{B\prec A}\mu(B) & \text{if } P^\circ \prec A \end{cases}$, and
- $\bar{u}$ is the topological closure of u.

Figure 1 gives an idea of how the inside-out reciprocity works.

With the reciprocity and by considering cases to deal with the half-openness, the generating function $M_c(z)$ for the cubical count $m_c(t)$ is computed to be [8]

$$\begin{aligned} M_c(z) &:= \sum_{t\geq 1} m_c(t) z^t \\ &= \frac{12z^3\begin{pmatrix} 1+4z+8z^2+22z^3+61z^4+125z^5+181z^6+210z^7+198z^8 \\ +181z^9+179z^{10}+211z^{11}+250z^{12}+260z^{13}+190z^{14}+79z^{15}\end{pmatrix}}{(1-z^2)^2(1-z^3)^2(1-z^4)(1-z^5)}, \end{aligned}$$

which gives us Table 7.

[b]Rational polytopes are polytopes with rational coordinate vertices

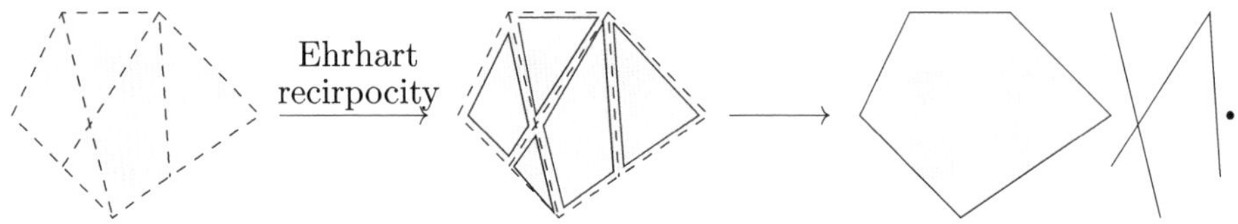

Fig. 1. The first arrow applies the Ehrhart reciprocity for rational polytopes to each region of the polytope, and the second arrow breaks the shape into different elements in $\mathcal{L}$.

Table 7. Number of Latin 3×3 semimagic squares with strict upper bound t.

t	3	4	5	6	7	8	9	10	11	12	13	14
$\frac{m_c(t)}{12}$	1	4	10	32	89	206	408	756	1293	2134	3358	5134

Similarly for affine count,

$$M_a(z) := \sum_{s\geq 1} m_a(s) z^s$$
$$= \frac{12z^3\begin{pmatrix} 1+3z+7z^2+15z^3+33z^4+65z^5+128z^6+208z^7+316z^8 \\ +434z^9+566z^{10}+676z^{11}+784z^{12}+852z^{13}+911z^{14}+936z^{15} \\ +967z^{16}+967z^{17}+1001z^{18}+995z^{19}+1000z^{20}+955z^{21} \\ +893z^{22}+752z^{23}+624z^{24}+456z^{25}+322z^{26}+174z^{27}+79z^{28} \end{pmatrix}}{(1+z)(1+z+z^2)(1+z^2)(1-z^3)(1-z^5)(1-z^6)(1-z^7)(1-z^8)},$$

and the counts for $s = 3, \ldots, 14$ is listed in Table 8.

Table 8. Number of Latin 3×3 semimagic squares with magic sum s.

s	3	4	5	6	7	8	9	10	11	12	13	14
$\frac{m_a(s)}{12}$	1	1	2	6	13	20	46	50	85	129	167	214

The counts for Latin $n \times n$ semimagic squares for $n \geq 4$ are not known.

4.4. *Latin Semimagic Rectangles*

We now revisit semimagic rectangles, this time under the Latin condition. The polytope associated with Latin $2 \times n$ semimagic rectangles is in $\mathbb{R}^n$ and has vertices of the form

$$\left(0, \ldots, 0, \tfrac{n-2n_2}{2(n-n_1-n_2)}, \ldots, \tfrac{n-2n_2}{2(n-n_1-n_2)}, 1, \ldots, 1\right),$$

where $0 \leq n_1, n_2 \leq \frac{n}{2}$, the first n_1 coordinates are 0, and the last n_2 coordinates are 1 [15]. These points, just like the ones in the hypersimplices of Section 4.2, represent a row of the semimagic rectangle. The interior points of this polytope has distinct coordinates, so half of the Latin condition checks off. However, we have to forbid some hyperplanes so that cells in the same column are not the same. The hyperplanes we forbid are the ones with $\frac{1}{2}$ as a coordinate. For any fixed j, the vertices of the intersection of the hyperplane $x_j = \frac{1}{2}$ and the polytope are

$$\left(0,\ldots,0,\frac{n-n_2-2n_3}{2(n-n_1-n_2-n_3)},\ldots,\frac{n-n_2-2n_3}{2(n-n_1-n_2-n_3)},\frac{1}{2},\ldots,\frac{1}{2},1,\ldots,1\right) \quad \text{or}$$
$$\left(0,\ldots,0,\frac{1}{2},\ldots,\frac{1}{2},\frac{n-n_2-2n_3}{2(n-n_1-n_2-n_3)},\ldots,\frac{n-n_2-2n_3}{2(n-n_1-n_2-n_3)},1,\ldots,1\right),$$

where $0 \le n_1, n_3 \le \frac{n}{2}$; $0 \le n_2 \le n$; the first n_1 coordinates are 0; n_2 consecutive coordinates, including x_j, are $\frac{1}{2}$; and the last n_3 coordinates are 1.

Using Gaußian binomial coefficients [2] and recursion, we get Table 9 to Table 12 [15].

Table 9. Number of Latin 2×2 semimagic rectangles with column sum c.

c	1	2	3	4	5	6	7	8	9	10	11	12	13	14
$\frac{\text{affine count}}{2}$	1	1	2	2	3	3	4	4	5	5	6	6	7	7

Table 10. Number of Latin 2×3 semimagic rectangles with column sum c.

c	6	8	10	12	14	16	18	20	22	24	26	28
$\frac{\text{affine count}}{6}$	2	4	8	12	18	24	32	40	50	60	72	84

Table 11. Number of Latin 2×4 semimagic rectangles with column sum c.

c	3	4	5	6	7	8	9	10	11	12	13	14	15	16
$\frac{\text{affine count}}{24}$	1	1	3	3	8	8	18	16	33	31	55	90	86	80

Table 12. Number of Latin 2×5 semimagic rectangles with column sum c.

c	8	10	12	14	16	18	20	22	24
$\frac{\text{affine count}}{120}$	4	16	42	90	172	296	482	740	1092

4.5. *Strong Semimagic Squares*

The inside-out polytopes for strong 3×3 semimagic squares is the same as the inside-out polytope for Latin 3×3 semimagic squares except instead of $0 \le \alpha$ and $0 \le \delta \le \beta$, it uses $0 < \alpha$ and $0 < \delta < \beta$. Table 13 and Table 14 show the magic counts of strong 3×3 semimagic squares in [8].

Table 13. Cubical count of strong 3×3 semimagic squares with strict upperbound t.

t	9	10	11	12	13	14	15	16	17
$\frac{\text{cubical count}}{72}$	1	4	13	36	80	160	291	496	794

Table 14. Affine count of strong 3×3 semimagic squares with column sum s.

s	14	15	16	17	18	19	20	21	22
$\frac{\text{affine count}}{72}$	1	2	4	8	12	20	29	42	54

5. Pandiagonal Magic Squares

As proved in [15], all 4×4 and 5×5 pandiagonal magic squares can be reduced to

0	$\alpha+\beta+\delta$	$\beta+\gamma$	$\alpha+\gamma+\delta$
$\alpha+\beta+\gamma$	$\gamma+\delta$	α	$\beta+\delta$
$\alpha+\delta$	β	$\alpha+\beta+\gamma+\delta$	γ
$\beta+\gamma+\delta$	$\alpha+\gamma$	δ	$\alpha+\beta$

and

0	$\alpha_3+\beta_1$	$\alpha_4+\beta_4$	$\alpha_2+\beta_2$	$\alpha_1+\beta_3$
$\alpha_4+\beta_2$	$\alpha_2+\beta_3$	α_1	β_1	$\alpha_3+\beta_4$
$\alpha_1+\beta_1$	β_4	$\alpha_3+\beta_2$	$\alpha_4+\beta_3$	α_2
$\alpha_3+\beta_3$	α_4	$\alpha_2+\beta_1$	$\alpha_1+\beta_4$	β_2
$\alpha_2+\beta_4$	$\alpha_1+\beta_2$	β_3	α_3	$\alpha_4+\beta_1$

Their magic counts can be found in [3], [1], and [15].

Though the structure of a 6×6 pandiagonal magic square is not known, we do know something about it. First, by adding diagonals

- $a_{11}, a_{22}, a_{33}, a_{44}, a_{55}, a_{66}$,
- $a_{14}, a_{23}, a_{32}, a_{41}, a_{56}, a_{65}$,
- $a_{11}, a_{25}, a_{34}, a_{43}, a_{52}, a_{61}$,
- $a_{14}, a_{25}, a_{36}, a_{41}, a_{52}, a_{63}$

and columns one and four, then subtract row two, three, five, and six, we get $a_{11} + a_{14} + a_{41} + a_{44} = \frac{2}{3}s$, where s is the magic sum. This tells us s must be a multiple of 3. Adding and subtracting different rows/columns/diagonals results in $a_{11}+a_{13}+a_{15}+a_{31}+a_{33}+a_{35}+a_{51}+a_{53}+a_{55} = \frac{3}{2}s$, which means s is a multiple of 2. Therefore, the smallest positive magic sum is 6.

Special thanks to Sebastian Manecke from Goethe-Universität Frankfurt for writing and running a program, we found that there are 117247 weak 6×6 pandiagonal magic squares with entries 0, 1, or 2 and magic sum 6. The equation $a_{11} + a_{14} + a_{41} + a_{44} = \frac{2}{3}s$ from earlier implies that if we use only nonnegative entries and fix magic sum 6, then the largest entry cannot exceed 4. Using entries 0 to 4, we found 1193431 magic squares. This is a surprisingly large amount of squares.

The next thing we want to know is how many of them are vertices if we treat each magic square as a 36-dimensional point. Intuitively, the vertices are squares with many 4's and 0's. Also, $a_{11}+a_{13}+a_{15}+a_{31}+a_{33}+a_{35}+a_{51}+a_{53}+a_{55} = \frac{3}{2}\cdot 6 = 9$ is an odd number, so we expect four 1's in the square. Here are some likely vertices:

0	0	0	0	4	2
0	4	2	0	0	0
0	0	1	2	2	1
4	0	0	0	0	2
0	0	2	4	0	0
2	2	1	0	0	1

0	0	2	0	2	2
2	0	2	2	0	0
4	0	1	0	0	1
0	0	0	4	2	0
0	4	0	0	0	2
0	2	1	0	2	1

0	2	2	0	0	2
0	0	2	2	0	2
4	0	1	0	0	1
0	0	0	4	2	0
2	4	0	0	0	0
0	0	1	0	4	1

2	2	0	0	2	0
2	2	0	2	0	0
0	0	1	4	0	1
2	0	0	0	0	4
0	2	4	0	0	0
0	0	1	0	4	1

.

Counting all pandiagonal squares of similar forms, we have 2232 potential vertices. To know whether they are actually vertices or if we have the complete list of vertices, further investigation is needed.

What makes 6×6 pandiagonal magic squares particularly challenging is that there is no pandiagonal 6×6 magic square with magic sum 1, which is because 6 is a multiple of 2 and 3 [11]. In fact, so let us look at $n \times n$ pandiagonal magic squares where n is a neither a multiple of 2 nor a multiple of 3. For $p = 5$, the spanning set consists of

1	0	0	0	0
0	0	1	0	0
0	0	0	0	1
0	1	0	0	0
0	0	0	1	0

0	1	0	0	0
0	0	0	1	0
1	0	0	0	0
0	0	1	0	0
0	0	0	0	1

0	0	1	0	0
0	0	0	0	1
0	1	0	0	0
0	0	0	1	0
1	0	0	0	0

0	0	0	1	0
1	0	0	0	0
0	0	1	0	0
0	0	0	0	1
0	1	0	0	0

0	0	0	0	1
0	1	0	0	0
0	0	0	1	0
1	0	0	0	0
0	0	1	0	0

1	0	0	0	0
0	0	0	1	0
0	1	0	0	0
0	0	0	0	1
0	0	1	0	0

0	1	0	0	0
0	0	0	0	1
0	0	1	0	0
1	0	0	0	0
0	0	0	1	0

0	0	1	0	0
1	0	0	0	0
0	0	0	1	0
0	1	0	0	0
0	0	0	0	1

0	0	0	1	0
0	1	0	0	0
0	0	0	0	1
0	0	1	0	0
1	0	0	0	0

0	0	0	0	1
0	0	1	0	0
1	0	0	0	0
0	0	0	1	0
0	1	0	0	0

.

There are two things to note here. First is that these are solutions to the n-queens problem played on a torus[c]. The n-queen problem is the riddle of putting queens on a $n \times n$ chess board so that no queen attacks each other. Second is that these matrices are all of the form

$$a_{ij} = \begin{cases} 1 & \text{if } i + mj \equiv k \mod 5 \\ 0 & \text{otherwise} \end{cases}$$

for some integers m and k. Specifically, $m \equiv 2 \mod 5$ for the first five matrices and $m \equiv 3 \mod 5$ for the rest. When $m \equiv 0, 1, 4 \mod 5$, all 1's are in the same

[c]The left-right boundaries connect and the top-bottom boundaries connect

column or diagonal. This reasonably let us conjecture that the spanning set of $n \times n$ pandiagonal magic squares consists of $n(n-3)$ matrices of the form

$$A = \left\{ \begin{pmatrix} a_{11} & \cdots & a_{1n} \\ \vdots & & \vdots \\ a_{n1} & \cdots & a_{nn} \end{pmatrix} : a_{ij} = \begin{cases} 1 & \text{if } i + mj \equiv k \mod n \\ 0 & \text{otherwise} \end{cases} \right\} \tag{2}$$

where $m \not\equiv 0, 1, n-1 \mod n$, and k is some integer. But alas, this is only the case for $n = 5, 7, 11$ [11]. There are 4524 pandiagonal magic squares with sum 1 for 13×13 and 140692 pandiagonal magic squares with sum 1 for 17×17 [10].

6. Conclusion

There are still a lot of open questions for magic squares. For example, how many traditional 6×6 magic square are there? Is there a way to compute the counts of weak $m \times n$ semimagic rectangles in general? What is the structure of a 6×6 pandiagonal magic square?

Recall that the spanning set of 13×13 pandiagonal squares consists of more than just matrices of the form in Equation (2). In light of the weirdness of the spanning set of the 13×13 pandiagonal square, let us define **pandirectional** magic squares to be $n \times n$ matrices where

$$\sum_{m_1 i + m_2 j \equiv k \mod n} a_{ij} = s$$

for all integer m_1, m_2, k where at least one of m_1, m_2 is relatively prime to n. Are there any nontrivial pandirectional magic squares? If so, what are the magic counts?

Finally, here is a special pandiagonal magic square for the workshop this proceeding is written for:

6	33	1	52	54
53	47	11	19	16
24	2	68	48	4
63	5	17	7	54
0	59	49	20	18

Fig. 2. Osaka, 2018

References

[1] M. Ahmed and J. De Loera and R. Hemmecke, Polyhedral cones of magic cubes and squares, Discrete and computational geometry, Algorithms Combin. (Springer, Berlin, 2003).

[2] G. E. Andrews, The theory of partitions, Cambridge Mathematical Library (Cambridge University Press, Cambridge, 1976).

[3] M. Beck and M. Cohen and J.Cuomo and P. Gribelyuk, Paul, The number of "magic" squares, cubes, and hypercubes, Amer. Math. Monthly (2003).

[4] M. Beck and D. Pixton, Ehrhart polynomials for n =1, ..., 9(2018), http://people.math.binghamton.edu/dennis/Birkhoff/polynomials.html.

[5] M. Beck and D. Pixton, The Ehrhart polynomial of the Birkhoff polytope, Discrete Comput. Geom. (2003).

[6] M. Beck and R. Sanyal, Combinatorial Reciprocity Theorems (AMS,2018)(to appear).

[7] M. Beck and T. Zaslavsky, Inside-out polytopes, Adv. Math. (2006).

[8] M. Beck and T. Zaslavsky, Six little squares and how their numbers grow, J. Integer Seq. (2010).

[9] A. Bellos, Macau's magic square stamps just made philately even more nerdy (The Guardian, 2014), https://www.theguardian.com/science/alexs-adventures-in-numberland/2014/nov/04/macaus-magic-square-stamps-just-made-philately-even-more-nerdy.

[10] M. Engelhardt, Number of ways of placing n nonattacking toroidal queens on an n X n chessboard(OEIS, 2012), https://oeis.org/A051906.

[11] M. Engelhardt, World records for the n queens problem(2016), http://nqueens.de/sub/WorldRecord.en.html.

[12] G. Jacob and A. Murugan, An Integrated approach for the Secure Transmission of Images based on DNA Sequences (CoRR, 2016).

[13] I. G. Macdonald, Polynomials associated with finite cell-complexes (J. London Math. Soc. (2),1971).

[14] National Gallery of Art, Melencolia I(2018), https://www.nga.gov/collection/art-object-page.6640.html.

[15] L. Ng, Magic Counting with Inside-Out Polytopes (San Francisco State University, 2018).

[16] N. K. Pareek, Design and Analysis of a Novel Digital Image Encryption Scheme (CoRR, 2012).

On local Dressians of matroids

Jorge Alberto Olarte

Institut für Mathematik, FU Berlin
Arnimallee 2, 14195 Berlin, Germany
E-mail: olarte@zedat.fu-berlin.de

Marta Panizzut

Institut für Mathematik, TU Berlin
Str. des 17. Juni 136, 10623 Berlin, Germany
E-mail: panizzut@math.tu-berlin.de

Benjamin Schröter

Department of Mathematical Sciences, Binghamton University
Binghamton, NY 13902-6000, USA
E-mail:schroeter@math.binghamton.edu

We study the fan structure of Dressians $\mathrm{Dr}(d,n)$ and local Dressians $\mathrm{Dr}(\mathcal{M})$ for a given matroid $\mathcal{M}$. In particular we show that the fan structure on $\mathrm{Dr}(\mathcal{M})$ given by the three term Plücker relations coincides with the structure as a subfan of the secondary fan of the matroid polytope $P(\mathcal{M})$. As a corollary, we have that a matroid subdivision is determined by its 3-dimensional skeleton. We also prove that the Dressian of the sum of two matroids is isomorphic to the product of the Dressians of the matroids. Finally we focus on indecomposable matroids. We show that binary matroids are indecomposable, and we provide a non-binary indecomposable matroid as a counterexample for the converse.

Keywords: Dressian, tropical Grassmannian, tropical linear space, hypersimplex, matroid subdivision.

1. Introduction

Let $\mathbb{K}_p$ be an algebraically closed field of characteristic p with a non-trivial, non-archimedean valuation. The *tropical Grassmannian* $\mathrm{TGr}_p(d,n)$ is a rational polyhedral fan parametrizing *realizable* $(d-1)$-dimensional tropical linear spaces in the tropical projective space $\mathbb{TP}^{n-1}$. These are contractible polyhedral complexes arising from the tropicalization of $(d-1)$-dimensional linear spaces in the projective space $\mathbb{P}^{n-1}_{\mathbb{K}_p}$. The tropical Grassmannian is the tropical variety obtained from the tropicalization of the Plücker ideal $I_{d,n}$ generated by the algebraic relations among the $d \times d$-minors of a $d \times n$-matrix of indeterminates. It is the tropicalization of the Grassmannian $\mathrm{Gr}(d,n)$. Its study has been initiated by Speyer and Sturmfels [28]. In the paper the authors focus in particular on the tropical Grassmannian $\mathrm{TGr}_p(2,n)$, exhibiting a bijection with the space of phylogenetic trees with n labeled leaves.

Herrmann, Jensen, Joswig and Sturmfels [11] studied the *Dressian* $\mathrm{Dr}(d,n)$, an outer approximation of the tropical Grassmannian which parametrizes all $(d-1)$-

dimensional tropical linear spaces in $\mathbb{TP}^{n-1}$. This is the tropical prevariety defined by the three term Plücker relations among the generators of $I_{d,n}$. These relations induce the *Plücker fan structure* on $\mathrm{Dr}(d,n)$. From work of Kapranov [17], restated by Speyer [26], it follows that a point is in the Dressian if and only if it induces a matroid subdivision of the hypersimplex $\Delta(d,n)$. This endows $\mathrm{Dr}(d,n)$ with a *secondary fan structure* as subfan of the secondary fan of $\Delta(d,n)$. In [11], the authors proved that for $d = 3$ the two fan structures coincide.

The Grassmannian $\mathrm{Gr}(d,n)$ can be stratified in strata consisting of points with coordinates equal to zero if and only if they are not indexed by a basis of a matroid. As remarked in [11], a similar stratification can be considered in the tropical setting. In particular, we can look at the intersection of the Dressian $\mathrm{Dr}(d,n)$ with each of the open faces of $\mathbb{TP}^{n-1}$. The intersection is not empty only if the face corresponds to a matroid of rank d on $[n]$. This motivates the authors to give a similar definition for the local Dressian $\mathrm{Dr}(\mathcal{M})$ of a matroid $\mathcal{M}$. In the article, they focused exclusively on this construction for the Pappus matroid. In our paper, we provide more examples and we analyze more deeply the properties of local Dressians.

Local Dressians can also be endowed with two fan structures: one coming from the Plücker relations, one as a subfan of the secondary fan. Our main contribution is Theorem 4.1 which states that the two fan structures coincide. The proof is based on a careful analysis of the subdivision induced by a point in the local Dressian on the 3-dimensional skeleton of the matroid polytope. From our study it follows that a matroid subdivision is completely determined by its restriction to the 3-skeleton.

We then focus on local Dressians of disconnected matroids. We show that the local Dressian of the direct sum of two matroids is the product of their local Dressians. Again, the key step in the proof is to look at the 3-dimensional skeleton of the matroid polytope.

Finally, we move our attention to indecomposable matroids, i.e., matroids which do not admits matroid subdivisions of their matroid polytopes. The local Dressians of such matroids are linear spaces. We prove that binary matroids are indecomposable. Moreover, we give a counterexample for the converse, exhibiting a indecomposable non-binary matroid.

Many questions related to the indecomposability of matroids arose during the work for this manuscript. We conclude with a short section collecting them, including a conjecture.

Acknowledgments

We are very grateful to Takayuki Hibi and Akiyoshi Tsuchiya for the kind hospitality and the organization of the *Summer Workshop on Lattice Polytopes 2018*. Furthermore, we are indebted to Alex Fink for suggesting us to look at the matroid of Proposition 5.2. We also thank Michael Joswig and Felipe Rincón for their useful comments. Research by the first and second author is supported by the Einstein Foundation Berlin. Research by third author is carried out in the framework of

Matheon (Project *MI6 - Geometry of Equilibria for Shortest Path*) supported by Einstein Foundation Berlin. The authors would also like to thank the Institute Mittag-Leffler for its hospitality during the program *Tropical Geometry, Amoebas and Polytopes*, where the collaboration started.

Notation

Before beginning, we fix some notation. Given $d \leq n$ non-negative integers, we define the sets $[n] := \{1, 2, \ldots, n\}$, and $\binom{[n]}{d} := \{\text{subset } S \text{ of } [n] \text{ with } d \text{ elements}\}$. Moreover, given $S \subset [n]$ and $i, j \in [n]$, we use the notation $Sij := S \cup \{i, j\}$. Furthermore, we denote the all ones vector by $\mathbf{1}$.

2. Tropical Grassmannians and tropical linear spaces

We begin with some basics about tropical geometry following Maclagan and Sturmfels [19], focusing in particular on the definition of tropical Grassmannian. We work over the tropical semiring $\big(\mathbb{T} = \mathbb{R} \cup \{\infty\}, \oplus, \odot\big)$, where arithmetic is defined by $a \oplus b = \min\{a, b\}$ and $a \odot b = a + b$.

Let $\mathbb{K}_p$ be an algebraically closed field of carachteristic $p \geq 0$ with a non-trivial non-archimedean valuation $\mathrm{val} : \mathbb{K}_\mathrm{p} \to \mathbb{R} \cup \{\infty\}$. Examples are the field of Puiseaux series and their generalizations with real exponents; see Markwig [20]. Given a polynomial $f \in \mathbb{K}_p[x_1, \ldots, x_n]$,

$$f(x_1, \ldots, x_n) = \sum_{u=(u_1,\ldots,u_n)} c_u x_1^{u_1} \cdot \cdots \cdot x_n^{u_n},$$

its *tropicalization trop*(f) is

$$\mathrm{trop}(f)(x_1, \ldots, x_n) = \bigoplus_{u=(u_1,\ldots,u_n)} \mathrm{val}(\mathrm{c_u}) \odot \mathrm{x_1^{\odot u_1}} \odot \cdots \odot \mathrm{x_n^{\odot u_n}}.$$

The *tropical hypersurface* $\mathrm{trop}(V(f))$ is defined as the set of points $w \in \mathbb{R}^n$ such that the minimum in $\mathrm{trop}(f)(w)$ is attained at least twice. Given an ideal $I \subseteq \mathbb{K}_p[x_1, \ldots, x_n]$, the *tropical variety* $\mathrm{trop}(V(I))$ is the intersection of the tropical hypersurfaces $\mathrm{trop}(V(f))$, with $f \in I$. A *tropical prevariety* is the intersection of finitely many tropical hypersurfaces. Any tropical variety is a tropical prevariety as it is the intersection of the hypersurfaces of a tropical basis; see Hept and Theobald [10] and [19, Section 2.6] for more details.

Any vector $w \in \mathbb{R}^n$ defines a partial term order on the polynomial ring $\mathbb{K}_p[x_1, \ldots, x_n]$. Given an homogeneous ideal I the set of initial ideals $\mathrm{in}_w(I)$ endows $\mathbb{R}^n$ with the structure of polyhedral complex called *Gröbner complex*. The following result, known as *Fundamental Theorem of Tropical Algebraic Geometry*, gives the connection between algebraic and tropical varieties, and subcomplexes of the Gröbner complex.

Theorem 2.1 ([19, Theorem 3.1.3]). *Let I be an ideal in $\mathbb{K}_p[x_1, \dots, x_n]$ and $V(I)$ its variety intersected with the torus $(\mathbb{K}_p \setminus \{0\})^n$. The following sets coincide in $\mathbb{R}^n$:*

(i) the tropical variety trop$(V(I))$;
(ii) the closure in $\mathbb{R}^n$ of set of vectors w such that $in_w(I)$ does not contain a monomial.
(iii) the closure in $\mathbb{R}^n$ of the set $\{(\mathrm{val}(w_1), \dots, \mathrm{val}(w_n)) \,|\, w \in V(I)\}$.

When working with homogeneous polynomials it makes sense to consider tropical hypersurfaces and varieties in the tropical torus $\mathbb{R}^n/\mathbb{R}\mathbf{1} \cong \mathbb{R}^{n-1}$ or in its compactification $\mathbb{TP}^{n-1} = (\mathbb{T}^n \setminus \{(\infty, \dots, \infty)\})/\mathbb{R}\mathbf{1}$. We will adopt both interpretations in this paper, making sure to specify which one we are considering.

Now, let R be the polynomial ring in $\binom{n}{d}$ variables

$$\mathbb{Z}\big[p_{i_1 i_2 \dots i_d} \,|\, i_1 < i_2 < \cdots < i_n\big].$$

We consider the *Plücker ideal* $I_{d,n}$ in $R \otimes \mathbb{K}_p$ generated by the algebraic relations among the $d \times d$-minors of any $d \times n$-matrix in any field. The Grassmannian $\mathrm{Gr}(d, n)$ is the variety $V(I_{d,n})$. The ideal $I_{d,n}$ is generated by quadrics. The *tropical Grassmannian* $\mathrm{TGr}_p(d, n)$ is the tropical variety trop$(V(I_{d,n}))$. It is a pure $d(n-d)$-dimensional rational polyhedral fan in $\mathbb{R}^{\binom{n}{d}-1} \cong \mathbb{R}^{\binom{n}{d}}/\mathbb{R}\mathbf{1}$.

The study of tropical Grassmannian was initiated by Speyer and Sturmfels [28]. The authors focused on $\mathrm{TGr}_p(2, n)$ and the special case $\mathrm{TGr}_p(3, 6)$. The fan structure and the homology of the tropical Grassmannian $\mathrm{TGr}_p(3, 7)$ is studied in [11].

Theorem 2.2 ([28, Theorem 3.4 and Corollary 4.5]). *The tropical Grassmannian $\mathrm{TGr}_p(2, n)$ is characteristic-free and coincides with the space of phylogenetic trees with n labeled leaves.*

Classically, the Grassmannian $\mathrm{Gr}(d, n)$ is an example of a moduli space. It parametrizes the d-dimensional subspaces of a n-dimensional $\mathbb{K}_p$-vector space. A d-dimensional linear subspace of $\mathbb{K}_p^n$ can be represented by a full rank $d \times n$ matrix. The $d \times d$-minors form the Plücker vector which is a point on the Grassmannian. This surjective map from linear spaces to Plücker vectors is called the Stiefel map.

Now, let v be a vector in the tropical Grassmanian $\mathrm{TGr}_p(d, n)$. For each subset $S \in \binom{[n]}{d+1}$, consider the tropical linear polynomial

$$f_S(v) := \bigoplus_{i \in S} v_{S \setminus i} \odot x_i \tag{1}$$

and define L_v as the intersection of the tropical hyperplanes defined by f_S, as S varies over all elements in $\binom{[n]}{d+1}$. Then, L_v is the tropicalization of a classical d-dimensional linear space.

Speyer in [26, Proposition 4.5.1] showed that every tropicalization of a linear space arises this way. Indeed, any classical linear space has Plücker coordinates.

Taking the valuation of each Plücker coordinate yields a vector $v \in \mathrm{TGr}_p(d,n)$ such that L_v coincides with the tropicalization of the linear space. We take the above as definition of *realizable* tropical linear space. The tropical Grassmannian is the moduli space of realizable tropical linear spaces.

Theorem 2.3 ([28, Theorem 3.8]). *The bijection between the classical Grassmannian* $\mathrm{Gr}(d,n)$ *and the set of d-planes in* $\mathbb{K}^n$ *induces a unique bijection* $v \mapsto L_v$ *between the tropical Grassmannian and the set of realizable tropical d-planes in n-space.*

Let $A \in \mathbb{R}^{d\times n}$ be a tropical matrix. The *tropical Stiefel map* $\pi : \mathbb{T}^{d\times n} \dashrightarrow \mathrm{TGr}(d,n)$ sends the matrix A to the vector of its tropical minors $\pi(A)$. More precisely, for any $B = \{b_1, \ldots, b_d\} \in \binom{[n]}{d}$ we have

$$\pi(A)_B = \bigoplus_{\sigma\in\mathcal{S}_d} \bigodot_{i=1}^{n} A_{b_\sigma(i),i}$$

The tropical linear spaces which lie in the image of the tropical Stiefel map are called Stiefel tropical linear spaces. These linear spaces have been studied by Ricón [25], Herrmann et al.[13], and Fink and Rincón [7]. Notice that not all tropical linear spaces arise in this way, i.e., the tropical Stiefel map is not surjective.

3. Matroids and Dressians

In this section we introduce the main characters of this paper: matroids and Dressians. The former are classical objects in discrete mathematics. They are an abstraction of the concept of linear independence. Nakasawa and Whitney introduced them independently in the 1930s. There are many cryptomorphic ways to define a matroid. We will present just one definition and focus on their relation to polyhedral structures. We refer the interested reader to Oxley [23] and White [30]. A *matroid* $\mathcal{M}$ is a pair $(E, \mathcal{B})$ where E is a finite set and $\mathcal{B}$ is a non-empty collection of subsets of E satisfying the *base exchange property*: whenever B and B' are in $\mathcal{B}$ and $e \in B \setminus B'$, there exists and element $f \in B' \setminus B$ such that $(B \setminus \{e\}) \cup \{f\}$ is also in $\mathcal{B}$. The sets in $\mathcal{B}$ are called *bases*. Each basis $B \in \mathcal{B}$ has the same number d of elements, called the *rank of* $\mathcal{M}$. Given a subset $A \subseteq E$, the *rank* $r(A)$ of A is $\max_{B\in\mathcal{B}} |B \cap A|$. A *flat* F of $\mathcal{M}$ is a subset of E such that for every $e \in E \setminus F$ we have $r(F \cup e) = r(F) + 1$.

Example 3.1. Given a matrix A with entries in a field $\mathbb{K}$, the pair $(E, \mathcal{B})$ consisting of the set E of columns of A and the collection of the maximally independent subsets of E is a matroid $\mathcal{M}[A]$.

Example 3.2. Given a (finite) graph, the pair $(E, \mathcal{B})$ consisting of the set of edges E and the collection of the maximally spanning subsets of E is a matroid.

One of the fundamental questions regarding matroid is about their representabilty. A matroid is *representable* over a field $\mathbb{K}$ if it is isomorphic to a matroid $\mathcal{M}[A]$ for a matrix A with entries in $\mathbb{K}$. A matroid representable over the finite field with two elements is called *binary*. A *ternary matroid* is one representable over the finite field with three elements. A matroid that can be obtained from a graph as described in Example 3.2 is called *graphical matroid*. Graphical matroids are *regular*, i.e., representable over any field. It is a recent result of Nelson [22] that almost no matroid is representable. The following example provides non-regular matroids.

Example 3.3. Let $E = [n]$ and $\mathcal{B}$ the collection of subsets of E with d elements. The matroid $(E, \mathcal{B})$ is the *uniform matroid* $\mathcal{U}_{d,n}$. The uniform matroid $U_{2,n}$ is not representable over a field with less than $n-1$ elements. In particular, the matroid $U_{2,4}$ is not binary.

Example 3.4. The *Fano matroid* $\mathcal{F}_7$ is an example of a binary matroid that is only representable over fields of characteristic two. It is represented by all seven non vanishing 0/1-vectors of length three over a field of characteristic 2.

The following operations on matroids are derived from taking minors of matrices. Let $M = (E, \mathcal{B})$ be a matroid and $e \in E$. The *deletion* of e from $\mathcal{M}$, denoted $\mathcal{M}\backslash e$, is the matroid $(E \setminus \{e\}, \{B \in \mathcal{B} \mid e \notin B\})$. The *contraction* of e from $\mathcal{M}$, denoted $\mathcal{M}/e$, is the matroid $(E \setminus \{e\}, \{B \setminus e \mid e \in B \in \mathcal{B}\})$. Any matroid that is the result of successive deletions and contractions of $\mathcal{M}$ is called a *minor* of $\mathcal{M}$. The *dual* of $\mathcal{M}$, denoted $\mathcal{M}^*$, is the matroid $(E, \{E \setminus B \mid B \in \mathcal{B}\})$. It is straightforward to verify that $(\mathcal{M}^*)^* = \mathcal{M}$, and that $\mathcal{M}^*\backslash e = (\mathcal{M}/e)^*$.

We are most interested in the polyhedral point of view of defining and studying matroids. We fix $E = [n]$ as ground set. Let $e_1, e_2, \ldots, e_n$ be the canonical basis of $\mathbb{R}^n$. For a collection $\mathcal{S}$ of subsets of $E = [n]$, we define the polytope

$$P_{\mathcal{S}} := \operatorname{conv}\{e_S \mid S \in \mathcal{S}\},$$

where $e_S := \sum_{i \in S} e_i$. The *d-th hypersimplex* in $\mathbb{R}^n$ is the polytope

$$\Delta(d,n) := P_{\binom{[n]}{d}}.$$

A subset $\mathcal{M} \subset \binom{[n]}{d}$ is a *matroid of rank d on n elements* if the edges of $P_{\mathcal{M}}$ are parallel to the edges of $\Delta(d,n)$, i.e., they are of the form $e_i - e_j$ for $i, j \in [n]$ distinct. The elements in $\mathcal{M}$ are the *bases* and $P_{\mathcal{M}}$ is a *matroid polytope*. The fact that this construction gives a matroid is a result of Edmonds [5]. See also Gelfand, Goresky, MacPherson and Serganova [8].

In terms of the matroid polytope, we have that

$$P_{\mathcal{M}\backslash i} \cong P_{f(\mathcal{M})} \cap \{x_i = 0\}, \text{ and } P_{\mathcal{M}/i} \cong P_{\mathcal{M}} \cap \{x_i = 1\}.$$

Moreover, $P_{\mathcal{M}^*} = f(P_{\mathcal{M}})$ where f is the affine involution that sends x_i to $1 - x_i$ for each coordinate i. In particular, the polytopes $P_{\mathcal{M}}$ and $P_{\mathcal{M}^*}$ are isomorphic.

We now move to the definition of Dressians. We will particularly highlight their connection with matroids and matroid polytopes. Among the quadric generators of the Plüker ideal $I_{d,n}$ are the *three term Plücker relations*

$$p_{Sij}\,p_{Skl} - p_{Sik}\,p_{Sjl} + p_{Sil}\,p_{Sjk}, \tag{2}$$

where $S \in \binom{[n]}{d-2}$ and $i,j,k,l \in [n] \setminus S$ pairwise distinct. The *Dressian* $\mathrm{Dr}(d,n)$ is the tropical prevariety in $\mathbb{R}^{\binom{n}{d}}/\mathbb{R}\mathbf{1} \cong \mathbb{R}^{\binom{n}{d}-1}$ defined by the Plücker relations. This means that for a vector w in the Dressian $\mathrm{Dr}(d,n)$ the minimum of

$$w_{Sij} + w_{Slm}\,, \quad w_{Sil} + w_{Sjm}\,, \quad w_{Sim} + w_{Sjl} \tag{3}$$

is achieved at least twice, where $S \in \binom{[n]}{d-2}$ and $i,j,l,m \in [n] \setminus S$ pairwise distinct. The name Dressian was proposed by Herrmann et al. [11] in honor of Andreas Dress who discovered these relations by looking at valuated matroids. We call a point in the Dressian *valuated matroid*. The three term Plücker relations endow $\mathrm{Dr}(d,n)$ with the *Plücker fan structure*. The Dressian can be also viewed as a subcomplex in the tropical projective space $\mathbb{TP}^{\binom{n}{d}-1}$. We will do this in the next section, when we introduce local Dressians.

It follows directly from the definition that the Dressian $\mathrm{Dr}(d,n)$ contains the tropical Grassmannian $\mathrm{TGr}_p(d,n)$ for any characteristic p. From the results in Maclagan–Sturmfels [19], it follows that $\mathrm{Dr}(2,n) = \mathrm{TGr}_p(2,n)$ as fans and $\mathrm{Dr}(3,6) = \mathrm{TGr}_p(3,6)$ only as sets. The tropical Grassmannian $\mathrm{TGr}_p(3,7)$ depends on the characteristic p of the field and $\mathrm{TGr}_p(3,7) \neq \mathrm{TGr}_2(3,7)$ for $p \neq 2$ due to the representability properties of the Fano matroid. This implies that the Dressian $\mathrm{Dr}(d,n)$ disagrees with the tropical Grassmannian $\mathrm{TGr}_p(d,n)$ for $d \geq 3$ and $n \geq 7$. This fact is even reflected in their dimensions. The dimension of $\mathrm{Dr}(d,n)$ is of order n^{d-1} for fixed d, while the dimension of $\mathrm{TGr}_p(d,n)$ grows linear in n, see [15, Corollary 32].

As we said, the Dressian $\mathrm{Dr}(d,n)$ is the intersection of $\binom{n}{d+2}\binom{d+2}{4}$ tropical hypersurfaces coming from the three term Plücker relations. Note that these relations do not generate the Plücker Ideal $I_{d,n}$ for $n \geq d+3 \geq 6$, but they generate its image in the Laurent polynomials ring $\mathbb{K}_p[p^{\pm}_{i_1\ldots i_d} \,|\, i_1 < i_2 < \cdots < i_n]$, see [11, Section 2]. The tropical variety defined by the ideal generated by the three term Plücker relations coincides with the tropical Grassmannian.

The following proposition provides an upper bound for the number of elements in a tropical basis for the tropical Grassmannian $\mathrm{TGr}_p(d,n)$, i.e., the number of tropical hypersurfaces defining $\mathrm{TGr}_p(d,n)$.

Proposition 3.1. *The tropical Grassmannian has a tropical basis of size:*

$$\binom{n}{d+1}\left(\binom{n}{d-1} - \binom{d+1}{2}\right) + \binom{n}{d} - d(n-d) \leq 2^{2n+1}\,.$$

Proof. This bound follows from Theorem 1 in Hept–Theobald [10] as $I_{d,n}$ is a prime ideal of codimension $\binom{n}{d} - d(n-d) - 1$ generated by $\binom{n}{d+1}\left(\binom{n}{d-1} - \binom{d+1}{2}\right)$ polynomials. The generators can be read off from the prove of Theorem 14.6 in Miller–Sturmfels [21]. □

Note that this is a better estimation of the minimal size of a tropical basis than the one that can be derived from a general degree bound given in [16, Example 9].

Our final goal for this section is to explain the relation of the Dressian to a general concept in polyhedral geometry. Let P be a polytope in $\mathbb{R}^n$ with m vertices and dimension k. Any vector $w \in \mathbb{R}^m$ induces a regular subdivision of P. We think w as a height function which lifts the vertex v_i to the height w_i. By projecting the lower faces of the convex hull $\operatorname{conv}\{(v_i, w_i) \mid v_i \text{ vertex of } P\} \subset \mathbb{R}^{m+1}$ we get a subdivision of P. Vectors inducing the same subdivision form a relatively open cone. The collection of all these cones is the secondary fan of the polytope P. The lineality space is the largest linear space contained in each cone of the fan. The secondary fan has a $(k+1)$-dimensional lineality space that contains $\mathbf{1} \in \mathbb{R}^m$. In particular we may consider its image in $\mathbb{R}^m/\mathbb{R}\mathbf{1}$.

A subdivision of $\Delta(d,n)$ is a *matroid subdivision* if each of its cell is a matroid polytope. Speyer proved a description of the Dressian in terms of matroid subdivisions.

Theorem 3.1 ([17, Chapter 1], [26, Proposition 2.2]). *A vector* $w \in \mathbb{R}^{\binom{n}{d}}$ *lies in the Dressian* $\operatorname{Dr}(d,n)$ *if and only if it induces a matroid subdivision of the hypersimplex* $\Delta(d,n)$.

This description sees the Dressian $\operatorname{Dr}(d,n)$ as a subfan of the secondary fan of the hypersimplex $\Delta(d,n)$, and define the *secondary fan structure* on $\operatorname{Dr}(d,n)$. Suppose that $d \geq 2$. For each $S \in \binom{[n]}{d}$ of cardinality $d-2$, and $i,j,l,m \in [n] \setminus S$, the points

$$e_{Sij},\ e_{Sil},\ e_{Sim},\ e_{Sjl},\ e_{Sjm} \text{ and } e_{Slm}$$

define the vertices of a octahedron O, which is a 3-dimensional face of the hypersimplex $\Delta(d,n)$. A point w in the Dressian $\operatorname{Dr}(d,n)$ induces a matroid subdivision of $\Delta(d,n)$. According to which of the three inequalities and equations in (3) are satisfied, the subdivision induced by w on O, determines one of the three possible subdivision of the octahedron in two quadrilateral pyramids or the trivial subdivision. Herrmann et al. [11] showed that for $d = 3$ the Plücker fan structure coincides with the secondary fan structure. In the next section we will prove that this holds in general.

For any valuated matroid $v \in \operatorname{Dr}(d,n)$, we can define L_v in the same way as we did for points in the Grassmanian in Section 2. We call such L_v a tropical linear space. Note that, as there are valuated matroids that are not in the tropical Grassmanian, there are tropical linear spaces which are not realizable. If $v \in \operatorname{TGr}_p(d,n)$

then all the faces of the subdivision induced by v must be polytopes of matroids representable in characteristic p; see [27, Example 4.5.4] and [15, Proposition 34].

Given a valuated matroid $v \in \mathrm{Dr}(d,n)$, any point $x \in \mathbb{R}^n$ defines a matroid $\mathcal{M}_x$ by taking the face from the regular subdivision of $\Delta(d,n)$ that is minimized in the direction of x. In other words, the bases of $\mathcal{M}_x$ are the sets $B \in \binom{[n]}{d}$ such that $v_B - \sum_{i \in B} x_i$ is minimal. A loop in a matroid is an element which is contained in no bases. The notation of a linear space in (1) generalizes to an arbitary Plücker vector $v \in \mathrm{Dr}(d,n)$ of its realizability. The following is a combinatorial description of such a tropical linear space.

Proposition 3.2 ([26, Proposition 2.3]). *The tropical linear space L_v consists of exactly all points $x \in \mathbb{R}^n$ such that $\mathcal{M}_x$ has no loops.*

For each loopless matroid $\mathcal{M}$ whose polytope appears in the regular subdivision induced by v, there is a corresponding polyhedral cell in L_v given by the closure of all the points x such that $\mathcal{M}_x = \mathcal{M}$. This way the tropical linear space L_v has the structure of a polyhedral complex. A cell is bounded in this complex if and only if $\mathcal{M}$ has no coloops. The subcomplex of bounded cells is the tight span. Note that the polyhedral structure of L_v is not unique. We will illustrate this by looking at the recession fan of L_v.

Whenever $v \in \mathrm{Dr}(d,n)$ in the tropical projective space has only 0 and ∞ as values, the tropical linear space L_v coincides with the Bergman fan of the matroid $\mathcal{M}$ whose bases are the coordinates where v is 0. This is also the recession fan of L_v. The Bergman fan has a natural fan structure by the above arguments. We can equip the Bergman fan with a finer structure in the following way: for each flat F of $\mathcal{M}$, i.e., closed set, let $e_F = \sum_{i \in F} e_i$ be a ray in the Bergman fan. And for every flag $F_1 \subset \cdots \subset F_k$ of flats we get a cone generated by $e_{F_1}, \ldots, e_{F_k}$. When v is any valuated matroid and $x \in L_v$, then L_v is locally near x the same as the Bergman fan of the matroid M_x. For further details see [6], [1] and [19, Chapter 4]. Moreover, the introduction of Hampe [9] gives a broad overview about properties and developments of tropical linear spaces.

4. Local Dressians

Let $\mathcal{M} = ([n], \mathcal{B})$ be a matroid on the set $[n]$ of rank d. Let $b = |\mathcal{B}|$ be the number of bases of $\mathcal{M}$. We consider the variety in $\mathbb{P}^{b-1}$ defined by the ideal $I_{\mathcal{M}}$ obtained from the Plücker ideal $I_{d,n}$ by setting all the variables p_B to zero, where B is not a basis of $\mathcal{M}$. The variety $V(I_{\mathcal{M}})$ is the realization space of the matroid $\mathcal{M}$. It parametrizes all the d-dimensional linear subspaces of $\mathbb{K}_p^n$ whose non-zero Plücker coordinates are the bases of $\mathcal{M}$. In particular $V(I_{\mathcal{M}}) = \emptyset$ if and only if the matroid is not representable over $\mathbb{K}_p$. This gives a stratification of the Grassmannian $\mathrm{Gr}(d,n)$, where the strata are defined as

$$\{p \in \mathrm{Gr}(d,n) \,|\, p_B = 0 \,\text{if and only if}\, B \notin \mathcal{B}\}.$$

A variation of Mnëv's Universality theorem implies that the strata can be complicated as any algebraic variety.

Remark 4.1. For the reader familiar with toric geometric, consider the Grassmannian $\mathrm{Gr}(d,n)$ over the complex numbers $\mathbb{C}$. The algebraic torus $T = (\mathbb{C}^*)^n$ acts on $\mathbb{C}^n$ by $(t_1,\ldots,t_n)\cdot(x_1,\ldots,x_n) = (t_1x_1,\ldots,t_nx_n)$. The action is linear so it maps subspaces to subspaces. Therefore, it induces an action on the Grassmannian $\mathrm{Gr}(d,n)$. Given a point $p \in \mathrm{Gr}(d,n)$, the closure of the orbit $T\cdot p$ is a toric variety. Let p be a point in the stratum defined by $\mathcal{M}$. The image of $\overline{T\cdot p}$ through the moment map is the matroid polytope $P_{\mathcal{M}}$. For further reading we refer to [8].

We now look at a similar local construction for the Dressian, i.e., we look at the Dressian $\mathrm{Dr}(\mathcal{M})$ of a matroid $\mathcal{M} = ([n],\mathcal{B})$. This construction has been introduced by Herrmann et al. [11]. In the article the authors focus just on a single example where $\mathcal{M}$ is the Pappus matroid of rank three on nine elements.

The *Dressian* $\mathrm{Dr}(\mathcal{M})$ *of a matroid* $\mathcal{M}$ is the tropical prevariety in $\mathbb{R}^{b-1} \cong \mathbb{R}^b/\mathbb{R}\mathbf{1}$ given by the set of quadrics obtained from the three term Plücker relations by setting the variables p_B to zero, where B is not a basis of $\mathcal{M}$. The Dressian $\mathrm{Dr}(d,n)$ contains the Dressians of all matroids of rank d on n elements as subcomplexes at infinity.

Let us be more precise. From the coordinatewise logarithmic map $-\log$ we get a homoemorphism $\mathrm{int}(\Delta_{n-1}) \to \mathbb{R}^n/\mathbf{1}\mathbb{R}$. The tropical projective space $\mathbb{TP}^{n-1}$ is a compactification of the tropical torus $\mathbb{R}^n/\mathbf{1}\mathbb{R}$, such that the pair $(\mathrm{int}(\Delta_{n-1}),\Delta_{n-1})$ is homeomorphic to $(\mathbb{R}^n/\mathbf{1}\mathbb{R}, \mathbb{TP}^{n-1})$.

Given Z a non-empty subset of $[n]$, we define the set

$$\mathbb{T}^n(Z) := \{(w_1,w_2,\ldots,w_n) \in \mathbb{T}^n \mid w_i = \infty \,\text{if and only if}\, i \in Z\}.$$

The image of the sets $\mathbb{T}^n(Z)$ through the quotient map give a stratification of the boundary of $\mathbb{TP}^{n-1}$. See Section 5 of Joswig [14] for further details.

The intersection of the closure of the Dressian $\mathrm{Dr}(d,n)$ in the tropical projective space $\mathbb{T}^{\binom{n}{d}-1}$ and the boundary stratum $\{w \in \mathbb{T}^{\binom{n}{d}-1} \mid w_S = \infty \text{ for } S \notin \mathcal{M}\}$ agrees with the local Dressian $\mathrm{Dr}(\mathcal{M})$. Therefore the Dressian $\mathrm{Dr}(d,n)$ contains the Dressians of the matroid $\mathcal{M}$ as subcomplex at infinity.

Remark 4.2. Our definition of the local Dressian $\mathrm{Dr}(\mathcal{U}_{d,n})$ of the uniform matroid agrees with the definition of the Dressian $\mathrm{Dr}(d,n)$ and it only takes only on the three term Plücker relations into account. While the definition of the local Dressian given in [11, Section 6] and [19, Definition 4.4.1] relies on all quadratic Plücker relations. These definitions agree, see [2, Example 2.32]. Note that in their paper, the authors call the elements in our definition of the Dressian *weak* matroids and the elements coming from all quadratic Plücker relations *strong* matroids over the tropical hyperfield.

The following statement follows from the definition of $\mathrm{Dr}(\mathcal{M})$ and Theorem 3.1.

Corollary 4.1. *A vector w lies in the Dressian* $\mathrm{Dr}(\mathcal{M})$ *if and only if it induces a matroid subdivision of the matroid polytope* $P_{\mathcal{M}}$.

Therefore we have again two fan structures on the Dressian of a matroid $\mathcal{M}$: one induced by the Plücker relations and one induced by the secondary fan.

Theorem 4.1. *Let $\mathcal{M}$ be a matroid of rank d on n elements. The Plücker fan structure coincides with the secondary fan structure on* $\mathrm{Dr}(\mathcal{M})$.

Proof. First, we take vectors v and w lying in the same cone of the secondary fan. They induce the same subdivision of the matroid polytope $P_{\mathcal{M}}$, in particular of the 3-dimensional skeleton. Therefore v and w satisfy the same three term Plücker relations and lie in the same cone of the local Dressian equipped with the Plücker structure.

Now we focus on the vice versa. We take v and w lying in the same Plücker cone C_v. This means that they satisfy the same equations and inequalitics coming from the three term Plücker relations. By Corollary 4.1, they induce two matroid subdivisions Σ_v and Σ_w of $P_{\mathcal{M}}$. We want to show that $\Sigma_v = \Sigma_w$. This will imply that v, w are in the same secondary cone. By the fact that they satisfy the same Plücker relations, we know that $\Sigma_v|_{3-\text{skeleton}} = \Sigma_w|_{3-\text{skeleton}}$ as the 3-faces are either tetrahedra or octahedra. We pick σ_v a maximal dimensional cell in Σ_v. We suppose that σ_v is not in Σ_w. It means without loss of generality there are vertices q_1 and q_k in the cell σ_v such that q_1 and q_k do not lie in a maximal dimensional cell of Σ_w. Let $q_1 q_2 \dots q_k$ be a path in the vertex-edge graph of the cell σ_v. We pick a cell σ_w in Σ_w that contains $q_1 \dots q_i$ for some $i \leq k$ and there is no cell in Σ_w containing $q_1 \dots q_{i+1}$.

Now we have that q_{i-1} and q_{i+1} are at most of distance two. So we can use the base exchange axiom in the definition of a matroid to construct up to six points giving the unique face F of σ_v spanned by q_{i-1} and q_{i+1}. The following situations may arise.

- Either F is a octahedron, then F is subdivided in Σ_w as q_{i-1}, q_i are in σ_w and q_{i+1} is not. This is a contradiction to the fact that the subdivisions agree on the 3-skeleton.
- If F is a pyramid, it cannot be subdivided, therefore F is a face of σ_w and hence q_{i+1} is a vertex of σ_w, and that contradicts our assumption.
- Similarly if F is 2-dimensional, i.e., a square or a triangle.

Hence we conclude that both points q_1 and q_k are in σ_w and hence the subdivisions Σ_v and Σ_w agree. □

Corollary 4.2. *The Plücker fan structure on the Dressian* $\mathrm{Dr}(d, n)$ *as a fan in* $\mathbb{R}^{\binom{n}{d}-1}$ *coincides with the secondary fan structure.*

Proof. It is enough to consider the uniform matroid $\mathcal{U}_{d,n}$ in the previous statement. □

Corollary 4.3. *Let $d \geq 2$, and Σ and Σ' be two matroid subdivisions of the hypersimplex $\Delta(d,n)$. If they induce the same subdivision on the 3-skeleton, or equivalently on the octahedral faces of $\Delta(d,n)$, then Σ and Σ' coincide.*

Remark 4.3. The above statement extends Proposition 4.3 and Theorem 4.4 by Herrmann et al. [11] and is the key in the algorithm in Section 6 of Herrmann et al. [13] for computing (local) Dressians. Note that the abstract tree arrangements in Section 4 of Herrmann et al. [11] are a cover of the 3-skeleton of the hypersimplex $\Delta(3,n)$ for $n \geq 6$ and the metric condition guarantees that the height functions agree on all three maps that contain a given vertex.

A *connected component* S of $\mathcal{M}$ is a minimal non empty subset with the property that $|S \cap B|$ is the same for every base B of $\mathcal{M}$. Connected components of $\mathcal{M}$ partition $[n]$. If $[n]$ is the only connected component, we say that $\mathcal{M}$ is *connected.* We derive the following characterization of the lineality space which follows from the characterization of the dimension of a matroid polytope in terms of connected components by Edmonds [5] or Feichtner and Sturmfels [6]. Together with the fact that the secondary fan of a set of vertices has a lineality space of the same dimension as the affine dimension of the set of vertices.

Corollary 4.4. *Let b be the number of bases of a matroid $\mathcal{M}$ on n elements and with c connected components. The lineality space of the Dressian* $\mathrm{Dr}(\mathcal{M})$ *in* $\mathbb{R}^b/\mathbb{R}\mathbf{1}$ *is of dimension* $dim P_{\mathcal{M}} = n - c$.

Proof. Adding a linear functions to the height function of a regular subdivision does not change the subdivision. Therefore the linealty space is the image of the map $\mathbb{R}^n \to \mathbb{R}^b$ with $e_i \mapsto \sum_{B \ni i} e_B$. □

Example 4.1. The local Dressian of the uniform matroid $\mathcal{U}_{2,4}$ coinsides with the Dressian $\mathrm{Dr}(2,4)$. This is a 5-dimensional pure balanced fan in $\mathbb{R}^6/\mathbb{R}\mathbf{1}$ consisting of three maximal cells and a 3-dimensional lineality space.

Example 4.2. The local Dressian of the matroid $\mathcal{U}_{1,2} \oplus \mathcal{U}_{1,2}$ is a 2-dimensional linear space in $\mathbb{R}^4/\mathbb{R}\mathbf{1}$ spanned by $e_{13} + e_{14}$ and $e_{13} + e_{23}$. The corresponding matroid polytope $P_{\mathcal{U}_{1,2}} \times P_{\mathcal{U}_{1,2}}$ is a square, which has no finer matroidal subdivision.

Let us discuss two examples of local Dressians of non-regular connected ternary $(3,6)$-matroids. These are matroids that are representable over the field with three elements, but are not representable over the field with two elements.

Example 4.3. Let $\mathcal{M}$ be the matroid on 6 elements and rank 3 whose bases are $\binom{[6]}{3} \setminus \{123, 145, 356\}$, see Figure 1. The polytope $P_{\mathcal{M}}$ is full dimensional so the local Dressian $\mathrm{Dr}(\mathcal{M})$ has a lineality space of dimension 5 in $\mathbb{R}^{16} = \mathbb{R}^{17}/\mathbb{R}\mathbf{1}$. The local Dressian is 6-dimensional and consists of three maximal cones. These cones correspond to the vertex split with the hyperplane $x_2 + x_4 + x_6 = 0$ and two

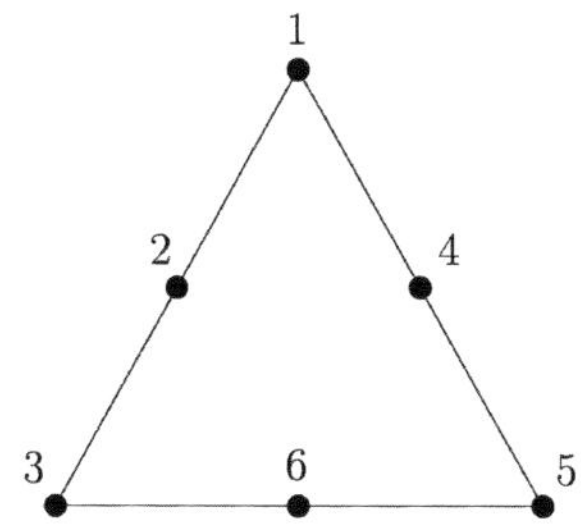

Fig. 1. The terrnary matroid of Example 4.3.

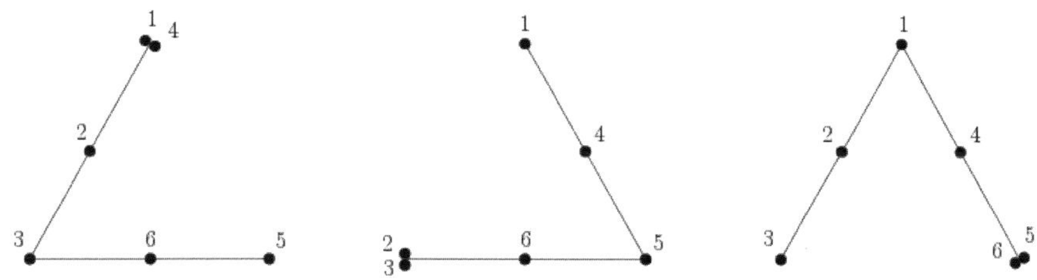

Fig. 2. The three matroids of one of the subdivisions of Example 4.3.

3-splits, i.e., a subdivision into three maximal cells that intersect in a common cell of codimension 2. The three maximal cells of one of those 3-splits is illustrated in Figure 2.

Example 4.4. Let $\mathcal{M}$ be the connected matroid given by the 14 bases:

$$135,\ 136,\ 145,\ 146,\ 156,\ 235,\ 236,\ 245,\ 246,\ 256,\ 345,\ 346,\ 356,\ 456\ .$$

The local Dressian $\mathrm{Dr}(\mathcal{M})$ consists of three maximal cones of dimension 6 and a 5-dimensional lineality space in $\mathbb{R}^{13}$. In other words the polytope $P_{\mathcal{M}}$ has four matroidal subdivisions. The trivial subdivision and three splits with respect to the hyperplanes $x_4 + x_5 + x_6 = 2$, $x_3 + x_5 + x_6 = 2$ or $x_3 + x_4 = 1$.

Remark 4.4. For any point w in the local Dressian $\mathrm{Dr}(\mathcal{M})$ we can construct a tropical linear space L_w, by taking the intersection over $S \in \mathcal{M}$ of the tropical hyperplanes defined by

$$f_S(w) = \bigoplus_{i \in S} w_{S \setminus i} \odot x_i \ .$$

Further details on tropical linear spaces can be found in [19, Section 4.4].

Proposition 4.1. *Let $\mathcal{M}$ and $\mathcal{M}'$ be matroids such that $P_{\mathcal{M}}$ is combinatorially isomorphic to $P_{\mathcal{M}'}$. Then,*

$$Dr(\mathcal{M}) \cong Dr(\mathcal{M}').$$

Proof. A matroid subdivision of the polytope $P_{\mathcal{M}}$ does not impose new edges. The isomorphism between the polytopes $P_{\mathcal{M}}$ and $P_{\mathcal{M}'}$ induces a subdivision of $P_{\mathcal{M}'}$ as images of cells. Moreover, this subdivision is matroidal as the 1-cells are edges of $P_{\mathcal{M}'}$. This subdivision is regular, as the map between $P_{\mathcal{M}}$ and $P_{\mathcal{M}'}$ is a concatenation of a coordinate permutation, an embedding and a reflection. This follows from the explicit description in Remark 4.5. □

Remark 4.5. It can be shown that the two matroid polytopes of $\mathcal{M}$ and $\mathcal{M}'$ are combinatorially isomorphic if and only if the matroids are isomorphic up to loops, coloops or dual connected components. This is part of the work by Pineda-Villavicencio and Schröter [24].

The following statement deals with Dressians of disconnected matroids. Let $\mathcal{M}_1$ and $\mathcal{M}_2$ be matroids $(E_1, \mathcal{B}_1)$ and $(E_2, \mathcal{B}_2)$ with E_1 and E_2 disjoint. We define the *direct sum* of $\mathcal{M}_1$ and $\mathcal{M}_2$ as

$$\mathcal{M}_1 \oplus \mathcal{M}_2 = (E_1 \cup E_2, B_1 \cup B_2 \,\text{with}\, B_1 \in \mathcal{B}_1 \,\text{and}\, B_2 \in \mathcal{B}_2).$$

Theorem 4.2. *Let $\mathcal{M}_1$ and $\mathcal{M}_2$ be matroids with disjoint element sets. Then*

$$\mathrm{Dr}(\mathcal{M}_1 \oplus \mathcal{M}_2) = \mathrm{Dr}(\mathcal{M}_1) \times \mathrm{Dr}(\mathcal{M}_2).$$

Remark 4.6. The statement was independently proved by Lopez de Medrano, Rincón and Shaw, see [18, Lemma 4.7 and Corollary 4.8].

Proof of Theorem 4.2. We have the map

$$\begin{aligned} \otimes : \ \mathrm{Dr}(\mathcal{M}_1) \times \mathrm{Dr}(\mathcal{M}_2) &\to \mathrm{Dr}(\mathcal{M}_1 \oplus \mathcal{M}_2) \\ (w, v) &\mapsto w \otimes v \end{aligned}$$

where $(w \otimes v)_{B_1 \sqcup B_2} := w_{B_1} + v_{B_2}$ for any $B_1 \in \mathcal{M}_1$ and $B_2 \in \mathcal{M}_2$. To check that $w \otimes v$ satisfies the tropical Plücker relations notice the following: any octahedron contained in $P_{\mathcal{M}_1 \oplus \mathcal{M}_2}$ must be of the form $\{e_{B_1}\} \times O_2$, with $B_1 \in \mathcal{M}_1$ and O_2 octahedron contained in $P_{\mathcal{M}_2}$, or $O_1 \times \{e_{B_2}\}$, with $B_2 \in \mathcal{M}_2$ and O_1 octahedron contained in $P_{\mathcal{M}_1}$. Then the Plücker relations follow from those of $\mathrm{Dr}(\mathcal{M}_1)$ and $\mathrm{Dr}(\mathcal{M}_2)$. In particular, the cone where $w \otimes v$ lies is determined by the cones where w and v lie, so $\otimes$ maps cones into cones.

To construct the inverse of $\otimes$, we fix a basis $B_1 \sqcup B_2 \in \mathcal{M}_1 \oplus \mathcal{M}_2$ and we define the map

$$\begin{aligned} \phi : \ \mathrm{Dr}(\mathcal{M}_1 \oplus \mathcal{M}_2) &\to \mathrm{Dr}(\mathcal{M}_1) \times \mathrm{Dr}(\mathcal{M}_2) \\ w &\mapsto (\phi_1(w), \phi_2(w)) \end{aligned}$$

where $\phi_1(w)_{A_1} := w_{A_1 \sqcup B_2}$ and $\phi_2(w)_{A_2} := w_{B_1 \sqcup A_2}$ for any $A_1 \in \mathcal{M}_1$ and any $A_2 \in \mathcal{M}_2$. It is straight forward to verify that the Plücker relations satisfied by w imply that the projections $\phi_1(w)$ and $\phi_2(w)$ satisfy them as well. In particular, ϕ maps cones to cones.

Now we prove that ϕ is independent of the choice of basis $B_1 \sqcup B_2$. We do this by contradiction. Suppose it is not, without loss of generality we can assume there exist $B_1 \sqcup B_2$ and $B_1 \sqcup B_2'$, with B_2 and B_2' of distance 1 such that ϕ does not agree for these two choices. Clearly ϕ_2 is the same for both choices, so we look at ϕ_1. Let $A, A' \in \mathcal{M}_1$ be bases at distance 1. We have that the points $e_{A\sqcup B_2}, e_{A\sqcup B_2'}, e_{A'\sqcup B_2'}, e_{A'\sqcup B_2}$ form a square face of $P_{\mathcal{M}_1\oplus\mathcal{M}_2}$. This square can not be subdivided, so

$$w_{A\sqcup B_2} - w_{A'\sqcup B_2} = w_{A\sqcup B_2'} - w_{A'\sqcup B_2'} \ .$$

But this means that the difference of ϕ_1 for A and A' is independent of the choice of B_2. By connectivity of the graph of $P_{\mathcal{M}_1}$, we can conclude that ϕ_1 is independent of the choice of B_2.

We are left with proving that ϕ is the inverse of $\otimes$. First we check that for any $(v, w) \in \mathrm{Dr}(\mathcal{M}_1) \times \mathrm{Dr}(\mathcal{M}_2)$ we have that $\phi(w \otimes v) = (w, v)$. To see this, notice that $\phi_1(w\otimes v)_A = (w\otimes v)_{A\sqcup B_2} = w_A + v_{B_2}$ for any $A \in \mathcal{M}_1$. But v_{B_2} is a constant independent of A, so $\phi_1(w \otimes v) = w$ in the tropical torus. Analogously, we get that $\phi_2(w \otimes v) = v$.

Now we check the other direction, that is, for any $w \in \mathrm{Dr}(\mathcal{M}_1 \oplus \mathcal{M}_2)$ we have $w = \phi_1(w) \otimes \phi_2(w)$. Consider two bases of $(\mathcal{M}_1 \oplus \mathcal{M}_2)$ at distance 1. Without loss of generality let them be $A_1 \sqcup A_2$ and $A_1 \sqcup A_2'$. We have that

$$\begin{aligned}(\phi_1(w) \otimes \phi_2(w))_{A_1\sqcup A_2} - (\phi_1(w)&\otimes\phi_2(w))_{A_1\sqcup A_2'} \\ &=\phi_1(w)_{A_1} + \phi_2(w)_{A_2} - \phi_1(w)_{A_1} - \phi_2(w)_{A_1'} \\ &=w_{A_1\sqcup B_2} + w_{B_1\sqcup A_2} - w_{A_1\sqcup B_2} - w_{B_1\sqcup A_2'} \\ &=w_{B_1\sqcup A_2} - w_{B_1\sqcup A_2'} \ .\end{aligned}$$

We have already shown that ϕ is independent of the choice of B_1, so we may assume $B_1 = A_1$. Hence, the above equals $w_{A_1\sqcup A_2} - w_{A_1\sqcup A_2'}$. By connectivity of the graph of $P_{\mathcal{M}_1\oplus\mathcal{M}_2}$, we get $w = \phi_1(w) \otimes \phi_2(w)$ as we wanted.

Therefore, the maps ϕ and $\otimes$ are bijective linear maps which send cones to cones, which implies $\mathrm{Dr}(\mathcal{M}_1 \oplus \mathcal{M}_2) = \mathrm{Dr}(\mathcal{M}_1) \times \mathrm{Dr}(\mathcal{M}_2)$. □

Remark 4.7. The statement above generalizes Theorem 4 by Chatelain and Ramírez [4] which deals with sequences of weakly compatible hyperplane splits. While the article by Joswig and Schröter [15] provides the case of sequences of strongly compatible hyperplane splits and the matroid polytopes that occur in these matroid subdivisions. We refer to Herrmann and Joswig [12] for the definitions.

Let $\mathcal{M}$ be a matroid $(E, \mathcal{B})$. Two elements e and e' in E are *parallel* if $\mathrm{rank}(\{e, e'\}) = 1$. We denote this by $e \parallel e'$. Remark that this implies that $\mathcal{M}\backslash e = \mathcal{M}\backslash e'$.

Theorem 4.3. *Let $\mathcal{M}$ be a matroid and $e \parallel e'$ in $\mathcal{M}$. Then*

$$\mathrm{Dr}(\mathcal{M}) \Big/ \mathrm{lin}\,\mathrm{Dr}(\mathcal{M}) \cong \mathrm{Dr}(\mathcal{M} \setminus e') \Big/ \mathrm{lin}\,\mathrm{Dr}(\mathcal{M} \setminus e')$$

and $\dim \operatorname{lin} \operatorname{Dr}(\mathcal{M}) = \dim \operatorname{lin} \operatorname{Dr}(\mathcal{M} \setminus e') + 1$.

Proof. Clearly, $\mathcal{M}$ contains the circuit $\{e, e'\}$. Hence, the number of connected components of $\mathcal{M}$ is the same as the number of connected components of $\mathcal{M} \setminus e'$. It follows that $\dim \operatorname{lin} \operatorname{Dr}(\mathcal{M}) = \dim \operatorname{lin} \operatorname{Dr}(\mathcal{M} \setminus e') + 1$.

The projection $\operatorname{Dr}(\mathcal{M}) \to \operatorname{Dr}(\mathcal{M} \setminus e')$ that forgets the coordinates that correspond to bases that contain e' is surjective. Our goal is to show that this projection is injective if we quoten by the lineality spaces. Let $w \in \operatorname{Dr}(\mathcal{M})$ and B_e be a basis of $\mathcal{M}$ that contains e and $B_{e'} = B_e \setminus \{e\} \cup \{e'\}$. We may assume that $w_{B_e} = w_{B_{e'}}$ as the lineality space of $\operatorname{Dr}(\mathcal{M})$ contains $\sum_{B \ni e'} e_B$. Let $B'_{e'}$ be a basis of $\mathcal{M}$ and $B'_e = B'_{e'} \setminus \{e'\} \cup \{e\}$ of distance $\# B'_e \setminus B_e = 1$. That is $e_{B_e}, e_{B'_e}, e_{B_{e'}}, e_{B'_{e'}}$, form a square in the vertex-edge graph of $P_{\mathcal{M}}$. The set $B_e \cap B'_{e'} \cup \{e, e'\}$ is a non-basis of distance 1 to those four bases. Therefore, the square is not subdivided by the regular subdivision induced by w. We conclude that $w_{B_e} + w_{B'_{e'}} = w_{B_{e'}} + w_{B'_e}$ and by our assumption $w_{B'_{e'}} = w_{B'_e}$. Iterating our argument shows that $w_B = w_{B \setminus \{e\} \cup \{e'\}}$ for any basis B that contains e. As the basis exchange graph of a matroid is connected. Therefore, we derive that the projection is injective up to lineality and therefore the desired isomorphism. □

The combination of Theorem 4.2 and Theorem 4.3 allows to deduce the local Dressian $\operatorname{Dr}(\mathcal{M})$ of an arbitrary matroid $\mathcal{M}$ from the simplifications of its connected components.

5. Indecomposable matroids

In this section, we begin with focusing on local Dressians of binary matroids, i.e., those matroids which are representable over the field with two elements. Recall that any matroid obtained from successive deletions and contractions form a matroid $\mathcal{M}$ is a minor of $\mathcal{M}$. The following is a useful characterization of binary matroids in terms of their minors.

Proposition 5.1 (Tutte [29]). *A matroid is binary if and only if it has no minor isomorphic to the uniform matroid* $\mathcal{U}_{2,4}$.

From Proposition 5.1 we derive a generalization of Theorem 3 by Chatelain and Ramírez [3] which states that a matroid polytope of a binary matroid can not be splited into two matroid polytopes.

Definition 5.1. A matroid is said to be indecomposable if and only if its polytope does not allow a non-trivial matroid subdivision.

Corollary 5.1. *Let* $\mathcal{M}$ *be a binary matroid. Then the local Dressian* $\operatorname{Dr}(\mathcal{M})$ *linear space. In particular, the matroid polytope* $P_{\mathcal{M}}$ *is indecomposable.*

Proof. Let $\mathcal{M}$ be a binary matroid and $P_{\mathcal{M}}$ its matroid polytope. The 3-skeleton of the polytope $P_{\mathcal{M}}$ does not contain a octahedral face as such a face corresponds to a minor isomorphic to the uniform matroid $\mathcal{U}_{2,4}$. From Corollary 4.3 we deduce that $P_{\mathcal{M}}$ only has a trivial matroid subdivision. That is the Dressian is a linear space and $\mathcal{M}$ is indecomposable. □

A matroid can be indecomposable even if its matroid polytope contains octahedral faces. Consider the simple matroid $\mathcal{P}$ on 13 elements and rank 3 given by the ternary projective plane. This matroid is not binary and its matroid polytope has 117 octahedral faces.

Proposition 5.2. *The local Dressian* $\mathrm{Dr}(\mathcal{P})$ *is a* 12*-dimensional linear space. In particular, the matroid* $\mathcal{P}$ *is indecomposable.*

Proof. We will show the indecomposability by contradiction. We assume that $\mathcal{M}$ is a proper connected submatroid of $\mathcal{P}$. Being a submatroid means that every basis of $\mathcal{M}$ is a basis of $\mathcal{P}$. In this proof we will use the *closure operators* of the matroids $\mathcal{M}$ and $\mathcal{P}$. Recall that the closure $cl(S)$ of a set S is the maximal set that contains S with $\mathrm{rank}(cl(S)) = \mathrm{rank}(S)$. Here with maximal we mean that for every element $e \notin cl(S)$ we have $\mathrm{rank}(e \cup cl(S)) > \mathrm{rank}(S)$. We denote by $cl_{\mathcal{P}}(S)$ the closure of S in $\mathcal{P}$ and by $cl_{\mathcal{M}}(S)$ the closure of S in $\mathcal{M}$. They satisfy $cl_{\mathcal{P}}(S) \subseteq cl_{\mathcal{M}}(S)$ whenever $\mathrm{rank}_{\mathcal{M}}(S) = \mathrm{rank}_{\mathcal{P}}(S)$.

The proof consists of five steps:

- There are two parallel elements in $\mathcal{M}$.
- Two lines in $\mathcal{P}$ collapse to a line in $\mathcal{M}$.
- A quadrilateral in $\mathcal{P}$ collapses to a point in $\mathcal{M}$.
- Three concurrent lines in $\mathcal{P}$ collapse into a line in $\mathcal{M}$.
- Contradiction.

After each step, for improving the exposition, we reset the labeling. We make sure to clarify the new assigned labels. We do this in order to assure that there is no loss of generality. Keep in mind that for $\mathcal{M}$ to be connected there is no line such that its complement is a single point. In particular there must be a least four points, i.e., four parallelism classes.

Our first step is to show that $\mathcal{M}$ contains a pair of parallel elements. Suppose that the set 123 is a basis of $\mathcal{P}$ but it is dependent in $\mathcal{M}$. Either 123 contains a parallel pair or $cl_{\mathcal{M}}(123)$ is of rank 2 as $\mathcal{M}$ is loop free. In the latter case, let 4 be not in the rank 2 flat $cl_{\mathcal{M}}(123)$. This implies that the intersection of the lines $cl_{\mathcal{M}}(14) \cap cl_{\mathcal{M}}(123)$ is of rank 1 in $\mathcal{M}$. As 2 is not parallel to 3 in $\mathcal{M}$, then $cl_{\mathcal{P}}(23) \subseteq cl_{\mathcal{M}}(23) = cl_{\mathcal{M}}(123)$ and, as 123 is independent in $\mathcal{P}$, there is an element 5 in $cl_{\mathcal{P}}(23) \cap cl_{\mathcal{P}}(14)$. This means that $5 \in cl_{\mathcal{M}}(14) \cap cl_{\mathcal{M}}(123)$ and hence it is parallel to 1 in $\mathcal{M}$.

Suppose now that 1 and 2 are two parallel elements in $\mathcal{M}$. Notice that there are at lest three elements not in $cl_{\mathcal{M}}(12)$. Moreover, $cl_{\mathcal{P}}(12)$ has four elements, at least two of which are in $cl_{\mathcal{M}}(12)$. Then there exists an element 3 such that 3 is not in $cl_{\mathcal{M}}(12) \cup cl_{\mathcal{P}}(12)$. Therefore, $cl_{\mathcal{P}}(13)$ and $cl_{\mathcal{P}}(23)$ are two different lines in $\mathcal{P}$ which are contained in $cl_{\mathcal{M}}(13) = cl_{\mathcal{M}}(23)$.

Suppose that the seven points on the two lines 1234 and 1567 in $\mathcal{P}$ span a line in $\mathcal{M}$. There must be at least two points 8 and 9 outside this line in the connected matroid $\mathcal{M}$. Each of the three lines $cl_{\mathcal{P}}(28)$, $cl_{\mathcal{P}}(38)$ and $cl_{\mathcal{P}}(48)$ intersects the line 1567 in a different point in the projective geometry $\mathcal{P}$. This induces a bijection between 234 and 567 where elements are mapped to parallel elements in $\mathcal{M}$. Similarly, a bijection can be constructed by considering the lines from 9. These bijections do not agree and hence, there are at least four parallel elements in $\mathcal{M}$ that span a quadrilateral in $\mathcal{P}$.

Suppose that 1234 is a quadrilateral in $\mathcal{P}$ which collapses to a point in $\mathcal{M}$. Let $5 \in cl_{\mathcal{P}}(12) \cap cl_{\mathcal{P}}(34)$, and $6 \in cl_{\mathcal{P}}(13) \cap cl_{\mathcal{P}}(24)$, and $7 \in cl_{\mathcal{P}}(14) \cap cl_{\mathcal{P}}(23)$. As $\mathcal{M}$ is connected, there are at least three elements outside $cl_{\mathcal{M}}(1234)$. Suppose that these points are exactly 5, 6 and 7. Then $cl_{\mathcal{P}}(56) \cap cl_{\mathcal{M}}(1234) \neq \emptyset$ forcing $cl_{\mathcal{M}}(1234)$, 5 and 6 to be colinear in $\mathcal{M}$, and $\mathcal{M}$ disconnected. So there is another point 8 outside $cl_{\mathcal{M}}(1234)$. In particular, three of the lines in $\mathcal{P}$ passing through 8 also pass through at least one point in the quadrilateral 1234. Therefore they collapse in a single line in $\mathcal{M}$.

Suppose three concurrent lines passing through 1 in $\mathcal{P}$ collapse to a single line in $\mathcal{M}$. Let S be the set of elements different from 1 forming these three lines. As $\mathcal{M}$ is connected there must be at least two elements outside $cl_{\mathcal{M}}(S)$. For each point, the lines passing through it and not 1 induces a partition of S in three subsets of size three, such that the elements in each subsets belong to the same parallelism class. The two partitions are transversal, therefore S is in the same parallelism class. As the complement of S is a line in $\mathcal{P}$, then $\mathcal{M}$ is disconnected and we obtain a contradiction. □

Remark 5.1. We actually proved a stronger statement, namely that the matroid $\mathcal{P}$ does not contain a proper connected submatroid.

We end this section by showing that a finest matroid subdivision of $\mathrm{Dr}(2, n)$ contains only indecomposable matroids.

Proposition 5.3. *The cells of a finest matroid subdivision of $\Delta(2, n)$ correspond to binary matroids. In particular, they are indecomposable.*

Proof. Every matroid subdivisions of the hypersimplex $\Delta(2, n)$ is regular, as it is a sequence of compatible splits, i.e., subdivisions that divide $\Delta(2, n)$ into two maximal cells. See [15] and [12] for further details. Every regular matroid subdivision of $\Delta(2, n)$ is representable in characteristic 0 as $\mathrm{Dr}(2, n) = \mathrm{TGr}_0(2, n)$. Moreover, it is a finest subdivision if and only if it has $n - 2$ maximal cells. The Main Theorem in [27] states that all cells in a subdivision of the hypersimplex $\Delta(d, n)$ are graphical and therefore binary whenever the subdivision is induced by a Plücker vector in $\mathrm{TGr}_0(d, n)$ and has $\binom{n-2}{d-1}$ maximal cells. This applies to the finest subdivisions of

$\Delta(2, n)$ and hence the maximal cells are matroid polytopes of binary matroids and indecomposable. □

6. Open questions

Several questions arise from the last section. We end this article by stating them and in particular, by making a conjecture.

A class of possible indecomposable matroids comes directly from the previous section. We conjecture the following generalization of Proposition 5.2:

Conjecture 6.1. *All matroids that arise from projective spaces over finite fields are indecomposable.*

Notice that a direct consequence would be examples of indecomposable matroids which are only representable over a particular characteristic. While the direct sum of the ternary projective plane and the binary projective plane, i.e., fano matroid, is an indecomposable matroid which is not representable over any field. We also want to remark that as the ternary projective plane $\mathcal{P}$ has a decomposable minor $\mathcal{U}_{2,4}$, a classification of indecomposable matroids can not relay on excluded minors.

Moreover, it would be interesting to find an efficient criterion to check indecomposability. In Proposition 5.2 we used that there does not exist a connected submatroid. We wonder whether a submatroid is in general a certificate of decomposability.

Question 6.1. Does there exist two connected matroids $\mathcal{M}$ and $\mathcal{M}'$ such that $P_{\mathcal{M}'}$ is strictly contained in $P_{\mathcal{M}}$ but no matroid subdivision of $P_{\mathcal{M}}$ has $P_{\mathcal{M}'}$ as a cell?

Notice that when $\mathcal{M}$ is a uniform matroid then the corank subdivision has $P_{\mathcal{M}'}$ as a cell. But the corank function of M' does not necessarily satisfy local Plücker relations.

Example 6.1. Let $\mathcal{M}$ be a matroid with bases 12, 13, 14, 23 and 24 and $\mathcal{M}'$ be the matroid with the two bases 12 and 13. Then the local corank lifting is $w = (0, 0, 1, 1, 1)$ and this vector is not in the local Dressian $\mathrm{Dr}(\mathcal{M})$ as it subdivides the square pyramid $P_{\mathcal{M}}$ into two tetrahedra.

Our last question is about finest matroid subdivisions of hypersimplicies and is derived from Proposition 5.3.

Question 6.2. Are all cells in a finest matroid subdivision of a hypersimplex matroid polytopes of indecomposable matroids?

References

[1] Federico Ardila and Caroline J. Klivans. The Bergman complex of a matroid and phylogenetic trees. *J. Combin. Theory Ser. B*, 96(1):38–49, 2006.

[2] Matthew Baker and Nathan Bowler. Matroids over partial hyperstructures. *Adv. Math.*, 343:821–863, 2019.
[3] Vanessa Chatelain and Jorge L. Ramírez Alfonsín. Matroid base polytope decomposition. *Adv. in Appl. Math.*, 47(1):158–172, 2011.
[4] Vanessa Chatelain and Jorge L. Ramírez Alfonsín. Matroid base polytope decomposition II: Sequences of hyperplane splits. *Adv. in Appl. Math.*, 54:121–136, 2014.
[5] Jack Edmonds. Submodular functions, matroids, and certain polyhedra. In *Combinatorial Structures and their Applications (Proc. Calgary Internat. Conf., Calgary, Alta., 1969)*, 69–87. Gordon and Breach, New York, 1970.
[6] Eva Maria Feichtner and Bernd Sturmfels. Matroid polytopes, nested sets and Bergman fans. *Port. Math. (N.S.)*, 62(4):437–468, 2005.
[7] Alex Fink and Felipe Rincón. Stiefel tropical linear spaces. *J. Combin. Theory Ser. A*, 135:291–331, 2015.
[8] I. M. Gelfand, R. M. Goresky, R. D. MacPherson, and V. V. Serganova. Combinatorial geometries, convex polyhedra, and Schubert cells. *Adv. in Math.*, 63(3):301–316, 1987.
[9] Simon Hampe. Tropical linear spaces and tropical convexity. *Electron. J. Combin.*, 22(4):Paper 4.43, 20, 2015.
[10] Kerstin Hept and Thorsten Theobald. Tropical bases by regular projections. *Proc. Amer. Math. Soc.*, 137(7):2233–2241, 2009.
[11] Sven Herrmann, Anders N. Jensen, Michael Joswig, and Bernd Sturmfels. How to draw tropical planes. *Electron. J. Combin.*, 16(2, Special volume in honor of Anders Björner): Research Paper 6, 26, 2009.
[12] Sven Herrmann and Michael Joswig. Splitting polytopes. *Münster J. Math.*, 1:109–141, 2008.
[13] Sven Herrmann, Michael Joswig, and David E Speyer. Dressians, tropical Grassmannians and their rays. *Forum Mathematicum*, pages 389–411, 2012.
[14] Michael Joswig. Essentials of tropical combinatorics. Book in preperation.
[15] Michael Joswig and Benjamin Schröter. Matroids from hypersimplex splits. *J. Combin. Theory Ser. A*, 151:254–284, 2017.
[16] Michael Joswig and Benjamin Schröter. The degree of a tropical basis. *Proc. Amer. Math. Soc.*, 146(3):961–970, 2018.
[17] Mikhail M. Kapranov. Chow quotients of Grassmannians. I, *I. M. Gelfand Seminar, Adv. Soviet Math.*, vol. 16, Amer. Math. Soc., Providence, RI, 29–110, 1993.
[18] Lucia Lopez de Medrano, Felipe Rincón and Kristin Shaw Chern-Schwartz-MacPherson cycles of matroids preprint `https://arxiv.org/abs/1707.07303`, 2017.
[19] Diane Maclagan and Bernd Sturmfels. *Introduction to Tropical Geometry*, volume 161 of *Graduate Studies in Mathematics*. American Mathematical Society, Providence, RI, 2015.
[20] Thomas Markwig. A field of generalised Puiseux series for tropical geometry. *Rend. Semin. Mat. Univ. Politec. Torino*, 68(1):79–92, 2010.
[21] Ezra Miller and Bernd Sturmfels. *Combinatorial commutative algebra*, volume 227 of *Graduate Texts in Mathematics*. Springer-Verlag, New York, 2005.
[22] Peter Nelson. Almost all matroids are nonrepresentable. *Bull. Lond. Math. Soc.*, 50(2):245–248, 2018.
[23] James Oxley. *Matroid theory*, volume 21 of *Oxford Graduate Texts in Mathematics*. Oxford University Press, Oxford, second edition, 2011.
[24] Guillermo Pineda-Villavicencio and Benjamin Schröter. Reconstruction of matroid polytopes. In preperation.
[25] Felipe Rincón. Local tropical linear spaces *Discrete Comput. Geom.*, 50:700–713, 2013.

[26] David E. Speyer. Tropical linear spaces. *SIAM J. Discrete Math.*, 22(4):1527–1558, 2008.

[27] David E. Speyer. A matroid invariant via the K-theory of the Grassmannian. *Adv. Math.*, 221(3):882–913, 2009.

[28] David Speyer and Bernd Sturmfels. The tropical Grassmannian. *Adv. Geom.*, 4(3):389–411, 2004.

[29] W. T. Tutte. A homotopy theorem for matroids. I, II. *Trans. Amer. Math. Soc.*, 88:144–174, 1958.

[30] Neil White, editor. *Theory of matroids*, volume 26 of *Encyclopedia of Mathematics and its Applications*. Cambridge University Press, Cambridge, 1986.

Polyhedral geometry for lecture hall partitions

McCabe Olsen

Department of Mathematics
The Ohio State University
231 W. 18th Ave., Columbus, OH 43210, USA
E-mail: olsen.149@osu.edu

Lecture hall partitions are fundamental combinatorial objects which have been studied extensively over the past two decades. These objects have produced new results, as well as reinterpretations and generalizations of classicial results, which are of interest in combinatorial number theory, enumerative combinatorics, and convex geometry. In a recent survey of Savage, a wide variety of these results are nicely presented. However, since the publication of this survey, there have been many new developments related to the polyhedral geometry and Ehrhart theory arising from lecture hall partitions. Subsequently, in this survey article, we focus exclusively on the polyhedral geometric results in the theory of lecture hall partitions in an effort to showcase these new developments. In particular, we highlight results on lecture hall cones, lecture hall simplices, and lecture hall order polytopes. We conclude with an extensive list of open problems and conjectures in this area.

Keywords: Lecture hall partitions, lattice polytopes, rational cones, Ehrhart theory, local Ehrhart theory.

1. Introduction

For $n \in \mathbb{Z}_{\geq 1}$, let $[n] := \{1, 2, \ldots, n\}$. Let $\boldsymbol{s} = (s_1, \ldots, s_n) \in \mathbb{Z}_{\geq 1}^n$ be a sequence of positive integers. The set of $\boldsymbol{s}$-*lecture hall partitions* is the set

$$L_n^{(\boldsymbol{s})} := \left\{ \lambda \in \mathbb{Z}^n \, : \, 0 \leq \frac{\lambda_1}{s_1} \leq \cdots \leq \frac{\lambda_n}{s_n} \right\}.$$

In the case that $\boldsymbol{s}$ is a weakly increasing sequence, $L_n^{(\boldsymbol{s})}$ refines the set of partitions with at most n parts. In other cases, one may actually have a composition rather than a partition, though much of the literature makes no distinction here. The study of lecture hall partitions was initiated by Bousquet-Mélou and Eriksson in a pair of seminal papers [5, 6]. In the two decades to follow, these objects have been studied in a variety of algebraic, combinatorial, geometric, and number theoretic mediums. Much of this work is summarized in the excellent survey of Savage [33].

The goal of this survey is to provide an update to the survey of Savage. In particular, we focus exclusively on results involving convex geometric objects arising from lecture hall partitions taking into account a significant number of results appearing after the publication of the Savage survey. The structure of this note is as follows. In Section 2, we provide a brief review of the necessary polyhedral geometry, Ehrhart theory, and local Ehrhart theory, as well as define the convex geometric objects arising from lecture hall partitions, namely *lecture hall cones*, *lec-*

ture hall simplices, *rational lecture hall simplices*, and *lecture hall order polytopes*. In Section 3, we outline known results of lecture hall cones. In Section 4, we discuss the known results regarding lecture hall simplices. Section 5 focuses on results of lecture hall order polytopes. Finally, Section 6 outlines some of the remaining open problems in the polyhedral geometry of lecture hall partitions.

2. Background

In this section, we provide the necessary background for the results on the polyhedral geometry of lecture hall partitions. First, we review the standard tools and definitions in Ehrhart theory and polyhedral geometry. We then discuss the notions of local Ehrhart theory and box unimodal triangulations. Finally, we briefly introduce the polyhedral geometric objects arising from lecture hall partitions. Subsequently, some or all of these sections may be safely skipped by the experts.

2.1. *Classical Ehrhart theory and polyhedral geometry*

We recall a few definitions from polyhedral geometry. A *polyhedral cone* C in $\mathbb{R}^n$ is the solution set to some finite collection of linear inequalities $Ax \geq 0$ for some real matrix A, or equivalently given some elements $w_1, w_2, \cdots, w_j \in \mathbb{R}^n$,

$$C = \text{span}_{\mathbb{R}_{\geq 0}}\{w_1, w_2, \cdots, w_j\}.$$

The minimal set of elements w_i to define the cone are called *ray generators*. The cone C is said to be *rational* if the matrix A contains rational entries (equivalently if each $w_i \in \mathbb{Q}^n$), it is said to be *simplicial* if it is defined by n independent inequalities (equivalently if $j = n$ and $\{w_i\}_{i=1}^n$ are linearly independent), and it is said to be *pointed* if it does not contain a linear subspace of $\mathbb{R}^n$. Let C° denote the interior of C.

If $C \subset \mathbb{R}^n$ is a pointed, rational cone, a *proper grading* of C is a function $g : C \cap \mathbb{Z}^n \to \mathbb{Z}^r_{\geq 0}$, for some r, satisfying (i) $g(\lambda + \mu) = g(\lambda) + g(\mu)$; (ii) $g(\lambda) = 0$ implies $\lambda = 0$; and (iii) for any $v \in \mathbb{Z}^r_{\geq 0}$, $g^{-1}(v)$ is finite. Moreover, the integer points $C \cap \mathbb{Z}^n$ form a semigroup. Semigroups of this type have unique minimal generating sets known as the *Hilbert basis* of C. Additionally, the lattice points in a pointed rational cone give rise to a semigroup algebra structure. We call this the *affine, normal semigroup algebra of* C, which is denoted $\mathbb{C}[C] := \mathbb{C}[C \cap \mathbb{Z}^n]$ For background and details see [3, 27].

We say that a pointed, rational cone $C \subset \mathbb{R}^n$ is *Gorenstein* if there exists a point $\boldsymbol{c} \in C^\circ$ such that $C^\circ \cap \mathbb{Z}^n = +(C \cap \mathbb{Z}^n)$. This point is known as the *Gorenstein point* of C. Due to theorems of Stanley [38], this notion of Gorenstein is equivalent to the commutative algebra notion of Gorenstein, as C is Gorenstein if and only if the algebra $\mathbb{C}[C]$ is Gorenstein. For reference and commutative algebra details, see [10, 41].

We say that $\mathcal{P} \subset \mathbb{R}^n$ is a *convex polytope* of dimension $d \leq n$ if it can be expressed as the convex hull of finitely many points in

$$\mathcal{P} = \text{conv}\{\boldsymbol{v}_0, \ldots, \boldsymbol{v}_m : \boldsymbol{v}_i \in \mathbb{R}^n\} := \left\{ \sum_{i=0}^{m} \mu_i \boldsymbol{v}_i : \boldsymbol{v}_i \in \mathbb{R}^n,\ \mu_i \geq 0, \text{ and } \sum_{i=0}^{m} \mu_i = 1 \right\},$$

which lies in a d-dimensional affine subspace. Let $V(\mathcal{P})$ be the minimal set of $\boldsymbol{v}_i$ such that $\mathcal{P} = \text{conv}\{\boldsymbol{v}_i : \boldsymbol{v}_i \in V(\mathcal{P})\}$. This set is called the *vertex set* of $\mathcal{P}$. If for all $\boldsymbol{v} \in V(\mathcal{P})$, we have $\boldsymbol{v} \in \mathbb{Z}^n$ (resp. $\boldsymbol{v} \in \mathbb{Q}^n$), we say that $\mathcal{P}$ is a *lattice* polytope (resp. *rational* polytope). Alternatively, $\mathcal{P}$ can be expressed as the intersection of finitely many halfspaces in $\mathbb{R}^n$.

The *lattice point enumerator* of $\mathcal{P}$, where $d = \dim(P) \leq n$, is the function

$$i(\mathcal{P}, t) = \#(t \cdot \mathcal{P} \cap \mathbb{Z}^n)$$

where $t \cdot \mathcal{P} = \{t \cdot \alpha \ : \ \alpha \in \mathcal{P}\}$ is the tth dilate of $\mathcal{P}$ with $t \in \mathbb{Z}_{\geq 0}$ and by convention $0 \cdot \mathcal{P} = \{\mathbf{0}\}$. By theorems of Ehrhart [14], if $\mathcal{P}$ is lattice, $i(\mathcal{P}, t)$ is a polynomial in the variable t of degree d and if $\mathcal{P}$ is rational, $i(\mathcal{P}, t)$ is a quasipolynomial in the variable t of degree d. Subsequently, we will call $i(\mathcal{P}, t)$ the *Ehrhart polynomial* of $\mathcal{P}$ or the *Ehrhart quasipolynomial* of $\mathcal{P}$ in each respective case. We say that $\mathcal{P}$ is *Ehrhart positive* if each coefficient of $i(\mathcal{P}, t)$ is a positive rational number. For background on Ehrhart positivity, see the survey of Liu [25].

We should note that given a rational or lattice polytope $\mathcal{P}$, we can associate a rational cone, and thus an affine, normal semigroup algebra, to $\mathcal{P}$. The *cone over* $\mathcal{P}$ is

$$\text{cone}(\mathcal{P}) := \text{span}_{\mathbb{R}_{\geq 0}} \{(\boldsymbol{p}, 1) \ : \ \boldsymbol{p} \in \mathcal{P} \cap \mathbb{Z}^n\} \subseteq \mathbb{R}^{n+1}$$

and the *affine, normal semigroup algebra* of $\mathcal{P}$ is $\mathbb{C}[\mathcal{P}] := \mathbb{C}[\text{cone}(\mathcal{P}) \cap \mathbb{Z}^{n+1}]$. We can alternatively view $i(\mathcal{P}, t)$ as the *Hilbert function* or $\mathbb{C}[\mathcal{P}]$.

Suppose that $\mathcal{P}$ is lattice. It is often more desirable to study the Ehrhart polynomial under a suitable change of basis. In particular, let $h^*(\mathcal{P}; z)$ be the Ehrhart polynomial expressed in the binomial basis $\left\{z^i := \binom{t+n-i}{n}\right\}_{i=0}^{n}$. The polynomial $h^*(\mathcal{P}; z)$ is called the *h^*-polynomial* of $\mathcal{P}$. Equivalently, we can view the h^*-polynomial as the numerator of the *Ehrhart series* of $\mathcal{P}$ which is

$$\text{Ehr}_{\mathcal{P}}(z) := \sum_{t \geq 0} i(\mathcal{P}, t) z^t = \frac{h^*(P; z)}{(1 - z)^{d+1}}.$$

We say that a lattice polytope $\Delta \subset \mathbb{R}^n$ is a *d-simplex* if it can be expressed as the convex hull of exactly $d + 1$ affinely independent lattice points. In the special case of a d-simplex $\Delta \subset \mathbb{R}^n$, we have a geometric interpretation of $h^*(\Delta; z)$. The *(half-open) fundamental parallelpiped* of Δ is

$$\Pi_\Delta := \left\{ \sum_{i=0}^{d} \mu_i (\boldsymbol{v}_i, 1) \in \mathbb{R}^{n+1} \ : 0 \leq \mu_i < 1 \, , i \in \{0, \ldots, d\} \right\}.$$

We then have

$$h^*(\Delta; z) = \sum_{(x_1,\ldots,x_n,x_{n+1})\in\Pi_\Delta\cap\mathbb{Z}^{n+1}} z^{x_{n+1}}.$$

We say that a lattice polytope $\mathcal{P}$ has the *integer decomposition property* (IDP) if for any $\boldsymbol{x} \in t\mathcal{P} \cap \mathbb{Z}^n$, there exist t lattice points $\boldsymbol{p}_1, \boldsymbol{p}_2, \ldots, \boldsymbol{p}_t \in P \cap \mathbb{Z}^n$ such that $\boldsymbol{p}_1 + \boldsymbol{p}_2 + \cdots + \boldsymbol{p}_t = \boldsymbol{x}$. In this case, we say that $\mathcal{P}$ has the IDP. This is equivalent to saying the semigroup algebra $\mathbb{C}[\mathcal{P}]$ is generated entirely in degree 1.

A *triangulation* $\mathcal{T}$ of a lattice polytope $\mathcal{P}$ of dimension d is a subdivision of $\mathcal{P}$ into d-dimensional lattice simplices framing a simplicial complex. We say that a triangulation $\mathcal{T}$ is *regular* if each simplex in $\mathcal{T}$ is a domains of linearity of a convex piecewise linear function. We say that a triangulation $\mathcal{T}$ is *unimodular* if each simplex in $\mathcal{T}$ is a unimodular simplex (i.e. a lattice simplex with minimal possible volume). If $\mathcal{P}$ has a unimodular triangulation, then $\mathcal{P}$ has the IDP.

Let $\mathcal{P}$ be a lattice polytope of dimension d such that $\mathcal{P} \setminus \partial\mathcal{P} \cap \mathbb{Z}^n = \{\mathbf{0}\}$. We say that $\mathcal{P}$ is *reflexive* if the dual polytope

$$\mathcal{P}^\vee := \{y \in \mathbb{R}^n \,:\, \langle x, y\rangle \leq 1 \text{ for all } x \in \mathcal{P}\}$$

is a lattice polytope. It is a well known result of Hibi [20] that the following are equivalent:

- $\mathcal{P}$ is reflexive (up to translation)
- $h^*(\mathcal{P}; z)$ is a palindromic polynomial of degree d.

Given $k \in \mathbb{Z}_{\geq 1}$, $\mathcal{P}$ is *Gorenstein of index c* if the polytope $k\mathcal{P}$ is reflexive. Equivalently, $\mathcal{P}$ is Gorenstein of index c (up to unimodular equivalence) if $h^*(\mathcal{P}; z)$ is a palindromic polynomial of degree $d - k + 1$ [13]. One should additionally note that this precisely means that the cone($\mathcal{P}$) is a Gorenstein cone. That is, that there exist some $\boldsymbol{c} \in \mathbb{Z}^{n+1}$ such that $\boldsymbol{c} + (\text{cone}(\mathcal{P}) \cap \mathbb{Z}^{n+1}) = \text{cone}(\mathcal{P})^\circ \cap \mathbb{Z}^{n+1}$. Hence, $\mathcal{P}$ is Gorenstein if and only if the algebra $\mathbb{C}[\mathcal{P}]$ is Gorenstein.

A generalization of the Gorenstein property is the level property. We say that $\mathcal{P}$ is *level* if there exists some $\boldsymbol{c}_1, \ldots, \boldsymbol{c}_k \in \mathbb{Z}^{n+1}$ such that

$$\text{cone}(\mathcal{P})^\circ \cap \mathbb{Z}^{n+1} = \sum_{i=1}^{k} (\boldsymbol{c}_i + (\text{cone}(\mathcal{P}) \cap \mathbb{Z}^{n+1})$$

and $\boldsymbol{c}_{1_{n+1}} = \cdots = \boldsymbol{c}_{k_{n+1}}$. Note that in the case of $k = 1$, this is precisely the definition of Gorenstein. This implies that the algebra $\mathbb{C}[\mathcal{P}]$ is a *level semigroup algebra*, which generalizes the notion of a Gorenstein algebra. While palidromicity of $h^*(\mathcal{P}; z)$ is lost with this generalization, additional inequalities on the coefficents remain. We recommend [41] for details on level algebras.

A commonly studied property in Ehrhart theory is h^*-unimodality. That is, determining under what conditions the polynomial $h^*(\mathcal{P}; z)$ has a unimodal sequence of coefficients. In general, one should not expect $h^*(\mathcal{P}; z)$ to be unimodal, even under the assumptions of Gorenstein (see, e.g., [31]). However, the following theorem on unimodality, due to Bruns and Römer, is known.

Theorem 2.1 ([11]). *If $\mathcal{P}$ is Gorenstein and has a regular unimodular triangulation, then $h^*(\mathcal{P}; z)$ is unimodal.*

The following conjecture of Hibi and Ohsugi, which is a weakening of the Bruns and Römer result, is the subject of much research.

Conjecture 2.1 ([29]). *Let $\mathcal{P}$ be Gorenstein and have the IDP, then $h^*(\mathcal{P}; z)$ is unimodal.*

There is even a weaker question, which does not have a negative answer as of yet.

Question 2.1 ([37]). *Does every IDP polytope have a unimodal h^*-polynomial?*

2.2. *Local Ehrhart Theory*

Building on the classical h^*-polynomials in Ehrhart theory is the theory of *local h^*-polynomials.* This theory has been developed in large part due to Betke and McMullen [4], Katz and Stapledon [23], Schepers and Van Langenhoven [37], and Stanley [40].

Suppose that Δ is a lattice d-simplex in $\mathbb{R}^n$ with vertex set $\{\boldsymbol{v}_0, \ldots, \boldsymbol{v}_d\}$. Then the *open-parallelpiped* of Δ is

$$\Pi_\Delta^\circ := \left\{ \sum_{i=0}^{d} \mu_i(\boldsymbol{v}_i, 1) \in \mathbb{R}^{n+1} \, : \, 0 < \mu_i < 1 \, , \, i \in \{0, \ldots, d\} \right\}.$$

Then the *local h^*-polynomial* for the simplex Δ is

$$\ell^*(\Delta; z) := \sum_{(x_1, \ldots, x_n, x_{n+1}) \in \Pi_\Delta^\circ \cap \mathbb{Z}^{n+1}} z^{x_{n+1}}.$$

We note that this is similar in nature to the geometric interpretation of $h^*(\Delta; z)$. The polynomial $\ell^*(\Delta; z)$ is sometimes referred to as the *box polynomial* of Δ. We should note that there is an involution on Π_Δ° which sends elements at height i to height $d - i$, which implies that $\ell^*(\Delta; z)$ is a palindromic polynomial. The following theorem relates the local h^*-polynomial and the classic h^*-polynomial. This theorem uses the notion of an *h-polynomial* and the *link* of a simplicial complex. For background and details, please consult [42].

Theorem 2.2 ([4]). *Let $\mathcal{P}$ be a lattice polytope with lattice triangulation T. Then*

$$h^*(\mathcal{P}; z) = \sum_{\Delta \in T} h(\mathrm{link}_T(\Delta); z)\ell^*(\Delta; z).$$

In the case of $\mathcal{P}$ a reflexive polytope, there is the following theorem.

Theorem 2.3 ([4]). *Let $\mathcal{P}$ be a reflexive lattice polytope with lattice triangulation T of the boundary $\partial\mathcal{P}$. Then*

$$h^*(\mathcal{P};z) = \sum_{\Delta\in T} h(\mathrm{link}_T(\Delta);z)\ell^*(\Delta;z).$$

We say that a lattice triangulation T of $\mathcal{P}$ is *box unimodal* if it is regular and $\ell^*(\Delta;z)$ is unimodal for $\Delta \in T$. Therefore, if a reflexive polytope $\mathcal{P}$ admits a boundary triangulation whose simplices all have unimodal local h^*-polynomials, then $h^*(\mathcal{P};z)$ is also unimodal.

In the case of $\mathcal{P}$ not a simplex, one can still define the local h^*-polynomial $\ell^*(\mathcal{P};z)$ with some additional work. Let $F(\mathcal{P})$ denote the set of faces of $\mathcal{P}$. Note that it is clear that $F(P)$ is a ranked poset under inclusion of faces where the rank is $\rho(F) = \dim(F)+1$. The *g-polynomial* of $\mathcal{P}$ is defined recursively as follows. If $\mathcal{P} = \varnothing$, then $g(\mathcal{P};z) = 1$. If $\mathcal{P}$ is dimension d, then $g(\mathcal{P};z)$ is the unique polynomial of degree strictly less than $\frac{d}{2}$ satisfying

$$z^d g\left(\mathcal{P};\frac{1}{z}\right) = \sum_{F\in F(\mathcal{P})} g(F;z)(t-1)^{d-\dim(F)-1}.$$

The reader should consult [42] for details on the g-polynomial. Then the *local h^*-polynomial* of $\mathcal{P}$ is

$$\ell^*(\mathcal{P};z) = \sum_{F\in F(\mathcal{P})} (-1)^{\dim(\mathcal{P})-\dim(F)} h^*(F;z) g([F,\mathcal{P}]^\vee;z),$$

where $[F,\mathcal{P}]^\vee$ is polytope obtained from the dual of the interval $[F,\mathcal{P}]$ in $F(\mathcal{P})$.

2.3. *Polyhedral objects arising from lecture hall partitions*

We now briefly define the common polyhedral objects arising from lecture hall partitions. We begin with the *$\boldsymbol{s}$-lecture hall cone*.

Definition 2.1. Let $\boldsymbol{s} \in \mathbb{Z}^n_{\geq 1}$. The *$\boldsymbol{s}$-lecture hall cone* is

$$\mathcal{C}_n^{(\boldsymbol{s})} := \left\{\lambda \in \mathbb{R}^n \,:\, 0 \leq \frac{\lambda_1}{s_1} \leq \cdots \leq \frac{\lambda_n}{s_n}\right\},$$

which may equivalently be expressed as

$$\mathcal{C}_n^{(\boldsymbol{s})} := \mathrm{span}_{\mathbb{R}_{\geq 0}} \{(0,\ldots,0,s_i,\ldots,s_n) \,:\, 1 \leq i \leq n\}.$$

It is straightforward to see that $\mathcal{C}_n^{(\boldsymbol{s})} \cap \mathbb{Z}^n = L_n^{(\boldsymbol{s})}$. Subsequently, studying this polyhedral object yields the same data as studying the lecture hall partitions directly. Results on lecture hall cones include a classification of the Gorenstein property, characterization of Hilbert bases in special cases, and triangulation results for the cone $\mathcal{C}_n^{(1,2,\ldots,n)}$. All of these results will be discussed in detail in Section 3.

A related object are the $\boldsymbol{s}$-*lecture hall simplices*[a], which arise as a truncation of the lecture hall cone. Particularly, they can be defined as follows.

Definition 2.2. Let $\boldsymbol{s} \in \mathbb{Z}^n_{\geq 1}$. The $\boldsymbol{s}$-*lecture hall simplex* is

$$\mathbf{P}_n^{(\boldsymbol{s})} := \left\{ \lambda \in \mathbb{R}^n \, : 0 \leq \frac{\lambda_1}{s_1} \leq \cdots \leq \frac{\lambda_n}{s_n} \leq 1 \right\},$$

which may equivalently expressed as

$$\mathbf{P}_n^{(\boldsymbol{s})} := \operatorname{conv}\left\{(0,\ldots,0),(0,\ldots,0,s_i,\ldots,s_n) \, : \, 1 \leq i \leq n\right\}.$$

This family of simplices has been very well-studied. There are results regarding the h^*- polynomials, existence of the IDP, existence of triangulations, classification of the Gorenstein property and the level property, and results on the local h^*-polynomial. These results will be outlined in detail in Section 4.

A less well-studied object arising from lecture hall partitions are the *rational* $\boldsymbol{s}$-*lecture hall simplices.*

Definition 2.3. Let $\boldsymbol{s} \in \mathbb{Z}^n_{\geq 1}$. The *rational* $\boldsymbol{s}$-*lecture hall simplex* is

$$R_n^{(\boldsymbol{s})} := \left\{ \lambda \in \mathbb{R}^n \, : 0 \leq \frac{\lambda_1}{s_1} \leq \cdots \leq \frac{\lambda_n}{s_n} \leq \frac{1}{s_n} \right\}.$$

The study of these objects was initiated by Pensyl and Savage [32]. This theory has been used while studying both the $\boldsymbol{s}$-lecture hall cones and $\boldsymbol{s}$-lecture hall simplices. Given the relative lack of known results regarding these rational polytopes, there is no section specifically dedicated to them in this survey, but they appear in Section 3 as well as Section 6.

Another important polyhedral object arising from lecture hall partitions are $\boldsymbol{s}$-*lecture hall order polytopes.* These objects were introduced by Brändén and Leander [8] as a generalization of the classical theory of *order polytopes* defined by Stanley [39] which have been well-studied.

In order to define $\boldsymbol{s}$-lecture hall order polytopes, we first need the notion of an $\boldsymbol{s}$-*lecture hall P-parition.* Let $P = ([n], \preceq)$ be a labeled poset on $[n]$. We say that P is *naturally-labeled* if $i \preceq j$ in P implies that $i \leq j$ in $\mathbb{Z}$.

Definition 2.4. Let $P = ([n], \preceq)$ be a labeled poset and let $\boldsymbol{s} : [n] \to \mathbb{Z}_{\geq 1}$ be an arbitrary map. A *lecture hall P-partition* is a map $f : [n] \to \mathbb{R}$ such that

(i) if $x \prec y$, then $f(x)/s(x) \leq f(y)/s(y)$, and
(ii) if $x \prec y$ and $x > y$, the $f(x)/s(x) < f(y)/s(y)$.

[a]Much of the literature refers to these as *lecture hall polytopes.* However, since these polytopes are simplices and since the development of lecture hall order polytopes, this is a better name in terms of specificity.

If we consider naturally-labeled posets, we need only consider the first condition of the definition. We now define the lecture hall order polytopes.

Definition 2.5. Let P be a naturally-labeled poset. The *$\boldsymbol{s}$-lecture hall order polytope* associated to $(P, \boldsymbol{s})$ is

$$O(P, \boldsymbol{s}) := \{f \in \mathbb{R}^{[n]} : f \text{ is a } (P, \boldsymbol{s})\text{-partition and } 0 \leq f(x)/s(x) \leq 1 \text{ for all } x \in [n]\}$$

Remark 2.1. Note that the naturally-labeled condition in Definition 2.5 ensures that the resulting polytopes are closed. One could define a similar notion for any lecture hall P-partition, however, the resulting polytope will be partially open without the naturally-labeled condition (i.e. missing certain facets).

We should note that if we let P be a totally ordered chain, we recover the $\boldsymbol{s}$-lecture hall simplex as $O(P, \boldsymbol{s}) = \mathbf{P}_n^{(\boldsymbol{s})}$. While these objects are relatively recent in development, significant results have been obtained. Specifically, there are known results about the h^* polynomial including some results real-rootedness. There are also known results regarding box unimodal traingulations and local h^*-polynomials. These results will be outlined in Section 5.

3. Results: Lecture hall cones

In this section, we discuss the results on $\boldsymbol{s}$-lecture hall cones. The known results fall into three categories: the classification of the Gorenstein property, the characterization of Hilbert bases for certain Gorenstein lecture hall cones, and the special case of the lecture hall cone $\mathcal{C}_n^{(1,2,\ldots,n)}$.

3.1. *Gorenstein characterization*

In combinatorial commutative algebra and polyhedral geometry, Gorenstein objects are well-behaved and it is desirable to classify which objects in the family have this property. In the case of $\boldsymbol{s}$-lecture hall cones, this property is entirely controlled by number theoretic properties of the $\boldsymbol{s}$-sequence.

Theorem 3.1 ([1, 6]). *For a positive integer sequence $\boldsymbol{s}$, the $\boldsymbol{s}$-lecture hall cone $\mathcal{C}_n^{(\boldsymbol{s})}$ is Gorenstein if and only if there exists some $\boldsymbol{c} \in \mathbb{Z}^n$ satisfying*

$$c_j s_{j-1} = c_{j-1} s_j + \gcd(s_j, s_{j+1})$$

for $j > 1$, with $c_1 = 1$.

This result is implicit from the seminal work of Bousquet-Mélou and Eriksson [6], but was formulated in this context by Beck et al. [1]. While this completely classifies when lecture hall cones are Gorenstein, the implicit nature of the definition is somewhat mysterious. Subsequently, it is of interest to study particular types of sequences, as the specificity may provide a more enlightening answer. Given a sequence $\boldsymbol{s} \in \mathbb{Z}_{\geq 1}^n$, we say that $\boldsymbol{s}$ is *$\boldsymbol{u}$-generated* if there is a sequence $\boldsymbol{u} =$

$(u_1, u_2, \cdots, u_{n-1}) \in \mathbb{Z}_{\geq 1}^{n-1}$ such that $s_2 = u_1 s_1 - 1$ and $s_{i+1} = u_i s_i - s_{i-1}$ for $i > 1$. Note that by construction a $\boldsymbol{u}$-generated sequence has the property $\gcd(s_i, s_{i+1}) = 1$ for $1 \leq i < n$. We can now state a characterization of the Gorenstein property for the special case where $\gcd(s_i, s_{i+1}) = 1$ for all i.

Theorem 3.2 ([1, 6]). *Let $\boldsymbol{s} = (s_1, \cdots, s_n)$ be a sequence of positive integers such that $\gcd(s_i, s_{i+1}) = 1$ for $1 \leq i < n$. Then $\mathcal{C}_n^{(\boldsymbol{s})}$ is Gorenstein if and only if $\boldsymbol{s}$ is $\boldsymbol{u}$-generated by some sequence $\boldsymbol{u} = (u_1, u_2, \cdots, u_{n-1})$ of positive integers. When such a sequence exists, the Gorenstein point $\mathbf{c}$ for $\mathcal{C}_n^{(\boldsymbol{s})}$ is defined by $c_1 = 1$, $c_2 = u_1$, and for $2 \leq i < n$, $c_{i+1} = u_i c_i - c_{i-1}$.*

Additionally, the two-term recurrence case has been studied. The following theorem specifies precisely when such a sequence will produce a Gorenstein lecture hall cone.

Theorem 3.3 ([1]). *Let $\ell > 0$ and $b \neq 0$ be integers such that $\ell^2 + 4b \geq 0$. Let $\boldsymbol{s} = (s_1, s_2, \ldots)$ defined by*

$$s_n = \ell s_{n-1} + b s_{n-2}$$

with $s_1 = 1$ and $s_0 = 0$. The $\mathcal{C}_n^{(\boldsymbol{s})}$ is Gorenstein for all $n \geq 0$ if and only if $b = -1$ If $b \neq -1$, there exists n_0 such that $\mathcal{C}_n^{(\boldsymbol{s})}$ fails to be Gorenstein for all $n \geq n_0$.

3.2. *Hilbert basis results*

Given a rational polyhedral cone, determining the Hilbert basis is a natural line of study. Understanding the Hilbert bases for $\boldsymbol{s}$-lecture hall cones is particularly of interest, because this is equivalent to determining a minimum additive generating set for the set of all $\boldsymbol{s}$-lecture hall partitions. In the special case of $\boldsymbol{s} = (1, 2, \ldots, n)$, the following elegant description of the Hilbert basis is known.

Theorem 3.4 ([2]). *For each $A = \{a_1 < a_2 < \cdots < a_k\} \subseteq [n-1]$, define the element v_A to be*

$$v_A = (0, \ldots, 0, a_1, a_2, \ldots, a_k, a_k + 1).$$

The Hilbert basis for $L_n^{(1,2,\ldots,n)}$ is

$$\mathcal{H}_n^{(1,2,\ldots,n)} := \{v_A \ : \ A \subseteq [n-1]\}.$$

As a corollary, the semigroup algebra $\mathbb{C}[\mathcal{C}_n^{(1,2,\ldots,n)}]$ is generated entirely by elements in degree 1 with respect to the grading given by $\lambda \mapsto (\lambda_n - \lambda_{n-1})$.

Given this surprisingly nice, combinatorial description for the Hilbert basis in this case, a natural follow-up is to determine which other $\boldsymbol{s}$-sequences have combinatorially nice Hilbert bases. It is unlikely that one could give a universal description for the Hilbert basis of any $\boldsymbol{s}$-lecture hall cone. Even in the case of 2-dimensional cones, one cannot even bound the cardinality of the Hilbert basis beyond the trivial bounds for simplicial cones. Subsequently, to hope for meaningful results, it is necessary to restrict our classes of $\boldsymbol{s}$-sequences.

The first more general family of $\boldsymbol{s}$-sequences considered in the literature on Hilbert bases are the 1 mod *k-sequences.* Given a $k \in \mathbb{Z}_{\geq 1}$, the 1 mod k-sequence of length n is

$$\boldsymbol{s} = (1, k+1, 2k+1, \ldots, (n-1)k+1).$$

For simplicity of notation, when $\boldsymbol{s}$ is as above, we will use $L_{k,n}$, $\mathcal{C}_{k,n}$, and $\mathbf{P}_{k,n}$ to denote the lecture hall partitions, lecture hall cone, and lecture hall simplex respectively. We have the following result for the Hilbert basis of $\mathcal{C}_{k,n}$.

Theorem 3.5 ([30]). *For all $k \geq 1$, the Hilbert basis for the* 1 mod k *cones in $\mathbb{R}^n$, denoted $\mathcal{C}_{k,n}$, consist of the following elements:*

- *The element $v_A := (0, 0, \ldots, 0, a_1, a_2, \ldots, a_k, a_k+1)$ for each $A \subseteq [n-2]$ where $A = \{a_1 < a_2 < \cdots < a_k\}$;*
- *Elements $w \in L_{k,n}$, where $w_{n-1} = (n-2)k+1$ and $w_n = (n-1)k+1$;*

where $L_{n,k}$ denotes the set of 1 mod k *lecture hall partitions.*

When $k = 1$, we recover the sequence $\boldsymbol{s} = (1, 2, \ldots, n)$ and moreover, this result agrees with the previous characterization of Beck et al. [2]. We can also enumerate the cardinality of the Hilbert basis of this cone with the following algebraic expression.

Corollary 3.1 ([30]).

$$|\mathcal{H}_{k,n}| = \frac{(k+1)^{n-2} + (k-1)}{k} + 2^{n-2}.$$

To see this result, it is clear that the 2^{n-2} appears from enumerating the elements indexed by subsets. The remaining elements can be enumerated by observing that each Hilbert element can be realized as a lattice point in $\mathbf{P}_{k,n-2}$ and applying Ehrhart theory results of Savage and Visawanathan [36].

The second more general family of $\boldsymbol{s}$-sequences studied are the *ℓ-sequences.* These sequences are recursively defined by $s_1 = 1$, $s_2 = \ell$, and $s_i = \ell s_{i-1} - s_{i-2}$ for $i \geq 2$. For ease of notation, given the above sequence let L_n^ℓ, $\mathcal{C}_n^\ell$, $\mathbf{P}_n^\ell$, and R_n^ℓ denote the lecture hall partitions, lecture hall cone, lecture hall simplex, and rational lecture hall simplex respectively. The Hilbert basis results for these lecture hall cones are given as follows.

Theorem 3.6 ([30]). *Let $\boldsymbol{s} = (s_1, s_2, \ldots, s_n)$ be an ℓ-sequence for some $\ell \geq 2$. The Hilbert basis $\mathcal{H}_n^\ell$ for the ℓ-sequence cone $\mathcal{C}_n^\ell$ is*

$$\mathcal{H}_n^\ell = \bigcup_{i=0}^{n} \{\lambda \in L_n^\ell \ : \ \lambda_{n-1} = s_i \, , \, \lambda_n = s_{i+1}\}$$

where L_n^ℓ denotes the set of ℓ-sequence lecture hall partitions.

We should note that when $\ell = 2$, we recover the sequence $\boldsymbol{s} = (1, 2, \ldots, n)$ and this characterization agrees with the Beck et al. result [2]. By identifying these Hilbert basis elements with lattice points in dilates of rational lecture hall simplices, one can derive the following expression for the cardinality of the Hilbert basis.

Corollary 3.2 ([30, Corollary 4.2]).

$$|\mathcal{H}_n^\ell| = 2 + \sum_{j=1}^{n-2} i(R_{n-2}^\ell, s_j)$$

where $i(R_{n-2}^\ell, t)$ denotes the Ehrhart quasipolynomial of the rational lecture hall simplices R_{n-2}^ℓ.

In addition to these Hilbert basis results, the author has also classified the Hilbert bases for $\boldsymbol{u}$-generated Gorenstein lecture hall cones when $n = 2, 3, 4$ [30].

3.3. *Triangulation of $L_n^{(1,2,\ldots,n)}$*

In the case of $\boldsymbol{s} = (1, 2, \ldots, n)$, one can make more specific observation of the geometric structure of the lecture hall cone. The result of Theorem 3.4 verifies that the Hilbert basis of $\mathcal{C}_n^{(1,2,\ldots,n)}$ is entirely at height 1 with respect to a chosen grading. This motivates asking whether or not that cone has a regular, unimodular triangulation. Beck et al. verify that this cone can be represented as the cone over a polytope. Specifically, $\mathcal{C}_n^{(1,2,\ldots,n)} \cong \operatorname{cone}(\mathscr{R}_n)$, where

$$\mathscr{R}_n = \operatorname{conv}\begin{bmatrix} 1 & 1 & 1 & \cdots & 1 \\ 0 & n-1 & 1 & \cdots & 1 \\ 0 & 0 & n-2 & \cdots & 1 \\ \vdots & \vdots & & \ddots & \vdots \\ 0 & 0 & 0 & \cdots & 1 \end{bmatrix}.$$

They then state the following theorem.

Theorem 3.7 ([2]). *For $n \geq 1$, $\mathscr{R}_n$ has a regular, flag, unimodular triangulation. Thus $\mathcal{C}_n^{(1,2,\ldots,n)}$ has a regular, unimodular triangulation.*

4. Results: Lecture hall simplices

We now discuss many of the known results for a lecture hall simplex $\mathbf{P}_n^{(\boldsymbol{s})}$. Perhaps the most pervasive question in Ehrhart theory is: given a lattice polytope $\mathcal{P}$, is there a combinatorial interpretation for $h^*(\mathcal{P}; z)$? To determine such an interpretation for $\mathbf{P}_n^{(\boldsymbol{s})}$, we must define some statistics related to lecture hall partitions. The set of *$\boldsymbol{s}$-inversion sequences* is

$$\mathbf{I}_n^{(\boldsymbol{s})} := \left\{ e \in \mathbb{Z}_{\geq 0}^n : 0 \leq e_i < s_i \text{ for } 1 \leq i \leq n \right\}.$$

Given $\boldsymbol{e} \in \mathbf{I}_n^{(\boldsymbol{s})}$, the *ascent set* of $\boldsymbol{e}$ is

$$\operatorname{Asc}(\boldsymbol{e}) := \left\{ i \in \{0, 1, \ldots, n-1\} : \frac{e_i}{s_i} < \frac{e_{i+1}}{s_{i+1}} \right\}$$

where by convention $s_0 = 1$ and $e_0 = 0$. The *ascent number* of $\boldsymbol{e} \in \mathbf{I}_n^{(\boldsymbol{s})}$ is $\operatorname{asc}(\boldsymbol{e}) := \# \operatorname{Asc}(\boldsymbol{e})$. Savage and Schuster show the following result.

Theorem 4.1 ([34]). *For any $\boldsymbol{s} \in \mathbb{Z}_{\geq 1}^n$,*

$$h^*\left(\mathbf{P}_n^{(\boldsymbol{s})}; z\right) = \sum_{\boldsymbol{e} \in \mathbf{I}_n^{(\boldsymbol{s})}} z^{\operatorname{asc}(\boldsymbol{e})}.$$

The proof of this result is essentially a bijection between lattice points in $\Pi_{\mathbf{P}_n^{(\boldsymbol{s})}}$ and $\boldsymbol{s}$-inversion sequences with respect to four statistics. When restricting to the ascent statistic alone, this corresponds to height of lattice points. In the special case of $\boldsymbol{s} = (1, 2, \ldots, n)$, one can show that

$$h^*(\mathbf{P}_n^{(1,\ldots,n)}; z) = \sum_{\boldsymbol{e} \in \mathbf{I}_n^{(\boldsymbol{s})}} z^{\operatorname{asc}(\boldsymbol{e})} = \sum_{\pi \in \mathfrak{S}_n} z^{\operatorname{des}(\pi)} = A_n(z)$$

where $\mathfrak{S}_n$ is the symmetric group and thus $A_n(z)$ is the *Eulerian Polynomial* (see, e.g., [42] for more details on Eulerian polynomials). Subsequently, the polynomials $h^*(\mathbf{P}_n^{(\boldsymbol{s})}; z)$ for arbitrary $\boldsymbol{s}$ generalize this family and are known as the *$\boldsymbol{s}$-Eulerian polynomials.*

The $\boldsymbol{s}$-Eulerian polynomials also possess many of the same nice properties of Eulerian polynomials, namely unimodality and real-rootedness due to the following theorem of Savage and Visontai [35].

Theorem 4.2 ([35]). *For any $\boldsymbol{s} \in \mathbb{Z}_{\geq 1}$, the $\boldsymbol{s}$-Eulerian polynomial $h^*(\mathbf{P}_n^{(\boldsymbol{s})}; z)$ has only real roots. Hence, the coefficients of $\boldsymbol{s}$-Eulerian polynomials are log-concave and unimodal.*

The proof of this theorem uses the idea of compatible polynomials and interlacing to prove real-rootedness. For background on real-rootedness in a combinatorial setting, the reader should consult [7]. Unfortunately, $\boldsymbol{s}$-Eulerian polynomials do not a priori possess the same palindromic properties as the Eulerian polynomial $A_n(z)$. The classification of this property is equivalent to the classification of which lecture hall simplices $\mathbf{P}_n^{(\boldsymbol{s})}$ are Gorenstein, which will be discussed in a later subsection.

While $h^*(\mathbf{P}_n^{(\boldsymbol{s})}; z)$ is unimodal for any choice of $\boldsymbol{s}$, it is not true in general that $\mathbf{P}_n^{(\boldsymbol{s})}$ is Ehrhart positive. For $a, b, k_2 \in \mathbb{Z}_{\geq 1}$ and $k_1, k_3 \in \mathbb{Z}_{\geq 0}$, let the vector $(1^{k_1}, a, 1^{k_2}, b, 1^{k_3}) = (\underbrace{1, \ldots, 1}_{k_1}, a, \underbrace{1, \ldots, 1}_{k_2}, b, \underbrace{1, \ldots, 1}_{k_3})$. The following theorem of Liu and Solus [26] provide examples where the Ehrhart polynomial of $\mathbf{P}_n^{(\boldsymbol{s})}$ has negative coefficients.

Theorem 4.3 **([26]).** *For every $n \geq 3$, there exist $\boldsymbol{s}$-lecture hall simplices of the form $\mathbf{P}_n^{(1^{k_1},a,1^{k_2},b,1^{k_3})}$ which are not Ehrhart positive.*

In addition to understanding this classical Ehrhart data of $\mathbf{P}_n^{(\boldsymbol{s})}$, it is also of interest to investigate geometric properties such as the existence of the integer decomposition property (IDP) and the existence of particular triangulations. Regarding the integer decomposition property, Hibi Tsuchiya, and the author showed some partial results [22], specifically that $\mathbf{P}_n^{(\boldsymbol{s})}$ has the IDP when $\boldsymbol{s}$ is a weakly monotonic sequence and that $\mathbf{P}_n^{(\boldsymbol{s},1,\boldsymbol{t})}$ has the IDP if and only if $\mathbf{P}_n^{(\boldsymbol{s})}$ and $\mathbf{P}_n^{(\boldsymbol{t})}$ both have the IDP. This results has since been generalized to all $\mathbf{P}_n^{(\boldsymbol{s})}$ in forthcoming work of Brändén and Solus [9], where they prove stronger results for the $\boldsymbol{s}$-lecture hall order polytopes.

Theorem 4.4 **([9]).** *Let $\boldsymbol{s} \in \mathbb{Z}_{\geq 1}^n$ be a sequence. The $\boldsymbol{s}$-lecture hall simplex $\mathbf{P}_n^{(\boldsymbol{s})}$ has the IDP.*

Given that all lecture hall simplices possess the IDP, a natural extension would be the classification of which lecture hall simplices possess regular, unimodular triangulations. There are two notable results in this direction which follow.

Theorem 4.5 **([9]).** *Suppose that $\boldsymbol{s}$ is weakly monotone sequence such that $|s_{i+1} - s_i| \leq 1$. Then $\mathbf{P}_n^{(\boldsymbol{s})}$ has a regular, unimodular triangulation.*

Theorem 4.6 **([22]).** *Suppose that $\mathbf{P}_n^{(\boldsymbol{s})}$ has a regular, unimodular triangulation. Then $\mathbf{P}_{n+1}^{(\boldsymbol{s},ks_n)}$ has a regular, unimodular triangulation for any $k \in \mathbb{Z}_{\geq 1}$.*

The first theorem above, due to Brändén and Solus [9], uses the theory of Gröbner bases. The second theorem, due to Hibi, Tsuchiya, and the author [22], uses the notion of a *chimney polytope*. The reader should consult [18] for details on triangulations, including these two methods. There are no known examples of lecture hall simplices which do not admit a regular, unimodular triangulation.

4.1. *Gorenstein and level classifications*

Both the Gorenstein property and the level property for lecture hall simplices have been studied and a classification for both exists. The Gorenstein property was studied in limited cases by Hibi Tsuchiya, and the author in [22]. These results have been fully generalized by Kohl and the author in [24].

Theorem 4.7 **([24]).** *Let $\boldsymbol{s} = (s_1, s_2, \ldots, s_n) \in \mathbb{Z}_{\geq 1}^n$. Then $\mathbf{P}_n^{(\boldsymbol{s})}$ is Gorenstein if and only if there exists a $\mathbf{c} \in \mathbb{Z}^{n+1}$ satisfying*

$$c_j s_{j-1} = c_{j-1} s_j + \gcd(s_{j-1}, s_j)$$

for $j > 1$ and

$$c_{n+1} s_n = 1 + c_n$$

with $c_1 = 1$.

While the next result is not as general, it guarantees that, under the condition that $\gcd(s_{i-1}, s_i) = 1$ for some $1 < i \leq n$, the vertex cones of $\mathbf{P}_n^{(s)}$ at $(0, 0, \ldots, 0)$ and at $(s_1, s_2, \ldots, s_n)$ being Gorenstein already implies that $\mathbf{P}_n^{(s)}$ is Gorenstein.

Theorem 4.8 ([24]). *Let $\boldsymbol{s} = (s_1, s_2, \ldots, s_n)$ be a sequence such that there exists some $1 < i \leq n$ such that $\gcd(s_{i-1}, s_i) = 1$ and define $\overleftarrow{\boldsymbol{s}} = (\overleftarrow{s_1}, \ldots, \overleftarrow{s_n}) := (s_n, s_{n-1}, \ldots, s_1)$. Then $\mathbf{P}_n^{(s)}$ is Gorenstein if and only if there exists $\mathbf{c}, \mathbf{d} \in \mathbb{Z}^n$ satisfying*

$$c_j s_{j-1} = c_{j-1} s_j + \gcd(s_{j-1}, s_j)$$

and

$$d_j \overleftarrow{s_{j-1}} = d_{j-1} \overleftarrow{s_j} + \gcd(\overleftarrow{s_{j-1}}, \overleftarrow{s_j})$$

for $j > 1$ with $c_1 = d_1 = 1$.

Note that Theorem 4.7 and Theorem 4.8 provide a complete description for when $\boldsymbol{s}$-Eulerian polynomials are palindromic. See Table 1 for some examples of palindromic $\boldsymbol{s}$-Eulerian polynomials which fall under the purview specifically of Theorem 4.8. Additionally, it is worth noting that in the special case of $\boldsymbol{s} = (2, 3, \ldots, n+1)$, Hibi, Tsuchiya, and the author [21] show that $\mathbf{P}_n^{(2,3,\ldots,n+1)}$ is a *self-dual* reflexive simplex. That is, $\mathbf{P}_n^{(2,3,\ldots,n+1)} \cong \mathbf{P}_n^{(2,3,\ldots,n+1)^\vee}$. Moreover, $h^*(\mathbf{P}_n^{(2,3,\ldots,n+1)}; z)$ is the usual Eulerian polynomial $A_{n+1}(z)$.

In addition, Kohl and the author [24] are able to also provide a characterization of the level property in terms of $\boldsymbol{s}$-inversion sequences.

Theorem 4.9 ([24]). *Let $\boldsymbol{s} = (s_1, s_2, \ldots, s_n)$ and let $r = \max\{\mathrm{asc}(\boldsymbol{e}) : \boldsymbol{e} \in \mathbf{I}_n^{(s)}\}$. Then $\mathbf{P}_n^{(s)}$ is level if and only if for any $\mathbf{e} \in \mathbf{I}_{n,k}^{(s)}$ with $1 \leq k < r$ there exists some $\mathbf{e}' \in \mathbf{I}_{n,1}^{(s)}$ such that $(\mathbf{e} + \mathbf{e}') \in \mathbf{I}_{n,k+1}^{(s)}$.*

The proof of this theorem relies heavily on the fact that $\mathbf{P}_n^{(s)}$ is a simplex and studying the additive properties of the lattice points in $\Pi_{\mathbf{P}_n^{(s)}}$ using the bijection that the lattice points at height i correspond to inversion sequences with i inversions. The authors use this classification to prove that certain classes of lecture hall simplices are level and that we can construct level lecture hall simplicies of arbitrarily high dimension. Additionally, levelness introduces new inequalities on the coefficients of the $\boldsymbol{s}$-Eulerian polynomials, as represented in following corollary.

Corollary 4.1 ([24]). *Let $\boldsymbol{s} = (s_1, s_2, \ldots, s_n)$ be a sequence such that $\mathbf{P}_n^{(s)}$ is level. Then the coefficients of the $\boldsymbol{s}$-Eulerian polynomial $h^*(\mathbf{P}_n^{(s)}; z) = 1 + h_1^* z + \cdots + h_r^* z^r$ satisfy the inequalities $h_i^* \leq h_j^* h_{i+j}^*$ for all pairs i and j such that $h_{i+j}^* > 0$.*

4.2. *Local h^*-polynomials for lecture hall simplices*

Given the breadth of knowledge on the traditional Ehrhart theory of $\boldsymbol{s}$-lecture hall simplices, a natural extension is to consider the local h^*-polynomials. Gustafsson

Table 1. Palindromic $\boldsymbol{s}$-Eulerian Polynomials [24]

	sequence $\boldsymbol{s}$	corresponding $\boldsymbol{c}$	corresponding $\boldsymbol{d}$	$\boldsymbol{s}$-Eulerian polynomial
(i)	$(2,1,3,2,1)$	$(1,1,4,3,2)$	$(1,3,5,2,5)$	$1+5z+5z^2+z^3$
(ii)	$(3,2,3,1,2)$	$(1,1,2,1,3)$	$(1,1,4,3,5)$	$1+9z+16z^2+9z^3+z^4$
(iii)	$(1,4,3,2,3)$	$(1,5,4,3,5)$	$(1,1,2,3,1)$	$1+16z+38z^2+16z^3+z^4$
(iv)	$(3,5,2,3,1)$	$(1,2,1,2,1)$	$(1,4,3,8,5)$	$1+20z+48z^2+20z^3+z^4$
(v)	$(1,2,3,4,5)$	$(1,3,5,7,9)$	$(1,1,1,1,1)$	$1+26z+66z^2+26z^3+z^4$
(vi)	$(1,2,5,8,3)$	$(1,3,8,13,5)$	$(1,3,2,1,1)$	$1+50z+138z^2+50z^3+z^4$
(vii)	$(4,3,2,5,3)$	$(1,1,1,3,2)$	$(1,2,1,2,3)$	$1+30z+149z^2+149z^3+30z^4+z^5$
(viii)	$(4,7,3,2,3)$	$(1,2,1,1,2)$	$(1,1,2,5,3)$	$1+43z+208z^2+208z^3+43z^4+z^5$
(ix)	$(5,9,4,3,2)$	$(1,2,1,1,1)$	$(1,2,3,7,6)$	$1+82z+457z^2+457z^3+82z^4+z^5$
(x)	$(3,5,12,7,2)$	$(1,2,5,3,1)$	$(1,4,7,3,2)$	$1+175z+1084z^2+1084z^3+175z^4+z^5$
(xi)	$(3,11,8,5,2)$	$(1,4,3,2,1)$	$(1,3,5,7,2)$	$1+180z+1139z^2+1139z^3+180z^4+z^5$
(xii)	$(2,7,5,10,4)$	$(1,4,3,7,3)$	$(1,3,2,3,1)$	$1+181z+1218z^2+1218z^3+181z^4+z^5$
(xiii)	$(3,8,13,5,2)$	$(1,3,5,2,1)$	$(1,3,8,5,2)$	$1+213z+1346z^2+1346z^3+213z^4+z^5$

and Solus in [15] do this in generality. To introduce these results, we need the following notation. Let

$$\tilde{I}_n^{(\boldsymbol{s})} := \left\{(e_0,\ldots,e_{n+1}) \in \mathbb{Z}^{n+2} \,:\, 0 \le e_i < s_i \text{ and } \frac{e_i}{s_i} \neq \frac{e_{i+1}}{s_{i+1}} \text{ for all } i \in \{0,\ldots,n\}\right\}$$

where $s_0 = s_{n+1} = 1$ and $e_0 = e_{n+1} = 0$.

Theorem 4.10 ([15, Proposition 3.3]). *Let $\boldsymbol{s} \in \mathbb{Z}^n_{\geq 1}$. The local h^*-polynomials* *for the* $\boldsymbol{s}$*-lecture hall simplex* $\mathbf{P}_n^{(\boldsymbol{s})}$ *is*

$$\ell^*(\mathbf{P}_n^{(\boldsymbol{s})}; z) = d_n^{\boldsymbol{s}}(z) := \sum_{\boldsymbol{e}\in \tilde{I}_n^{(\boldsymbol{s})}} z^{\mathrm{asc}(\boldsymbol{e})}.$$

The polynomials $d_n^{\boldsymbol{s}}(z)$ are called the *$\boldsymbol{s}$-derangement polynomials*, because in the case of $\boldsymbol{s} = (2, 3, \ldots, n+1)$, this polynomial agrees with the Eulerian polynomial computed only over $\pi \in \mathfrak{S}_n$ such that π is a derangement. Moreover, there is a parallel to the real-rootedness result of Savage and Visontai as follows.

Theorem 4.11 ([15]). *Let* $\boldsymbol{s} \in \mathbb{Z}_{\geq 1}$. *Then the* $\boldsymbol{s}$*-derangement polynomial* $d_n^{\boldsymbol{s}}(z)$ *is real-rooted and hence log-concave and unimodal.*

Suppose that $p(z)$ is a palindromic polynomial of degree at most d, then we can express p in the basis $\{x^i(x+1)^{d-2i}\}_{i=0}^{\lfloor \frac{d}{2}\rfloor}$. That is,

$$p(z) = \sum_{i=0}^{\lfloor \frac{d}{2}\rfloor} \gamma_i x^i (x+1)^{d-2i}.$$

We say that $p(z)$ is *γ-nonnegative* if $\gamma_0, \ldots, \gamma_{\lfloor \frac{d}{2}\rfloor}$ are all nonnegative. It is a straight forward observation to see that if $p(z)$ is γ-nonnegative, then $p(z)$ is unimodal. The $\boldsymbol{s}$-derangement polynomials can be shown to have this property by the following theorem.

Theorem 4.12 ([15]). *All* $\boldsymbol{s}$*-derangement polynomials are* γ*-nonnegative.*

5. Results: Lecture hall order polytopes

The study of lecture hall order polytopes was initiated by Brändén and Leander [8] by way of introducing *lecture hall* P*-partitions*, which generalize the P-partition theory of Stanley. This subject has since been studied by Brändén and Solus [9] and Gustafsson and Solus [15] from a polyhedral geometry perspective, as well as by Corteel and Kim [12] from a partition theoretic framework. This section is organized as follows. First, we discuss general results regarding the integer decomposition property and the h^*-polynomials of $\boldsymbol{s}$-lecture hall order polytopes. We then discuss results regarding unimodality and real-rootedness of h^*-polynomials. We conclude by discussing local Ehrhart theory results on these polytopes.

We begin by discussing the results on the integer decomposition property. The following theorem, due to Brändén and Solus, further implies that both order polytopes and $\boldsymbol{s}$-lecture hall simplices have the IDP.

Theorem 5.1 ([9]). *Let* $P = ([n], \preceq)$ *be a labeled poset and* $\boldsymbol{s} : [n] \to \mathbb{Z}_{\geq 1}$ *be any function. Then* $O(P, \boldsymbol{s})$ *has the IDP.*

To introduce some of the initial Ehrhart results on $\boldsymbol{s}$-lecture hall order polytopes, we must review some theory of P-partitions and some generalizations. For additional background on P-partitions beyond that reviewed here, the reader should consult [42]. Let $P = ([n], \preceq)$ be a labeled poset. Then the set

$$\mathcal{L}(P) = \{\pi \in \mathfrak{S}_n \,:\, \text{if } \pi_i \preceq \pi_j, \text{ then } i \leq j, \text{ for all } i, j \in [n]\}$$

are the *linear extensions* of P or the *Jordan-Hölder set* of P. If $O(P)$ denotes the *order polytope* of P, then a classical result of Stanley [39] says that

$$h^*(O(P); z) = \sum_{\pi \in \mathcal{L}(P)} z^{\operatorname{des}(\pi)}.$$

A similar result holds in the case of $\boldsymbol{s}$-lecture hall order polytopes. Given $\boldsymbol{s} : [n] \to \mathbb{Z}_{\geq 1}$, an $\boldsymbol{s}$*-colored permutation* is a pair $\tau = (\pi, \boldsymbol{r})$, where $\pi \in \mathfrak{S}_n$ and $\boldsymbol{r} : [n] \to \mathbb{Z}_{\geq 0}$ such that $\boldsymbol{r}(\pi_i) \in \{0, 1, \ldots, \boldsymbol{s}(\pi_i) - 1\}$ for all $1 \leq i \leq n$. Now, given a labeled poset P,

$$\mathcal{L}(P, \boldsymbol{s}) := \{\tau \,:\, \tau = (\pi, \boldsymbol{r}) \text{ such that } \pi \in \mathfrak{S}_n \text{ and } \tau \text{ is an } \boldsymbol{s}\text{-colored permutation}\}.$$

Given $\tau = (\pi, \boldsymbol{r})$, we say that $i \in [n-1]$ is a *descent* of τ if

$$\begin{cases} \pi_i < \pi_{i+1} \text{ and } \boldsymbol{r}(\pi_i)/\boldsymbol{s}(\pi_i) > \boldsymbol{r}(\pi_{i+1})/\boldsymbol{s}(\pi_{i+1}), \text{ or} \\ \pi_i > \pi_{i+1} \text{ and } \boldsymbol{r}(\pi_i)/\boldsymbol{s}(\pi_i) \geq \boldsymbol{r}(\pi_{i+1})/\boldsymbol{s}(\pi_{i+1}) \end{cases}$$

and we let

$$D_1(\tau) := \{i \in [n-1] \,:\, i \text{ is a descent}\}.$$

Moreover, let

$$D(\tau) := \begin{cases} D_1(\tau), & \text{if } \boldsymbol{r}(\pi_n) = 0 \\ D_1(\tau) \cup \{n\}, & \text{if } \boldsymbol{r}(\pi_n) \neq 0 \end{cases}$$

Given $\tau \in \mathcal{L}(P, \boldsymbol{s})$, let $\operatorname{des}_{\boldsymbol{s}}(\tau) = |D(\tau)|$. We can now state the following Ehrhart series result for $\boldsymbol{s}$-lecture hall order polytopes.

Theorem 5.2 ([8, Corollary 3.7]). *Let P be a labeled poset and $\boldsymbol{s} : [n] \to \mathbb{Z}_{\geq 1}$. Then*

$$h^*(O(P, \boldsymbol{s}); z) = \sum_{\tau \in \mathcal{L}(P, \boldsymbol{s})} z^{\operatorname{des}_{\boldsymbol{s}}(\tau)}.$$

We should note that this simultaneously generalizes both the known Ehrhart series results for order polytopes and $\boldsymbol{s}$-lecture hall simplices. Specifically, if $\boldsymbol{s}(i) = 1$ for all i, then we recover the standard order polytope $O(P)$ and the Ehrhart series given by Stanley. Given any choice of $\boldsymbol{s}$, if P is chosen to be the totally ordered chain, we recover $\mathbf{P}_n^{(\boldsymbol{s})}$ and an alternative description of its h^*-polynomial.

When assuming certain conditions on $(P, \boldsymbol{s})$, one can make palindromic statements. We say that a poset $P = ([n], \preceq)$ is *graded* if it is equipped with a rank function $\rho : [n] \to \mathbb{Z}$ that has the properties that $\rho(x) < \rho(y)$ whenever $x \prec y$ and $\rho(y) = \rho(x) + 1$ if y covers x in P. We can now state the following theorem of Brändén and Leander for such posets.

Theorem 5.3 ([8]). *Suppose that $P = ([n], \preceq)$ is a naturally-labeled, graded poset with nonnegative rank function ρ and $\boldsymbol{s} = \rho + 1$. Then $h^*(O(P, \boldsymbol{s}); z)$ is a palindromic polynomial of degree $n-1$. Equivalently, $O(P, \boldsymbol{s})$ is a Gorenstein polytope of index 2.*

5.1. *Unimodality and real-rootedness*

Given that the h^*-polynomials for $O(P, \boldsymbol{s})$ can be combinatorially described, it is natural to consider when these polynomials are unimodal. Additionally, since all of $O(P, \boldsymbol{s})$ are known to have the IDP and since there is a natural family of Gorenstein $O(P, \boldsymbol{s})$ arising from sign-ranked posets, unimodality questions are especially of interest in terms of the Hibi and Ohsugi Conjecture (see Conjecture 2.1) and the question of Schepers and Van Langenhoven (See Question 2.1).

A powerful technique for showing unimodality of polynomials with positive coefficients is showing real-rootedness. The following two results of Brändén and Leander show real-rootedness in certain cases.

Theorem 5.4 ([8]). *Let P, Q be labeled posets on $[p]$ and $[q]$ respectively and let $\boldsymbol{s}_P : [p] \to \mathbb{Z}_{\geq 1}$ and $\boldsymbol{s}_Q : [q] \to \mathbb{Z}_{\geq 1}$. Let $P \sqcup Q$ denote the disjoint union of P and Q and let $\boldsymbol{s}_{P\sqcup Q}$ be the unique function on $P \sqcup Q$ which agrees with $\boldsymbol{s}_P$ and $\boldsymbol{s}_Q$ on the components P and Q respectively. If $h^*(O(P, \boldsymbol{s}_P); z)$ and $h^*(O(Q, \boldsymbol{s}_Q); z)$ are real-rooted, then $h^*(O(P \sqcup Q, \boldsymbol{s}_{P\sqcup Q}); z)$ is real-rooted.*

Theorem 5.5 ([8]). *Let $P = A_{p_1} \oplus \cdots \oplus A_{p_m}$ be an ordinal sum of anti-chains and let $\boldsymbol{s} : P \to \mathbb{Z}_{\geq 1}$ be a function which is constant on A_{p_i} for $i \leq i \leq m$. Then $h^*(O(P, \boldsymbol{s}); z)$ is real-rooted.*

The first theorem follows quickly as it can be shown that

$$h^*(O(P \sqcup Q, \boldsymbol{s}_{P\sqcup Q}; z) = h^*(O(P, \boldsymbol{s}_P); z) \cdot h^*(O(Q, \boldsymbol{s}_Q); z).$$

The second theorem requires more sophisticated techniques. In particular, they show that $h^*(O(P, \boldsymbol{s}); z)$ can be expressed as a sum of related polynomials which are an *interlacing sequence.* We should note that this is similar to the method used by Savage and Visontai [35] to prove Theorem 4.2. The reader should consult [7] for background and details on this method of proving real-rootedness.

In the special case that P is a naturally-labeled poset with a unique minimal element and nonnegative rank function and where we let $\boldsymbol{s} = \rho+1$, which is a subcase of that covered by Theorem 5.3, unimodality can be shown using triangulation results. In particular, the following theorem of Gustafsson and Solus follows with some work from Theorem 5.7 in the next subsection.

Theorem 5.6 ([15]). *Suppose that P is a naturally-labeled, graded poset with a unique minimal element, a nonnegative rank function ρ, and $\boldsymbol{s} = \rho + 1$. Then $h^*(O(P, \boldsymbol{s}); z)$ is unimodal.*

5.2. *Box unimodal triangulations and local h^*-polynomials*

To conclude our discussion of results on $\boldsymbol{s}$-lecture hall order polytopes, we will consider work regarding box unimodal triangulations and local Ehrhart theory. This work is due entirely to Gustafsson and Solus [15]. The *$\boldsymbol{s}$-canonical triangulation* of $O(P,\boldsymbol{s})$ is the analogue to the canonical triangulation of the order polytope $O(P)$ introduced by Stanley [39] with the primary difference being that rather than decomposing the polytope into unimodular simplicies indexed by linear extensions of P, $O(P,\boldsymbol{s})$ is decomposed into $\boldsymbol{s}$-lecture hall simplices, which are in general not unimodular, in accordance with linear extensions of P. We should note that both the canonical and $\boldsymbol{s}$-canonical triangulations are regular. We begin with a theorem on the existence of box unimodal triangulations.

Theorem 5.7 ([15, Theorem 5.6]). *Let $O(P,\boldsymbol{s})$ be an $\boldsymbol{s}$-lecture hall order polytope. Then the $\boldsymbol{s}$-canonical triangulation of $O(P,\boldsymbol{s})$ is box unimodal. Moreover, if $O(P,\boldsymbol{s})$ is reflexive, then the h^*-polynomial $h^*(O(P,\boldsymbol{s});z)$ is unimodal.*

This theorem combined with Theorem 5.1 provides further support of Conjecture 2.1. While this statement does not guarantee the existence of a regular, unimodular triangulation, which is the case in the order polytope case, it is nevertheless a very strong statement which gives some evidence for h^*-polynomial unimodality results in generality.

Given that the local h^*-polynomials of $\boldsymbol{s}$-lecture hall simplices are both real-rooted and γ-positive, it is a natural extension to consider such properties of the local h^*-polynomials of $O(P,\boldsymbol{s})$. In general, these strong conditions cannot currently be verified. However, Theorem 5.7 can be leveraged, along with results of Katz and Stapledon [23], to show the following unimodality result.

Theorem 5.8 ([15, Theorem 5.9]). *The local h^*-polynomial of an $\boldsymbol{s}$-lecture hall order polytope is unimodal.*

6. Open problems

We conclude this survey with a collection of conjectures and questions which have accumulated in the literature, so that the majority of open problems regarding the polyhedral geometry of lecture hall partitions appear in a single location. We will follow the order of the survey by first addressing open problems for $\boldsymbol{s}$-lecture hall cones, then $\boldsymbol{s}$-lecture hall simplices, and culminating with $\boldsymbol{s}$-lecture hall order polytopes.

6.1. *Lecture hall cones*

Given that the Gorenstein property for $\mathcal{C}_n^{(\boldsymbol{s})}$ has been completely characterized, a natural extension is to consider the level property. This generates the following question:

Question 6.1. Given $\boldsymbol{s} \in \mathbb{Z}^n_{\geq 1}$, what conditions on $\boldsymbol{s}$ guarantee that $\mathcal{C}_n^{(\boldsymbol{s})}$ is level?

For this question, there is a notable obstruction; there is no canonical choice of grading for $\mathcal{C}_n^{(\boldsymbol{s})}$. Choices for gradings which have been used in the literature include the standard "height" grading used in the polytope case, as well as the difference of last two coordinates which was used heavily in the case of $\mathcal{C}_n^{(1,2,\ldots,n)}$. After choosing a grading, it seems plausible that one may be able to adapt the strategy of Kohl and the author in the $\mathbf{P}_n^{(\boldsymbol{s})}$ case [24], though one would need to understand the Hilbert basis for any given cone considered.

Additionally, there are many potential extensions of the Hilbert basis classifications. This leads to the following problem and question.

Problem 6.1. *Identify the Hilbert basis elements of additional combinatorial families of lecture hall cones.*

Question 6.2. Is there a reasonable description of the Hilbert bases of $\boldsymbol{u}$-generated Gorenstein lecture hall cones of dimension 5?

For Problem 6.1, a possible candidate would be to consider the (k, ℓ)-sequences, or even a special case such as the $(4, 1)$-sequences. The reader should consult [33] for an overview of these sequences and the deep mathematics behind them. Regarding Question 6.2, the Hilbert bases for $\boldsymbol{u}$-generated Gorenstein lecture hall cones have been classified for dimension ≤ 4. Continuing the classification to higher dimensions should certainly be possible, though there are likely many subcases and enumeration of the cardinality of the Hilbert bases may be significantly more challenging.

For the special case of $\mathcal{C}_n^{(1,2,\ldots,n)}$ where a triangulation of the cone is known, the following conjecture stated by Beck et al. would provide additional geometric information about the cone.

Conjecture 6.1 ([2]). *There exists a regular, flag, unimodular triangulation of* $\mathscr{R}_n$ *that admits a shelling order such that the maximal simplices of the triangulation are indexed by* $\pi \in \mathfrak{S}_{n-1}$ *and each such simplex is attached along* $\operatorname{des}(\pi)$ *many of its facets.*

6.2. *Lecture hall simplices*

While many results are known for the $\boldsymbol{s}$-lecture hall simplices, there are several open questions in well-studied areas such as triangulations and the levelness property. To begin, we consider a conjecture stated by Hibi, Tsuchiya, and the author.

Conjecture 6.2 ([22]). *Let* $\boldsymbol{s} \in \mathbb{Z}^n_{\geq 1}$ *be given. Then* $\mathbf{P}_n^{(\boldsymbol{s})}$ *has a regular, unimodular triangulation.*

This has been shown in very particular cases as detailed in the survey. However, there is no known counterexample in other cases. Verifying this conjecture is also crucial for showing results regarding the s-canonical triangulation of s-lecture hall order polytopes.

While the level property is completely classified in terms of s-inversion sequences, the characterization is often difficult to apply. Subsequently, it is of interest to determine alternative characterizations for restricted families of s-sequences or to determine conditions which are sufficient though perhaps not necessary for levelness. This motivates the following question and conjecture.

Question 6.3. Is there an alternative characterization of levelness if s is a strictly monotone sequence?

Conjecture 6.3 ([24]). *Let $s \in \mathbb{Z}^n_{\geq 1}$ be a sequence such that there exists some $c \in \mathbb{Z}^n$ satisfying*

$$c_j s_{j-1} = c_{j-1} s_j + \gcd(s_{j-1}, s_j)$$

for $j > 1$ with $c_1 = 1$. Then $\mathbf{P}_n^{(s)}$ is level.

While there are some results on the Ehrhart positivity for s-lecture hall simplices, the property is largely unstudied. This motivates the following question.

Question 6.4. What conditions on s ensure that $\mathbf{P}_n^{(s)}$ is Ehrhart positive? What conditions on s, besides those in Theorem 4.3, ensure that $\mathbf{P}_n^{(s)}$ is not Ehrhart positive?

One property which has yet to be studied regarding the s-lecture hall simplices is *reflexive dimension.* Given a polytope $\mathcal{P}$, the *reflexive dimension* of $\mathcal{P}$ is the smallest m such that $\mathcal{P}$ is the face of an m-dimensional reflexive polytope. By a result of Haase and Melnikov [17], this is known to be finite for any $\mathcal{P}$. This yields the following question.

Question 6.5. Let $s \in \mathbb{Z}^n_{\geq 1}$ be given. What is the reflexive dimension of $\mathbf{P}_n^{(s)}$?

6.3. *Lecture hall order polytopes*

We begin the discussion of open problems for s-lecture hall polytopes by mentioning two conjectures of Brändén and Leander [8] in the seminal paper.

Conjecture 6.4 ([8]). *Suppose P is a naturally-labeled, graded poset with nonnegative rank function ρ and $s = \rho + 1$, then $h^*(O(P, s); z)$ is γ-nonnegative.*

Conjecture 6.5 ([8]). *Suppose that P is a naturally-labeled, graded poset with nonnegative rank function ρ and $s = \rho + 1$, then $O(P, s)$ has a regular, unimodular triangulation.*

Both of these conjectures are primarily of interest because they imply that $h^*(O(P); z)$ is unimodal. Gustafsson and Solus [15] proved a slightly weaker unimodality result to that of Conjecture 6.4 in the case that P has a unique minimal element as discussed previously. However, this question is still open in the general setting. Conjecture 5.7 is also open in generality. In fact, this is only known to hold in very specific cases, such as the case where P is the totally ordered chain with minimal element x_0 and $\rho(x_0) \in \{0, 1\}$ where it overlaps with known results for lecture hall simplices.

We conclude with three general questions regarding $\boldsymbol{s}$-lecture hall order polytopes.

Question 6.6. Let $P = ([n], \preceq)$ be a labeled poset and $\boldsymbol{s} : [n] \to \mathbb{Z}_{\geq 1}$. Can we determine conditons of P and $\boldsymbol{s}$ different than those given in Theorem 5.3 which guarantee that $O(P, \boldsymbol{s})$ is Gorenstein?

Question 6.7. Let $P = ([n], \preceq)$ be a labeled poset and $\boldsymbol{s} : [n] \to \mathbb{Z}_{\geq 1}$. What conditions on P and $\boldsymbol{s}$ guarantee that $O(P, \boldsymbol{s})$ is level?

Question 6.8. Let $P = ([n], \preceq)$ be a labeled poset and $\boldsymbol{s} : [n] \to \mathbb{Z}_{\geq 1}$. When does $O(P, \boldsymbol{s})$ admit a regular, unimodular triangulation?

Regarding Questions 6.6 and 6.7, characterizations for the Gorenstein and level properties are known for $\boldsymbol{s}$-lecture hall simplices as discussed in Section 4, as well as for order polytopes (see [19] for the Gorenstein property and see [16, 28] for levelness). It seems reasonable that one may be able to adapt the methods and techniques to find suitable characterizations in this context. One possible approach to Question 6.8 is to first consider the case of lecture hall simplices and see if the $\boldsymbol{s}$-canonical triangulation can be exploited.

References

[1] Matthias Beck, Benjamin Braun, Matthias Köppe, Carla D. Savage, and Zafeirakis Zafeirakopoulos. $\boldsymbol{s}$-lecture hall partitions, self-reciprocal polynomials, and Gorenstein cones. *Ramanujan J.*, 36(1-2):123–147, 2015.

[2] Matthias Beck, Benjamin Braun, Matthias Köppe, Carla D. Savage, and Zafeirakis Zafeirakopoulos. Generating functions and triangulations for lecture hall cones. *SIAM J. Discrete Math.*, 30(3):1470–1479, 2016.

[3] Matthias Beck and Sinai Robins. *Computing the continuous discretely: Integer-point enumeration in polyhedra.* Undergraduate Texts in Mathematics. Springer, New York, 2007.

[4] U. Betke and P. McMullen. Lattice points in lattice polytopes. *Monatsh. Math.*, 99(4):253–265, 1985.

[5] Mireille Bousquet-Mélou and Kimmo Eriksson. Lecture hall partitions. *Ramanujan J.*, 1(1):101–111, 1997.

[6] Mireille Bousquet-Mélou and Kimmo Eriksson. Lecture hall partitions. II. *Ramanujan J.*, 1(2):165–185, 1997.

[7] Petter Brändén. Unimodality, log-concavity, real-rootedness and beyond. In *Handbook of enumerative combinatorics*, Discrete Math. Appl. (Boca Raton), pages 437–483. CRC Press, Boca Raton, FL, 2015.

[8] Petter Brändén and Madeleine Leander. Lecture hall p-paritions. *ArXiv eprints,* 2016. arxiv:1609.02790.

[9] Petter Brändén and Liam Solus. *Private Communication.*
[10] Winfried Bruns and Jürgen Herzog. *Cohen-Macaulay rings*, volume 39 of *Cambridge Studies in Advanced Mathematics.* Cambridge University Press, Cambridge, 1993.
[11] Winfried Bruns and Tim Römer. h-vectors of Gorenstein polytopes. *J. Combin. Theory Ser. A*, 114(1):65–76, 2007.
[12] Sylvie Corteel and Jang Soo Kim. Lecture hall tableaux. *ArXiv eprints,* 2018. arXiv:1804.02489.
[13] Emanuela De Negri and Takayuki Hibi. Gorenstein algebras of Veronese type. *J. Algebra,* 193(2):629–639, 1997.
[14] Eugène Ehrhart. Sur les polyèdres homothétiques bordés à n dimensions. *C. R. Acad. Sci. Paris*, 254:988–990, 1962.
[15] Nils Gustafsson and Liam Solus. Derangements, Ehrhart theory, and local h-polynomials. *ArXiv eprints*, 2018. arXiv:1807.05246.
[16] Christian Haase, Florian Kohl, and Akiyoshi Tsuchiya. Levelness of order polytopes. *ArXiv eprints,* 2018. arXiv:1805.02967.
[17] Christian Haase and Ilarion V. Melnikov. The reflexive dimension of a lattice polytope. *Ann. Comb.*, 10(2):211–217, 2006.
[18] Christian Haase, Andreas Paffenholz, Lindsay C. Piechnik, and Francisco Santos. Existence of unimodular triangulations–positive results. *ArXiv eprints,* 2014. arXiv:1405.1687.
[19] Takayuki Hibi. Distributive lattices, affne semigroup rings and algebras with straightening laws. In *Commutative algebra and combinatorics (Kyoto, 1985)*, volume 11 of *Adv. Stud. Pure Math.*, pages 93–109. North-Holland, Amsterdam, 1987.
[20] Takayuki Hibi. Dual polytopes of rational convex polytopes. *Combinatorica,* 12(2):237–240, 1992.
[21] Takayuki Hibi, McCabe Olsen, and Akiyoshi Tsuchiya. Self dual reflexive simplices with Eulerian polynomials. *Graphs Combin.* 33(6): 1401–1404, 2017.
[22] Takayuki Hibi, McCabe Olsen, and Akiyoshi Tsuchiya. Gorenstein properties and integer decomposition properties of lecture hall polytopes. *Mosc. Math. J.* 18(4):667–679, 2018.
[23] Eric Katz and Alan Stapledon. Local h-polynomials, invariants of subdivisions, and mixed Ehrhart theory. *Adv. Math.*, 286:181–239, 2016.
[24] Florian Kohl and McCabe Olsen. Level algebras and lecture hall polytopes. *ArXiv eprints*, 2017. arXiv:1710.10892.
[25] Fu Liu. On positivity of Ehrhart polynomials. *ArXiv eprints*, 2017. arxiv:1711.09962
[26] Fu Liu and Liam Solus. On the Relationship Between Ehrhart Unimodality and Ehrhart Positivity. *ArXiv eprints*, 2018. arXiv:1804:08258.
[27] Ezra Miller and Bernd Sturmfels. *Combinatorial commutative algebra*, volume 227 of *Graduate Texts in Mathematics.* Springer-Verlag, New York, 2005.
[28] Mitsuhiro Miyazaki. On the generators of the canonical module of a Hibi ring: a criterion of level property and the degrees of generators. *J. Algebra*, 480:215–236, 2017.
[29] Hidefumi Ohsugi and Takayuki Hibi. Special simplices and Gorenstein toric rings. *J. Combin. Theory Ser. A*, 113(4):718–725, 2006.
[30] McCabe Olsen. Hilbert bases and lecture hall partitions. *Ramaujan J.*, To Appear.
[31] Sam Payne. Ehrhart series and lattice triangulations. *Discrete Comput. Geom.*, 40(3):365–376, 2008.
[32] Thomas W. Pensyl and Carla D. Savage. Rational lecture hall polytopes and infl ated Eulerian polynomials. *Ramanujan J.*, 31(1-2):97–114, 2013.
[33] Carla D. Savage. The mathematics of lecture hall partitions. *J. Combin. Theory Ser. A*, 144:443–475, 2016.

[34] Carla D. Savage and Michael J. Schuster. Ehrhart series of lecture hall polytopes and Eulerian polynomials for inversion sequences. *J. Combin. Theory Ser. A*, 119(4):850–870, 2012.

[35] Carla D. Savage and Mirkó Visontai. The $\boldsymbol{s}$-Eulerian polynomials have only real roots. *Trans. Amer. Math. Soc.*, 367(2):1441–1466, 2015.

[36] Carla D. Savage and Gopal Viswanathan. The $1/k$-Eulerian polynomials. *Electron. J. Combin.*, 19(1):Paper 9, 21, 2012.

[37] Jan Schepers and Leen Van Langenhoven. Unimodality questions for integrally closed lattice polytopes. *Ann. Comb.*, 17(3):571–589, 2013.

[38] Richard P. Stanley. Hilbert functions of graded algebras. *Advances in Math.*, 28(1):57–83, 1978.

[39] Richard P. Stanley. Two poset polytopes. *Discrete Comput. Geom.*, 1(1):9–23, 1986.

[40] Richard P. Stanley. Subdivisions and local h-vectors. *J. Amer. Math. Soc.*, 5(4):805–851, 1992.

[41] Richard P. Stanley. *Combinatorics and commutative algebra*, volume 41 of *Progress in Mathematics*. Birkhauser Boston, Inc., Boston, MA, second edition, 1996.

[42] Richard P. Stanley. *Enumerative combinatorics*. Volume 1, volume 49 of *Cambridge Studies in Advanced Mathematics*. Cambridge University Press, Cambridge, second edition, 2012.

A note on deformations and mutations of fake weighted projective planes

Irem Portakal
Mathematics Department, Otto-von-Guericke University
Magdeburg, Germany
E-mail: irem.portakal@ovgu.de

It has been shown by Hacking and Prokhorov that if the projective surface X with quotient singularities and self-intersection number 9 has a smoothing to the projective plane, then X is the general fiber of a $\mathbb{Q}$-Gorenstein deformation of the weighted projective plane with weights giving solutions to the Markov equation. This result has been understood and generalized by combinatorial mutations of Fano triangles by Akhtar, Coates, Galkin, and Kasprzyk. In this note, we study this result by utilizing polarized T-varieties and describe the associated deformation explicitly in terms of certain Minkowski summands of so-called divisorial polytopes.

Keywords: Toric degenerations, Markov equation, Minkowski summands, weighted projective plane, T-variety, $\mathbb{Q}$-Gorenstein deformation, combinatorial mutation, divisorial polytope.

1. Introduction

In their paper [8], Hacking and Prokhorov generalize Manetti's classification of surfaces X with self intersection number $K_X \geq 5$. They classify all surfaces which admit a $\mathbb{Q}$-Gorenstein smoothing. As a corollary to their theorem, they show that if X is a projective surface with quotient singularities and with $K_X^2 = 9$ which admits a smoothing to the projective plane, then X is a $\mathbb{Q}$-Gorenstein deformation of the weighted projective plane $\mathbb{P}(a^2, b^2, c^2)$ where (a, b, c) is a solution for the Markov equation, i.e. $a^2 + b^2 + c^2 = 3abc$ (Theorem 1.1). The Markov equation also appears in the context of derived categories. The solutions of the equation are in one-to-one correspondence with the exceptional bundles on the projective plane with slopes in the interval $[0, 1/2]$ ([13]).

It turns out that we can understand this smoothing in terms of deformations of T-varieties. A normal variety X is called a complexity-d T-variety if it admits an effective algebraic torus action $T \cong (\mathbb{C}^*)^n$ such that $\dim(X) - \dim(T) = d$. The deformations of T-varieties have been studied by Altmann in [3] and by Ilten and Vollmert in [10]. The key idea of the theory is to associate certain Minkowski decompositions to deformations. In this note, we consider the deformations of projective toric surfaces by interpretating them as complexity-one polarized T-varieties. To a polarized T-variety, one can associate a so-called divisorial polytope consisting of a lattice polytope and piecewise affine concave function. In [9], a technique has been introduced to deform these varieties by using admissible Minkowski summands of the associated piecewise affine concave function. The latest improvement on $\mathbb{Q}$-Gorenstein smoothings has been done combinatorially by Akhtar, Coates, Galkin and, Kasprzyk in [1] in terms of combinatorial mutations of Fano polytopes. This construction generalizes the result of Hacking and Prokhorov to fake projective planes satisfying a certain type of Diophantine equation. In fact, for this combinatorial mutation, one can relate a birational transformation. Let f be a Laurent polynomial in n variables. A mutation of f is a birational transformation $\varphi\colon (\mathbb{C}^*)^n \dashrightarrow (\mathbb{C}^*)^n$ such that it preserves the logarithmic volume form of the algebraic torus $(\mathbb{C}^*)^n$ and $\varphi^* f$ is again a Laurent polynomial. In [11], Ilten shows that for given such a mutation, there exists a flat projective family over $\mathbb{P}^1$ such that zero fiber is $\mathrm{TV}(\triangle(f))$ and the infinity fiber is $\mathrm{TV}(\triangle(\varphi^* f))$. The proof involves the mentioned language of affine T-varieties via Minkowski decompositions by taking the cone over these projective varieties.

It is natural to try to relate the combinatorial mutations to deformations of projective varieties using lattice polytopes and T-varieties. Indeed, we associate a one-parameter deformation to a mutation of Fano triangle by utilizing divisorial polytopes. It gives an explicit exposition of the Minkowski summands of the associated deformation. In [6], Baytrev proposed a method to construct Landau-Ginzburg model mirror dual to a smooth Fano variety X as a Laurent polynomial f such that the Newton polytope $\mathrm{Newt}(f)$ is the fan polytope of the small toric degeneration of X. The choice of the Laurent polynomial f is not unique, however one can always transform f to another Laurent polynomial via mutations which is also mirror dual to X. Although the notion of small degenerations is restrictive, this model has been extended to any $\mathbb{Q}$-Gorenstein degeneration of the complex projective plane in [7]. In particular, this is the case of Prokhorov-Hacking's deformations and it is another reason that we focus our attention into two dimensional case.

Remark that by toric degenerations we mean that the variety degenerates to a toric variety as one can guess. We say that X degenerates to a toric variety X_0 if there exists a flat morphism over a smooth curve germ such that the general fiber is isomorphic to X and the zero fiber is isomorphic to the toric variety X_0. This may be also expressed as the deformation of X_0 over a smooth curve germ.

Definition 1.1. Let X be a normal surface with quotient singularities and Picard rank 1. We say that X is a $\mathbb{Q}$-Gorenstein deformation of X_0 if there exists a deformation $\mathcal{X}$ over a smooth curve germ such that X_t is isomorphic to X for all $t \neq 0$ and $K_{\mathcal{X}}$ is $\mathbb{Q}$-Cartier. If X is smooth, the deformation is called $\mathbb{Q}$-Gorenstein smoothing.

Prokhorov and Hacking prove the following result.

Theorem 1.1 ([8], Corollary 1.2). *Let X be a projective surface with quotient singularities and $K_X^2 = 9$. If X admits a smoothing to the complex projective plane,* *then X is a $\mathbb{Q}$-Gorenstein deformation of $\mathbb{P}(a^2, b^2, c^2)$ where $a^2 + b^2 + c^2 = 3abc$.*

We denote the one-parameter subgroups of an algebraic torus $T \cong (\mathbb{C}^*)^n$ by N and the characters of T by M. There is a natural bilinear pairing between these two lattices which is the usual dot product

$$\langle \bullet, \bullet \rangle \colon M \times N \to \mathbb{Z}.$$

We let $N_{\mathbb{Q}} := N \otimes_{\mathbb{Z}} \mathbb{Q}$ and $M_{\mathbb{Q}} := M \otimes_{\mathbb{Z}} \mathbb{Q}$ be the corresponding vector spaces to the lattices N and M.

2. Mutations of Fano polytopes

The notion of mutations of lattice polytopes has been introduced by Akhtar, Galkin, Coates, and Kaspryzk in their paper [1]. This construction is motivated by certain birational transformations of Laurent polynomials which have been studied in Galkin and Usnich's paper [7]. Let $f \in \mathbb{C}[x_1^{\pm}, \dots, x_n^{\pm}]$ be a Laurent polynomial. The period of f is defined as

$$\pi_f(t) := \left(\frac{1}{2\pi i}\right)^n \int_{|x_1|=\dots|x_n|=1} \frac{1}{1 - tf(x_1, \dots x_n)} \frac{dx_1}{x_1} \cdots \frac{dx_n}{x_n}$$

where $t \in \mathbb{C}$ and $|t| \ll \infty$. A mutation of f, as in Definition 7 of [7], is a birational transformation $\varphi \in \mathrm{Aut}(\mathbb{C}(x_1, \dots, x_n))$ preserving the period π_f such that $\varphi(f)$ is again a Laurent polynomial. We are in particular interested in the following mutation.

Example 2.1. Let $g \in \mathbb{C}[x_2^{\pm}, \dots, x_n^{\pm}]$ be a Laurent polynomial. The birational transformation $\varphi : (x_1, \dots, x_n) \mapsto (x_1/g, x_2, \dots, x_n)$ is a mutation of f if and only if f can be written as $\sum_{i=k}^{l} f_i x_1^i$ where $f_i \in \mathbb{C}[x_2^{\pm}, \dots, x_n^{\pm}]$ such that for $i > 0$, $f_i/g^i \in \mathbb{C}[x_2^{\pm}, \dots, x_n^{\pm}]$.

Mutations of Laurent polynomials induce transformations of their associated Newton polytopes, which we call *combinatorial mutations.* Moreover, if $0 \in \mathrm{Newt}(f)$, then above-presented type of mutation of f has been associated to the deformation of the toric variety of $\mathrm{Newt}(f)$. Note that in this case, we require that $k < 0$ and $l > 0$. We will explain this relation in Section 4.

2.1. *Combinatorial mutations*

Let $P \subset N_{\mathbb{Q}}$ be a full dimensional convex lattice polytope. We say P is *Fano*, if the origin lies in the strict interior of P and the vertices of P are primitive lattice points. The dual of P is defined as

$$P^* = \{u \in M_{\mathbb{Q}} \mid \langle u, v \rangle \geq -1, \ \forall v \in P\} \subset M_{\mathbb{Q}}.$$

For the construction of combinatorial mutations, we follow [1, 2].

Let $w \in M$ be a primitive lattice vector. We set

$$h_{\min} := \min\{\langle w, v\rangle | v \in P\}, \quad h_{\max} := \max\{\langle w, v\rangle | v \in P\}.$$

Since P is a Fano polytope, we obtain that $h_{\min} < 0$ and $h_{\max} > 0$. Note that the lattice height function $w : N \to \mathbb{Z}$ naturally extends to $w_{\mathbb{Q}} : N_{\mathbb{Q}} \to \mathbb{Q}$.

Definition 2.1. We say that a lattice point $v \in P$ is at height m with respect to $w \in M$, if $\langle w, v\rangle = m$. For each height $h \in \mathbb{Z}$, we define a hyperplane

$$H_{w,h} := \{x \in N_{\mathbb{Q}} \mid \langle w, x\rangle = h\}.$$

We let $w_h(P) = \operatorname{conv}(H_{w,h} \cap P \cap N)$.

Definition 2.2. A factor of P with respect to w is a lattice polytope $F \subset N_{\mathbb{Q}}$ satisfying:

(1) $\langle w, v\rangle = 0$ for all $v \in F$.
(2) For every $h \in \mathbb{Z}$ such that $h_{\min} \leq h < 0$, there exists a (possibly empty) lattice polytope $G_h \subset N_{\mathbb{Q}}$ at height h such that

$$H_{w,h} \cap \operatorname{vert}(P) \subseteq G_h + (-h)F \subseteq w_h(P).$$

Remark 2.1. A factor does not need to exist for every $w \in M$. If it exists, then we are in a position to define a combinatorial mutation of P. For a given combinatorial mutation of P, there exists a Laurent polynomial f with its Newton polytope $\operatorname{Newt}(f)$ to be equal to P and the combinatorial mutation arises from an algebraic mutation.

Definition 2.3. The combinatorial mutation of P with respect to height function w, factor F, and lattice polytopes $\{G_h\}$ is defined as the convex lattice polytope

$$\operatorname{mut}_w(P, F; \{G_h\}) := \operatorname{conv}\left(\bigcup_{h=h_{\min}}^{-1} G_h \cup \bigcup_{h=0}^{h_{\max}} (w_h(P) + hF)\right) \subset N_{\mathbb{Q}}.$$

This definition might appear technical at first glance. Let us construct an explicit combinatorial mutation induced by the birational transformation in Example 2.1.

Example 2.2. Let $f = \sum_{i=k}^{l} f_i x_2^i \in \mathbb{C}[x_1^{\pm}, x_2^{\pm}]$ with $k < 0$ and $l > 0$, and let $f_i, g \in \mathbb{C}[x_1^{\pm}]$ be Laurent polynomials for $i \in [k, l]$. Suppose that $g^i \mid f_i$ for $i > 0$. Consider the mutation

$$\varphi \colon (x_1, x_2) \to (x_1, x_2/g).$$

The Laurent polynomial $\varphi^* f$ is the algebraic mutation of f and can be written as $\sum_{i=k}^{l}(f_i/g^i)x_2^i \in \mathbb{C}[x_1^{\pm}, x_2^{\pm}]$. Let $P := \text{Newt}(f) \subset N_{\mathbb{Q}}$ and $Q := \text{Newt}(\varphi^* f) \subset N_{\mathbb{Q}}$. We set the height function for the induced combinatorial mutation $w = [0,1] \in M$ and the factor with respect to w as $F = \text{conv}(0,(1,0)) \subset N_{\mathbb{Q}}$. First infer that, for $k \leq h \leq -1$, the lattice polytopes G_h are $\text{Newt}(f_h g^h x_2^h)$. Moreover, by the definition of the factor, we have the condition $\text{Newt}(f_h g^h x_2^h) + [0,h] \subseteq w_h(P)$. Hence, we obtain that $g = 1 + x_1$ and observe that $F = \text{Newt}(g)$.

In fact, a combinatorial mutation $P \subset N_{\mathbb{Q}}$ induces a piecewise linear transformation φ of $P^* \subset M_{\mathbb{Q}}$ such that $(\varphi(P^*))^* = \text{mut}_w(P, F; \{G_h\})$ and it is given by

$$u \mapsto u - u_{\min} w$$

where $u_{\min} = \min\{\langle u, v_F\rangle \mid v_F \in \text{vert}(F)\}$.

2.2. *Mutations of Fano triangles*

In this section, we bring our attention to the two dimensional case, more particularly to combinatorial mutations of Fano triangles. Let $P := \text{conv}(v_1, v_2, v_3) \subset N_{\mathbb{Q}}$ be a Fano triangle. Since the origin is contained in the interior of P, there exists a unique choice of coprime positive integers $(\lambda_1, \lambda_2, \lambda_3)$ such that $\lambda_1 v_1 + \lambda_2 v_2 + \lambda_3 v_3 = 0$. Let $N' \subset N$ be the sublattice generated by the lattice points v_i. The associated projective toric surface $\text{TV}(P)$ is the *fake projective plane* with weights $\lambda_1, \lambda_2, \lambda_3$, i.e. it is the quotient of $\mathbb{P}(\lambda_1, \lambda_2, \lambda_3)$ by the action of the finite group N/N' acting free in codimension one.

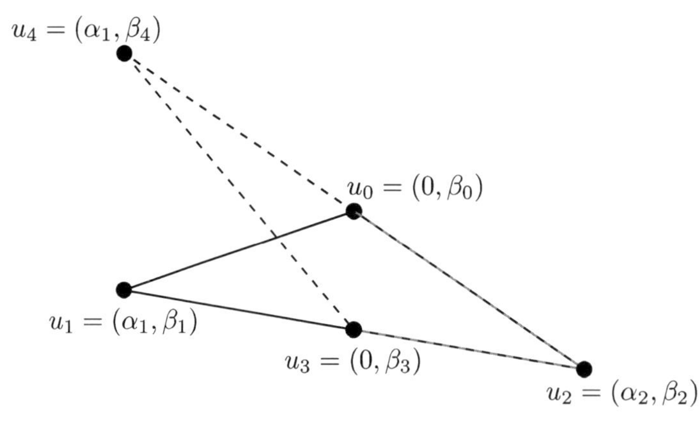

Fig. 1. Mutation Ψ of a dual of a Fano triangle in $M_{\mathbb{Q}}$

Proposition 2.1 ([2], Proposition 3.3). *Let $P \subset N_{\mathbb{Q}}$ be a Fano triangle and $TV(P)$ be the fake weighted projective plane with weights $(\lambda_1, \lambda_2, \lambda_3)$. Suppose that* *there exists a mutation of P as $Q = mut_w(P, F; \{G_h\})$ for some height function w and factor F such that $TV(Q)$ is a fake weighted projective plane. Then, up to relabelling, $\lambda_1 \mid (\lambda_2 + \lambda_3)^2$ and $TV(Q)$ has weights*

$$\left(\lambda_2, \lambda_3, \frac{(\lambda_2 + \lambda_3)^2}{\lambda_1}\right).$$

As a consequence to this proposition, one can show that the weights of $\mathrm{TV}(P)$ and $\mathrm{TV}(Q)$ belong to the solution set of the Diophantine equation

$$mx_1x_2x_3 = k(c_1x_1^2 + c_2x_2^2 + c_3x_3^2) \tag{1}$$

where $m, k, c_i \in \mathbb{Z}_{>0}$ and c_i is square free. Remark that a special case of this equation is the Markov equation which has been mentioned in the introduction.

Proposition 2.2 ([2], Proposition 3.12). *Let P and Q be Fano triangles as in Proposition 2.1. Then the weights of $TV(P)$ and $TV(Q)$ give solutions to the same Diophantine equation (1).*

In the next sections, we study how these mutations give rise to deformations and hence this previous result can be considered as a generalization of Theorem 1.1 by Hacking and Prokhorov in [8].

Example 2.3. Consider the combinatorial mutation Ψ of the triangle $P^* \subset M_{\mathbb{Q}}$ to the dashed triangle $Q^* \subset M_{\mathbb{Q}}$ in Figure 1. It is induced by the mutation of the type of Example 2.1 with $w = (0, 1) \in M$ and $F = \mathrm{conv}(0, (1, 0))$. In particular, if $\mathrm{TV}(P)$ has the weights (a^2, b^2, c^2), then $\mathrm{TV}(Q)$ has weights $(b^2, c^2, \frac{(b^2+c^2)^2}{a^2})$. Suppose now (a, b, c) is a solution to the Markov equation. Since the triple $(b, c, 3bc - a)$ is also a solution to Markov equation, we obtain

$$(b^2, c^2, \frac{(b^2 + c^2)^2}{a^2}) = (b^2, c^2, \frac{(3abc - a^2)^2}{a^2}) = (b^2, c^2, (3bc - a)^2).$$

and thus the weights of $\mathrm{TV}(Q)$ also give solution to the Markov equation.

3. Deformations of T-varieties

T-varieties naturally appear during the study of deformations of toric varieties preserving the torus action of the embedded torus. In this case, the total space of a deformation of a toric variety is not always toric, but they admit a lower dimensional effective torus action. The natural question to ask is if these varieties also admit a nice combinatorial description as in the toric case. A such combinatorial construction of T-varieties has been introduced in [4]. The affine T-varieties are in one-to-one correspondence with the so-called p-divisors. A p-divisor is a formal product of a set of polyhedra and divisors from the "good" quotient of the T-variety by the effective torus action. For a detailed read, one can refer to the survey on the language of T-varieties [5]. In this section, we restrict our attention to the combinatorial construction for polarized complexity-one T-varieties. Next, we study one-parameter deformations of these varieties in terms of admissible Minkowski decompositions.

3.1. *Polarized T-varieties and Divisorial Polytopes*

The correspondence between polarized toric varieties and lattice polytopes in toric geometry has been generalized to the complexity-one T-varieties case in terms of so-called divisorial polytopes in [12]. In this section, we present this construction and in particular we apply this construction explicitly to fake weighted projective planes.

Definition 3.1. A divisorial polytope $(\square, \Phi)$ consists of a lattice polytope $\square \subset M'_{\mathbb{Q}}$ and a piecewise linear concave function

$$\Phi = \sum_{P \in \mathbb{P}^1} \Phi_P.P \colon \square \to \mathrm{Div}_{\mathbb{Q}}\mathbb{P}^1$$

such that

(i) $\deg \Phi(u) > 0$ for u in the interior of $\square$.
(ii) $\deg \Phi(u) > 0$ or $\Phi(u) \sim 0$ for u a vertex of $\square$.
(iii) For all $P \in \mathbb{P}^1$, the graph of Φ_P has its vertices in $M' \times \mathbb{Z}$.

Theorem 3.1 ([12], Theorem 3.2). *There is a one-to-one correspondence between divisorial polytopes and complexity-one T-varieties $(X, \mathcal{L})$ where $\mathcal{L}$ is an equivariant ample line bundle.*

For more details of this construction, we refer the reader to [9, 12]. One can always consider a polarized toric variety with a one-dimension lower subtorus action. Let $\triangle \subset M_{\mathbb{Q}}$ be a lattice polytope in M. Then there exists the exact sequence of lattices

$$0 \to \mathbb{Z} \xrightarrow{\mathrm{F}} M \xrightarrow{\deg} M' \to 0.$$

We choose a section $s \colon M' \to M$, i.e. $\deg(s) = \mathrm{id}_{M'}$. Consider the map $\Phi_\triangle \colon \deg(\triangle) \to \mathrm{Div}_{\mathbb{Q}}(\mathbb{P}^1)$ given by

$$(\Phi_\triangle)_0(u) = \max\{x \in \mathbb{Q} \mid F_{\mathbb{Q}}(x) + s(u) \in \triangle \cap \deg_{\mathbb{Q}}^{-1}(u)\},$$

$$(\Phi_\triangle)_\infty(u) = -\min\{x \in \mathbb{Q} \mid F_{\mathbb{Q}}(x) + s(u) \in \triangle \cap \deg_{\mathbb{Q}}^{-1}(u)\}.$$

Then, $(\Phi_\triangle, \deg(\triangle))$ is a divisorial polytope with respect to the restricted effective torus action $T_{N'}$.

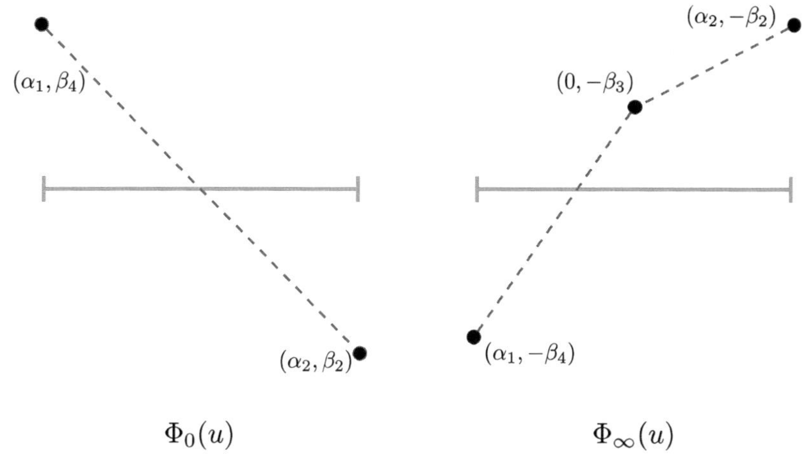

Fig. 2. The divisorial polytope associated to fake projective plane with weights $(\lambda_1, \lambda_2, \lambda_3)$

Example 3.1. Consider the dashed triangle $P^* \subset M_{\mathbb{Q}}$ from Figure 1. Set $\deg\colon M \xrightarrow{(1,0)} M'$ to be the restricted torus action and choose the section $s\colon M' \xrightarrow{(1,0)^\intercal} M$. Note that P^* does not need to be a lattice polytope. Let us suppose for the moment that P^* is lattice polytope. Then we obtain the lattice polytope $\deg(P^*) = \Box \subset M'_{\mathbb{Q}}$ to be the interval $[\alpha_1, \alpha_2]$ and the piecewise linear concave functions Φ_0 and Φ_∞ as in Figure 2.

3.2. *One-parameter deformations of polarized T-varieties*

A *deformation* of a projective algebraic variety X_0 is a flat map $\pi\colon \mathcal{X} \longrightarrow S$ with $0 \in S$ such that $\pi^{-1}(0) = X_0$. The variety $\mathcal{X}$ is called the total space and S is called the base space of the deformation. If S is a open subset of $\mathbb{A}^1$, we call π a one-parameter deformation. Since we consider polarized T-varieties, we will study embedded deformations in this section. Let $X_0 \hookrightarrow \mathbb{P}^N$ be a projectively embedded variety. An *embedded deformation* π of X_0 consists of $\mathcal{X} \hookrightarrow \mathbb{P}^N_S$ such that the projection to S is a deformation of X_0 and the embedding of $\mathcal{X}$ restricts to the embedding of X_0. Our aim is to construct certain embedded one-parameter deformations of polarized T-varieties in terms of Minkowski summands.

Definition 3.2. Let $(\Box, \Phi)$ be a divisorial polytope. An admissible one-parameter Minkowski decomposition of $(\Box, \Phi)$ consists of two piecewise affine functions $\Phi^0_P, \Phi^1_P\colon \Box \to \mathbb{Q}$ for some $P \in \mathrm{Div}_{\mathbb{Q}}\mathbb{P}^1$ such that:

(1) The graph of Φ^i_P has lattice vertices for $i = 0, 1$.
(2) $\Phi_P(u) = \Phi^0_P(u) + \Phi^1_P(u)$ for all $u \in \Box$.
(3) For any full-dimensional polytope $\triangle \subset \Box$ on which Φ_P is affine, Φ^i_P has non-integral slope for at most one $i \in \{0, 1\}$.

Given a one-parameter Minkowski decomposition of $(\square, \Phi)$, for any $s \in S$, define the divisorial polytope $\Phi^{(s)} : \square \to \mathrm{Div}_{\mathbb{Q}}\mathbb{P}^1$ by

$$\Phi^{(s)}(u) = \sum_{P' \neq P} \Phi_{P'}(u) \otimes \mathbb{V}(y_{P'}) + \Phi_P^0(u) \otimes \mathbb{V}(y_P) + \Phi_P^1(u) \otimes \mathbb{V}(y_P - s)$$

where $y_P \in \mathbb{C}(\mathbb{P}^1)$ is a rational function with its sole zero at P. Denote $X(\Phi)$ and $X(\Phi^{(s)})$ to be the associated polarized T-varieties to divisorial polytopes $(\square, \Phi)$ and $(\square, \Phi^{(s)})$.

Theorem 3.2 ([9], Theorem 7.3.2). *There exists a flat family of T-varieties $\{X(\Phi^{(s)})\}$ over S. Moreover, if Φ is very ample and gives a projectively normal embedding, this deformation can be realized as an embedded deformation of $X(\Phi^{(0)}) \cong X(\Phi)$.*

The deformations of above form are called *T-deformations*. These deformations admit the torus T_N of $X(\Phi^{(0)})$ acting on the total space $\mathcal{X}$ and (trivially) on the base space S. Moreover, the maps $\pi \colon \mathcal{X} \to S$ and $X_0 \hookrightarrow X$ are T_N-equivariant.

4. Comparison of mutations and deformations

We now recall the mutation of the type of Example 2.1. Suppose that Newt(f) contains the origin in its interior.

Theorem 4.1 ([11], Theorem 1.3). *There is a flat projective family $\pi \colon \mathcal{X} \to \mathbb{P}^1$ such that $\pi^{-1}(0) = TV(Newt(f))$ and $\pi^{-1}(\infty) = TV(Newt(\varphi^* f))$.*

This deformation is constructed by taking the affine cone over the projective toric varieties and applying the techniques for the deformations of affine complexity-one T-varieties. Note also that, as we have shown in Example 2.2, this mutation induces a combinatorial mutation. We utilize this combinatorial mutation to understand Proposition 2.2 in terms of divisorial polytopes.

Theorem 4.2. *For the combinatorial mutation Ψ, there exists an one-parameter embedded deformation of the polarized variety $TV(P)$ such that the general fiber is isomorphic to $TV(Q)$.*

Proof. Let us consider the mutation Ψ from the lattice polygon $P^* \subset M_{\mathbb{Q}}$ to the lattice polygon $Q^* \subset M_{\mathbb{Q}}$. A lattice polygon in $M_{\mathbb{Q}}$ is very ample. Since P^* is not necessarily a lattice polytope, we consider a dilation of P^*, i.e. the polarized toric variety $(\mathrm{TV}(P), \mathcal{L}_{aP^*})$ for some $a \in \mathbb{Z}_{>0}$. As we explained in Example 2.3, the height function is $w = (0, 1)$ and the factor is the line segment $F = \mathrm{conv}(0, (x, 0))$. Thus, we obtain the following piecewise linear transformation $v \mapsto v\Psi$ where $v = (v_1, v_2)$ and,

$$\Psi = \begin{cases} \begin{pmatrix} 1 & 0 \\ 0 & 1 \end{pmatrix} & \text{if } v_1 > 0 \\ \begin{pmatrix} 1 & x \\ 0 & 1 \end{pmatrix} & \text{otherwise.} \end{cases}$$

Next, we utilize the associated divisorial polytope of $\mathrm{TV}(P)$ from Example 3.1. The subtorus action is constructed by the map $M \xrightarrow{(1,0)} M'$. Without loss of generality, we assume that (u_0, u_1, u_2) are lattice points of the lattice polytope $aP^* \in M_{\mathbb{Q}}$. In Figure 2, we observe that the divisorial polytope with respect to the restricted subtorus action consists of the line segment $[\alpha_1, \alpha_2]$ and two piecewise linear functions Φ_0 and Φ_∞.

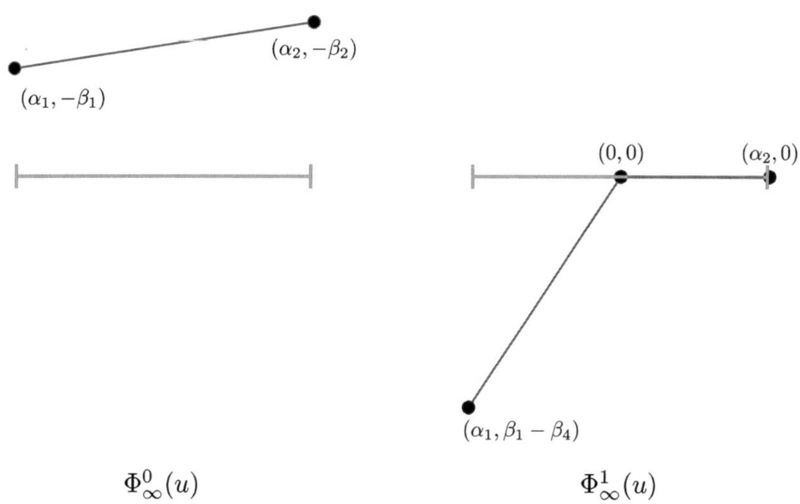

Fig. 3. An admissible one-parameter Minkowski decomposition of Φ_∞

In order to construct an embedded deformation of $\mathrm{TV}(P)$, we examine the following admissible one-parameter Minkowski decomposition of $\Phi_\infty(u) = \Phi^0_\infty(u) + \Phi^1_\infty(u)$ as in Figure 3. This decomposition is indeed admissible, since we obtain an integral slope for Φ^1_∞ on interval $[\alpha_1, 0]$:

$$\frac{\beta_1 - \beta_4}{\alpha_1} = \frac{\beta_1 - \alpha_1 x - \beta_1}{\alpha_1} = -x.$$

Since we would like to obtain a toric variety as the general fiber of this deformation, the associated divisorial polytope $\Phi^{(s)}$ may have at most two nontrivial coefficient. Recall that the general fiber of this deformation is

$$\Phi^{(s)}(u) = \Phi_0(u) \otimes \{0\} + \Phi^0_\infty(u) \otimes \mathbb{V}(y_\infty) + \Phi^1_\infty(u) \otimes \mathbb{V}(y_\infty - s).$$

We get rid of the additional coefficient by adding the slope of $\frac{\beta_2-\beta_4}{\alpha_2-\alpha_1}$ of Φ_0 to Φ^1_∞. Thus we obtain the coefficient for the divisor $\{0\} \in \mathrm{Div}_{\mathbb{Q}}\mathbb{P}^1$ as the piecewise linear function with slopes $\frac{\beta_1-\beta_0}{\alpha_1}$ on $[\alpha_1, 0]$ and $\frac{\beta_2-\beta_0}{\alpha_2}$ on $[0, \alpha_2]$. Hence, we conclude that $X(\Phi^{(s)})$ corresponds to the projective toric plane $\mathrm{TV}(Q)$.

□

One can interpret these embedded deformations as deformations of so-called divisorial fans in lattice N. This construction can be found in [12]. The T-deformations of divisorial polytopes can be also seen as the T-deformations of affine cone over $\mathrm{TV}(P)$ which uses a generalized method of deformations of toric varieties in [3]. In fact, this is the method which has been used to prove Theorem 4.1. Moreover, if the coefficients Φ_0 and Φ^i_∞ form an admissible one-parameter Minkowski decomposition for some $i \in \{0,1\}$, then we can extend the deformation constructed in Theorem 4.2 to a deformation over $\mathbb{P}^1$.

From the proof of Theorem 4.2, one deduces the following interesting fact.

Corollary 4.1. *Let $\Phi_\infty = \Phi^0_\infty + \Phi^1_\infty$ be the admissible one-parameter Minkowski decomposition associated to the combinatorial mutation Ψ as in Theorem 4.2. Then Φ^0_∞ is a linear function and Φ^1_∞ has exactly two affine pieces with integral slopes. In particular, the decomposition of the slopes appear as*

$$\frac{\beta_2-\beta_1}{\alpha_1-\alpha_2} + \left\{0, \frac{\beta_1-\beta_4}{\alpha_1}\right\}.$$

Example 4.1. Let P^* be the dual polytope $\mathrm{conv}((-3,1),(3,1),(0,-1/2))$. Then $\mathrm{TV}(P)$ is the weighted projective plane with weights $(1,1,4)$. Consider the ample

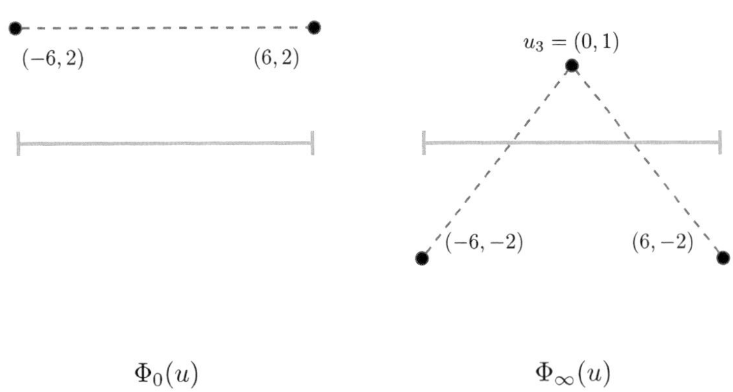

Fig. 4. A divisorial polytope for the weighted projective plane $\mathbb{P}(1,1,4)$

line bundle $\mathcal{O}_{\mathbb{P}(1,1,4)}(4)$. We choose the restricted torus action again as $M \xrightarrow{(1,0)} M'$. Then the associated divisorial polytope consists of lattice polytope $[-6,6] \subset M'_{\mathbb{Q}}$ and the piecewise linear function as in Figure 4. To obtain a deformation with its general fiber isomorphic to $\mathbb{P}^2$, we set the decomposition of Φ_∞ as Φ^0_∞ with slope $-\frac{1}{2}$ and Φ^1_∞ with slopes $\{0,1\}$. It gives us a $\mathbb{Q}$-Gorenstein smoothing of $\mathbb{P}(1,1,4)$ to $\mathbb{P}^2$.

Now consider another admissible decomposition of Φ_∞ as in Figure 5. Note that this decomposition has slopes $\{1/2, 0\}$ and $\{0, -1/2\}$. As in the proof of Theorem 4.2, we construct the divisorial polytope $([-6,6], \Phi^{(s)})$. Thus, we obtain the general fiber $X(\Phi^{(s)})$ being isomorphic to $\mathbb{P}^1 \times \mathbb{P}^1$. In particular, this is not a $\mathbb{Q}$-Gorenstein deformation.

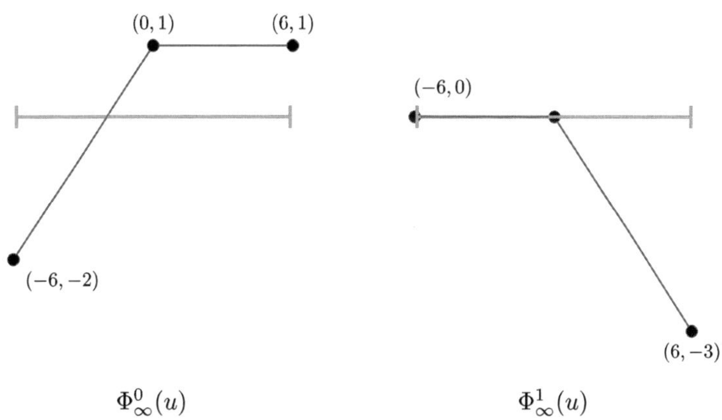

Fig. 5. A decomposition of Φ_∞ giving a deformation to $\mathbb{P}^1 \times \mathbb{P}^1$

References

[1] Mohammad E. Akhtar, Tom Coates, Sergey Galkin and Alexander M. Kasprzyk. *Minkowski polynomials and mutations*, SIGMA Symmetry Integrability Geom. Methods Appl. **8** (2012), 094, pp. 707.

[2] Mohammad E. Akhtar. Alexander M. Kasprzyk. *Mutations of fake weighted projective planes*, Proceedings of the Edinburgh Mathematical Society, Volume **59**, Issue 2, May 2016 , pp. 271-285.

[3] Klaus Altmann. *Minkowski sums and homogenous deformations of toric Varieties*, Tohoku Math. J. **47** (1995), 151-184.

[4] Klaus Altmann and Jürgen Hausen. *Polyhedral divisors and algebraic torus actions.* Math. Ann., 334:557-607, 2006.

[5] Klaus Altmann, Nathan Owen Ilten, Lars Petersen and, Hendrik Süß, and Robert Vollmert. *The geometry of T-varieties*, Contributions to algebraic geometry, EMS Ser. Congr. Rep., Eur. Math. Soc., Zürich, 2012, pp. 17-69.

[6] Victor Batyrev. *Toric degenerations of Fano varieties and constructing mirror manifolds*, The Fano Conference, 109-122, Univ. Torino, Turin, 2004.
[7] Sergey Galkin and Alexandr Usnich. *Mutations of potentials*, Preprint IPMU 10-0100, 2012.
[8] Paul Hacking and Yuri Prokhorov. *Smoothable del Pezzo Surfaces with quotient Singularities*, Compos. Math. **146** (2010), no.1, 169-192.
[9] Nathan Ilten. *Deformations of rational varieties with codimension one torus action*, Doctoral Thesis, FU Berlin, 2010. [urn:nbn:de:kobv:188-fudissthesis000000018440-0].
[10] Nathan Ilten and Robert Vollmert. *Deformation of rational T-varieties*, J. Algebraic Geometry **21** (2012) pp. 473-493).
[11] Nathan Ilten. *Mutations of Laurent polynomials and flat families with toric fibers*, Symmetry, Integrability and Geometry: Methods and Applications **8** (2012), 047.
[12] Nathan Ilten and Hendrik Süß. *Polarized complexity-one T-Varieties*, Michigan Mathematical Journal **60** (2011) no. 3 pp. 561-578.
[13] A. N. Rudakov. *The Markov numbers and exceptional bundles on* $\mathbb{P}^2$, Izv. Akad. Nauk SSSR Ser. Mat., 1988, Volume 52, Issue 1, 100-112.

Special cases and a dual view on the local formulas for Ehrhart coefficients from lattice tiles

Maren H. Ring
Institute for Mathematics, University of Rostock
18051 Rostock, Germany
E-mail: maren.ring@uni-rostock.de
https://www.geometrie.uni-rostock.de/people/maren/

McMullen's formulas or local formulas for Ehrhart coefficients are functions on rational cones that determine the i-th coefficient of the Ehrhart polynomial as a weighted sum of the volumes of the i-dimensional faces of a polytope. This work focuses on the RS-μ-construction as given in [13]. We give an explicit description of the construction from the dual point of view, i.e. given the cone of feasible directions instead of the normal cone as input value. We further show some properties of the construction in special cases, namely in case of symmetry and for the codimension one case.

Keywords: Ehrhart coefficients, local formula, lattice tiling.

1. Introduction

For a lattice polytope P in a Euclidean space V with lattice Λ, the *Ehrhart polynomial* E_P counts the number of lattice points in the t-th dilate of P,

$$\mathrm{E}_P(t) = |\Lambda \cap tP| = e_d t^d + e_{d-1}t^{d-1} + \ldots + e_1 t + e_0,$$

for all $t \in \mathbb{Z}_{\geq 0}$ and with $d = \dim(P)$ (cf. Ehrhart 1962 [7]).

In 1983, McMullen [10] showed the existence of so-called *McMullen's formulas* or *local formulas* for Ehrhart polynomials.

Definition 1. A real valued function μ on rational cones in V is called a *McMullen's formula* or *local formula* for Ehrhart coefficients, if for any lattice polytope P with Ehrhart polynomial $\mathrm{E}_P(t) = e_d t^d + e_{d-1}t^{d-1} + \cdots + e_1 t + e_0$, we have

$$e_i = \sum_{\mathcal{F}_i} \mu(N_f)\mathrm{vol}(f),$$

for all $i \in \{0, \ldots, d\}$ with $\mathcal{F}_i$ the set of all i-dimensional faces of P.

Here, N_f is the *(outer) normal vector* of the face f, i.e. the cone over the outer normal vectors of all facets meeting in f. It is a *convex rational cone*, which means that it can be written as $\{a_1v_1 + \ldots + a_kv_k \mid a_i \in \mathbb{R}_{\geq 0}\}$ for some vectors $a_1, \ldots, a_k \in \Lambda$. The volume $\mathrm{vol}(f)$ denotes the *relative volume* of f with respect to the induced lattice in the affine span of f, as defined in Section 2.

There have been several nice constructions of McMullen's formulas, for example the ones by Pommersheim and Thomas, see [12], and Berline and Vergne, see [4]. This work will focus on the construction given in [13] by the author together with

Achill Schürmann. Following the notation of Federico Castillo in his talk at Osaka University in 2018, we will refer to our construction as the RS-μ-construction, in analogy to the BV-α-construction by Berline and Vergne (cf. [5]). Each construction has different advantages. This construction, for instance, does not require a prior decomposition into unimodular cones and can be described using elementary geometric means. In contrast to [13], it might sometimes be more convenient to consider μ not as a function on the normal cones of a polytope but from the dual point of view, given *cones of feasible directions* as input values. Note that both cones, the normal and the cone of feasible directions of a face, hold the same local information about the face and can be recovered from one another. In Section 2 we give a detailed construction of the RS-μ-values given a full dimensional cone. To give more insight on how to compute the values, we then follow the construction given a polytope with cones of feasible directions and alongside compute the values in an example. In Section 3 we focus on the connection between the RS-μ-values and the number of lattice points, thus giving a motivation for the idea behind the construction and an outline of the proof. In Section 4 we focus on symmetry and show a new result about the codimension one case under central symmetry.

2. Construction

Fix an ambient Euclidean space V and a lattice Λ of full rank. Let Q be a polyhedron and f a face of Q. We define the *cone of feasible directions* of Q in the face f, *fcone* of Q in f for short, as

$$\text{fcone}(Q, f) = \{x \in V \mid \exists \varepsilon > 0\colon s + \varepsilon x \in Q\}$$

for any point s in the relative interior $\text{int}(f)$ of f. Let further $\text{lineal}(Q)$ be the *lineality space* of Q, i.e. the biggest linear subspace contained in Q, and denote $\text{lat}(Q) := \Lambda \cap \text{lineal}(Q)$.

The construction we will give relies on a choice of certain lattice tiles, namely *fundamental domains* as defined below. Different choices lead to different values and thus give an infinite family of constructions.

Definition 2. For a polyhedron $Q \subseteq V$ with induced sublattice $L = \text{lineal}(Q) \cap \Lambda$ in the lineality space of Q, a *fundamental domain* $T(Q)$ is a subset of $\text{lineal}(Q)$ with the following properties:

- $\bigcup_{x \in L} x + T(Q) = \text{lineal}(Q)$,
- $(x + T(Q)) \cap (y + T(Q)) = \varnothing$, for $x, y \in L$ with $x \neq y$ and
- every intersection of $T(Q)$ with a linear subspace is measureable.

Examples of fundamental domains are given in Section 4.

The *relative volume* of a subset $A \subseteq V$ denotes the volume of A in the affine span $\text{aff}(A)$ normalized such that any fundamental domain in $\text{aff}(A)$ has volume 1. Note that it is a lower dimensional volume if the affine span of A is.

We now give an explicit description of the construction of regions that determine the RS-μ-values defined on full dimensional cones. In the subsequent section, we will give the whole construction in the way it occurs when starting with a polytope and determining the RS-μ-values on its fcones.

2.1. *Construction on cones*

We will inductively define a map R from rational cones to subsets of V, associating a *region* to each cone. From these regions, the RS-μ-values can be computed via volume computations. The author is well aware that this is a formal and very compact description of the RS-μ-construction. For a step-by-step construction with examples and pictures, see Section 2.2. The aim of this subsection is to give a clear, short and formally precise definition of the construction.

Let C be a full dimensional rational cone in V. For non-full dimensional cones we intersect V with the linear span of C and consider that as our ambient space. For each subspace $A \subseteq V$ we assume to have chosen a fundamental domain $T(A)$ and denote $T(B) := T(\text{lineal}(B))$ for arbitrary subsets $B \subseteq V$.

If $C = V$ is the whole space, we set

$$R(V) := T(V).$$

Otherwise, if $C \subsetneq V$, we assume we have constructed the regions $R(\text{fcone}(C, F))$ for all faces $F < C$. Let X_F^C be the set of all points x in $\text{lat}(\text{fcone}(C, F))$ that fulfill the conditions:

- $[x + (R(\text{fcone}(C, F)) \cap \text{int}(\text{fcone}(C, F)))] \subseteq \text{int}(C)$ and
- $(x + R(\text{fcone}(C, F))) \cap (x' + R(\text{fcone}(C, F'))) = \varnothing$ for all $F' < C$, with F' incomparable to F and $x' \in \text{lat}(\text{fcone}(C, F'))$.

Then we define

$$R(C) := \left(T(C) + \text{lineal}(C)^{\perp}\right) \setminus \bigcup_{F<C} \left(X_F^C + R(\text{fcone}(C, F))\right).$$

From this we can compute the values for the *relative domain volume* in each region $R(C)$ for the cone C:

$$v_C := \text{vol}\big(R(C) \cap ((C \cap \Lambda) + T)\big).$$

And further the *correction volumes* for each $F < C$:

$$w_{\text{fcone}(C,F)}^C := \text{vol}(R(C) \cap (\text{lineal}(\text{fcone}(C, F)) \cap C))$$

for all faces $F < C$.

Then we get the value for C as

$$\mu(C) := v_C - \sum_{F<C} w_{\text{fcone}(C,F)}^C \cdot \mu(F).$$

Remark. Most recently, the construction of the regions as given in [13] has changed in the way that $R(C)$ is additionally intersected with the union of all lattice point translates that intersect C. For facets, this corresponds to the *relevant part* of the region defined in Section 4.2 and it does not change the value of μ. It does, however, make the region itself bounded and thus better constructable.

2.2. *Construction on fcones of polytopes*

To give a feeling for how the construction works given an actual polytope, we go through the construction from the perspective of a general, full dimensional lattice polytope P. Simultaneously, we compute the values for one particular example. Again, for each subspace $A \subseteq V$ we choose and fix a fundamental domain $T(A)$ and set $T := T(V)$.

To avoid unreadable expressions, we expand μ, R, v and w to functions on the faces of P by setting $\mu(f) := \mu(\text{fcone}(P, f))$, $R(f) := R(\text{fcone}(P, f))$, $v_g^f := v_{\text{fcone}(P,f)}$ and $w_g := w_{\text{fcone}(P,g)}^{\text{fcone}(P,f)}$. In the example we use the same notation and keep in mind that all fcones are taken with respect to the example simplex S.

Example. Let $S = \text{conv}(v_1, v_2, v_3) \subseteq \mathbb{R}^2$ be the simplex with vertices $v_1 = (1, 0)$, $v_2 = (2, 1)$ and $v_3 = (0, 2)$. We consider it as a lattice polytope in $\mathbb{R}^2$ with respect to the lattice $\mathbb{Z}^2$.

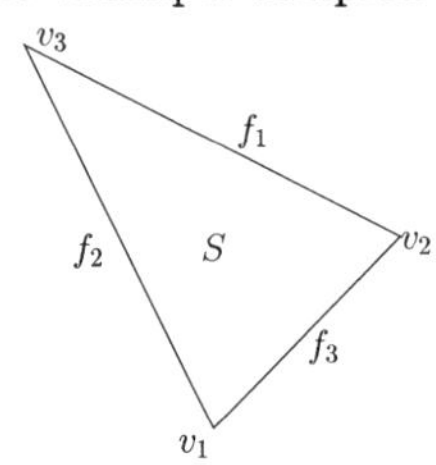

The first value we compute is $\mu(P)$ for P as a face of itself. Since $\text{fcone}(P, P) = V$, we get

$$R(P) = R(V) = T$$

and

$$v_P = \text{vol}\big(R(P) \cap ((V \cap \Lambda) + T)\big) = \text{vol}(T) = 1,$$

which directly determines the RS-μ-value as

$$\mu(P) = v_P = 1.$$

Remark. Using this to compute the d-th Ehrhart coefficient, we get

$$e_d = \sum_{f \in \mathcal{F}_d} \mu(f)\text{vol}(f) = \mu(P)\text{vol}(P) = 1 \cdot \text{vol}(P) = \text{vol}(P)$$

as desired, since the highest Ehrhart coefficient is known to be the relative volume of the polytope.

Example. The region $R(S) = R(\mathbb{R}^2)$ is given as the fundamental domain T of $\mathbb{Z}^2$, which we choose to be the square with edgelength 1 and the origin as barycentre. Then we have $\mu(S) = v_S = \mathrm{vol}(T) = 1$.

Now, let $F < P$ be a facet of P. Then $H^+ := \mathrm{fcone}(P, F)$ is a halfspace containing a hyperplane H and $R(H^+)$ is defined as

$$R(H^+) = \left(T(F) + H^\perp\right) \setminus \left(X_V^{H^+} + T\right),$$

where

$$X_V^{H^+} = \{x \in \Lambda \mid (x + T_V) \subseteq \mathrm{int}(H^+)\ \}.$$

That means that $R(H^+)$ equals the strip $\left(T(F) + H^\perp\right)$ minus all fundamental domains that lie entirely inside of H^+.

Then we have

$$v_{H^+} = \mathrm{vol}(R(H^+) \cap ((H^+ \cap \Lambda) + T))$$

and

$$w_V^{H^+} = \mathrm{vol}(R(H^+) \cap H^+).$$

That yields the RS-μ-value as

$$\mu(H^+) = v_{H^+} - w_V^{H^+} \cdot \mu(V) = v_{H^+} - w_V^{H^+}.$$

Example. S has three facets, namely the edges f_1, f_2 and f_3. Then for $i \in \{1, 2, 3\}$, $R(f_i) = \mathrm{Strip}(f_i) \setminus \left(X_S^{f_i} + T\right)$, where $\mathrm{Strip}(f_i) := T(f_i) + \mathrm{lineal}(\mathrm{fcone}(S, f_i))^\perp$. An illustration of the construction of the region $R(f_1)$ is given in Figure 1, top. The area of v_{f_1} is depicted in Figure 2 and the area of $w_S^{f_1}$ in Figure 3. Altogether we get

$$\mu(f_1) = v_{f_1} - w_S^{f_1} = 2 - 3/2 = 1/2.$$

Using the results from Section 4.2, we know that the values for all facets equal 1/2.

If we now consider a codimension two face f of P, then we have exactly two facets F_1 and F_2 of P that meet in f. That means that $W := \mathrm{fcone}(P, f)$ is a wedge defined by the intersection of the halfspaces $H_1^+ := \mathrm{fcone}(P, F_1)$ and $H_2^+ := \mathrm{fcone}(P, F_2)$, whose lineality spaces are the hyperplanes $H_1 := \mathrm{lineal}(\mathrm{fcone}(P, F_1))$ and $H_2 := \mathrm{lineal}(\mathrm{fcone}(P, F_2))$, respectively. Hence, the lineality space of $\mathrm{fcone}(P, f)$ is the line $\mathrm{lineal}(\mathrm{fcone}(P, f)) = H_1 \cap H_2$.

As above, for $\mathrm{fcone}(P, P) = V$ we have

$$X_V^W = \{x \in \Lambda \mid (x + T_V) \subseteq \mathrm{int}(W)\ \},$$

for H_1^+ we have

$$\begin{aligned} X_{H_1^+}^W = \{x \in \mathrm{lat}(H_1^+) \mid &(x + R(H_1^+) \cap \mathrm{int}(H_1^+)) \subseteq \mathrm{int}(W) \text{ and} \\ &(x + R(H_1^+)) \cap (x' + R(H_2^+)) = \varnothing \text{ for all } x' \in \mathrm{lat}(H_2^+)\} \end{aligned}$$

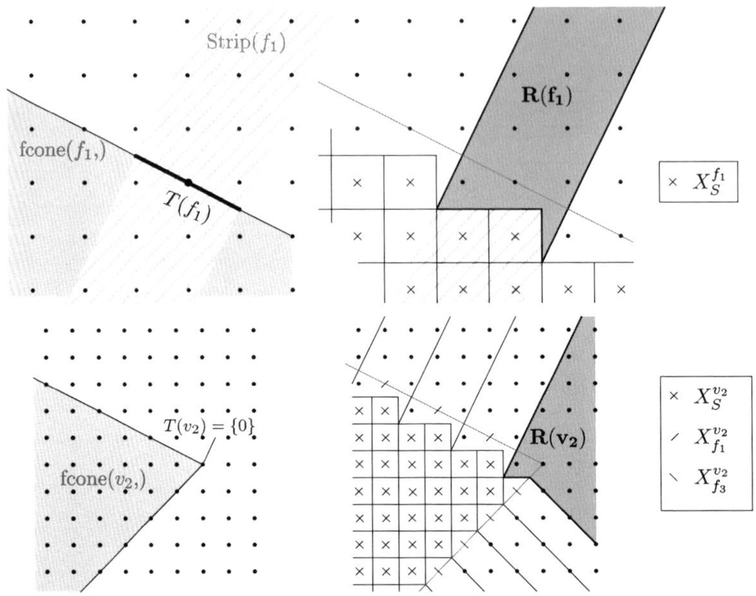

Fig. 1. Construction of the regions $R(f_1)$ and $R(v_2)$ for the edge f_1 (above) and the vertex v_2 (below).

and analogously for H_2^+

$$X_{H_2^+}^W = \{x \in \text{lat}(H_2^+) \mid (x + R(H_2^+) \cap \text{int}(H_2^+)) \subseteq \text{int}(W) \text{ and}$$
$$(x + R(H_2^+)) \cap (x' + R(H_1^+)) = \varnothing \text{ for all } x' \in \text{lat}(H_1^+)\}.$$

Then the region $R(W)$ is given by

$$R(W) = \big(T(W) + \text{lineal}(W)^{\perp}\big)$$
$$\setminus \Big((X_V^W + T) \cup (X_{H_1^+}^W + R(H_1^+)) \cup (X_{H_2^+}^W + R(H_2^+))\Big).$$

The relative domain volume is

$$v_W = \text{vol}(R(W) \cap ((W \cap \Lambda) + T))$$

and we have to consider three correction volumes:

$$w_{H_1^+}^W = \text{vol}(R(W) \cap (H_1 \cap C)),$$
$$w_{H_2^+}^W = \text{vol}(R(W) \cap (H_2 \cap C)) \text{ and}$$
$$w_V^W = \text{vol}(R(W) \cap W).$$

The RS-μ-value for W then is

$$\mu(W) = v_W - w_{H_1^+}^W \cdot \mu(H_1^+) - w_{H_2^+}^W \cdot \mu(H_2^+) - w_V^W \cdot \mu(V)$$

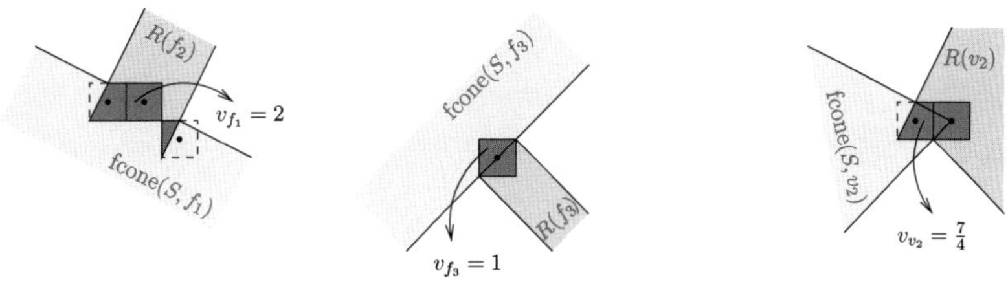

Fig. 2. Values for the relative domain volume v_f for certain faces $f < g$ of S.

Example. To compute the value corresponding to the vertex v_2 of S, we construct the region

$$R(v_2) = \mathbb{R}^2 \setminus \left((X_S^{v_2} + T) \cup (X_{f_1}^{v_2} + T(f_1)) \cup (X_{f_3}^{v_2} + T(f_3)) \right)$$

as shown in Figure 1, bottom. We then get $v_{v_2} = 7/4$ (cf. Figure 2) and as correction volumes we have the 1-dim. relative volumes $w_{f_1}^{v_2} = 1/2$ and $w_{f_2}^{v_2} = 1/2$ and the 2-dim. relative volume $w_S^{v_2} = 7/8$ (cf. Figure 3). Altogether we get the RS-μ-value for the fcone of S at the vertex v_2 as

$$\begin{aligned} \mu(v_2) &= v_{v_2} - w_{f_1}^{v_2} \cdot \mu(f_1) - w_{f_3}^{v_2} \cdot \mu(f_3) - w_S^{v_2} \cdot \mu(S) \\ &= \frac{7}{4} - \frac{1}{2} \cdot \frac{1}{2} - \frac{1}{2} \cdot \frac{1}{2} - \frac{7}{8} \cdot 1 = \frac{3}{8}. \end{aligned}$$

Analogously, we can compute $\mu(v_3) = 1/4$ and $\mu(v_1) = 3/8$. For reasons of symmetry, the latter has to equal the value $\mu(v_2)$, see Section 4.1 for an explanation.

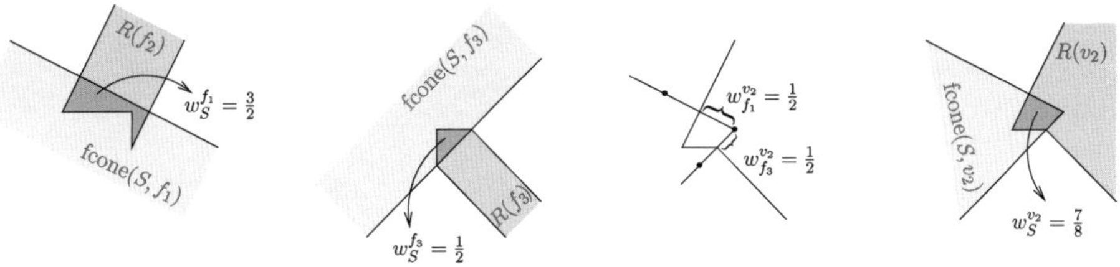

Fig. 3. Values for the correction volume w_g^f for certain faces $f < g$ of S.

For the general faces f of P the region of f is computed as

$$R(f) := \left(T(f) + \text{lineal}(\text{fcone}(P, f))^{\perp}\right) \setminus \bigcup_{g > f} \left(X_g^f + R(g)\right),$$

the relative domain volume is

$$v_f := \text{vol}\big(R(f) \cap ((f \cap \Lambda) + T)\big)$$

and the correction volume for each face g of P with $g > f$ is

$$w_g^f := \mathrm{vol}(R(f) \cap (\mathrm{lineal}(\mathrm{fcone}(P, g)) \cap f)).$$

Then we get the value for $\mathrm{fcone}(P, f)$ as

$$\mu(f) := v_f - \sum_{g>f} w_g^C \cdot \mu(g).$$

3. The geometrical connection between the RS-μ-values and the number of lattice points

This section gives a motivation behind the definition of the RS-μ-values and, along the way, an outline of the proof. Complete proofs can be found in [13].

Let $P = \mathrm{conv}(v_1, \ldots, v_m)$ be a full dimensional lattice polytope in d-dimensional Euclidean space V. We use the same notation as in Section 2.2, where we write $\mu(f)$ for $\mu(\mathrm{fcone}(P, f))$ and the same for R, v and w.

For a face f of P and $t \in \mathbb{Z}_{\geq 0}$ define the set $\mathcal{X}(tf)$ of all *feasible lattice points* in $t \cdot f$ as

$$\mathcal{X}(tf) := \bigcap_{i=1}^{m} \left(X_f^{v_i} + tv_i \right).$$

After the construction of all regions for P, we get the following tiling of V

Theorem 3.1 ([13, Theorem 1]). *Let $P \subseteq V$ be a full dimensional lattice polytope. There exists a $t_0 \in \mathbb{Z}_{>0}$ such that for each $t \geq t_0$ we have a tiling of V into translated regions of the form*

$$\{x + R(f) : \ f \leq P, \, x \in \mathcal{X}(tf)\}.$$

Example The tiling of $8 \cdot S$ for the example polygon S from Section 2.2 can be seen in Figure 4 on the left; in the picture, each region is translated by a unique lattice point marked • and the union of all these lattice points within a certain face f form the set $\mathcal{X}(8f)$.

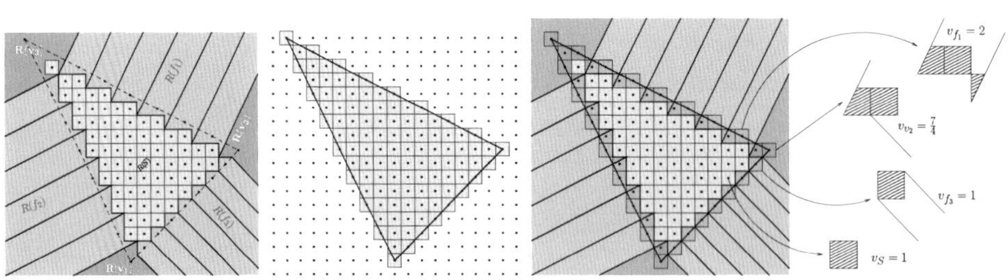

Fig. 4. **Left:** Tiling of $\mathbb{R}^2$ from Thm. 3.1 for the triangle S with $t = 8$; **middle:** The (fundamental) domain complex of $8S$; **right:** Values of v_f for some faces $f < S$

The crucial step between counting the number of lattice points in P and the RS-μ-value given by volumes is to interpret the number of lattice points in tP as the volume of all translates of T by the lattice points in P:

$$|\Lambda \cap tP| = \sum_{x\in\Lambda\cap tP} \mathrm{vol}(x+T) = \mathrm{vol}\underbrace{((\Lambda \cap tP)+T)}_{=:DC},$$

The first part of the equation holds, since by definition $\mathrm{vol}(T) = 1$ for any fundamental domain T of Λ, and the second part follows from $(x+T)\cap(y+T) = \varnothing$ for all $x, y \in \Lambda$ with $x \neq y$. We call the set $DC := (\Lambda\cap tP)+T$ a *(fundamental) domain complex* of tP (cf. Figure 4, middle). By taking the volume of the respective part of the domain complex in each region of the tiling in Theorem 3.1, we get

$$|\Lambda \cap tP| = \mathrm{vol}(DC) = \sum_{f\le P}\sum_{x\in\mathcal{X}(tf)} \underbrace{\mathrm{vol}\big((x+R(f))\cap DC\big)}_{(*)}.$$

It turns out that $(*)$ is exactly the value v_f as defined in Section 2, which thus yields

$$|\Lambda \cap tP| = \sum_{f\le P}\sum_{x\in\mathcal{X}(tf)} v_f = \sum_{f\le P} v_f \cdot |\mathcal{X}(tf)|, \tag{1}$$

see Figure 4, right.

Example Equation (1) for $4S$ gives

$$\begin{aligned}|\mathbb{Z}^2 \cap 4S| &= 109 = 1\cdot 70 + 2\cdot 6 + 2\cdot 6 + 1\cdot 7 + \frac{7}{4}\cdot 1 + \frac{7}{4}\cdot 1 + \frac{9}{2}\cdot 1\\ &= \sum_{f\le S} v_f \cdot |\mathcal{X}(8f)|,\end{aligned}$$

where the sum runs over the faces $S, f_1, f_2, f_3, v_1, v_2, v_3$ in that order.

To extract the value for the local formula from Equation (1), we need to determine the difference between $|\mathcal{X}(tf)|$ and $\mathrm{vol}(f)$. It can be shown (see [13, Lemma 8]) that by the construction of the correction volumes w we have that

$$\mathrm{vol}(tf) = \sum_{g\le f} w_f^g \cdot |\mathcal{X}(tf)|, \tag{2}$$

where we formally define $w_f^f = 1$.

We want the RS-μ-value to be defined such that

$$|\Lambda \cap tP| = \sum_{f\le P} \mu(f)\cdot \mathrm{vol}(f)$$

holds. Using Equation (2), we get

$$\sum_{f\le P} \mu(f)\cdot\mathrm{vol}(f) = \sum_{f\le P}\mu(f)\cdot\left(\sum_{g\le f} w_f^g\cdot|\mathcal{X}(tf)|\right).$$

By combinatorially reordering the right hand side, we get

$$\begin{aligned}\sum_{f \leq P} \mu(f) \cdot \mathrm{vol}(f) &= \sum_{f \leq P} \left[\mu(f) \cdot |\mathcal{X}(tf)| + \mu(f) \cdot \sum_{h>f} w_h^f \cdot |\mathcal{X}(tf)| \right] \\ &= \sum_{f \leq P} \left[\left(\mu(f) + \sum_{h>f} \cdot w_h^f \right) \cdot |\mathcal{X}(tf)| \right]\end{aligned}$$

Hence, we want to define the RS-μ-value such that the following equation holds

$$\sum_{f \leq P} v_f \cdot |\mathcal{X}(tf)| = |\Lambda \cap tP| = \sum_{f \leq P} \left[\left(\mu(f) + \sum_{h>f} \cdot w_h^f \right) \cdot |\mathcal{X}(tf)| \right].$$

Comparing coefficients we get

$$v_f = \mu(f) + \sum_{h>f} \cdot w_h^f,$$

which directly leads to the definition of the RS-μ-values as

$$\mu(f) = v_f - \sum_{h>f} \cdot w_h^f.$$

4. Symmetry and special cases

Symmetry of the RS-μ-values can be achieved by choosing symmetric fundamental domains, for example by taking *Dirichlet–Voronoi cells*, as we will show in Section 4.1. We will finish this article by proving that given central symmetry, the values on halfspaces are determined to be 1/2. That implies that the value is always 1/2 for fcones of a facet of a polytope.

4.1. *Dirichlet–Voronoi cells and symmetry of RS-μ*

Possibly the most natural choice of fundamental domains are *Dirichlet–Voronoi cells*. Given a space V and an inner product $\langle \cdot, \cdot \rangle$ with induced norm $\| \cdot \|$, the Dirichlet–Voronoi cell of a sublattice $L \subseteq \Lambda$ is defined as

$$\mathrm{DV}(L, \langle \cdot, \cdot \rangle) := \{x \in \mathrm{lineal}(L) : \|x\| \leq \|x - a\| \text{ for all } a \in L\}.$$

In this definition, it is not yet a fundamental domain of the lattice L, but by considering the Dirichlet–Voronoi cell half open, it can be seen as a fundamental domain of the lattice. Dirichlet–Voronoi cells are naturally centrally symmetric and can be forced to have certain symmetries by choosing a suitable inner product.

Let P be a lattice polytope and $\mathcal{G}$ a subgroup of all lattice symmetries of P, i.e. $\mathcal{G}$ is a finite matrix group with $A \cdot P := \{A \cdot x : x \in P\} = P$ and $A \cdot \Lambda = \Lambda$ for all $A \in \mathcal{G}$. Then we can define a $\mathcal{G}$-invariant inner product by taking

$$\langle x, y \rangle_{\mathcal{G}} := x^t G y \qquad \text{for all } x, y \in V, \tag{3}$$

with the Gram matrix G given by

$$G := \frac{1}{|\mathcal{G}|}\sum_{A\in\mathcal{G}} A^t A. \tag{4}$$

Let $\|\cdot\|_{\mathcal{G}}$ be the induced norm and let D be the Dirichlet–Voronoi cell for Λ given by the that particular inner product,

$$D := \mathrm{DV}(\Lambda, \langle\cdot,\cdot\rangle_{\mathcal{G}}) = \{x \in V : \|x\|_{\mathcal{G}} \leq \|x - p\|_{\mathcal{G}} \text{ for all } p \in \Lambda\}.$$

Then D is invariant under the action of $\mathcal{G}$: Let $x \in D$, then for $A \in \mathcal{G}$ we have

$$\|Ax\|_{\mathcal{G}} = \|x\|_{\mathcal{G}} \leq \|x - p\|_{\mathcal{G}} = \|Ax - Ap\|_{\mathcal{G}} \qquad \text{for all } p \in \Lambda.$$

Since $A\Lambda = \Lambda$, we get $AD \subseteq D$ for all $A \in \mathcal{G}$. Substituting A by A^{-1}, we get $A^{-1}D \subseteq D$ which yields $D \subseteq AD$ and hence $AD = D$. Similarly, we see that for all faces f in the same $\mathcal{G}$-orbit the fcones and Dirichlet–Voronoi cells in $\Lambda \cap N_f^{\perp}$ are mapped onto each other. Hence, the used regions are invariant under the action of $\mathcal{G}$ and μ is constant on $\mathcal{G}$-orbits.

4.2. *Codimension one faces under central symmetry*

It is known that the second highest Ehrhart coefficient always equals 1/2 times the sum over the relative volumes of the facets of a polytope. A natural conjecture would be that all values of McMullen's formulas corresponding to facets (in this case all values on halfspaces) have the value 1/2. This is not true in general for the RS-μ-values, but we show here that it does hold for the RS-μ-values when all fundamental domains are centrally symmetric. Therefore, in the following let $T(A)$ be a centrally symmetric fundamental domain for each $A \subseteq V$. Again, we denote $T := T(V)$.

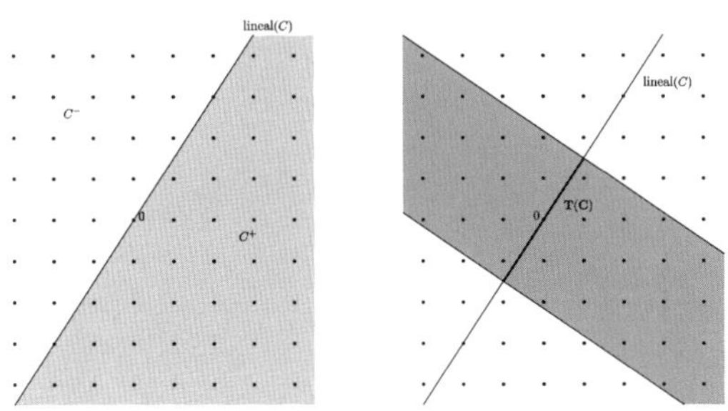

Fig. 5. **left**: C with C^+, C^- and lineal(C); **right**: Strip(C) $= T(C) + \mathrm{lineal}(C)^{\perp}$

Now, let F be a facet of P with fcone $C := \mathrm{fcone}(P, F)$. That means lineal($C$) is a hyperplane in V. Let C^+ be the open halfspace inside of C and C^- the open halfspace on the other side, i.e. the complement of C.

Then

$$R(C) = \underbrace{(T(C) + \text{lineal}(C)^{\perp})}_{=:\text{Strip}(C)} \setminus \left(X_V^C + T\right)$$
$$= \text{Strip}(C) \cap \left((\Lambda \setminus X_V^C) + T\right),$$

where

$$X_V^C = \{x \in \Lambda \mid (x + T) \subseteq C^+ \}.$$

Taking a closer look at the construction in Section 2, we note that for computing v_C and w_K^C and hence $\mu(C)$ it is only necessary to consider $R(C)$ intersected with the union of all fundamental domains that have a nonempty intersection with lineal(C). We therefore consider the *relevant part* $\tilde{R}(C)$ of the region $R(C)$:

$$\tilde{R}(C) = (\underbrace{\{p \in \text{lat}(W) \mid (p + T) \cap \text{lineal}(C) \neq \varnothing\}}_{:=X} + T) \cap \text{Strip}(C)$$

This relevent part of $R(C)$ can be partitioned into three parts

$$\begin{aligned}\tilde{R}(C) =& (X + T) \cap \text{Strip}(C)\\ =& ((\underbrace{X \cap \text{lineal}(C)}_{:=X_0} + T) \cap \text{Strip}(C)\\ & \cup ((\underbrace{X \cap C^+}_{:=X_+} + T) \cap \text{Strip}(C)\\ & \cup ((\underbrace{X \cap C^-}_{:=X_-} + T) \cap \text{Strip}(C),\end{aligned}$$

where the unions are disjoint, since X_0, X_+, X_- are. For an illustration see Figure 6.

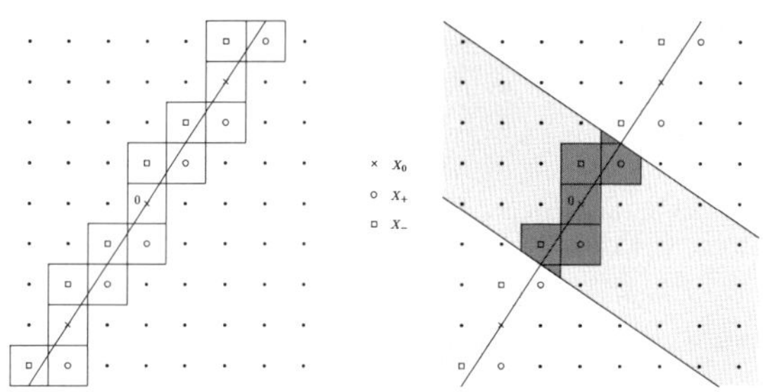

Fig. 6. $(X + T)$ and $\tilde{R}(C)$

Recall that $\mu(C)$ is defined as

$$\begin{aligned}\mu(C) &= v_C - \mu(W) \cdot w_W^C\\ &= \text{vol}(\tilde{R}(C) \cap ((C \cap \text{lat}(W))) + T) - 1 \cdot \text{vol}(\tilde{R}(C) \cap (W \cap C))\end{aligned}$$

To show that $\mu(C) = 1/2$, we show that everything but half the lattice cell around the origin cancels out nicely. We use the fact that everything is centrally symmetric in the following sense:

Let σ_0 be the point reflection at the origin:

$$\sigma_0 \colon V \to V$$
$$v \mapsto -v$$

Then $\sigma_0(T) = T$ by assumption and since C is a halfspace, we also have

$$\begin{aligned}
\sigma_0(\mathrm{lineal}(C)) &= \mathrm{lineal}(C), \\
\sigma_0(C^+) &= C^-, \\
\sigma_0(C^-) &= C^+, \\
\sigma_0(\mathrm{Strip}(C)) &= \mathrm{Strip}(C) \\
\sigma_0(\tilde{R}(C)) &= \tilde{R}(C)
\end{aligned}$$

Now observe that $\mathrm{vol}((X_0 + T) \cap \mathrm{Strip}(C)) = \mathrm{vol}(0 + T) = 1$, since the intersection of a centrally symmetric fundamental domain with a linear subspace is always contained in the centrally symmetric fundamental domain of that subspace.

Since σ_0 does not change the volume, we can use it to gain information on the occuring volumes:

$$\sigma_0((T \cap \mathrm{Strip}(C)) \cap C^+) = T \cap \mathrm{Strip}(C) \cap C^-$$

and hence,

$$\mathrm{vol}(((X_0 + T) \cap \mathrm{Strip}(C)) \cap C^-) = \frac{1}{2}.$$

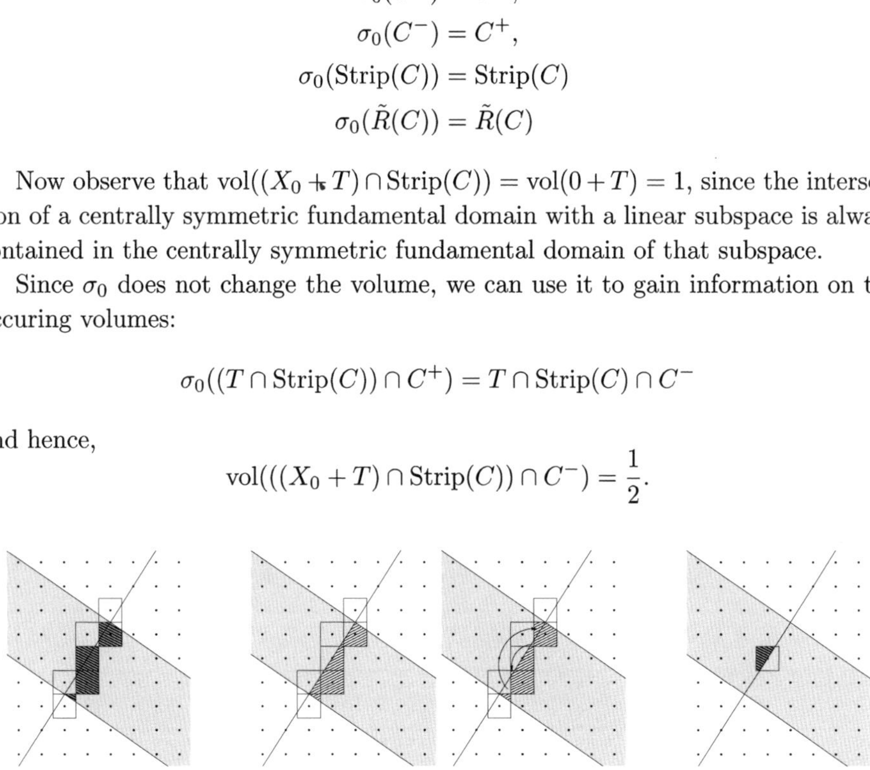

Fig. 7. **Left:** v_C, **middle:** w_W^C and **right:** $w_W^C - v_C$

Moreover, we have the disjoint union:

$$\begin{aligned}
\tilde{R}(C) \cap C^+ =& ((X_0 + T) \cap \mathrm{Strip}(C)) \cap C^+ \\
& \cup (X_+ + T) \cap \mathrm{Strip}(C)) \cap C^+ \\
& \cup (X_- + T) \cap \mathrm{Strip}(C)) \cap C^+
\end{aligned}$$

and since $\sigma_0(((X_- + T) \cap \mathrm{Strip}(C)) \cap C^+) = ((X_+ + T) \cap \mathrm{Strip}(C)) \cap C^-$, the two have equal volume and we have

$$\begin{aligned} w_V^C =& \mathrm{vol}(\tilde{R}(C) \cap (V \cap C^+)) \\ =& \mathrm{vol}(((X_0 + T) \cap \mathrm{Strip}(C)) \cap C^+) \\ &+ \mathrm{vol}((X_+ + T) \cap \mathrm{Strip}(C)) \cap C^+) \\ &+ \mathrm{vol}((X_+ + T) \cap \mathrm{Strip}(C)) \cap C^-) \\ =& \mathrm{vol}(((X_0 + T) \cap \mathrm{Strip}(C)) \cap C^+) \\ &+ \mathrm{vol}((X_+ + T) \cap \mathrm{Strip}(C))), \end{aligned}$$

see Figure 7, middle. Together with

$$v_C = \mathrm{vol}(\tilde{R}(C) \cap ((C \cap \Lambda) + T)) = \mathrm{vol}((X_0 + T) \cap \mathrm{Strip}(C)) + \mathrm{vol}((X_+ + T) \cap \mathrm{Strip}(C)),$$

(cf. Figure 7, left), we finally get

$$\begin{aligned} \mu(C) &= v_C - w_V^C \\ &= \mathrm{vol}(\tilde{R}(C) \cap ((C \cap \Lambda)) + T) - \mathrm{vol}(\tilde{R}(C) \cap (V \cap C)) \\ &= \mathrm{vol}((X_0 + T) \cap \mathrm{Strip}(C)) - \mathrm{vol}(((X_0 + T) \cap \mathrm{Strip}(C)) \cap C^+) \\ &= \mathrm{vol}(((X_0 + T) \cap \mathrm{Strip}(C)) \cap C^-) \\ &= \frac{1}{2} \end{aligned}$$

as we wanted to show (cf. Figure 7, right).

Acknowledgements

The author would like to thank Professor Takayubi Hibi and Akiyoshi Tuchiya for the invitation and amiable organization of the summer workshop on lattice polytopes 2018, and also all other participants for making it a wonderful and very productive stay. Further thanks go to *The Birth of Modern Trends on Commutative Algebra and Convex Polytopes with Statistical and Computational Strategies* (Grant-in-Aid for Scientific Research (S) 26220701) for funding the workshop.

References

[1] Alexander I. Barvinok (2008), *Integer points in polyhedra*, Zurich Lectures in Advanced Mathematics, European Mathematical Society, Zürich.

[2] Alexander I. Barvinok and James E. Pommersheim (1999), *An algorithmic theory of lattice points in polyhedra*, in New perspectives in algebraic combinatorics (Berkeley, CA, 1996-97), Math. Sci. Res. Inst. Publ., vol. 38, Cambridge Univ. Press, Cambridge, 91–147.

[3] Matthias Beck and Sinai Robins (2015), *Computing the continuous discretely*, 2nd. ed, Springer, New York.

[4] Nicole Berline and Michèle Vergne (2007), *Local Euler McLaurin formula for polytopes*, Mosc. Math. J., Volume 7, Number 3, 355–386.

[5] Federico Castillo and Fu Liu (2015), *Berline-Vergne valuation and generalized permutohedra*, preprint: arXiv:1509.07884.
[6] Vladimir I. Danilov (1978), *The Geometry of Toric Varieties*, Russian Math. Surveys 33 (2), 97–154.
[7] Eugène Ehrhart (1962), *Sur les polyèdres rationnels homothétiques à n dimensions*, C. R. Acad. Sci. Paris. 254, 616–618.
[8] Robert Morelli (1993), *Pick's theorem and the Todd class of a toric variety*, Adv. Math. 100, 183–231.
[9] Peter McMullen (1978), *Lattice invariant valuations on rational polytopes*, Arch. Math. 31, 509–516.
[10] Peter McMullen (1983), *Weakly continuous valuations on convex polytopes*, Arch. Math. 41, 555–564.
[11] Georg Pick (1899), *Geometrisches zur Zahlenlehre*, Sitzungsberichte des deutschen naturwissenschaftlich-medicinischen Vereines für Böhmen Lotos in Prag, 311–319.
[12] James E. Pommersheim and Hugh Thomas (2004), *Cycles representing the Todd class of a toric variety*, J. Amer. Math. Soc. 17, 983–994.
[13] Maren H. Ring and Achill Schürmann (2017), *Local Formulas for Ehrhart Coefficients from Lattice Tiles*, preprint: arXiv:1709.10390.
[14] Günter M. Ziegler (1997), *Lectures on polytopes*, 2nd printing, Springer, New York.

Local h^*-polynomials of some weighted projective spaces

Liam Solus

Institutionen för Matematik
KTH Royal Institute of Technology
SE-100 44 Stockholm, Sweden
E-mail: solus@kth.se

There is currently a growing interest in understanding which lattice simplices have unimodal local h^*-polynomials (sometimes called box polynomials); specifically in light of their potential applications to unimodality questions for Ehrhart h^*-polynomials. In this note, we compute a general form for the local h^*-polynomial of a well-studied family of lattice simplices whose associated toric varieties are weighted projective spaces. We then apply this formula to prove that certain such lattice simplices, whose combinatorics are naturally encoded using common systems of numeration, all have real-rooted, and thus unimodal, local h^*-polynomials. As a consequence, we discover a new restricted Eulerian polynomial that is real-rooted, symmetric, and admits intriguing number theoretic properties.

Keywords: Box polynomial, local h^*-polynomial, Eulerian polynomial, Ehrhart theory, simplex, weighted projective space, numeral systems, simplices for numeral systems, factoradics, real-rooted, unimodal, symmetric, log-concave.

1. Introduction

For a positive integer $n \in \mathbb{Z}_{\geq 0}$ we set $[n] := \{1, \ldots, n\}$ and $[n]_0 := \{0, \ldots, n\}$. Let $p(z) = p_0 + p_1z + \cdots + p_nz^n$ be the generating polynomial for a combinatorial sequence $(p_0, \ldots, p_n)$; in particular, assume that p_k is a nonnegative integer for all $k \in [n]_0$. A popular endeavour in combinatorics is to understand the distributional properties of the polynomial $p(z)$, and thus of the sequence $(p_0, \ldots, p_n)$. We say that $p(z)$ is **unimodal** if $p_0 \leq p_1 \leq \cdots \leq p_t \geq \cdots \geq p_{n-1} \geq p_n$ for some $t \in [n]_0$. We say it is **log-concave** if $p_i^2 \geq p_{i-1}p_{i+1}$ for all $i \in [n-1]$, and we say it is **symmetric** with respect to $m \in \mathbb{Z}_{\geq 0}$ if $p_i = p_{m-i}$ for all $i \in [m]_0$. As stated, these properties seem mysteriously independent of the generating polynomial $p(z)$, and only really pertaining to the combinatorial sequence $(p_0, \ldots, p_n)$. However, the structure of the polynomial $p(z)$, and particularly it roots, can help us deduce these various distributional properties. We say that $p(z)$ is **real-rooted** if all of its roots are real numbers. If $p(z)$ is real-rooted then it is also log-concave and unimodal (see Theorem 1.2.1 in [11]). Thus, it is particularly desirable if the generating polynomial $p(z)$ is both real-rooted and symmetric.

In algebraic combinatorics, a well-studied family of generating polynomials are the (Ehrhart) h^*-polynomials of lattice polytopes. Let $P \subset \mathbb{R}^n$ be a d-dimensional convex lattice polytope; i.e. a convex polytope whose affine span has dimension d and all of whose vertices lie in $\mathbb{Z}^n$. For a nonnegative integer $t \in \mathbb{Z}_{\geq 0}$, the tth **dilate**

of P is $tP := \{tp \in \mathbb{R}^n : p \in P\}$, and the **Ehrhart series of** P is

$$\mathrm{Ehr}_P(z) := \sum_{t\geq 0} |tP \cap \mathbb{Z}^n| z^t = \frac{h_0^* + h_1^* z + \cdots + h_d^* z^d}{(1-z)^{d+1}}.$$

When written in its closed rational form, $\mathrm{Ehr}_P(z)$ appears as in the right-most expression above. The polynomial in the numerator is called the **(Ehrhart)** h^***-polynomial** of P, and it is known to have only nonnegative integer coefficients [24]. In recent decades, the distributional properties of h^*-polynomials have been investigated extensively (see for instance [4]).

In the case of lattice simplices, the h^*-polynomial has a closely related local-invariant known as its local h^*-polynomial [17], or box polynomial [2, 4, 22]. Let

$$\Delta := \mathrm{conv}(v^{(0)}, \ldots, v^{(d)}) \subset \mathbb{R}^n$$

be a lattice d-simplex defined as the convex hull of the $d+1$ affinely independent points $v^{(0)}, \ldots, v^{(d)} \in \mathbb{Z}^n$. The **open parallelpiped** of Δ is

$$\Pi_\Delta^\circ := \left\{ \sum_{i=0}^{d} \lambda_i (v^{(i)}, 1) \in \mathbb{R}^{n+1} : 0 < \lambda_i < 1,\, i \in [d]_0 \right\},$$

and the **local** h^***-polynomial**, or **box polynomial**, of Δ is

$$\ell^*(\Delta; z) := \sum_{(x_1,\ldots,x_{n+1})\in \Pi_\Delta^\circ \cap \mathbb{Z}^n} z^{x_{n+1}}.$$

We additionally define the **half-open parallelpiped** of Δ, denoted Π_Δ, by replacing the inequalities $0 < \lambda_i$ in the definition of Π_Δ° with $0 \leq \lambda_i$ for all $i \in [d]_0$. As outlined in [4] and [22] and applied in [16], unimodality of local h^*-polynomials of lattice simplices can sometimes be used to recover unimodality of h^*-polynomials by way of a theorem shown in [3]. Moreover, as demonstrated in [16], the distributional properties of local h^*-polynomials can also be used to answer questions on the distributional properties of a related family of generating polynomials; namely, the local h-polynomials of subdivisions of simplices [1]. The applications of local h^*-polynomials to unimodality of h^*-polynomials prompted the following general question, posed in [4]:

Question 1.1 (Question 5 in [4]). *Which lattice simplices have unimodal local h^*-polynomials?*

It can be seen from the definition of $\ell^*(\Delta; z)$ that local h^*-polynomials are always symmetric with respect to $d+1$. Thus, if we can further show $\ell^*(\Delta; z)$ is real-rooted, it will have all of the desirable distributional properties we previously outlined, including the unimodality requested by Question 1.1.

One well-studied family of simplices for which we may consider Question 1.1 are those whose associated toric varieties are **weighted projective spaces** [12]. For convenience we may refer to a simplex in this family simply as a weighted projective space. Recently, the distributional properties of the h^*-polynomials of a particularly nice family of weighted projective spaces, denoted $\mathcal{Q}$, have been intensely investigated [5–7, 18, 23]. Within the family $\mathcal{Q}$ live the **simplices for numeral systems**, examples of which were shown in [23] to have real-rooted h^*-polynomials. In this note, we first state a general formula for the local h^*-polynomials for each simplex in $\mathcal{Q}$. We then use this formula to show that all simplices for numeral systems studied in [23] also have real-rooted local h^*-polynomials, thereby answering Question 1.1 for this family of weighted projective spaces.

In Section 2, we recall the definition of the weighted projective spaces $\mathcal{Q}$ and compute a general formula for their associated local h^*-polynomials. We additionally recall some necessary preliminaries on numeral systems [15] and real-rooted polynomials in subsections 2.1 and 2.2, respectively. In Section 3, we prove that the local h^*-polynomial of the weighted projective spaces associated to the factoradic numeral system are all real-rooted, and thus unimodal. In doing so, we discover a previously unknown restricted version of the Eulerian polynomial that is both symmetric and real-rooted. In Section 4, we prove that the local h^*-polynomials of the base-r-simplices for numeral systems are all real-rooted as well. In Section 5, we then outline some future directions motivated by this work pertaining to γ-nonnegativity and local h-polynomials of subdivisions of simplicial complexes.

2. Some Weighted Projective Spaces

Let $e^{(1)}, \ldots, e^{(n)}$ denote the standard basis vectors in $\mathbb{R}^n$. Given a weakly increasing vector of positive integers $q := (q_1, \ldots, q_n)$, define an n-simplex $\Delta_{(1,q)}$ as the convex hull

$$\Delta_{(1,q)} := \operatorname{conv}\left(e^{(1)}, \ldots, e^{(n)}, -\sum_{i=1}^{n} q_i e^{(i)}\right) \subset \mathbb{R}^n .$$

The projective toric variety associated to $\Delta_{(1,q)}$ is the weighted projective space in which the weight q_i is assigned to the ith coordinate [12]. Recently, the simplices $\Delta_{(1,q)}$ have been considered for their combinatorial properties, particularly in regards to the unimodality of their h^*-polynomials [5–7, 18, 23]. Thus, in light of Question 1.1, it is natural to ask which $\Delta_{(1,q)}$ have unimodal local h^*-polynomials. To this end, we first determine a general formula for the local h^*-polynomial of $\Delta_{(1,q)}$.

Theorem 2.1. *Let $q := (q_1, \ldots, q_n)$ be a sequence of positive integers, set $Q := 1 + q_1 + \cdots + q_n$, and define the set*

$$T_q := \{b \in [Q-1] \,:\, Q \nmid q_i b \text{ for all } i \in [n]\} .$$

Then the local h^*-polynomial of the n-simplex $\Delta_{(1,q)}$ is

$$\ell^*(\Delta_{(1,q)}; z) = \sum_{b \in T_q} z^{\omega(b)}, \qquad \text{where} \qquad \omega(b) = b - \sum_{i=1}^{n} \left\lfloor \frac{q_i b}{Q} \right\rfloor .$$

Proof. In the following we let $\Pi_q := \Pi_{\Delta_{(1,q)}}$ denote the half-open parallelpiped of $\Delta_{(1,q)}$, and we let $\Pi_q^\circ := \Pi_{\Delta_{(1,q)}}^\circ$ denote the associated open parallelpiped. To prove the desired result, we take an analogous approach to the proof of Theorem 2.2 in [6], in which the h^*-polynomial of $\Delta_{(1,q)}$ is computed by way of the heights of the lattice points in Π_q. In this fashion, we let $(v^{(1)},1)$, $(v^{(2)},1)$, $\ldots$, $(v^{(n)},1)$, $(v^{(0)},1)$ denote the columns of the matrix

$$\begin{pmatrix} 1 & 1 & 1 & \cdots & 1 & 1 \\ 1 & 0 & 0 & \cdots & 0 & -q_1 \\ 0 & 1 & 0 & \cdots & 0 & -q_2 \\ 0 & 0 & 1 & \cdots & 0 & -q_3 \\ \vdots & \vdots & \vdots & \ddots & \vdots & \vdots \\ 0 & 0 & 0 & \cdots & 1 & -q_n \end{pmatrix}$$

as read from left-to-right. Then every lattice point in the open parallelpiped Π_q° is of the form

$$p = \sum_{i=0}^{n} \lambda_i (v^{(i)}, 1),$$

where $0 < \lambda_i < 1$ for all $0 \leq i \leq n$. Moreover, by Proposition 4.4 in [19] and Cramer's Rule we know that $\lambda_i = \frac{b_i}{Q}$, for some $0 \leq \lambda_i < Q$. However, since we are assuming $p \in \mathbb{Z}^{n+1}$, then for $1 \leq i \leq n$ it must be that $\lambda_i = q_i\lambda_0 - \lfloor q_i\lambda_0 \rfloor$. So it follows that $p \in \Pi_q^\circ$ if and only if $0 < \lambda_0 < 1$ and $0 < q_i\lambda_0 - \lfloor q_i\lambda_0 \rfloor < 1$ for all $1 \leq i \leq n$. Since $\lambda_0 = \frac{b}{Q}$ for some $0 \leq b < Q$, this is equivalently stated as $p \in \Pi_q^\circ$ if and only if $Q \nmid q_i b$ for all $i \in [n]$ and $0 < b < Q$. Since, for each $0 < b < Q$, the height of the corresponding lattice point in Π_q° is given by

$$\sum_{i=0}^{n} \lambda_i = b - \sum_{i=1}^{n} \left\lfloor \frac{q_i b}{Q} \right\rfloor,$$

the proof is complete. □

As a first example, we can use Theorem 2.1 to compute the local h^*-polynomial for the n-simplex whose associated projective toric variety is $\mathbb{P}^n$; i.e. n-dimensional projective space.

Corollary 2.1. *Let $\Delta_{(1,q^{(n)})}$ be the n-simplex with $q^{(n)} := (1,1,\ldots,1) \in \mathbb{R}^n$. Then the local h^*-polynomial of $\Delta_{(1,q^{(n)})}$ is*

$$\ell^*(\Delta_{(1,q^{(n)})}; z) = z + z^2 + \cdots + z^n = zh^*(\Delta_{(1,q^{(n-1)})}; z).$$

Proof. Since $Q = n+1$ and $q_i^{(n)} = 1$ for all $1 \leq i \leq n$ for $\Delta_{(1,q^{(n)})}$, then $b \in [Q-1]$ is in the set $T_{q^{(n)}}$ if and only if $n+1$ divides b. Therefore, $T_{q^{(n)}} = [Q-1]$, and the first inequality follows by checking that $\omega(b) = b$ for all $b \in [Q-1]$. The second inequality follows from well-known results. □

In a recent paper [16], the authors showed that the local h^*-polynomials of the s-**lecture hall simplices** are all real-rooted, and therefore affirmatively answered Question 1.1 for this family of simplices. They additionally showed that they contain the local h-polynomials for some well-studied subdivisions of simplices which are closely related to derangement polynomials. The h^*-polynomials of the s-lecture hall simplices, called the s-**Eulerian polynomials**, are also real-rooted and unimodal [21]. The results of [16] then suggest that these nice properties of the h^*-polynomial of a simplex can be inherited by its local h^*-polynomial. In [23] the author indexed families of $\Delta_{(1,q)}$ via the place-values of positional numeral systems and used this stratification to deduce numeral systems for which the associated h^*-polynomials $h^*(\Delta_{(1,q)}; z)$ were all real-rooted. In the following, we will use Theorem 2.1 to show that these properties of the h^*-polynomial are again inherited by the associated local h^*-polynomials. We first recall the necessary details from [23] in subsections 2.1 and 2.2.

2.1. *Numeral Systems and Weighted Projective Spaces*

The **normalized volume** of a lattice polytope is the value given by evaluating its h^*-polynomial at 1. In this paper, the normalized volume of the simplex $\Delta_{(1,q)}$ is denoted Q. When studying properties of lattice polytopes it can often be useful to stratify the collection of all lattice polytopes by their normalized volume rather than their dimension. In special cases, this can even allow us to discover recursive formulae that can be used to prove desirable properties of the associated h^*-polynomials, such as real-rootedness. To do exactly this, in [23] the author studied when one simplex $\Delta_{(1,q)}$ in each dimension $n \geq 1$ with normalized volume Q_n can be found so that the sequence $(Q_n)_{n=0}^{\infty}$ are the place values of some (positional) numeral system.

A (positional) numeral system is a method for expressing numbers. A **numeral** is a string of nonnegative integers $\eta = \eta_{n-1}\eta_{n-2}\cdots\eta_0$ and the location of η_i is called the **place** of i. The **digits** are the numbers allowable in the place i and the ith **base** is the number of digits for place i. A **(positional) numeral system** is a sequence of positive integers $(a_n)_{n=0}^{\infty}$ satisfying $a_0 = 1 < a_1 < a_2 < \cdots$, and the term a_n is called its nth **place value**. Any nonnegative integer b can be expressed uniquely as a numeral with respect to the numeral system $(a_n)_{n=0}^{\infty}$ by repeatedly performing Euclidean division starting with the largest a_{n-1} such that $b > a_{n-1}$ and repeating this process with the remainder (see for instance Theorem 1 in [15]). Specifically, if $b = p_{n-1}a_{n-1} + r_{n-1}$ and $r_{n-1} = p_{n-2}a_{n-2} + r_{n-2}$ then p_{n-1} is the digit in the $(n-1)^{st}$ place and p_{n-2} is the digit in the $(n-2)^{nd}$ place in the numeral representing b, and so on.

Example 2.1. The following numeral systems will be of special interest in the coming sections.

(i) (The Binary Numeral System). Let $(a_n)_{n=0}^{\infty} := (2^n)_{n=0}^{\infty}$. Then the nth place value is 2^n and the possible digits in each place are only 0 or 1. The number 13 is represented by the numeral 1101 since

$$13 = 1 \cdot 2^3 + 1 \cdot 2^2 + 0 \cdot 2^1 + 1 \cdot 2^0.$$

(ii) (The Base-r Numeral System). Generalizing the previous example, let $r \geq 2$ and $(a_n)_{n=0}^{\infty} := (r^n)_{n=0}^{\infty}$. Then the nth place value is r^n and the possible digits in each place are $[r-1]_0$. For example, when $r = 10$, we get the typical base-10 numeral system which we all use every day.

(iii) (The Factoradic Numeral System). Let $(a_n)_{n=0}^{\infty} := ((n+1)!)_{n=0}^{\infty}$. Then the nth place value is $(n+1)!$ and the possible digits in each place are only $[n]_0$. This numeral system is used in computer science to efficiently encode permutations in a lexicographic order, and it will be the focus of Section 3.

Thinking geometrically, given a numeral system $(a_n)_{n=0}^{\infty}$ we can ask for a sequence of $\Delta_{(1,q)}$, one in each dimension $n \geq 1$, with normalized volume Q_n such that $Q_n = a_n$ for all $n \geq 1$. As demonstrated in [23], if $\Delta_{(1,q)}$ are all well-chosen, then the h^*-polynomials of these simplices can be nicely expressed in terms of the combinatorics of the representations of integers in the numeral system $(a_n)_{n=0}^{\infty}$. Even more, these expressions can help us recover real-rootedness of the h^*-polynomials by revealing recursions that preserve interlacing of polynomials. We now summarize the theory we will need on real-rootedness and interlacing polynomials in the next subsection.

2.2. *Real-rooted and Interlacing Polynomials*

Let $p(z) = p_0 + p_1 z + \cdots + p_d z^d \in \mathbb{R}[z]$ be a real-rooted polynomial with degree d, denoted $\deg(p) = d$, and let $\alpha_1 \geq \alpha_2 \geq \cdots \geq \alpha_d$ denote its roots. If $q(z) = q_0 + q_1 z + \cdots + q_n z^n \in \mathbb{R}[z]$ is a second real-rooted polynomial with $\deg(q) = n$ and roots $\beta_1 \geq \beta_2 \geq \cdots \geq \beta_n$. We say that q **interlaces** p, denoted $q \preceq p$, if the roots of p and q can be ordered such that

$$\alpha_1 \geq \beta_1 \geq \alpha_2 \geq \beta_2 \geq \alpha_3 \geq \beta_3 \geq \cdots.$$

A sequence of polynomials $(f_i)_{i=0}^{m}$ with $f_i \in \mathbb{R}[z]$ of polynomials is called **interlacing** if $f_i \preceq f_j$ for all $0 \leq i \leq j \leq m$. Let $\mathcal{F}_m^+$ denote the space of all interlacing sequences $(f_i)_{i=0}^{m}$ for which f_i has only nonnegative coefficients for all $0 \leq i \leq m$. In [9], the author gave a complete characterization of matrices $(G_{i,j}(z))_{i,j=1}^{m,n}$ that map $\mathcal{F}_n^+$ to $\mathcal{F}_m^+$. These matrices are said to **preserve interlacing**, and by iteratively applying the same matrix we produce recursions that preserve interlacing. In the coming sections, we will use two recursions that preserve interlacing.

Lemma 2.1 (Corollary 8.7 in [9]). *Let $(f_i)_{i=0}^{n}$ be an interlacing sequence of polynomials. For $m \in \mathbb{Z}_{\geq 0}$ and $\varphi : [m]_0 \longrightarrow \mathbb{Z}_{\geq 0}$ satisfying $\varphi(i) \leq \varphi(i+1)$ for all $i \in [m-1]_0$, define the polynomials*

$$g_i := z \sum_{j<\varphi(i)} f_i + \sum_{\varphi(i)\geq j} f_i.$$

Then the sequence of polynomials $(g_i)_{i=0}^m$ is interlacing.

A closely related, but distinct, recursion that also preserves interlacing is the following:

Lemma 2.2 (Lemma 4.4 in [23]). *Let $(f_i)_{i=0}^n$ be an interlacing sequence of polynomials. For $i \in [n]_0$ define the polynomial*

$$g_i := z \sum_{j \leq \varphi(i)} f_i + \sum_{\varphi(i) \geq j} f_i$$

Then the sequence of polynomials $(g_i)_{i=0}^n$ is interlacing.

In [23], the above lemmas were used to prove that families of $\Delta_{(1,q)}$ associated via their normalized volumes to the factoradic and base-r numeral systems described in Example 2.1 all have real-rooted Ehrhart h^*-polynomials. In the coming sections, we prove these families of simplices also have real-rooted local h^*-polynomials.

3. The Factoradic Simplex

Given a permutation $\pi = \pi_1\pi_2\cdots\pi_n \in \mathfrak{S}_n$ we say that an index $i \in [n-1]$ is a **descent** of π if $\pi_i > \pi_{i+1}$. We then let

$$\mathrm{des}(\pi) := |\{i \in [n-1] \,:\, \pi_i > \pi_{i+1}\}|,$$

and we call the polynomial

$$A_n(z) := \sum_{\pi \in \mathfrak{S}_n} z^{\mathrm{des}(\pi)}$$

the nth **Eulerian polynomial**. It is well-known that $A_n(z)$ is symmetric and real-rooted of degree $n-1$ for all $n \geq 1$. We additionally define

$$\mathrm{maxDes}(\pi) := \max\{i \in [n-1] \,:\, \pi_i > \pi_{i+1}\},$$

and we define a second polynomial

$$B_n(z) := \sum_{\pi \in \mathfrak{S}_n} z^{\mathrm{maxDes}(\pi)}.$$

We further let $a_{n,k}$ and $b_{n,k}$ denote the kth coefficient of $A_n(z)$ and $B_n(z)$, respectively. It follows from Lemma 3.4 in [23] that $B_n(z)$ is also a unimodal polynomial of degree $n-1$, and in [23], the author used the simplices $\Delta_{(1,q)}$ to offer a geometric interpolation between the two generating polynomials $A_n(z)$ and $B_n(z)$.

Definition 3.1. For $n \geq 1$, let $q^{(n)} := (b_{n+1,1}, b_{n+1,2}, \ldots, b_{n+1,n})$ be the coefficient sequence of $B_{n+1}(z)$ excluding its constant term. The **factoradic n-simplex** is the n-simplex

$$\Delta_n^! := \Delta_{(1,q^{(n)})}.$$

In [23] the author proved the following theorem which shows that if one takes the coefficient vector of the polynomial $B_n(z)$ as the vector q, then the polynomial $A_n(z)$ is returned as the h^*-polynomial of the simplex $\Delta_{(1,q)}$.

Theorem 3.1 (Theorem 3.5 in [23]). *The factoradic n-simplex $\Delta_n^!$ has h^*-polynomial*

$$h^*(\Delta_n^!; z) = A_{n+1}(z).$$

Since Theorem 3.1 immediately implies that $h^*(\Delta_n^!; z)$ is real-rooted and unimodal, it is natural to ask if these same properties hold for the associated local h^*-polynomial $\ell^*(\Delta_n^!; z)$. Indeed, we will prove that these properties do in fact carry over to $\ell^*(\Delta_n^!; z)$. To do this, we must first identify a nice combinatorial expression for this polynomial in terms of descent statistics for permutations. This will require a few more definitions, which will in turn shed a bit more light on the choice of the name "factoradic n-simplex."

Given a permutation $\pi \in \mathfrak{S}_n$, its associated **Lehmer code** (or **inversion sequence**) is the sequence $\ell(\pi) := (\ell_{n-1}, \ell_{n-2}, \ldots, \ell_1)$, where

$$\ell_i := |\{0 \leq j < i \,:\, \pi_{n-i} > \pi_{n-j}\}|.$$

The collection of all Lehmer codes for $\mathfrak{S}_n$ consists of the collection of lattice points

$$L_n := \left\{(\ell_{n-1}, \ell_{n-2}, \ldots, \ell_1) \in \mathbb{Z}^{n-1} \,:\, \ell_k \in [k]_0,\, k \in [n-1]\right\},$$

and there is a natural bijection between $\mathfrak{S}_n$ and L_n. In particular, to construct a permutation π from its Lehmer code start with the zero-indexed sequence of numbers $s = (s_i)_{i=0}^{n-1} := (1, 2, \ldots, n)$. Beginning with $k = 1$, set $\pi_k := s_{\ell_{n-k}}$, drop $s_{\ell_{n-k}}$ from the list s and repeat with $k := k + 1$. Finally, end by setting π_n equal to the last remaining number.

It turns out that Lehmer codes offer computer scientists an efficient way to encode the permutations $\mathfrak{S}_n$ in a lexicographic order. Given two strings of positive integers $a := a_1a_2\cdots a_n$ and $b := b_1b_2\cdots b_n$, we say that a is **lexicographically larger** than b if and only if the leftmost nonzero number in the string $(a_1 - b_1)(a_2 - b_2)\cdots(a_n - b_n)$ is positive. To each Lehmer code $\ell := (\ell_{n-1}, \ell_{n-2}, \ldots, \ell_1) \in L_n$, we associate a unique number $0 \leq b^{(\ell)} < n!$

$$b^{(\ell)} := \sum_{i=1}^{n-1} \ell_{n-i}(n-i)!, \tag{1}$$

and we note that every such number $0 \leq b < n!$ has such a unique representation. The Lehmer code ℓ associated to b via equation (1) is the representation of b in the factoradic numeral system, which was introduced in Example 2.1. Since $0 \leq b^{(m)} < b^{(\ell)} < n!$ if and only if ℓ is lexicographically larger than m, this provides us with a natural lexicographic ordering of the permutations via their Lehmer codes. In particular, given $0 \leq b < n!$, we let $\pi^{(b)}$ denote the associated b^{th} lexicographically largest permutation in $\mathfrak{S}_n$.

Similar to the permutations $\pi \in \mathfrak{S}_n$, we can also define descents in Lehmer codes: assuming that $\ell_0 := 0$, we say that an index $i \in [n-1]$ is a **descent** of a Lehmer code $\ell := (\ell_{n-1}, \ell_{n-2}, \ldots, \ell_1)$ if $\ell_i > \ell_{i-1}$, and we let

$$\mathrm{des}(\ell) := |\{i \in [n-1] : \ell_i > \ell_{i-1}\}|.$$

It is a short exercise to check that $n-k$ is a descent of $\ell(\pi)$ if and only if k is a descent in π. In order to prove Theorem 3.1, in [23] the author showed that, in the context of Theorem 2.2 of [6] and $\Delta_n^!$, for all $0 \leq b < n!$,

$$\omega(b) = \mathrm{des}(\pi^{(b)}).$$

That is, in Theorem 3.1, the Eulerian polynomial $A_n(z)$ is computed with respect to the lexicographic ordering of the permutations. By Theorem 2.1, to compute the local h^*-polynomial of $\Delta_n^!$, we then need only compute which values of $0 \leq b < n!$ satisfy the necessary divisibility conditions with respect to $q^{(n)} := (b_{n+1,1}, b_{n+1,2}, \ldots, b_{n+1,n})$. As we will see in the next theorem, these are, intriguingly, exactly those values $0 \leq b < n!$ that are congruent to 1 or 5 modulo 6. In the following, we let $\overline{k}_m$ denote the equivalence class of all integers congruent to k modulo m.

Theorem 3.2. *The factoradic n-simplex $\Delta_n^!$ has local h^*-polynomial*

$$\ell^*(\Delta_n^!; z) = \sum_{b \in [(n+1)!] \cap \left(\overline{1}_6 \cup \overline{5}_6\right)} z^{\mathrm{des}(\pi^{(b)})}.$$

In particular,

$$\ell^*(\Delta_n^!; 1) = \frac{n!}{3}.$$

Proof. To prove the statement we apply Theorem 2.1. In particular, for $q = (b_{n+1,1}, b_{n+1,2}, \ldots, b_{n+1,n})$, we would like to show that

$$T_q = [n+1] \cap \left(\overline{1}_6 \cup \overline{5}_6\right).$$

To see this, note that by Lemma 3.4 of [23], $0 < b < (n+1)!$ satisfies $b \in T_q$ if and only if $(n+1)!$ does not divide

$$\frac{(n+1)!b}{(n-k+1)! + (n-k)!}$$

for all $k \in [n]$. Equivalently, $b \in T_q$ if and only if $(n-k+1)! + (n-k)!$ does not divide b for all $k \in [n]$. To see that the latter condition is equivalent to $b \in \overline{1}_6 \cup \overline{5}_6$, notice that when $k = n$ we have

$$(n-k+1)! + (n-k)! = 1! + 0! = 2,$$

and when $k = n-1$ we have that

$$(n-k+1)! + (n-k)! = 2! + 1! = 3.$$

$$
\begin{array}{ccccccccccccc}
 & & & & & & 1 & & & & & & \\
 & & & & & 1 & & 1 & & & & & \\
 & & & & 1 & & 6 & & 1 & & & & \\
 & & & 1 & & 19 & & 19 & & 1 & & & \\
 & & 1 & & 48 & & 142 & & 48 & & 1 & & \\
 & 1 & & 109 & & 730 & & 730 & & 109 & & 1 & \\
1 & & 234 & & 3087 & & 6796 & & 3087 & & 234 & & 1
\end{array}
$$

Fig. 1. The triangle of coefficients for $\ell^*(\Delta_n^!; z)$ for $2 \leq n \leq 8$.

Moreover, for every $k < n-1$, both 2 and 3 divide $(n-k+1)!+(n-k)!$. Thus, $b \in T_q$ if and only if it is never divisible by 2 or 3, and this happens only if $b \in \overline{1}_6 \cup \overline{5}_6$. Since, by the proof of Theorem 3.5 in [23], we know that $\omega(b) = \mathrm{des}(\pi^{(b)})$ for all $0 \leq b < (n+1)!$, then we can apply Theorem 2.1 to recover that

$$\ell^*(\Delta_n^!; z) = \sum_{b \in [(n+1)!] \cap \left(\overline{1}_6 \cup \overline{5}_6\right)} z^{\mathrm{des}(\pi^{(b)})}.$$

To see the final statement, notice that $\ell^*(\Delta_n^!; 1)$ is the number of integers $0 \leq b < n!$ that are congruent to 1 or 5 modulo 6, and this can be computed as

$$n! - 4\left(\frac{n!}{6}\right) = \frac{n!}{3}.$$

This completes the proof. □

Remark 3.1. As noted in Theorem 3.2, $\ell^*(\Delta_n^!; 1) = \frac{n!}{3}$ and thus the open parallelpiped $\Pi_!^\circ$ of $\Delta_n^!$ contains one third of the lattice points within the half-open parallelpiped $\Pi_!$ of $\Delta_n^!$. Since the local h^*-polynomial $\Delta_n^!$ enumerates these lattice points via the descent statistics of their associated permutations, it is interesting to note that the sequence $\left(\frac{n!}{3}\right)_{n=0}^{\infty}$ also has other known connections to permutation statistics. This sequence is number A002301 on the *Online Encyclopedia of Integer Sequences (OEIS)*,[20] where more details of these connections can be found.

Example 3.1 (Some Factoradic local h^*-Polynomials). *The triangle of coefficients of $\ell^*(\Delta_n^!; z)$ for $2 \leq n \leq 8$ is given in Figure 1. We see that these are indeed not the* ***derangement polynomials****, which were shown to be the local h^*-polynomials of the s-lecture hall simplices P_n^s for $s = (2, 3, \ldots, n)$ [16]. The simplices $\Delta_n^!$ and P_n^s are both* ***reflexive*** *(see [23] for the definition of reflexive) with h^*-polynomial $A_{n+1}(z)$, and the combinatorics of both are intimately tied to that of inversion sequences [21, 23]. Despite this, the distinction in their local h^*-polynomials further highlights their fundamentally different geometry and combinatorics.*

Using the formula from Theorem 3.2 we can now affirmatively answer Question 1.1 for the factoradic simplex.

Theorem 3.3. *The local h^*-polynomial $\ell^*(\Delta_n^!; z)$ of the factoradic n-simplex $\Delta_n^!$ is real-rooted, and thus unimodal.*

Proof. To prove that the polynomials $\ell^*(\Delta_n^!; z)$ are real-rooted we will identify a polynomial recursion that will allow us to apply Lemma 2.1. For convenience, we first recall that the descents in a permutation $\pi \in \mathfrak{S}_n$ are in bijection with the descents in the associated Lehmer code $\ell(\pi)$. Thus, we may also express the polynomial $\ell^*(\Delta_n^!; z)$ as

$$\ell^*(\Delta_n^!; z) = \sum_{b \in [(n+1)!] \cap \left(\overline{1}_6 \cup \overline{5}_6\right)} z^{\mathrm{des}(\ell(b))},$$

where $\ell(b) = (\ell_{b,n}, \ell_{b,n-1}, \ldots, \ell_{b,1}) := \ell(\pi^{(b)})$ for all $0 \leq b < (n+1)!$. Recall that $0 \leq \ell_{b,n} < n$, and let $\chi(p)$ denote the boolean function yielding 1 if p is a true statement and 0 otherwise. For $0 \leq k < n$, we then define the polynomial

$$B_{n,k}(z) := \sum_{b \in [(n+1)!] \cap \left(\overline{1}_6 \cup \overline{5}_6\right)} \chi\left(\ell_{b,n} = k\right) z^{\mathrm{des}(\ell(b))}. \tag{2}$$

Notice first that

$$\ell^*(\Delta_n^!; z) = \sum_{k=0}^{n} B_{n,k}(z), \quad \text{and that} \quad \ell^*(\Delta_n^!; z) = B_{n+1,0}(z).$$

Moreover, it is quick to check that we have the recursion

$$B_{n,k}(z) = z \sum_{t<k} B_{n-1,t}(z) + \sum_{t \geq k} B_{n-1,t}(z), \tag{3}$$

with initial conditions $B_{3,0}(z) = z$, $B_{3,1}(z) = 0$, and $B_{3,2}(z) = z^2$. We note here that we must take the initial conditions with $n = 3$ since if we express b as in equation (1)

$$b = \sum_{i=1}^{n-1} \ell_{n-i}(n-i)!,$$

and set $b' := \sum_{i=1}^{n-2} \ell_{n-i}(n-i)!$, then we require that b is congruent to 1 or 5 modulo 6 if and only if b' is congruent to 1 or 5 modulo 6. However, this fails for $n < 3$. (Note however, that $\ell^*(\Delta_1^!; z) = 0$ and $\ell^*(\Delta_2^!; z) = z$, so real-rootedness will still hold in all cases.) Given the recursion in equation 3, the local h^*-polynomial of the factoradic n-simplex $\Delta_n^!$ is then real-rooted, and thus unimodal, since we observed that $\ell^*(\Delta_n^!; z) = B_{n+1,0}(z)$. □

Remark 3.2. The same type of recursion as given in equation (3) has recently been used in order to recover real-rootedness results for a number of polynomial generating functions in combinatorics [10, 16, 21, 23]. In particular, this recursion as applied to refined descent polynomials over inversion sequences, defined analogously to those in equation (2), can be viewed as the “reverse” of the recursion applied to inversion sequences to recover real-rootedness of the Ehrhart h^*-polynomials of the s-lecture hall simplices [21].

Since the local h^*-polynomials $\ell^*(\Delta_n^!; z)$ are both symmetric and real-rooted, it follows that they also have only nonnegative integer coefficients in another well-studied polynomial basis. A polynomial $p(z) \in \mathbb{R}[z]$ of degree at most n is called **γ-nonnegative** (or **γ-positive**) if it has only nonnegative coefficients when expressed in the basis $\left\{z^i(z+1)^{n-2i}\right\}_{i=0}^{\lfloor n/2 \rfloor}$. It is well-known that if a $p(z)$ is real-rooted and symmetric with nonnegative coefficients in the standard basis, then $p(z)$ is γ-nonnegative (see for instance Remark 3.1 in [9]). Thus, we have the following corollary.

Corollary 3.1. *The local h^*-polynomial $\ell^*(\Delta_n^!; z)$ of the factoradic n-simplex $\Delta_n^!$ is γ-nonnegative.*

Proof. Since $\ell^*(\Delta_n^!; z)$ is symmetric and real-rooted with nonnegative coefficients it follows that it is also γ-nonnegative. □

Remark 3.3. In Corollary 3.3 of [9], the author presents a sufficient condition for a restricted Eulerian polynomial

$$A(T; z) := \sum_{\pi \in T \subset \mathfrak{S}_n} z^{\mathrm{des}(\pi)}$$

to be γ-nonnegative. This sufficient condition requires that the subset $T \subset \mathfrak{S}_n$ be invariant under the **valley–hopping** action of $\mathbb{Z}_2^n$ on $\mathfrak{S}_n$ first considered in [14] Since the polynomial $\ell^*(\Delta_n^!; z)$ is an example of restricted Eulerian polynomial (as seen by Theorem 3.2), it is natural to ask whether or not its γ-nonnegativity arises as a consequence of Corollary 3.3 from [9]. In fact, it does not. To see this, suppose that

$$T := \left\{\pi^{(b)} \in \mathfrak{S}_n \,:\, b \in \overline{1}_6 \cup \overline{5}_6\right\},$$

and notice that $\pi^{((n+1)!-1)} = n \cdots 321 \in T$. The unique element of the orbit of $\pi^{((n+1)!-1)}$ under the valley–hopping action that has no double descents (i.e. no indices $k \in [n]$ for which $\pi_{k-1} > \pi_k > \pi_{k+1}$) is $\pi^{(0)} = 123 \cdots n$. However, $\pi^{(0)} \notin T$, and thus T cannot be invariant under the valley–hopping action.

Since the valley–hopping action cannot be used to give a direct combinatorial proof of the γ-nonnegativity of $\ell^*(\Delta_n^!; z)$ that is implied by its real-rootedness, it would therefore be of interest to provide an alternate combinatorial interpretation of the coefficients of $\ell^*(\Delta_n^!; z)$ when it is written in the γ-basis.

4. The Base-r Simplex

A second family of $\Delta_{(1,q)}$'s that are known to have real-rooted h^*-polynomials are those with

$$q^{(r,n)} := \left((r-1), (r-1)r, \ldots, (r-1)r^{i-1}, \ldots, (r-1)r^{n-1}\right),$$

for $r \geq 2$ and $n \geq 1$ [23]. For simplicity, we will set $\mathcal{B}_{(r,n)} := \Delta_{(1,q^{(r,n)})}$ throughout this section. In [23], these simplices are referred to as the **base-r n-simplices** since the normalized volume of $\mathcal{B}_{(r,n)}$ is equal to r^n, and this is the nth place value in the base-r numeral system.

In [16] the authors observed that the s-lecture hall simplices, which have real-rooted h^*-polynomials also have real-rooted local h^*-polynomials. In [23] and Section 3 we, respectively, saw that the same holds true for the factoradic simplex $\Delta_n^!$. In this section, we continue this trend by further observing that all base-r simplices have real-rooted local h^*-polynomials as well. To do this, we must first define an additional family of polynomials.

For $r \geq 2$ and $n \geq 1$ define the polynomial

$$f_{(r,n)}(z) := (1 + z + z^2 + \cdots + z^{r-1})^n.$$

By writing each $0 \leq m \leq n(r-1)$ as $m = p(r-1) + q$ where $0 \leq q < r-1$, the uniqueness of this expression guarantees that for any $r \geq 2$ and polynomial $f(z) \in \mathbb{R}[z]$ there exist unique polynomials $f^{(0)}, \ldots, f^{(r-2)} \in \mathbb{R}[z]$ such that

$$f(z) = f^{(0)}(z^{r-1}) + zf^{(1)}(z^{r-1}) + z^2 f^{(2)}(z^{r-1}) + \cdots + z^{r-2} f^{(r-2)}(z^{r-1}).$$

For $\ell \in [r-2]_0$ we then let $f^{\langle r-1,\ell\rangle} := f^{(\ell)}$. In [23], the author gave the following formula for $h^*(\mathcal{B}_{(r,n)}; z)$, from which they recovered that this h^*-polynomial is real-rooted.

Theorem 4.1 (Theorem 4.2 in [23]). *For $r \geq 2$ and $n \geq 1$, the base-r n-simplex $\mathcal{B}_{(r,n)}$ has h^*-polynomial*

$$h^*(\mathcal{B}_{(r,n)}; z) = f_{(r,n)}^{\langle r-1,0\rangle} + z \sum_{\ell=1}^{r-2} f_{(r,n)}^{\langle r-1,\ell\rangle}.$$

In a similar fashion, one can show that the local h^*-polynomial can also be computed in terms of the polynomials $f_{(r,n)}^{\langle r-1,\ell\rangle}$.

Theorem 4.2. *For $r \geq 2$ and $n \geq 1$, the base-r n-simplex has local h^*-polynomial*

$$\ell^*(\mathcal{B}_{(r,n)}; z) = z \sum_{i=0}^{r-2} f_{(r,n-1)}^{\langle r-1,i\rangle} + z \sum_{\ell=1}^{r-2} \left(\sum_{i=0}^{\ell-1} f_{(r,n-1)}^{\langle r-1,i\rangle} + z \sum_{i=\ell}^{r-2} f_{(r,n-1)}^{\langle r-1,i\rangle} \right).$$

Proof. To deduce the desired formula for $\ell^*(\mathcal{B}_{(r,n)}; z)$ we first apply Theorem 2.1. We have that $Q = r^n$, and so each lattice point in the half-open parallelpiped for $\mathcal{B}_{(r,n)}$ corresponds to an integer $0 \leq b < r^n$. As such, we will denote the lattice point corresponding to b by $x^{(b)}$. Letting $\Pi_{(r,n)}^\circ$ denote the open parallelpiped of $\mathcal{B}_{(r,n)}$, Theorem 2.1 tells us that $x^{(b)} \in \Pi_{(r,n)}^\circ$ if and only if $r^n \nmid (r-1)r^{i-1}b$ for all $i \in [n]$. Equivalently, $x^{(b)} \in \Pi_{(r,n)}^\circ$ if and only if $r^{n-i+1} \nmid (r-1)b$ for all $i \in [n]$. Furthermore, since r and $r-1$ are always coprime, then so are r^{n-i+1} and $(r-1)$

for all $i \in [n]$. Therefore, $x^{(b)} \in \Pi^\circ_{(r,n)}$ if and only if $r^{n-i+1} \nmid b$ for all $i \in [n]$. Equivalently, we then have that $x^{(b)} \in \Pi^\circ_{(r,n)}$ if and only if $r \nmid b$. It follows that,

$$\ell^*(\mathcal{B}_{(r,n)}; z) = h^*(\mathcal{B}_{(r,n)}; z) - \sum_{0<b<r^n \,:\, r\nmid b} z^{\omega(b)}. \tag{4}$$

To compute the latter sum in equation (4), notice that for $0 < b < r^n$, if $r \big| b$ such that $b = rb'$ for $0 < b' < r^{n-1}$ then

$$\begin{aligned}
\omega(b) &= b - \sum_{i=1}^{n} \left\lfloor \frac{(r-1)b}{r^i} \right\rfloor, \\
&= rb' - (r-1)b' - \sum_{i=1}^{n-1} \left\lfloor \frac{(r-1)b'}{r^i} \right\rfloor, \\
&= b' - \sum_{i=1}^{n-1} \left\lfloor \frac{(r-1)b'}{r^i} \right\rfloor, \\
&= \omega(b').
\end{aligned}$$

Therefore, we have that the latter sum in equation (4) is actually the h^*-polynomial of $\mathcal{B}_{(r,n-1)}$, the base-r simplex of one dimension less. That is,

$$\sum_{0<b<r^n \,:\, r\nmid b} z^{\omega(b)} = h^*(\mathcal{B}_{(r,n-1)}; z).$$

Consequently, we have that

$$\ell^*(\mathcal{B}_{(r,n)}; z) = h^*(\mathcal{B}_{(r,n)}; z) - h^*(\mathcal{B}_{(r,n-1)}; z). \tag{5}$$

In the proof of Theorem 4.2 in [23], it is shown that the polynomials $f^{\langle r-1,\ell\rangle}_{(r,n)}$ for $0 \le \ell \le r-2$ satisfy the recursion

$$f^{\langle r-1,\ell\rangle}_{(r,n)} = \sum_{i=0}^{\ell} f^{\langle r-1,i\rangle}_{(r,n-1)} + z\sum_{i=\ell}^{r-2} f^{\langle r-1,i\rangle}_{(r,n-1)}. \tag{6}$$

Using the recursion in equation (6) together with Theorem 4.1 we can deduce from equation (5) that

$$\begin{aligned}
\ell^*(\mathcal{B}_{(r,n)}; z) &= h^*(\mathcal{B}_{(r,n)}; z) - h^*(\mathcal{B}_{(r,n-1)}; z), \\
&= \left(f^{\langle r-1,0\rangle}_{(r,n)} - f^{\langle r-1,0\rangle}_{(r,n-1)}\right) + z\sum_{\ell=1}^{r-2}\left(f^{\langle r-1,\ell\rangle}_{(r,n)} - f^{\langle r-1,\ell\rangle}_{(r,n-1)}\right), \\
&= z\sum_{i=0}^{r-2} f^{\langle r-1,i\rangle}_{(r,n-1)} + z\sum_{\ell=1}^{r-2}\left(\sum_{i=0}^{\ell-1} f^{\langle r-1,i\rangle}_{(r,n-1)} + z\sum_{i=\ell}^{r-2} f^{\langle r-1,i\rangle}_{(r,n-1)}\right),
\end{aligned}$$

which completes the proof. □

It then follows from the formula given in Theorem 4.2 that $\ell^*(\mathcal{B}_{(r,n)};z)$ is also a real-rooted, and therefore unimodal, polynomial for all $r \geq 2$ and $n \geq 1$.

Corollary 4.1. *The local h^*-polynomial $\ell^*(\mathcal{B}_{(r,n)};z)$ of the base-r n-simplex is real-rooted, and thus unimodal, for all $r \geq 2$ and $n \geq 1$.*

Proof. To prove that $\ell^*(\mathcal{B}_{(r,n)};z)$ is real-rooted we again apply the theory of interlacing sequences and the interlacing-preserving recursions presented in Lemmas 2.1 and 2.2. By Lemma 2.2, we know that the recursion for the polynomials $f_{(r,n)}^{\langle r-1,\ell\rangle}$ given in equation (6) preserves interlacing. Since the polynomials

$$g_{\ell-1} := \sum_{i=0}^{\ell-1} f_{(r,n-1)}^{\langle r-1,i\rangle} + z\sum_{i=\ell}^{r-2} f_{(r,n-1)}^{\langle r-1,i\rangle}$$

for $1 \leq \ell \leq r-1$ are produced by applying the recursion in Lemma 2.1 once to the interlacing sequence

$$\left(f_{(r,n-1)}^{\langle r-1,r-2\rangle}, \ldots, f_{(r,n-1)}^{\langle r-1,1\rangle}, f_{(r,n-1)}^{\langle r-1,0\rangle}\right),$$

and since this recursion also preserves interlacing, then we know that the sequence of polynomials $(g_{r-2}, \ldots, g_1, g_0)$ is also an interlacing sequence. Moreover, by Theorem 4.2 we know that

$$\ell^*(\mathcal{B}_{(r,n)};z) = z\sum_{\ell=0}^{r-1} g_{\ell-1}.$$

It follows that $\ell^*(\mathcal{B}_{(r,n)};z)$ is real-rooted (by, for example, Lemma 2.2 in [8]). $\square$

Similar to the case of the local h^*-polynomials for the factoradic simplices $\Delta_n^!$, the real-rootedness of $\ell^*(\mathcal{B}_{(r,n)};z)$ immediately implies that it is also γ-nonnegative.

Corollary 4.2. *For all $r \geq 2$ and $n \geq 1$ the local h^*-polynomial $\ell^*(\mathcal{B}_{(r,n)};z)$ of the base-r n-simplex is γ-nonnegative.*

Example 4.1 (The Base-2 Simplex). *In the case when $r = 2$, the normalized volume of $\mathcal{B}_{(2,n)}$ is equal to 2^n, the nth place value in the binary numeral system. In Theorem 3.6 of [23] it is shown that the h^*-polynomial of $\mathcal{B}_{(2,n)}$ is a familiar symmetric, real-rooted and unimodal polynomial; namely,*

$$h^*(\mathcal{B}_{(2,n)};z) = (1+z)^n.$$

By applying Corollary 4.1 we can similarly deduce that

$$\ell^*(\mathcal{B}_{(2,n)};z) = z(1+z)^{n-1}. \tag{7}$$

In particular, when $r = 2$, we have that $f_{(2,n)} = (1+z)^n$, $r-2 = 0$, and

$$f_{(2,n)}^{\langle 1,0\rangle} = (1+z)^n.$$

Since $r-2=0$, the second sum in Theorem 4.2 is empty, which leaves us with the formula for $\ell^*(\mathcal{B}_{(2,n)};z)$ in equation (7).

Finally, we remark that the formula for $\ell^*(\mathcal{B}_{(2,n)}; z)$ given in equation (7) *also has a natural combinatorial interpretation. In Theorem 3.6 of [23], it is shown that*

$$h^*(\mathcal{B}_{(2,n)}; z) = \sum_{b=0}^{2^n-1} z^{\mathrm{supp}_2(b)},$$

where $\mathrm{supp}_2(b)$ *denotes the number of* 1*'s in the binary representation of the integer* b*. Equation* (7) *simply says that*

$$\ell^*(\mathcal{B}_{(2,n)}; z) = \sum_{0<b<2^n \,:\, b \text{ is odd}} z^{\mathrm{supp}_2(b)}.$$

5. Final Remarks

We end with a few comments on the possible future directions of research relating to the results in this note. First, we make a remark on the general relationship between real-rootedness of h^*-polynomials of simplices and that of their local h^*-polynomials: The results of [16] and those of this paper demonstrate that certain families of lattice simplices with real-rooted h^*-polynomials also admit real-rooted local h^*-polynomials. It is therefore natural to ask if real-rootedness of the h^*-polynomial implies real-rootedness of the local h^*-polynomial and/or vice versa. Using the observations in this paper, we can already show that the converse of this statement does not hold: By Corollary 2.1, we know that the 2-simplex $\Delta_{(1,q)}$ with $q = (1,1)$ has a real-rooted local h^*-polynomial, but not a real-rooted h^*-polynomial. On the other hand, the author currently is not aware of a lattice simplex with a real-rooted h^*-polynomial but a non-real-rooted local h^*-polynomial. Examples of such simplices and an understanding of their lattice point combinatorics would be interesting to see.

As noted in Corollaries 3.1 and 4.2, it follows immediately from the real-rootedness of the local h^*-polynomials studied in this note that they are also γ-nonnegative. The well-known technique of valley-hopping does not seem to offer a second proof of this fact (see Remark 3.3 in the case of the factoradic simplex). Thus, it would be interesting to see a purely combinatorial proof of the γ-nonnegativity of the local h^*-polynomials studied in this paper, as well as those in [16].

It was also recently shown that the local h^*-polynomial of a simplex satisfies a monotonicity property with respect the local h-polynomials of its geometric subdivisions (see Remark 7.18 in [17]). In particular, if Δ is a lattice simplex with a lattice triangulation T, and $\ell_\Delta(T_\Delta; z)$ is the local h-polynomial of the (abstract) subdivision of Δ defined by T, then

$$\ell^*(\Delta; z) \geq \ell_\Delta(T_\Delta; z).$$

Thus, it would be interesting to know if the local h^*-polynomials computed here can be realized as local h-polynomials for some (possibly only abstract) subdivision of a simplex. One way to accomplish this would be to prove that the simplices studied here all admit unimodular triangulations (see Example 7.19 in [17]).

Acknowledgements. The author was supported by an NSF Mathematical Sciences Postdoctoral Research Fellowship (DMS - 1606407). This manuscript was prepared as a contribution to the conference proceedings of the 2018 *Workshop on Lattice Polytopes* at Osaka University. The author thanks the organizers, Dr. Takayuki Hibi and Dr. Akiyoshi Tsuchiya, for their support during this workshop.

References

[1] C. A. Athanasiadis. *A survey of subdivisions and local h-vectors.* The Mathematical Legacy of Richard P. Stanley (2016): 39-52.

[2] M. Beck and S. Robins. *Computing the continuous discretely.* Springer Science+ Business Media, LLC, 2007.

[3] U. Betke and P. McMullen. *Lattice points in lattice polytopes.* Monatshefte für Mathematik 99.4 (1985): 253-265.

[4] B. Braun. *Unimodality problems in Ehrhart theory.* Recent trends in combinatorics. Springer, Cham, 2016. 687-711.

[5] B. Braun and R. Davis. *Ehrhart series, unimodality, and integrally closed reflexive polytopes.* Annals of Combinatorics 20.4 (2016): 705-717.

[6] B. Braun, R. Davis, and L. Solus. *Detecting the integer decomposition property and Ehrhart unimodality in reflexive simplices.* Advances in Applied Mathematics 100 (2018): 122-142.

[7] B. Braun and F. Liu. *h^*-polynomials with roots on the unit circle.* Preprint available at `https://arxiv.org/pdf/1807.00105.pdf` (2018).

[8] P. Brändén. *On linear transformations preserving the Pólya frequency property.* Transactions of the American Mathematical Society 358.8 (2006): 3697-3716.

[9] P. Brändén. *Unimodality, log-concavity, real-rootedness and beyond.* Handbook of Enumerative Combinatorics (2015): 437-483.

[10] P. Brändén and M. Leander. *Lecture hall P-partitions.* Preprint available at `https://arxiv.org/abs/1609.02790` (2016).

[11] F. Brenti. *Unimodal log-concave and Pólya frequency sequences in combinatorics.* No. 413. American Mathematical Soc., 1989.

[12] H. Conrads. *Weighted projective spaces and reflexive simplices.* Manuscripta Mathematica 107.2 (2002): 215-227.

[13] E. Ehrhart. *Sur les polyhèdres rationnels homothètiques à n dimensions.* C. R. Acad Sci. Paris, 254:616-618, 1962.

[14] D. Foata and V. Strehl. *Euler numbers and variations of permutations.* Colloquio Internazionale sulle Teorie Combinatorie 1 (1973): 119-131.

[15] A. S. Fraenkel. *Systems of numeration.* The American Mathematical Monthly, Vol. 92, No. 2 (Feb., 1985), pp. 105-114.

[16] N. Gustafsson and L. Solus. *Derangements, Ehrhart theory, and local h-polynomials.* Preprint available at `https://arxiv.org/abs/1807.05246` (2018).

[17] E. Katz and A. Stapledon. *Local h-polynomials, invariants of subdivisions, and mixed Ehrhart theory.* Advances in Mathematics 286 (2016): 181-239.

[18] F. Liu and L. Solus. *On the relationship between Ehrhart unimodality and Ehrhart positivity.* Preprint available at `https://arxiv.org/abs/1804.08258` (2018).

[19] B. Nill. *Volume and lattice points of reflexive simplices.* Discrete & Computational Geometry 37.2 (2007): 301-320.

[20] N. J. Sloane. *The On-Line Encyclopedia of Integer Sequences.* (2003).

[21] C. D. Savage, M. Visontai, *The s-Eulerian polynomials have only real roots,* Trans. Amer. Math. Soc., **367**(2), (2015), 1441-1466.

[22] J. Schepers and L. Van Langenhoven. *Unimodality questions for integrally closed lattice polytopes.* Annals of Combinatorics 17.3 (2013): 571-589.

[23] L. Solus. *Simplices for numeral systems* Trans. Amer. Math. Soc. DOI: https://doi.org/10.1090/tran/7424 (2017).

[24] R. P. Stanley. *Decompositions of rational convex polytopes.* Annals of discrete mathematics 6 (1980): 333-342.

On the faces of simple polytopes

Johanna Steinmeyer

Institut für Mathematik
Freie Universität Berlin
Arnimallee 2, 14195 Berlin, Germany
E-mail: steinmeyer@math.fu-berlin.de

For a general polytope, the graph as well as higher-dimensional skeletons of the polytope are not enough to determine the combinatorial structure. In the case of simple polytopes however, Blind and Mani showed in a celebrated result that the graph already uniquely determines the combinatorial structure of the polytope. To reconstruct it, the entire graph is used at once, while the question of locally determining which subgraph corresponds to a face of the polytope remains unanswered. We show how the faces of a simple polytope may be locally characterized by the 2-skeleton and survey related progress and results since the theorem of Blind and Mani.

Keywords: Simple polytopes; reconstruction; skeleton; graph of a polytope.

1. Introduction

The question how to reconstruct a polytope from partial information of face incidences is a central problem in polytope theory. One of the earliest comprehensive instances of its analysis is found in Grünbaum's classical book of polytope theory [7]. Polytope reconstruction has seen many developments since then, see for example [2] for a recent survey on this and related topics.

We consider a d-dimensional convex polytope P, up to combinatorial equivalence. That is, we consider two polytopes equivalent, whenever they have the same *face lattice* – the set of faces, partially ordered by inclusions. We refer to [15] for a general introduction to the theory of convex polytopes. For a non-negative number $k \leq d$, the *k-skeleton* $\Delta^{\leq k}P$ of P is the collection of the faces of P of dimension at most k, considered as a polyhedral complex. We say that two polytopes P and Q are *k-equivalent* if $\Delta^{\leq k}P$ is combinatorially equivalent to $\Delta^{\leq k}Q$. The 1-skeleton is also called the *vertex-edge graph* $G(P)$ of the polytope, defined by the vertices (0-dimensional faces) and edges (1-dimensional faces) of P. We call $N_P(x) := \{v \in V(P), x \sim y\}$ the *neighborhood* of x.

In Chapter 12 of his book [7], Grünbaum analyzed in what cases k-equivalence of polytopes implies the combinatorial equivalence of the polytopes. While the facet-vertex incidences alone fully describe the face lattice of a polytope, the k-skeleton may be far from enough information to determine the polytope.

Example 1.1. Consider a cyclic polytope on n vertices in dimension d. It has a complete $\lfloor d/2 \rfloor$-skeleton, the same as the $(n-1)$-dimensional simplex. For a general cyclic polytope, this shows the possibility of non-equivalent polytopes which do not even have the same dimension, despite being $\lfloor d/2 \rfloor$-equivalent.

For $k > \lfloor d/2 \rfloor$, one gets the first non-trivial positive reconstruction result.

Theorem 1.1 (Grünbaum, 1967). *Any two $\lfloor d/2 \rfloor$-equivalent polytopes are combinatorially equivalent.*

Any two $(d-2)$-equivalent polytopes are combinatorially equivalent.

The second assertion follows from the fact that every subset of the d-dimensional sphere which is a homeomorphic image of the $(d-1)$-dimensional sphere divides the d-sphere into two connected components.

Grünbaum already gave examples showing the sharpness of this statement. The following easy example is due to Doolittle, Nevo, Pineda-Villavicencio, Ugon, and Yost [5].

Example 1.2. For $d \geq 4$, consider the bipyramid with the $(d-1)$-dimensional simplex as a base, and the pyramid over the bipyramid with the $(d-2)$-dimensional simplex as a base, see Figure 1. Both of these polytopes are of dimension d. It is easy to see that they have isomorphic $(d-3)$-skeletons. However, their respective f-vectors, which list the number of faces in each dimension, are $(1, 6, 14, 16, 8, 1)$ and $(1, 6, 14, 15, 7, 1)$, showing that the polytopes are non-isomorphic.

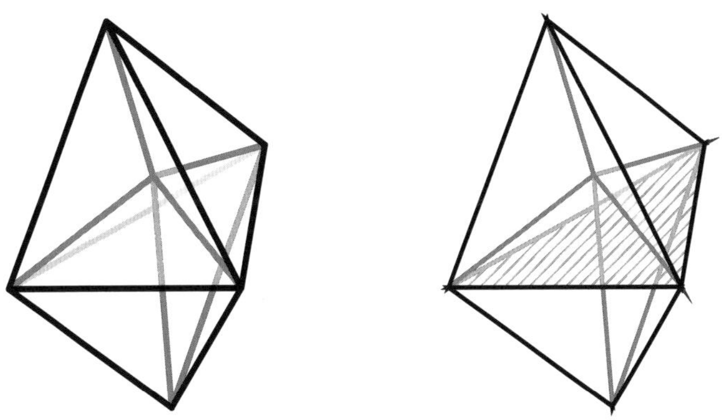

Fig. 1. Schlegel diagrams of the two 4-dimensional polytopes $\mathrm{bipyr}(\Delta_3)$ and $\mathrm{pyr}(\mathrm{bipyr}(\Delta_2))$, with the missing 2-face in $\mathrm{pyr}(\mathrm{bipyr}(\Delta_2))$ marked.

For most of this note, we focus on simple polytopes. A d-polytope P is *simple*, if its graph $G(P)$ is d-regular. In the context of reconstruction, simple polytopes are of particular interest, as they turn out to have a much more regular structure than general polytopes, making them more accessible to reconstruction.

The following special role of the 2-skeleton of simple polytopes is well-known. See for example Murty [12], who proves the it in the more general setting of abstract polytopes and Kalai [11], who accredits it to Perles.

Proposition 1.1. *The 2-skeleton of a simple polytope determines its combinatorial structure.*

To see this, first observe the following simple, but often crucial fact of simple polytopes.

Observation 1.1. Let P be a simple d-polytope, v a vertex of P, and $0 \leq k \leq d$, then for any choice $S \subseteq N_P(v)$ of k neighboring vertices of v, those k vertices lie in a unique k-face F of P together with v.

We say that S spans F at v. When it is clear at which vertex F is spanned, we may also say that F is spanned by the set $S \cup \{v\}$ of vertices or the set of edges from v to S.

Proof of Proposition 1.1. Let x, y be adjacent vertices in P. We consider a bijection between their respective sets of neighboring vertices,

$$\varphi_{x,y} : N_P(x) \setminus \{y\} \to N_P(y) \setminus \{x\},$$

where $\varphi_{x,y}$ sends a vertex v to the unique vertex w such that $\{x, y, v\}$ and $\{x, y, w\}$ span the same 2-face of P. On the other hand, let F be the unique facet spanned at x by all neighbors except for v. Sending v to the only neighbor of y that is not contained in F again yields a bijection between $N_P(x) \setminus \{y\}$ and $N_P(y) \setminus \{x\}$. This map coincides with $\varphi_{x,y}$. Thus we may determine the vertex-facet incidences with $\varphi_{x,y}$, which give the entire combinatorial structure of P. □

2. Reconstructing a simple polytope from its graph

While for general polytopes the graph does not determine the combinatorial structure of the polytope, Perles conjectured and Blind and Mani proved in [3] that this is indeed the case for simple polytopes.

Theorem 2.1 (Blind–Mani, 1987). *The graph of a simple polytope fully determines its combinatorial structure.*

Their proof is based in the dual setting of facet adjacencies of simplicial polytopes, making use of shellings and homology groups.

Shortly after, Kalai [11] gave a very short and elegant proof of the same result. His proof resides in the original simple setting, using shellings only in the disguise of abstract objective functions. We give a version of Kalai's proof, as it is the basis for several more recent related developments.

Let $\mathcal{O}$ be an acyclic orientation on the edges of $G(P)$. A *sink* of a subgraph $H \subseteq G(P)$ is a vertex v with all incident edges in H oriented towards v. $\mathcal{O}$ is an *abstract objective function* on $G(P)$, if the graph of every non-empty face of P contains precisely one sink. We may also describe the abstract objective function by the induced partial order on the vertices, where $v \leq w$ whenever there is an oriented path from v to w. Abstract objective functions generalize linear objective functions in general position with respect to the polytope.

Proof of Theorem 2.1. For a simple d-polytope P and $0 \leq k \leq d$, let $h_k(\mathcal{O})$ be the number of vertices of the graph $G(P)$ of P with indegree k under the orientation $\mathcal{O}$. Note that when $\mathcal{O}$ is an abstract objective function, these are the h-numbers of the simplicial polytope dual to P, and independent of the choice of the specific abstract objective function. See for example [15, Chapter 8.3] for an introduction to h-numbers. Define

$$\mathcal{H}(\mathcal{O}) := \sum_{k=0}^{d} h_k(\mathcal{O})\, 2^k.$$

Any vertex v of P with indegree k is a sink in precisely 2^k faces of P, since any subset of these k edges span a unique face. As every face contains at least one sink, we know that $\mathcal{H}(\mathcal{O})$ is no less than the total number of non-empty faces, with equality if and only if $\mathcal{O}$ is an abstract objective function. This characterizes the abstract objective functions among all acyclic orientations without any knowledge of the combinatorial structure of the polytope beyond the graph $G(P)$.

We now claim that an induced connected k-regular subgraph $H \subseteq G(P)$ corresponds to a k-face of P if and only if the set of vertices of H is initial with respect to (the partial order given by) some abstract objective function. For any face of the polytope, a linear objective function in general position may be chosen such that the vertices of the face are initial under the induced abstract objective function. On the other hand, consider a k-regular connected induced subgraph $H \subseteq G(P)$ whose vertices are initial under a given abstract objective function $\mathcal{O}$. Let v be a sink in H. It has k incident edges in H, all oriented towards it. Let F be the unique k-face F of P spanned by the k edges at v. Under the abstract objective function $\mathcal{O}$, the vertex v is the unique sink of F. Thus for all vertices w of F, we get $w \leq v$ implying $w \in H$, since the vertex set of H is initial with respect to $\mathcal{O}$. With $G(F)$ and H both being k-regular connected subgraphs of $G(P)$, we get $G(F) = F$.

The faces of the polytope thus may be identified directly from $G(P)$. □

In contrast to the proof of Blind and Mani, this proof given by Kalai is constructive. The implied algorithm however gives an exponential running time, computing all acyclic orientations of the graph.

3. Extending the result of Blind–Mani

Joswig, Kaibel, and Körner [9, 10] extended Kalai's approach of abstract objective functions, refining the function of Kalai.

$$\mathcal{H}(\mathcal{O}) = \sum_{k=0}^{d} h_k(\mathcal{O})2^k = \sum_{k=0}^{d} \mathcal{H}_k(\mathcal{O}), \quad \text{where} \quad \mathcal{H}_k(\mathcal{O}) := \sum_{i=0}^{d} h_k(\mathcal{O})\binom{i}{k}$$

They proposed the notion of a k-system, a set of induced subgraphs satisfying simple combinatorial conditions always fulfilled by those sets of subgraphs given by the set of k-faces. They thus provide certificates for the correctness of the reconstruction which may be checked in polynomial time. Among the k-systems, the set of k-faces is characterized by maximal cardinality, while $\mathcal{H}_k(\mathcal{O})$ is minimal if and only if $\mathcal{O}$ induces exactly one sink on every k-face. In their work, Joswig, Kaibel, and Körner emphasize the special role played by the 2-faces of the polytope.

Theorem 3.1 (Joswig–Kaibel–Körner, 2002). *Let $\mathcal{O}$ be an acyclic orientation on the graph $G(P)$ of a simple polytope P. Then $\mathcal{O}$ is an abstract objective function if and only if it induces precisely one sink on every 2-face of P.*

Building on the work of Joswig, Kaibel, and Körner, Friedman [6] showed that a simple polytope may be reconstructed from its graph in polynomial time, making use of linear programming.

Theorem 3.2 (Friedman, 2009). *A simple polytope can be reconstructed from its graph in polynomial time.*

Recently, Doolittle, Nevo, Pineda-Villavicencio, Ugon and Yost considered in [5] a generalization of the problem, reconstructing d-polytopes with a limited number of vertices which have more than d incident edges.

Theorem 3.3 (Doolittle–Nevo–Pineda-Villavicencio–Ugon–Yost, 2018). *The graph of a d-polytope P with at most two non-simple vertices determines fully determines the combinatorial structure of P.*

They gave examples of non-isomorphic polytopes with $d-2$ non-simple vertices and isomorphic graphs, showing that this theorem is tight for $d = 4$. For polytopes with only one non-simple vertex, they further showed that the reconstruction may be done in polynomial time in the number of vertices of the polytope.

Again highlighting the importance of the 2-skeleton of the polytopes, Doolittle, Nevo, Pineda-Villavicencio, Ugon and Yost further characterized the reconstruction of a nearly simple polytope from its 2-skeleton.

Theorem 3.4 (Doolittle–Nevo–Pineda-Villavicencio–Ugon–Yost, 2018). *The 2-skeleton of a d-polytope P with at most $d-2$ non-simple vertices in each facet determines the combinatorial structure of P.*

Furthermore, all facets can be reconstructed in linear time in the number of vertices of the polytope.

4. How to locally tell faces of simple polytopes

A direct proof of an efficient reconstruction algorithm would have been given, if a question of Perles (see [11]) had turned out to be true. It asks to locally characterize those subgraphs of the graph of the polytope that belong to a facet of the polytope.

Question 4.1 (Perles). *Does every $(d-1)$-regular, induced, connected, non-separating subgraph of the graph of a simple polytope correspond to a facet of P?*

The conditions proposed are all immediate properties of facet subgraphs.

For $d = 3$ this is equivalent to a result of Whitney [14] which holds for (not necessarily simple) polytopes.

Theorem 4.1 (Whitney, 1932). *The graphs of the 2-faces of a 3-polytope P are precisely the induced cycles of $G(P)$ that do not separate the graph.*

In dimensions 4 and up, not all induced non-separating cycles correspond to 2-faces, even for simple polytopes. To see this, consider the following graph of a 4-dimensional hypercube and the cycle marked in red. It is induced, and non-separating but not the graph of a 2-face of the hypercube.

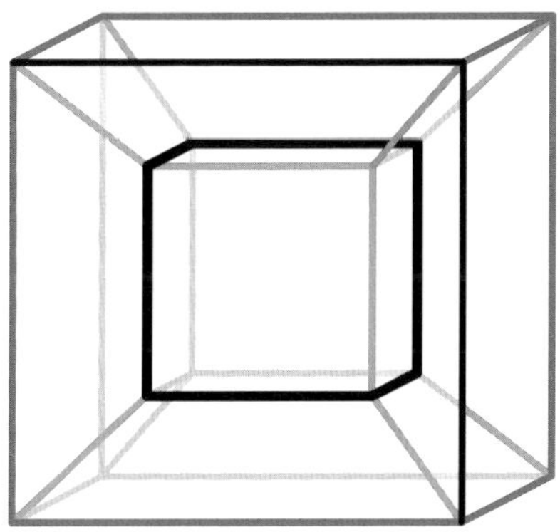

Fig. 2. A hypercube with a non-separating induced cycle which is not a face marked.

For $d \geq 4$ Perles' Conjecture was disproved by Haase and Ziegler [8]. They also gave a counterexample with the subgraph in question $(d-1)$-connected, a property of the graph of any $(d-1)$-polytope proved by Balinksi [1]. Haase and Ziegler asked whether every induced, non-separating subgraph isomorphic to the graph of a simple $(d-1)$-polytope is the graph of a facet.

During the time of this writing, Doolittle [4] found a new 4-dimensional counterexample to Perles' Conjecture. This example contains an induced, non-separating subgraph isomorphic to the graph of a simple 3-polytope which is *not* the graph of a facet, thus giving a negative answer to the question posed by Haase and Ziegler.

This however leaves open the question of locally determining the faces of a polytope. While every previous attempt of doing this from the 1-skeleton has been disproved, we show how one may locally determine the faces of a simple polytope from its 2-skeleton.

For a polytope P and its k-skeleton $\Delta^{\leq k}P$, we say that a subskeleton $S \subseteq \Delta^{\leq k}P$ (viewed as polytopal complexes) is *induced*, if S contains all faces of $\Delta^{\leq k}P$ spanned by the vertices of S, i.e.

$$S = \{f \in \Delta^{\leq k}P : V(f) \subseteq V(S)\}.$$

Theorem 4.2. *Let P, Q be simple polytopes. Q is a face of P if and only if $\Delta^{\leq 2}Q$ is an induced subskeleton of $\Delta^{\leq 2}P$.*

Let x, y be adjacent vertices in P and recall from the proof of Proposition 1.1 the bijection between the respective sets of neighboring vertices given by

$$\varphi_{x,y} : N_P(x) \setminus \{y\} \to N_P(y) \setminus \{x\},$$

where $\varphi_{x,y}$ sends a vertex v to the unique vertex w such that $\{x, y, v\}$ and $\{x, y, w\}$ span the same 2-face of P. We extend this map to subsets of the neighborhoods.

Lemma 4.1. *For a subset $S \subset N_P(x) \setminus \{y\}$, $\varphi(S)$ is the unique subset of $N_P(y) \setminus \{x\}$ such that the face spanned at x by $S \cup \{y\}$ and the face spanned at y by $\varphi(S) \cup \{x\}$ coincide.*

Proof. Let $k = |S| + 1$ be the dimension of the unique face F spanned at x by $S \cup \{y\}$. There is a unique set $T \subseteq N_P(y) \setminus \{x\}$ of cardinality $k - 1$ such that F is the face spanned at y by $T \cup \{x\}$. Let $w \in S$ and let $E \subseteq F$ be the 2-face spanned at x by y and w. Then $\{x, y, w, \varphi_{x,y}(w)\}$ are vertices of $E \subseteq F$, showing that $\varphi_{x,y}$ maps vertices of F to vertices of F. Thus $\varphi_{x,y}(S) \subseteq T$, with equality since $\varphi_{x,y}$ is bijective. □

Proof of Theorem 4.2. Let d be the dimension of P, let k be the dimension of Q. Let x, y be adjacent vertices of $\Delta^{\leq 2}Q \subseteq \Delta^{\leq 2}P$ and consider the maps

$$\varphi^P_{x,y} : N_P(x) \setminus \{y\} \to N_P(y) \setminus \{x\} \quad \text{and} \quad \varphi^Q_{x,y} : N_Q(x) \setminus \{y\} \to N_Q(y) \setminus \{x\}$$

as described above. As the maps are defined by the 2-skeletons of P and Q, they coincide by assumption when restricted to the neighborhoods in $G(Q)$, that is $\varphi^P_{x,y}|_{N_Q(x)\setminus\{y\}} = \varphi^Q_{x,y}$. By Lemma 4.1, the face of P spanned by $N_Q(x) \setminus \{y\}$ at $x \in P$ is the same as the face spanned at y by $\varphi^P_{x,y}(N_Q(x) \setminus \{y\}) = N_Q(y) \setminus \{x\}$. Since $G(Q) \subseteq G(P)$ is connected and k-regular, we get that the face spanned in P is isomorphic to Q. □

We may combine this result with the following Proposition of Pfeifle, Pilaud and Santos [13].

Proposition 4.1 (Pfeifle–Pilaud–Santos, 2012). *For a simple d-polytope P, all induced cycles of length $3, 4$ and 5 in $G(P)$ correspond to 2-faces of P.*

Corollary 4.1. *Let P be a simple d-polytope P. Every induced subgraph of $G(P)$ that is isomorphic to the graph of a simple polytope with at most 5 vertices in any 2-face is the graph of a face of P.*

Acknowledgments

The author is very grateful to Takayuki Hibi for his great hospitality and Akiyoshi Tsuchiya for the organization of the Summer Workshop on Lattice Polytopes 2018.

Furthermore, the author would like to thank Christian Haase for suggesting the question and for valuable discussions. The author was supported by DFG via the Collaborative Research Center TRR 109 "Discretization in Geometry and Dynamics" and via the Berlin Mathematical School BMS.

References

[1] M. L. Balinski, On the graph structure of convex polyhedra in n-space, *Pacific Journal of Mathematics* **11**, 431 (1961).

[2] M. M. Bayer, Graphs, skeleta and reconstruction of polytopes, *Acta Mathematica Hungarica* **155**, 61 (2018).

[3] R. Blind and P. Mani-Levitska, Puzzles and polytope isomorphisms, *aequationes mathematicae* **34**, 287 (1987).

[4] J. Doolittle, A minimal counterexample to a strengthening of Perles' conjecture, Preprint, September 2018, 8 pages, arXiv:1809.00662v2.

[5] J. Doolittle, E. Nevo, G. Pineda-Villavicencio, J. Ugon and D. Yost, On the reconstruction of polytopes, *Discrete & Computational Geometry* (2018).

[6] E. J. Friedman, Finding a simple polytope from its graph in polynomial time, *Discrete & Computational Geometry* **41**, 249 (2009).

[7] B. Grünbaum, *Convex polytopes, 2nd ed.*, Graduate Text in Mathematics, Vol. 221 (Springer-Verlag, New York, 2003).

[8] C. Haase and G. M. Ziegler, Examples and counterexamples for the Perles conjecture, *Discrete & Computational Geometry* **28**, 29 (2002).

[9] M. Joswig, V. Kaibel and F. Körner, On the k-systems of a simple polytope, *Israel Journal of Mathematics* **129**, 109 (2002).

[10] V. Kaibel, Reconstructing a simple polytope from its graph, in *Combinatorial optimization – Eureka, You Shrink!*, eds. M. Jünger, G. Reinelt and G. Rinaldi, Lecture Notes in Computer Science, Vol. 2570 (Springer, Berlin, 2002) pp. 105–118.

[11] G. Kalai, A simple way to tell a simple polytope from its graph, *Journal of Combinatorial Theory, Series A* **49**, 381 (1988).

[12] K. G. Murty, The graph of an abstract polytope, *Mathematical Programming* **4**, 336 (1973).

[13] J. Pfeifle, V. Pilaud and F. Santos, Polytopality and cartesian products of graphs, *Israel Journal of Mathematics* **192**, 121 (2012).

[14] H. Whitney, Non-separable and planar graphs, *Transactions of the American Mathematical Society* **34**, 339 (1932).

[15] G. M. Ziegler, *Lectures on Polytopes*Graduate texts in mathematics, Graduate texts in mathematics (Springer-Verlag, 1995).

Notes on toric Fano varieties associated to building sets

Yusuke Suyama

Department of Mathematics, Graduate School of Science, Osaka University
1-1 Machikaneyama-cho, Toyonaka, Osaka 560-0043 Japan
E-mail: y-suyama@cr.math.sci.osaka-u.ac.jp

This article gives an overview of toric Fano and toric weak Fano varieties associated to graphs and building sets. We also study some properties of such toric Fano varieties and discuss related topics.

Keywords: Toric geometry; Fano varieties; weak Fano varieties; building sets; nested sets; graph associahedra; reflexive polytopes; graph cubeahedra; root systems.

1. Introduction

An n-dimensional *toric variety* is a normal algebraic variety X over $\mathbb{C}$ containing the algebraic torus $(\mathbb{C}^*)^n$ as an open dense subset, such that the natural action of $(\mathbb{C}^*)^n$ on itself extends to an action on X. It is well known that the category of toric varieties is equivalent to the category of fans, see [11] for details.

A nonsingular projective algebraic variety is said to be *Fano* if its anticanonical divisor is ample. It is known that there are a finite number of isomorphism classes of toric Fano varieties in any given dimension. The classification of toric Fano varieties is a fundamental problem and has been studied by many researchers. In particular, Øbro [10] gave an explicit algorithm that classifies all toric Fano varieties for any dimension. A nonsingular projective variety is said to be *weak Fano* if its anticanonical divisor is nef and big. Since the fan of a toric weak Fano variety determines a reflexive polytope, there are a finite number of isomorphism classes of toric weak Fano varieties of fixed dimension as in the case of toric Fano varieties [13]. Sato [14] classified toric weak Fano 3-folds that are not Fano but are deformed to Fano under a small deformation, which are called toric *weakened Fano* 3-folds.

There is a construction of nonsingular projective toric varieties from building sets, which are formed by subsets of a finite set. Since a finite simple graph determines a building set, we can also associate to the graph a toric variety. Toric varieties associated to building sets were first studied by De Concini–Procesi [4] as smooth completions of hyperplane arrangement complements in a projective space with normal crossing boundary divisor and they are now called *wonderful models*. Building sets were originally defined as subspace arrangements with some suitable properties. In this article, we give an overview of toric Fano and toric weak Fano varieties associated to finite simple graphs and building sets. Furthermore, we provide some new results and discuss related topics.

This note is organized as follows: In Section 2, we recall the construction of a toric variety from a building set and a description of the intersection number of the anticanonical divisor with a torus-invariant curve. In Section 3, we survey characterizations of graphs and building sets whose associated toric varieties are Fano or weak Fano. In Section 4, we survey the relationship between the class of reflexive polytopes determined by toric weak Fano varieties associated to building sets, and that of reflexive polytopes associated to finite directed graphs. Section 5 contains new results. We study some properties of toric Fano varieties associated to building sets. Finally, in Section 6, we discuss characterizations of graph cubeahedra and root systems whose associated toric varieties are Fano or weak Fano.

2. Building sets

2.1. *Toric varieties associated to building sets*

Definition 2.1. A *building set* on a nonempty finite set S is a finite set B of nonempty subsets of S satisfying the following conditions:

(1) If $I, J \in B$ and $I \cap J \neq \varnothing$, then we have $I \cup J \in B$.
(2) For every $i \in S$, we have $\{i\} \in B$.

We denote by $B_{\max}$ the set of all maximal (with respect to inclusion) elements of B. An element of $B_{\max}$ is called a B-*component* and B is said to be *connected* if $B_{\max} = \{S\}$. For a nonempty subset C of S, we call $B|_C = \{I \in B \mid I \subset C\}$ the *restriction* of B to C. The restriction $B|_C$ is a building set on C. Note that $B|_C$ is connected if and only if $C \in B$. For any building set B, we have $B = \bigsqcup_{C \in B_{\max}} B|_C$. In particular, any building set is a disjoint union of connected building sets.

Definition 2.2. Let B be a building set. A *nested set* of B is a subset N of $B \setminus B_{\max}$ satisfying the following conditions:

(1) If $I, J \in N$, then we have either $I \subset J$ or $J \subset I$ or $I \cap J = \varnothing$.
(2) For any integer $k \geq 2$ and for any pairwise disjoint $I_1, \ldots, I_k \in N$, the union $I_1 \cup \cdots \cup I_k$ is not in B.

The empty set is a nested set for any building set. The set $\mathcal{N}(B)$ of all nested sets of B is called the *nested complex*. The nested complex $\mathcal{N}(B)$ is in fact an abstract simplicial complex on $B \setminus B_{\max}$. We denote by $\mathcal{N}(B)_{\max}$ the set of all maximal (with respect to inclusion) nested sets of B.

Proposition 2.1 ([20, Proposition 4.1]). *Let B be a building set on S. Then every maximal nested set of B has the same cardinality $|S| - |B_{\max}|$. In particular, the cardinality of every maximal nested set of a connected building set on S is $|S|-1$.*

We construct a toric variety from a building set. First, suppose that B is a connected building set on $S = \{1, \ldots, n+1\}$. We denote by $e_1, \ldots, e_n$ the standard basis for $\mathbb{R}^n$ and we put $e_{n+1} = -e_1 - \cdots - e_n$. For a nonempty subset I of S, we denote by e_I the vector $\sum_{i \in I} e_i$ in $\mathbb{R}^n$. Note that $e_S = 0$. For $N \in \mathcal{N}(B) \setminus \{\varnothing\}$, we denote by $\mathbb{R}_{\geq 0}N$ the $|N|$-dimensional rational strongly convex polyhedral cone $\sum_{I \in N} \mathbb{R}_{\geq 0} e_I$ in $\mathbb{R}^n$, and we define $\mathbb{R}_{\geq 0}\varnothing$ to be $\{0\} \subset \mathbb{R}^n$. Then $\Delta(B) = \{\mathbb{R}_{\geq 0}N \mid N \in \mathcal{N}(B)\}$ forms a fan in $\mathbb{R}^n$ and thus we have an n-dimensional toric variety $X(\Delta(B))$. If B is disconnected, then we define $X(\Delta(B)) = \prod_{C \in B_{\max}} X(\Delta(B|_C))$.

Theorem 2.1 ([20, Corollary 5.2 and Theorem 6.1]). *Let B be a building set. Then the associated toric variety $X(\Delta(B))$ is nonsingular and projective.*

Example 2.1. Let $S = \{1, 2, 3\}$ and $B = \{\{1\}, \{2\}, \{3\}, \{2, 3\}, \{1, 2, 3\}\}$. Then the nested complex $\mathcal{N}(B)$ is

$$\{\varnothing, \{\{1\}\}, \{\{2\}\}, \{\{3\}\}, \{\{2,3\}\},$$
$$\{\{1\}, \{2\}\}, \{\{1\}, \{3\}\}, \{\{2\}, \{2,3\}\}, \{\{3\}, \{2,3\}\}\}.$$

Hence we have the fan $\Delta(B)$ in Fig. 1. Therefore the associated toric variety $X(\Delta(B))$ is $\mathbb{P}^2$ blown-up at one point.

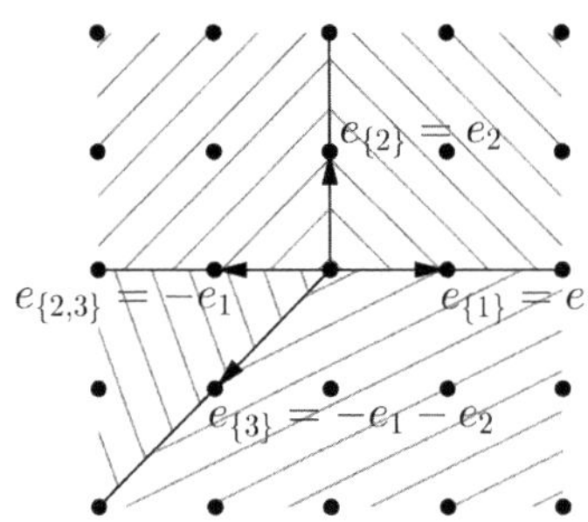

Fig. 1. The fan $\Delta(B)$.

Definition 2.3. Let G be a finite simple graph, that is, a finite undirected graph with no loops and no multiple edges. We denote by $V(G)$ and $E(G)$ its node set and edge set respectively. For a subset I of $V(G)$, the *induced subgraph* $G|_I$ is defined by $V(G|_I) = I$ and $E(G|_I) = \{\{v, w\} \in E(G) \mid v, w \in I\}$. The *graphical building set* $B(G)$ of G is defined to be $\{I \subset V(G) \mid I \neq \varnothing, G|_I \text{ is connected}\}$. The graphical building set $B(G)$ is in fact a building set on $V(G)$. The graphical building set $B(G)$ is connected if and only if G is connected.

Remark 2.1. Carr–Devadoss [3] introduced *graph associahedra* for finite simple graphs. Graph associahedra include many important families of polytopes such as associahedra (or Stasheff polytopes), cyclohedra (or Bott–Taubes polytopes), stellohedra and permutohedra. A graph associahedron can be realized as a smooth polytope and $\Delta(B(G))$ coincides with the normal fan of the graph associahedron of G.

2.2. *Intersection numbers*

Let Δ be a nonsingular complete fan in $\mathbb{R}^n$ and let $X(\Delta)$ be the associated toric variety. We denote by $\Delta(r)$ the set of r-dimensional cones in Δ for $r = 0, 1, \ldots, n$, and denote by $V(\sigma)$ the orbit closure corresponding to $\sigma \in \Delta$. The codimension of $V(\sigma)$ in $X(\Delta)$ equals the dimension of σ.

Proposition 2.2 (e.g., [11]). *Let Δ be a nonsingular complete fan in $\mathbb{R}^n$ and $\tau = \mathbb{R}_{\geq 0}v_1 + \cdots + \mathbb{R}_{\geq 0}v_{n-1} \in \Delta(n-1)$, where $v_1, \ldots, v_{n-1}$ are primitive vectors in $\mathbb{Z}^n$. Let v and v' be the distinct primitive vectors in $\mathbb{Z}^n$ such that $\tau + \mathbb{R}_{\geq 0}v, \tau + \mathbb{R}_{\geq 0}v' \in \Delta(n)$. Then there exist unique integers $a_1, \ldots, a_{n-1}$ such that $v + v' + a_1v_1 + \cdots + a_{n-1}v_{n-1} = 0$. Furthermore, the intersection number $(-K_{X(\Delta)}.V(\tau))$ of the anticanonical divisor $-K_{X(\Delta)}$ with the torus-invariant curve $V(\tau)$ equals $2 + a_1 + \cdots + a_{n-1}$.*

Proposition 2.3. *Let $X(\Delta)$ be an n-dimensional nonsingular projective toric variety. Then the following hold:*

(1) $X(\Delta)$ is Fano if and only if the intersection number $(-K_{X(\Delta)}.V(\tau))$ is positive for every $(n-1)$-dimensional cone τ in Δ ([11, Lemma 2.20]).
(2) $X(\Delta)$ is weak Fano if and only if $(-K_{X(\Delta)}.V(\tau))$ is nonnegative for every $(n-1)$-dimensional cone τ in Δ ([13, Proposition 6.17]).

To compute intersection numbers for the toric variety associated to a building set, we need the following proposition:

Proposition 2.4 ([20, Proposition 4.5]). *Let B be a building set on S and let $I_1, I_2 \in B$ with $I_1 \neq I_2$ and $N \in \mathcal{N}(B)$ such that $N \cup \{I_1\}, N \cup \{I_2\} \in \mathcal{N}(B)_{\max}$. Then the following hold:*

(1) We have $I_1 \not\subset I_2$ and $I_2 \not\subset I_1$.
(2) If $I_1 \cap I_2 \neq \varnothing$, then $(B|_{I_1 \cap I_2})_{\max} \subset N$.
(3) There exists a family $\{I_3, \ldots, I_k\} \subset N$ such that $I_1 \cup I_2, I_3, \ldots, I_k$ are pairwise disjoint and $I_1 \cup \cdots \cup I_k \in N \cup B_{\max}$ (the family $\{I_3, \ldots, I_k\}$ can be empty).

Let B be a connected building set on S and let $I_1, I_2 \in B$ with $I_1 \neq I_2$ and $N \in \mathcal{N}(B)$ such that $N \cup \{I_1\}, N \cup \{I_2\} \in \mathcal{N}(B)_{\max}$. Then by Proposition 2.4 (3), there exists $\{I_3, \ldots, I_k\} \subset N$ such that $I_1 \cup I_2, I_3, \ldots, I_k$ are pairwise disjoint and $I_1 \cup \cdots \cup I_k \in N \cup B_{\max} = N \cup \{S\}$. Since

$$e_{I_1} + e_{I_2} - \sum_{C \in (B|_{I_1 \cap I_2})_{\max}} e_C + e_3 + \cdots + e_k - e_{I_1 \cup \cdots \cup I_k} = 0,$$

Proposition 2.2 gives

$$(-K_{X(\Delta(B))}.V(\mathbb{R}_{\geq 0}N)) = \begin{cases} k - |(B|_{I_1 \cap I_2})_{\max}| - 1 & (I_1 \cup \cdots \cup I_k \in N), \\ k - |(B|_{I_1 \cap I_2})_{\max}| & (I_1 \cup \cdots \cup I_k = S). \end{cases}$$

If $I_1 \cap I_2 = \varnothing$, then $(B|_{I_1 \cap I_2})_{\max}$ is understood to be empty. The theorems in Section 3 are proved by using Proposition 2.3 and the equation above.

3. Toric Fano varieties associated to building sets

3.1. *Toric Fano varieties associated to finite simple graphs*

The following theorems characterize finite simple graphs whose associated toric varieties are Fano or weak Fano.

Theorem 3.1 ([16, Theorem 3.1]). *Let G be a finite simple graph. Then the following are equivalent:*

(1) The associated nonsingular projective toric variety $X(\Delta(B(G)))$ is Fano.
(2) Each connected component of G has at most three nodes.

Theorem 3.2 ([16, Theorem 3.4]). *Let G be a finite simple graph. Then the following are equivalent:*

(1) The toric variety $X(\Delta(B(G)))$ is weak Fano.
(2) For any connected component G' of G and for any proper subset I of $V(G')$, the induced subgraph $G'|_I$ is not isomorphic to any of the following:
  (a) A cycle with ≥ 4 nodes.
  (b) A diamond graph, that is, the graph obtained by removing an edge from a complete graph with four nodes.

Table 1. The number of connected graphs whose associated toric varieties are weak Fano.

# of nodes	# of connected graphs	# of connected graphs whose associated toric varieties are weak Fano
1	1	1
2	1	1
3	2	2
4	6	6
5	21	10
6	112	23

Example 3.1.

(1) Toric varieties associated to trees, cycles, complete graphs, and a diamond graph are weak Fano.
(2) The toric variety associated to the left graph in Fig. 2 is weak Fano, but the toric variety associated to the right graph is not weak Fano.

3.2. *Toric Fano varieties associated to building sets*

Theorem 3.3 ([17, Theorem 2.5]). *Let B be a building set. Then the following are equivalent:*

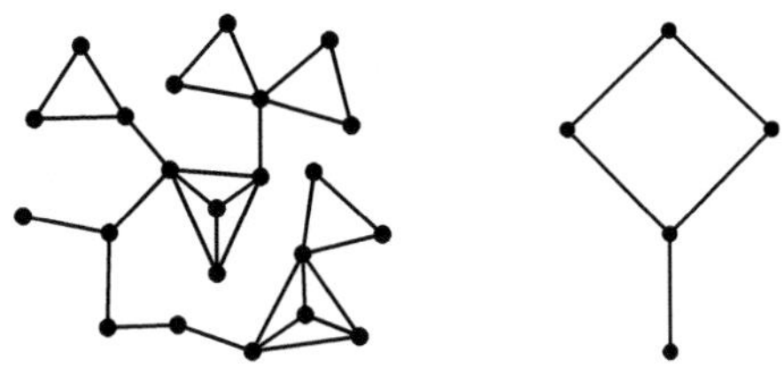

Fig. 2. Examples.

(1) The toric variety $X(\Delta(B))$ is Fano.

(2) For any B-component C and for any $I_1, I_2 \in B|_C$ such that $I_1 \cap I_2 \neq \varnothing, I_1 \not\subset I_2$ and $I_2 \not\subset I_1$, we have $I_1 \cup I_2 = C$ and $I_1 \cap I_2 \in B|_C$.

Table 2. The number of toric Fano varieties associated to building sets.

dimension	# of toric Fano varieties	# of toric Fano varieties associated to building sets
1	1	1
2	5	5
3	18	14
4	124	50
5	866	161

Example 3.2. If $|S| \leq 3$, then a connected building set B on S is isomorphic to one of the following six types:

(1) $\{\{1\}\}$: a point, which is understood to be Fano.
(2) $\{\{1\}, \{2\}, \{1,2\}\}$: $\mathbb{P}^1$.
(3) $\{\{1\}, \{2\}, \{3\}, \{1,2,3\}\}$: $\mathbb{P}^2$.
(4) $\{\{1\}, \{2\}, \{3\}, \{1,2\}, \{1,2,3\}\}$: $\mathbb{P}^2$ blown-up at one point.
(5) $\{\{1\}, \{2\}, \{3\}, \{1,2\}, \{1,3\}, \{1,2,3\}\}$: $\mathbb{P}^2$ blown-up at two points.
(6) $\{\{1\}, \{2\}, \{3\}, \{1,2\}, \{1,3\}, \{2,3\}, \{1,2,3\}\}$: $\mathbb{P}^2$ blown-up at three non-collinear points.

Thus $X(\Delta(B))$ is Fano in every case. Since the disconnected building set $\{\{1\}, \{2\}, \{1,2\}, \{3\}, \{4\}, \{3,4\}\}$ yields $\mathbb{P}^1 \times \mathbb{P}^1$, it follows that all toric Fano varieties of dimension ≤ 2 are obtained from building sets.

Theorem 3.4 ([18, Theorem 2.4]). *Let B be a building set. Then the following are equivalent:*

(1) The toric variety $X(\Delta(B))$ is weak Fano.

(2) For any B-component C and for any $I_1, I_2 \in B|_C$ such that $I_1 \cap I_2 \neq \varnothing, I_1 \not\subset I_2$ and $I_2 \not\subset I_1$, we have at least one of the following:
   (a) $I_1 \cap I_2 \in B|_C$.
   (b) $I_1 \cup I_2 = C$ and $|(B|_{I_1 \cap I_2})_{\max}| \leq 2$.

In particular, all toric varieties of dimension ≤ 3 associated to building sets are weak Fano.

Theorem 3.1 follows immediately from Theorem 3.3. However, it is unclear whether Theorem 3.2 can be obtained from Theorem 3.4.

4. Reflexive polytopes associated to building sets

4.1. *Reflexive polytopes and directed graphs*

An n-dimensional integral convex polytope P in $\mathbb{R}^n$ is said to be *reflexive* if the origin is in the interior of P and the dual $P^* = \{u \in \mathbb{R}^n \mid \langle u, v\rangle \geq -1 \text{ for any } v \in P\}$ is also an integral convex polytope, where $\langle \cdot, \cdot \rangle$ denotes the standard inner product on $\mathbb{R}^n$. An n-dimensional integral convex polytope in an n-dimensional real vector space with respect to a lattice is said to be *smooth Fano* if the origin is the only lattice point in the interior and the vertices of every facet form a basis for the lattice. Note that any smooth Fano polytope in $\mathbb{R}^n$ is a reflexive polytope.

Let Δ be a nonsingular complete fan in $\mathbb{R}^n$. If the associated toric variety $X(\Delta)$ is weak Fano, then the convex hull of the primitive generators of the rays in $\Delta(1)$ is a reflexive polytope. This correspondence induces a bijection between isomorphism classes of toric Fano varieties and isomorphism classes of smooth Fano polytopes. For a building set B whose associated toric variety $X(\Delta(B))$ is weak Fano, we denote by P_B the corresponding reflexive polytope.

Higashitani [8] gave a construction of integral convex polytopes from finite connected directed graphs (with no loops and no multiple arrows). Let G be a finite connected directed graph whose node set is $V(G) = \{1, \ldots, n+1\}$ and whose arrow set is $A(G) \subset V(G) \times V(G)$. For $\overrightarrow{e} = (i, j) \in A(G)$, we define $\rho(\overrightarrow{e}) \in \mathbb{R}^{n+1}$ to be $e_i - e_j$. We define P_G to be the convex hull of $\{\rho(\overrightarrow{e}) \mid \overrightarrow{e} \in A(G)\}$ in $\mathbb{R}^{n+1}$. Then P_G is an integral convex polytope in the hyperplane $H = \{(x_1, \ldots, x_{n+1}) \in \mathbb{R}^{n+1} \mid x_1 + \cdots + x_{n+1} = 0\}$.

Higashitani also characterized finite directed graphs whose associated integral convex polytopes are smooth Fano, see [8] for details.

4.2. *Reflexive polytopes associated to building sets*

The following theorem is proved by using Theorem 3.3 (2).

Theorem 4.1 ([17, Theorem 4.2]). *Let B be a building set. If the associated toric variety $X(\Delta(B))$ is Fano, then there exists a finite connected directed graph G such that $P_G \subset H$ is a smooth Fano polytope (with respect to the lattice $H \cap \mathbb{Z}^{n+1}$)*

and its associated fan is isomorphic to $\Delta(B)$*, that is, there is a linear isomorphism* $F : H \to \mathbb{R}^n$ *such that* $F(H \cap \mathbb{Z}^{n+1}) = \mathbb{Z}^n$ *and* $F(P_G) = P_B$*, where* n *is the dimension of* $X(\Delta(B))$*.*

Remark 4.1. Higashitani showed that any pseudo-symmetric smooth Fano polytope is obtained from a finite directed graph (see [8, Theorem 3.3]). On the other hand, there exists a building set B such that $X(\Delta(B))$ is Fano and the corresponding smooth Fano polytope is not pseudo-symmetric. For example, the toric variety in Example 2.1 is Fano while the corresponding smooth Fano polytope is not pseudo-symmetric.

Example 4.1. The converse of Theorem 4.1 is not true. The finite directed graphs in Fig. 3 yield smooth Fano 3-polytopes. However, these polytopes cannot be obtained from building sets.

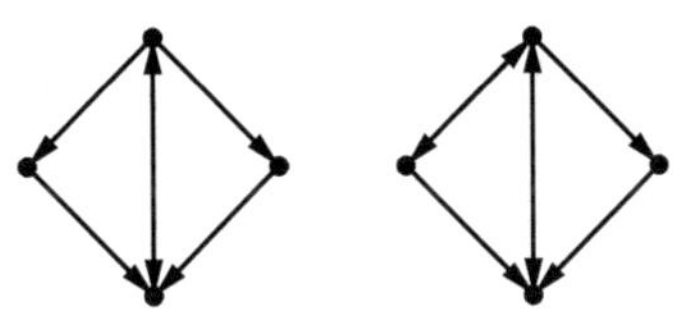

Fig. 3. Directed graphs whose smooth Fano polytopes cannot be obtained from building sets.

Remark 4.2. In contrast to Theorem 4.1, there exist infinitely many reflexive polytopes associated to building sets that cannot be obtained from finite directed graphs (note that such reflexive polytopes are not smooth Fano). On the other hand, there also exists a reflexive polytope associated to a finite directed graph that is not smooth Fano and cannot be obtained from any building set, see [18] for details.

5. Properties of toric Fano varieties associated to building sets

5.1. *Extremal contractions*

Let B be a connected building set on S whose associated toric variety $X(\Delta(B))$ is Fano. Let $I_1, I_2 \in B$ with $I_1 \neq I_2$ and $N \in \mathcal{N}(B)$ such that $N \cup \{I_1\}, N \cup \{I_2\} \in \mathcal{N}(B)_{\max}$ and $V(\mathbb{R}_{\geq 0}N)$ is an extremal curve. We find the wall relation associated to $V(\mathbb{R}_{\geq 0}N)$. We refer to [12] for details on extremal contractions of toric varieties.

Suppose $I_1 \cap I_2 = \varnothing$. By Proposition 2.4 (3), there exists $\{I_3, \ldots, I_k\} \subset N$ such that $I_1 \cup I_2, I_3, \ldots, I_k$ are pairwise disjoint and $I_1 \cup \cdots \cup I_k \in N \cup B_{\max} = N \cup \{S\}$. If $I_1 \cup \cdots \cup I_k \in N$, then the wall relation is $e_{I_1} + e_{I_2} + e_{I_3} + \cdots + e_{I_k} - e_{I_1 \cup \cdots \cup I_k} = 0$. If $I_1 \cup \cdots \cup I_k = S$, then the wall relation is $e_{I_1} + e_{I_2} + e_{I_3} + \cdots + e_{I_k} = 0$.

Suppose $I_1 \cap I_2 \neq \varnothing$. By Proposition 2.4 (1), we have $I_1 \not\subset I_2$ and $I_2 \not\subset I_1$. Applying Theorem 3.3 (2) to I_1 and I_2, we have $I_1 \cup I_2 = S$ and $I_1 \cap I_2 \in B$. By

Proposition 2.4 (2), we have $\{I_1 \cap I_2\} = (B|_{I_1 \cap I_2})_{\max} \subset N$. Since $e_{I_1 \cup I_2} = e_S = 0$, the wall relation is $e_{I_1} + e_{I_2} - e_{I_1 \cap I_2} = 0$.

In every case, at most one of the coefficients in the wall relation associated to $V(\mathbb{R}_{\geq 0}N)$ is negative. This implies the following corollary:

Corollary 5.1. *Let B be a building set. If the associated toric variety $X(\Delta(B))$ is Fano, then there are no small extremal contractions of $X(\Delta(B))$.*

5.2. *Toric 2-Fano varieties associated to building sets*

2-Fano varieties were introduced by de Jong–Starr [5].

Definition 5.1. A Fano variety X is said to be **2-Fano** if the second Chern character $\mathrm{ch}_2(X) = (c_1(X)^2 - 2c_2(X))/2$ is ample, that is, the intersection number $(\mathrm{ch}_2(X).S)$ is positive for any surface S on X.

At present, the only known examples of toric 2-Fano varieties are projective spaces. In this subsection, we prove the following theorem:

Theorem 5.1. *Let B be a building set whose associated toric variety $X(\Delta(B))$ is Fano. Then $X(\Delta(B))$ is 2-Fano if and only if it is a projective space.*

Let Δ be a nonsingular complete fan in $\mathbb{R}^n$ and $v_1, \ldots, v_m$ be the primitive generators of the rays in $\Delta(1)$. We put $D_i = V(\mathbb{R}_{\geq 0}v_i)$ for $i = 1, \ldots, m$. For a torus-invariant subvariety $Y \subset X(\Delta)$ of codimension l, we put

$$I_{Y/X(\Delta)} = \sum_{(i_1,\ldots,i_l)\in\{1,\ldots,m\}^l} (D_{i_1} \cdots D_{i_l}.Y)X_{i_1} \cdots X_{i_l} \in \mathbb{Z}[X_1, \ldots, X_m].$$

Example 5.1. Let $v_{i_1}, \ldots, v_{i_{n-1}}, v_{j_1}, v_{j_2}$ be distinct primitive generators such that $\tau = \mathbb{R}_{\geq 0}v_{i_1} + \cdots + \mathbb{R}_{\geq 0}v_{i_{n-1}} \in \Delta(n-1)$ and $\tau + \mathbb{R}_{\geq 0}v_{j_1}, \tau + \mathbb{R}_{\geq 0}v_{j_2} \in \Delta(n)$. If $v_{j_1} + v_{j_2} + a_1 v_{i_1} + \cdots + a_{n-1}v_{i_{n-1}} = 0$ for $a_1, \ldots, a_{n-1} \in \mathbb{Z}$, then $I_{V(\tau)/X(\Delta)} = X_{j_1} + X_{j_2} + a_1 X_{i_1} + \cdots + a_{n-1}X_{i_{n-1}}$.

Theorem 5.2 ([15, Theorem 3.5]). *Let $X(\Delta)$ be a nonsingular projective toric variety and let S be a torus-invariant surface on $X(\Delta)$ isomorphic to a Hirzebruch surface of degree $a \geq 0$. Then $I_{S/X(\Delta)} = aI^2_{C_{\mathrm{fib}}/X(\Delta)} + 2I_{C_{\mathrm{fib}}/X(\Delta)}I_{C_{\mathrm{neg}}/X(\Delta)}$, where C_{fib} is a fiber of the projection $S \to \mathbb{P}^1$ and C_{neg} is the negative section of S.*

For a nonsingular complete toric variety $X(\Delta)$, it is known that $\mathrm{ch}_2(X(\Delta)) = (D_1^2 + \cdots + D_m^2)/2$. Hence if $I_{S/X(\Delta)}$ has an expression of the form $\sum_{1\leq i\leq j\leq m} a_{ij}X_iX_j$, then we have $(\mathrm{ch}_2(X(\Delta)).S) = (a_{11} + \cdots + a_{mm})/2$.

Proof of Theorem 5.1. We only need to prove the necessity. Suppose that $X(\Delta(B))$ is not a projective space. If $X(\Delta(B))$ is a surface, then Noether's theorem implies that $\mathrm{ch}_2(X(\Delta(B)))$ is not ample. Since a product of nonsingular projective varieties is not 2-Fano (see [1, Lemma 17]), we may assume that

$n \geq 3$ and B is a connected building set on $S = \{1, \ldots, n+1\}$. By the assumption, $B' = B \setminus \{\{1\}, \ldots, \{n+1\}, S\}$ is nonempty. For a torus-invariant subvariety Y of $X(\Delta(B))$, we regard $I_{Y/X(\Delta(B))}$ as a polynomial in the polynomial ring $\mathbb{Z}[X_I \mid I \in B \setminus \{S\}]$.

Case 1. Suppose that every maximal element of B' has cardinality n. We pick a maximal element I of B'. We may assume $I = \{1, \ldots, n\}$.

Subcase 1.1. Suppose that there is no element of B' containing $n+1$. Let $K_1, \ldots, K_r$ be all maximal elements of $B|_I \setminus \{I\}$. Note that $r \geq 2$. Since $X(\Delta(B))$ is Fano, Theorem 3.3 (2) implies that I is the disjoint union of $K_1, \ldots, K_r$. We pick $N_i \in \mathcal{N}(B|_{K_i})_{\max}$ for each $i = 1, \ldots, r$. Let

$$N = N_1 \cup \cdots \cup N_r \cup \{K_3, \ldots, K_r\}.$$

Claim 1. *The families $N \cup \{I, K_1\}, N \cup \{K_1, \{n+1\}\}, N \cup \{\{n+1\}, K_2\}, N \cup \{K_2, I\}$ are maximal nested sets of B.*

Proof. For each family, condition (1) of Definition 2.2 is obviously satisfied. We check condition (2). Suppose that $I_1, \ldots, I_s \in N_1 \cup \cdots \cup N_r \cup \{K_1, K_3, \ldots, K_r\}$ such that $I_1, \ldots, I_s$ are pairwise disjoint. We have $I_1 \cup \cdots \cup I_s \subsetneq I$, since if $I_1 \cup \cdots \cup I_s = I$ then K_2 is the disjoint union of $I_{i_1}, \ldots, I_{i_t}$ for some $1 \leq i_1 < \cdots < i_t \leq s$ with $\{I_{i_1}, \ldots, I_{i_t}\} \subset N_2$, which contradicts that N_2 is a nested set.

Assume that $N \cup \{I, K_1\}$ does not satisfy condition (2) for contradiction. Since I is a unique maximal element of $N \cup \{I, K_1\}$, we may assume that $s \geq 2$ and $I_1 \cup \cdots \cup I_s \in B$. If $K_i = I_k$ for some i, k, then $K_i \subsetneq I_1 \cup \cdots \cup I_s \in B|_I \setminus \{I\}$, a contradiction. If $I_1, \ldots, I_s \in N_1 \cup \cdots \cup N_r$, then $K_i \supset I_1$ for some i and thus $K_i \subsetneq I_1 \cup \cdots \cup I_s \cup K_i \in B|_I \setminus \{I\}$, a contradiction. Therefore $N \cup \{I, K_1\}$ is a nested set.

We show that $N \cup \{K_1, \{n+1\}\}$ satisfies condition (2). If $s \geq 2$ and $I_1 \cup \cdots \cup I_s \in B$, then a similar argument leads to a contradiction. If $I_1 \cup \cdots \cup I_s \cup \{n+1\} \in B$, then it contradicts that there is no element of B' containing $n+1$. Therefore $N \cup \{K_1, \{n+1\}\}$ is a nested set.

Similarly, $N \cup \{\{n+1\}, K_2\}$ and $N \cup \{K_2, I\}$ are nested sets. Each nested set is maximal since it has cardinality n. This completes the proof of the claim. □

Since $e_{K_1} + e_{K_2} + e_{K_3} + \cdots + e_{K_r} - e_I = 0$ and $e_I + e_{\{n+1\}} = 0$, the surface $V(\mathbb{R}_{\geq 0}N)$ is isomorphic to a Hirzebruch surface of degree one and $C_{\text{neg}} = V(\mathbb{R}_{\geq 0}(N \cup \{I\}))$. We choose $C_{\text{fib}} = V(\mathbb{R}_{\geq 0}(N \cup \{K_1\}))$. Since

$$\begin{aligned} I_{C_{\text{neg}}/X(\Delta(B))} &= X_{K_1} + X_{K_2} + X_{K_3} + \cdots + X_{K_r} - X_I, \\ I_{C_{\text{fib}}/X(\Delta(B))} &= X_I + X_{\{n+1\}}, \\ I_{V(\mathbb{R}_{\geq 0}N)/X(\Delta(B))} &= (X_I + X_{\{n+1\}})^2 + 2(X_I + X_{\{n+1\}})(X_{K_1} + \cdots + X_{K_r} - X_I), \end{aligned}$$

we have $(\text{ch}_2(X(\Delta(B))).V(\mathbb{R}_{\geq 0}N)) = (1+1-2)/2 = 0$.

Subcase 1.2. Suppose that there exists an element of B' containing $n+1$. We pick a maximal element J of B' containing $n+1$. We may assume $J = \{1, \ldots, n-1, n+1\}$.

Since $X(\Delta(B))$ is Fano, Theorem 3.3 (2) implies $\{1,\ldots,n-1\} = I\cap J \in B$. Let $K_1,\ldots,K_r$ be all maximal elements of $B|_{I\cap J}\setminus\{I\cap J\}$. Note that $r\geq 2$. Theorem 3.3 (2) implies that $I\cap J$ is the disjoint union of $K_1,\ldots,K_r$. We pick $N_i\in\mathcal{N}(B|_{K_i})_{\max}$ for each $i=1,\ldots,r$. Let

$$N = N_1\cup\cdots\cup N_r\cup\{K_3,\ldots,K_r,\{n\}\}.$$

Claim 2. *The families* $N\cup\{I,K_1\}, N\cup\{K_1,\{n+1\}\}, N\cup\{\{n+1\},K_2\}, N\cup\{K_2,I\}$ *are maximal nested sets of* B.

Proof. It suffices to show that $N\cup\{I,K_1\}$ and $N\cup\{K_1,\{n+1\}\}$ satisfy condition (2) of Definition 2.2. Suppose that $I_1,\ldots,I_s\in N_1\cup\cdots\cup N_r\cup\{K_1,K_3,\ldots,K_r\}$ such that $I_1,\ldots,I_s$ are pairwise disjoint. Note that $I_1\cup\cdots\cup I_s\subsetneq I\cap J$.

We show that $N\cup\{I,K_1\}$ satisfies condition (2). If $s\geq 2$ and $I_1\cup\cdots\cup I_s\in B$, then an argument similar to the proof of Claim 1 leads to a contradiction. Suppose $I_1\cup\cdots\cup I_s\cup\{n\}\in B$. Applying Theorem 3.3 (2) to $I_1\cup\cdots\cup I_s\cup\{n\}$ and $I\cap J$, we have $S=(I_1\cup\cdots\cup I_s\cup\{n\})\cup(I\cap J)=I$, a contradiction. Since I is a unique maximal element of $N\cup\{I,K_1\}$, it follows that $N\cup\{I,K_1\}$ is a nested set.

We show that $N\cup\{K_1,\{n+1\}\}$ satisfies condition (2). If $I_1\cup\cdots\cup I_s\in B$ with $s\geq 2$ or $I_1\cup\cdots\cup I_s\cup\{n\}\in B$, then a similar argument leads to a contradiction. Suppose $I_1\cup\cdots\cup I_s\cup\{n+1\}\in B$. Applying Theorem 3.3 (2) to $I_1\cup\cdots\cup I_s\cup\{n+1\}$ and $I\cap J$, we have $S=(I_1\cup\cdots\cup I_s\cup\{n+1\})\cup(I\cap J)=J$, a contradiction. Suppose that $\{n,n+1\}\in B$ or $I_1\cup\cdots\cup I_s\cup\{n,n+1\}\in B$. Then there exists a maximal element L of B' containing it. Theorem 3.3 (2) implies $I\cap L, J\cap L\in B$. Since $n\geq 3$, we have $(I\cap L)\cap(J\cap L)=(I\cap J)\cap L\neq\varnothing$ and thus Theorem 3.3 (2) implies $S=(I\cap L)\cup(J\cap L)\subset L$, a contradiction. Therefore $N\cup\{K_1,\{n+1\}\}$ is a nested set. □

Since $e_{K_1}+e_{K_2}+e_{K_3}+\cdots+e_{K_r}+e_{\{n\}}-e_I=0$ and $e_I+e_{\{n+1\}}=0$, the surface $V(\mathbb{R}_{\geq 0}N)$ is isomorphic to a Hirzebruch surface of degree one and $C_{\mathrm{neg}}=V(\mathbb{R}_{\geq 0}(N\cup\{I\}))$. We choose $C_{\mathrm{fib}}=V(\mathbb{R}_{\geq 0}(N\cup\{K_1\}))$. Since

$$\begin{aligned}
I_{C_{\mathrm{neg}}/X(\Delta(B))} &= X_{K_1}+X_{K_2}+X_{K_3}+\cdots+X_{K_r}+X_{\{n\}}-X_I,\\
I_{C_{\mathrm{fib}}/X(\Delta(B))} &= X_I+X_{\{n+1\}},\\
I_{V(\mathbb{R}_{\geq 0}N)/X(\Delta(B))} &= (X_I+X_{\{n+1\}})^2\\
&\quad +2(X_I+X_{\{n+1\}})(X_{K_1}+\cdots+X_{K_r}+X_{\{n\}}-X_I),
\end{aligned}$$

we have $(\mathrm{ch}_2(X(\Delta(B))).V(\mathbb{R}_{\geq 0}N))=(1+1-2)/2=0$.

Case 2. Suppose that there exists a maximal element I of B' such that $|I|\leq n-1$. Let $K_1,\ldots,K_r$ be all maximal elements of $B|_I\setminus\{I\}$ and $L_1,\ldots,L_s$ be all maximal elements of $B|_{S\setminus I}\setminus\{S\setminus I\}$. Note that $r,s\geq 2$. Since $X(\Delta(B))$ is Fano, Theorem 3.3 (2) implies that I is the disjoint union of $K_1,\ldots,K_r$ and $S\setminus I$ is the disjoint union of $L_1,\ldots,L_s$. We pick $N_i\in\mathcal{N}(B|_{K_i})_{\max}$ for each $i=1,\ldots,r$ and $N'_j\in\mathcal{N}(B|_{L_j})_{\max}$ for each $j=1,\ldots,s$. Let

$$N=N_1\cup\cdots\cup N_r\cup N'_1\cup\cdots\cup N'_s\cup\{K_3,\ldots,K_r,L_3,\ldots,L_s,I\}.$$

Claim 3. *The families* $N\cup\{K_1,L_1\}, N\cup\{L_1,K_2\}, N\cup\{K_2,L_2\}, N\cup\{L_2,K_1\}$ *are maximal nested sets of* B.

Proof. It suffices to show that $N \cup \{K_1, L_1\}$ satisfies condition (2) of Definition 2.2. Suppose that $I_1,\dots,I_t \in N_1 \cup\cdots\cup N_r \cup \{K_1, K_3,\dots,K_r, I\}$ and $J_1,\dots,J_u \in N'_1\cup\cdots\cup N'_s \cup\{L_1, L_3,\dots,L_s\}$ such that $I_1,\dots,I_t, J_1,\dots,J_u$ are pairwise disjoint. Note that $J_1\cup\cdots\cup J_u \subsetneq S\setminus I$.

If $t \geq 2$ and $I_1 \cup\cdots\cup I_t \in B$, then an argument similar to the proof of Claim 1 leads to a contradiction. If $I_1\cup\cdots\cup I_t\cup J_1\cup\cdots\cup J_u \in B$, then we have $I \subsetneq I\cup J_1\cup\cdots\cup J_u \in B'$, which contradicts that I is a maximal element of B'. Suppose that $u \geq 2$ and $J_1\cup\cdots\cup J_u \in B$. If $L_j = J_k$ for some j,k, then $L_j \subsetneq J_1\cup\cdots\cup J_u \in B|_{S\setminus I}\setminus\{S\setminus I\}$, a contradiction. If $J_1,\dots,J_u \in N'_1\cup\cdots\cup N'_s$, then $L_j \supset J_1$ for some j and thus $L_j \subsetneq J_1\cup\cdots\cup J_u\cup L_j \in B|_{S\setminus I}\setminus\{S\setminus I\}$, a contradiction. Therefore $N\cup\{K_1,L_1\}$ is a nested set. □

Since $e_{K_1}+e_{K_2}+e_{K_3}+\cdots+e_{K_r}-e_I = 0$ and $e_{L_1}+e_{L_2}+e_{L_3}+\cdots+e_{L_s}+e_I = 0$, the surface $V(\mathbb{R}_{\geq 0}N)$ is isomorphic to $\mathbb{P}^1\times\mathbb{P}^1$. We choose $C_{\mathrm{neg}} = V(\mathbb{R}_{\geq 0}(N\cup\{L_1\}))$ and $C_{\mathrm{fib}} = V(\mathbb{R}_{\geq 0}(N\cup\{K_1\}))$. Since

$$\begin{aligned} I_{C_{\mathrm{neg}}/X(\Delta(B))} &= X_{K_1}+X_{K_2}+X_{K_3}+\cdots+X_{K_r}-X_I,\\ I_{C_{\mathrm{fib}}/X(\Delta(B))} &= X_{L_1}+X_{L_2}+X_{L_3}+\cdots+X_{L_s}+X_I,\\ I_{V(\mathbb{R}_{\geq 0}N)/X(\Delta(B))} &= 2(X_{L_1}+\cdots+X_{L_s}+X_I)(X_{K_1}+\cdots+X_{K_r}-X_I), \end{aligned}$$

we have $(\mathrm{ch}_2(X(\Delta(B))).V(\mathbb{R}_{\geq 0}N)) = (-2)/2 = -1$.

In every case, $\mathrm{ch}_2(X(\Delta(B)))$ is not ample. This completes the proof of the theorem. □

6. Related topics

6.1. *Toric Fano varieties associated to graph cubeahedra*

Devadoss–Heath–Vipismakul [6] introduced *graph cubeahedra* for finite simple graphs and proved that graph associahedra and graph cubeahedra appear as some compactified moduli spaces of marked bordered Riemann surfaces. A graph cubeahedron can also be realized as a smooth polytope and thus we can obtain a nonsingular projective toric variety. Let G be a finite simple graph on $V(G) = \{1,\dots,n\}$, and let $\square^n$ be an n-cube whose facets are labeled by $1,\dots,n$ and $\overline{1},\dots,\overline{n}$, where the two facets labeled by i and $\overline{i}$ are on opposite sides. Then every face of $\square^n$ is labeled by a subset I of $\{1,\dots,n,\overline{1},\dots,\overline{n}\}$ such that $I\cap\{1,\dots,n\}$ and $\{i\in\{1,\dots,n\}\mid \overline{i}\in I\}$ are disjoint, that is, the face corresponding to I is the intersection of the facets labeled by the elements of I. The *graph cubeahedron* $\square_G$ is obtained from $\square^n$ by truncating the faces labeled by the elements of the graphical building set $B(G)$ in increasing order of dimension. We have a bijection between the set of facets of $\square_G$ and $B(G)\cup\{\{\overline{1}\},\dots,\{\overline{n}\}\}$. For $I \in B(G)\cup\{\{\overline{1}\},\dots,\{\overline{n}\}\}$, we denote by F_I the corresponding facet.

Theorem 6.1 ([6, Theorem 12]). *Let G be a finite simple graph. Then the two facets F_I and F_J of the graph cubeahedron $\square_G$ intersect if and only if one of the following holds:*

(1) $I, J \in B(G)$ and we have either $I \subset J$ or $J \subset I$ or $I \cup J \notin B(G)$.
(2) One of I and J, say I, is in $B(G)$ and $J = \{\bar{j}\}$ for some $j \in \{1, \ldots, n\} \setminus I$.
(3) $I = \{\bar{i}\}$ and $J = \{\bar{j}\}$ for some $i, j \in \{1, \ldots, n\}$.

Furthermore, $\square_G$ is a flag polytope.

We can realize $\square_G$ as a smooth polytope such that the outward-pointing primitive normal vector e_I of the facet F_I is

$$\begin{cases} \sum_{i \in I} e_i & (I \in B(G)), \\ -e_i & (I = \{\bar{i}\}, i \in \{1, \ldots, n\}), \end{cases}$$

for any $I \in B(G) \cup \{\{\bar{1}\}, \ldots, \{\bar{n}\}\}$. Let

$$\mathcal{N}^{\square}(G) = \{N \subset B(G) \cup \{\{\bar{1}\}, \ldots, \{\bar{n}\}\} \mid F_I \text{ and } F_J \text{ intersect for any } I, J \in N\}.$$

For $N \in \mathcal{N}^{\square}(G) \setminus \{\varnothing\}$, we denote by σ_N the $|N|$-dimensional cone $\sum_{I \in N} \mathbb{R}_{\geq 0} e_I$ in $\mathbb{R}^n$, and we define $\sigma_\varnothing$ to be $\{0\} \subset \mathbb{R}^n$. Then $\{\sigma_N \mid N \in \mathcal{N}^{\square}(G)\}$ coincides with the normal fan $\Delta_{\square_G}$. Note that if $G_1, \ldots, G_m$ are the connected components of G, then $X(\Delta_{\square_G})$ is isomorphic to the product $X(\Delta_{\square_{G_1}}) \times \cdots \times X(\Delta_{\square_{G_m}})$.

Remark 6.1. Manneville–Pilaud proved that for connected graphs G and G', the graph associahedron of G and the graph cubeahedron of G' are combinatorially equivalent if and only if G is a tree with at most one node of degree more than two and G' is its line graph (see [9, Proposition 64]). Hence we can see that there exist many toric varieties associated to graph cubeahedra that cannot be obtained from graph associahedra.

Recently, the author proved the following:

Theorem 6.2 ([19, Theorem 5]). *Let G be a finite simple graph. Then the following are equivalent:*

(1) The nonsingular projective toric variety $X(\Delta_{\square_G})$ associated to the graph cubeahedron $\square_G$ is Fano.
(2) Each connected component of G has at most two nodes.

In particular, any toric Fano variety associated to a graph cubeahedron is a product of copies of $\mathbb{P}^1$ and $\mathbb{P}^1 \times \mathbb{P}^1$ blown-up at one point.

Theorem 6.3 ([19, Theorem 6]). *Let G be a finite simple graph. Then the following are equivalent:*

(1) The toric variety $X(\Delta_{\square_G})$ is weak Fano.
(2) For any subset I of $V(G)$, the induced subgraph $G|_I$ is not isomorphic to any of the following:

(a) A cycle with ≥ 4 nodes.
(b) A diamond graph.
(c) A claw, that is, a star with three edges.

In particular, if $X(\Delta_{\square_G})$ is weak Fano, then G is chordal.

Table 3. The number of connected graphs whose graph cubeahedra correspond to toric weak Fano varieties.

# of nodes	# of connected graphs	# of connected graphs whose graph cubeahedra correspond to toric weak Fano varieties
1	1	1
2	1	1
3	2	2
4	6	3
5	21	6
6	112	11

Remark 6.2. A connected graph satisfies condition (2) of Theorem 6.3 if and only if it is the line graph of a tree (see, for example [7, Theorem 8.5]), and such a graph is called a *claw-free block graph.* Therefore for a connected graph G, the toric variety $X(\Delta_{\square_G})$ is weak Fano if and only if G is the line graph of a tree.

6.2. *Toric Fano varieties associated to Weyl chambers*

The collection of Weyl chambers for a root system determines a nonsingular complete fan. Let Φ be a root system in a Euclidean space V. Let $M(\Phi)$ be the root lattice $\sum_{\alpha\in\Phi}\mathbb{Z}\alpha$ and $N(\Phi)$ be its dual lattice $\mathrm{Hom}_{\mathbb{Z}}(M(\Phi),\mathbb{Z})$. For a set of simple roots $S\subset\Phi$, we define a cone $\sigma_S=\{v\in N(\Phi)_{\mathbb{R}}\mid\langle u,v\rangle\geq 0 \text{ for any } u\in S\}$, where $N(\Phi)_{\mathbb{R}}=N(\Phi)\otimes_{\mathbb{Z}}\mathbb{R}$. The set $\Delta(\Phi)$ of all such cones and their faces forms a nonsingular complete fan in the lattice $N(\Phi)$. The associated toric variety $X(\Delta(\Phi))$ is nonsingular and projective [2].

Proposition 6.1. *Let $S=\{\alpha_1,\ldots,\alpha_n\}\subset\Phi$ be a set of simple roots and $\omega_1,\ldots,\omega_n$ be the dual basis of $\alpha_1,\ldots,\alpha_n$. We put $a_{ij}=2(\alpha_i,\alpha_j)/(\alpha_j,\alpha_j)$. Then we have $(-K_{X(\Delta(\Phi))}.V(\sum_{i\neq j}\mathbb{R}_{\geq 0}\omega_i))=\sum_{i=1}^n a_{ij}$ for any $j=1,\ldots,n$.*

Proof. For $\alpha\in\Phi$, we denote by $s_\alpha:V\to V$ the reflection through the hyperplane associated to α, that is, the map defined by $s_\alpha(\lambda)=\lambda-(2(\lambda,\alpha)/(\alpha,\alpha))\alpha$. For $j=1,\ldots,n$, we consider the set of simple roots $s_{\alpha_j}S=\{s_{\alpha_j}(\alpha_1),\ldots,s_{\alpha_j}(\alpha_n)\}$. The dual basis of $s_{\alpha_j}(\alpha_1),\ldots,s_{\alpha_j}(\alpha_n)$ is $\omega_1 s_{\alpha_j},\ldots,\omega_n s_{\alpha_j}$. We see that $\omega_i s_{\alpha_j}=\omega_i$ on $M(\Phi)$ for any $i\neq j$. On the other hand, we have

$$(\omega_j s_{\alpha_j})(\alpha_i)=\omega_j\left(\alpha_i-\frac{2(\alpha_i,\alpha_j)}{(\alpha_j,\alpha_j)}\alpha_j\right)=\begin{cases}-1 & (i=j),\\ -a_{ij} & (i\neq j).\end{cases}$$

Thus $\omega_j s_{\alpha_j} = -\omega_j - \sum_{i\neq j} a_{ij}\omega_i$ on $M(\Phi)$. Hence we have

$$\sigma_{s_{\alpha_j}S} = \sum_{i=1}^{n} \mathbb{R}_{\geq 0}(\omega_i s_{\alpha_j}) = \sum_{i\neq j} \mathbb{R}_{\geq 0}\omega_i + \mathbb{R}_{\geq 0}(-\omega_j - \sum_{i\neq j} a_{ij}\omega_i)$$

and $\sigma_S \cap \sigma_{s_{\alpha_j}S} = \sum_{i\neq j} \mathbb{R}_{\geq 0}\omega_i$. Since $\omega_j + (-\omega_j - \sum_{i\neq j} a_{ij}\omega_i) + \sum_{i\neq j} a_{ij}\omega_i = 0$, Proposition 2.2 gives

$$(-K_{X(\Delta(\Phi))}.V(\sum_{i\neq j} \mathbb{R}_{\geq 0}\omega_i)) = 2 + \sum_{i\neq j} a_{ij} = \sum_{i=1}^{n} a_{ij}.$$

This completes the proof. □

Example 6.1. The toric variety associated to the root system of type A_n is isomorphic to $X(\Delta(B))$ for the building set $B = 2^{\{1,\dots,n+1\}} \setminus \{\varnothing\}$. On the other hand, the toric varieties associated to the root systems of type B and C cannot be obtained from building sets since $a_{ij} = -2$ for some i, j.

Finally, we characterize root systems whose associated toric varieties are Fano or weak Fano.

Theorem 6.4. *Let Φ be a root system. Then the associated toric variety $X(\Delta(\Phi))$ is Fano (resp. weak Fano) if and only if each irreducible component of Φ is of type A_1 or A_2 (resp. type A_n or B_n for some positive integer n).*

Proof. Suppose that Φ is an irreducible root system of rank n. We fix a set of simple roots $S = \{\alpha_1, \dots, \alpha_n\} \subset \Phi$ and we put $a_{ij} = 2(\alpha_i, \alpha_j)/(\alpha_j, \alpha_j)$. For any $\tau \in \Delta(\Phi)(n-1)$, there exists a set of simple roots $S' \subset \Phi$ such that τ is a face of $\sigma_{S'}$. Proposition 6.1 gives $(-K_{X(\Delta(\Phi))}.V(\tau)) = \sum_{i=1}^{n} a_{ij}$ for some $j = 1, \dots, n$. Conversely, the proof of Proposition 6.1 shows that for any $j = 1, \dots, n$, there exists $\tau \in \Delta(\Phi)(n-1)$ such that $(-K_{X(\Delta(\Phi))}.V(\tau)) = \sum_{i=1}^{n} a_{ij}$. Therefore by Proposition 2.3, $X(\Delta(\Phi))$ is Fano (resp. weak Fano) if and only if $\sum_{i=1}^{n} a_{ij} > 0$ (resp. $\sum_{i=1}^{n} a_{ij} \geq 0$) for any $j = 1, \dots, n$. Since Φ is irreducible and (a_{ij}) is the Cartan matrix, $\sum_{i=1}^{n} a_{ij} > 0$ (resp. $\sum_{i=1}^{n} a_{ij} \geq 0$) for any $j = 1, \dots, n$ if and only if Φ is of type A_1 or A_2 (resp. type A_n or B_n for some positive integer n). The theorem follows from the fact that the direct sum of root systems corresponds to the product of the associated toric varieties. □

Acknowledgments

The author would like to thank the organizers of the conference "Algebraic and Geometric Combinatorics on Lattice Polytopes", Takayuki Hibi and Akiyoshi Tsuchiya, for their invitation and their kind hospitality. Professor Hidefumi Ohsugi gave him valuable comments. This work was supported by Grant-in-Aid for JSPS Fellows 18J00022.

References

[1] C. Araujo and A-M. Castravet, Classification of 2-Fano manifolds with high index, *A Celebration of Algebraic Geometry* **18**, 1–36 (2013).

[2] V. Batyrev and M. Blume, The functor for toric varieties associated with Weyl chambers and Losev–Manin moduli spaces, *Tohoku Math. J.* **63**, 581–604 (2011).

[3] M. Carr and S. L. Devadoss, Coxeter complexes and graph-associahedra, *Topology Appl.* **153**, 2155–2168 (2006).

[4] C. De Concini and C. Procesi, Wonderful models of subspace arrangements, *Selecta Math. (N.S.)* **1**, 459–494 (1995).

[5] A. J. de Jong and J. Starr, Higher Fano manifolds and rational surfaces, *Duke Math. J.* **139**, 173–183 (2007).

[6] S. L. Devadoss, T. Heath and W. Vipismakul, Deformations of bordered surfaces and convex polytopes, *Notices Amer. Math. Soc.* **58**, 530–541 (2011).

[7] F. Harary, *Graph Theory* (Addison-Wesley, Reading, MA, 1969).

[8] A. Higashitani, Smooth Fano polytopes arising from finite directed graphs, *Kyoto J. Math.* **55**, 579–592 (2015).

[9] T. Manneville and V. Pilaud, Compatibility fans for graphical nested complexes, *J. Combin. Theory Ser. A* **150**, 36–107 (2017).

[10] M. Øbro, An algorithm for the classification of smooth Fano polytopes, preprint, 2007. `arXiv:0704.0049`

[11] T. Oda, *Convex Bodies and Algebraic Geometry. An Introduction to the Theory of Toric Varieties*, Ergeb. Math. Grenzgeb. (3) **15** (Springer-Verlag, Berlin, 1988).

[12] M. Reid, Decomposition of toric morphisms, Arithmetic and Geometry, Vol. II, *Progr. Math.* **36**, 395–418 (Birkhäuser, Boston, 1983).

[13] H. Sato, Toward the classification of higher-dimensional toric Fano varieties, *Tohoku Math. J.* **52**, 383–413 (2000).

[14] H. Sato, The classification of smooth toric weakened Fano 3-folds, *Manuscripta Math.* **109**, 73–84 (2002).

[15] H. Sato, The numerical class of a surface on a toric manifold, *Int. J. Math. Math. Sci.* **2012**, Art. ID 536475.

[16] Y. Suyama, Toric Fano varieties associated to finite simple graphs, *Tohoku Math. J.*, to appear. `arXiv:1604.08440`

[17] Y. Suyama, Toric Fano varieties associated to building sets, *Kyoto J. Math.*, to appear. `arXiv:1611.01636`

[18] Y. Suyama, Toric weak Fano varieties associated to building sets, *Osaka J. Math.*, to appear. `arXiv:1705.07304`

[19] Y. Suyama, Toric Fano varieties associated to graph cubeahedra, preprint, 2018. `arXiv:1802.07059`

[20] A. Zelevinsky, Nested complexes and their polyhedral realizations, *Pure Appl. Math. Q.* **2**, 655–671 (2006).

A Reider-type result for smooth projective toric surfaces

Bach Le Tran

School of Mathematics
University of Edinburgh
James Clerk Maxwell Building
Peter Guthrie Tait Road
Edinburgh EH9 3FD, UK
E-mail: b.tran@sms.ed.ac.uk

Let L be an ample line bundle over a smooth projective toric surface X. Then L corresponds to a very ample lattice polytope P that encodes many geometric properties of L. In this article, by studying P, we will give some necessary and sufficient numerical criteria for the adjoint series $|K_X + L|$ to be either nef or (very) ample.

Keywords: Reider theorem, smooth toric surface, adjoint line bundle, lattice polygon, Pick theorem, Fujita's conjecture.

1. Introduction

The problem of determining whether a line bundle is nef or (very) ample is an important question in algebraic geometry. The Nakai-Moishezon criterion [10, 12] states that a Cartier divisor D on a proper scheme X over an algebraically closed field is ample if and only if $D^{\dim(Y)} \cdot Y > 0$ for every closed integral subscheme Y of X. For toric varieties, a special form of the criterion holds: if $D \cdot C > 0$ for every torus-invariant curve $C \subset X$ then D is ample. Furthermore, if $D \cdot C \geq 0$ for every torus-invariant curve $C \subset X$ then D is globally generated [7, 9, 11]. However, the question is more complicated when we consider the adjoint bundle $D + K_X$. Namely, are there numerical conditions for $D \cdot C$ so that $D + K_X$ is globally generated or ample? Fujita conjectured the following:

Conjecture 1.1 ([3]). *Let X be an n-dimensional projective algebraic variety, smooth or with mild singularities, and D an ample divisor on X. Then*

(i) For $t \geq n + 1$, $tD + K_X$ is basepoint free.
(ii) For $t \geq n + 2$, $tD + K_X$ is very ample.

The conjecture is true for toric varieties [4, 13]. For smooth surfaces, Fujita's conjecture follows from Reider's theorem [15].

In this article, we will present a combinatorial proof for a Reider-type result for smooth projective toric surfaces.

Proposition 1.1. *Let X be a smooth projective toric surface not isomorphic to $\mathbb{P}^2$, and let L be an ample line bundle on X.*

(i) *The adjoint series $|K_X + L|$ is not base point free if and only if there exists an effective torus-invariant divisor $D \subset X$ such that*

$$D \cdot L = 1 \text{ and } D^2 = 0.$$

(ii) *The adjoint series $|K_X + L|$ is not ample if and only if there exists an effective torus-invariant divisor $D \subset X$ such that either*

$$\begin{aligned} &D \cdot L = 1 \text{ and } D^2 = -1 \text{ or } D^2 = 0; \text{ or} \\ &D \cdot L = 2 \text{ and } D^2 = 0; \text{ or} \\ &D \cdot L = 3 \text{ and } D^2 = 1. \end{aligned}$$

Furthermore, if $L^2 \geq 10$, then $|K_X + L|$ is not ample if and only if there exists an effective torus-invariant divisor $D \subset X$ such that either

$$\begin{aligned} &D \cdot L = 1 \text{ and } D^2 = -1 \text{ or } D^2 = 0; \text{ or} \\ &D \cdot L = 2 \text{ and } D^2 = 0. \end{aligned}$$

As a convention, in this article, we will follow the notations in [2]. In particular, we will always use M to denote the ambient lattice if there is no confusions.

Acknowledgments

We would like to thank Milena Hering for suggesting the problem and for her invaluable guidance. We also want to thank Ivan Cheltsov for some of the comments.

2. Toric Surfaces Reviewed

Let A be an ample line bundle over a projective toric variety X corresponding to a polytope $P \subset M_{\mathbb{R}}$. Then we have a combinatorial interpretation of the intersection number $A \cdot C$ where $C \subset X$ is any torus-invariant curve as follows.

Lemma 2.1 ([7, (1.4) and Page 457]). *Let A be an ample line bundle on a projective toric variety X corresponding to a polytope P. For a torus invariant curve C, let E be the corresponding edge on P. Then $A \cdot C$ is equal to the lattice length of E, i.e.,*

$$A \cdot C = |E \cap M| - 1.$$

For our purpose, we will need to use the classification of smooth projective toric surfaces: every smooth complete toric surfaces is a finite blowup of either $\mathbb{P}^2$, $\mathbb{P}^1 \times \mathbb{P}^1$, or the Hirzebruch surface $\mathscr{F}_a$, where $a \geq 2$ ([2, Theorem 10.4.3]). Another important fact that we will use is that every ample line bundle on a smooth projective toric surface is also very ample.

Lemma 2.2 ([2, Theorem 6.1.15]). *A line bundle on a smooth complete toric variety is ample if and only if it is very ample.*

Smooth toric surfaces are interesting objects to work with; partially because of their computability. For example, we have the following lemma.

Lemma 2.3 ([2, Proposition 10.4.11]). *Let $u_0, \ldots, u_r$ be ray generators of a smooth complete fan Σ in $N_{\mathbb{R}} \cong \mathbb{R}^2$. Let $X = X_\Sigma$ be the smooth projective toric surface from Σ and $D_i = V(u_i)$ for $0 \le i \le r$. Let K_X be the canonical divisor $K_X = -\sum_{i=0}^r D_i$. Then*

$$K_X \cdot D_i = b_i - 2,$$

where the $b_1, \ldots, b_{r-1}$ are integers such that $u_{i-1} + u_{i+1} = b_i u_i$ for all $0 \le i \le r$, where $u_{-1} = u_r$ and $u_{r+1} = u_0$.

The following corollary follows directly from [2, Lemma 10.4.1] and Lemma 2.3.

Corollary 2.1. *Let $u_0, \ldots, u_r$ be ray generators of a smooth complete fan Σ in $N_{\mathbb{R}} \cong \mathbb{R}^2$. Let $X = X_\Sigma$ be the smooth projective toric surface from Σ and $D_i = V(u_i)$ for $0 \le i \le r$. Let K_X be the canonical divisor $K_X = -\sum_{i=0}^r D_i$. Then for $0 \le i \le r$,*

$$(L + K_X) \cdot D_i = L \cdot D_i - D_i^2 - 2.$$

We also know that the blowup of a toric variety corresponds to a subdivision of fan. Thus the number of generating rays of the fan corresponding to a toric surface increases after a blowup ([2, Proposition 3.3.15]).

Example 2.1. Consider the Hirzebruch surface $\mathscr{F}_r = \mathbb{P}(\mathcal{O}_{\mathbb{P}^1} \oplus \mathcal{O}_{\mathbb{P}^1}(r))$, $r \ge 1$, whose fan Σ given by the following figure

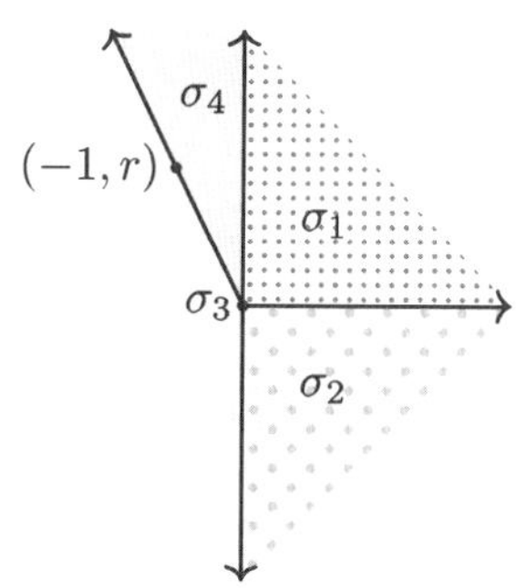

Fig. 1. The Hirzebruch fan

The ray generators of Σ are $v_1 = (1, 0)$, $v_2 = (0, 1)$, $v_3 = (-1, r)$, and $v_4 = (0, -1)$. Let the associated divisors be D_1, D_2, D_3, and D_4, respectively. By [2,

Proposition 4.1.2],

$$0 \sim div(\chi^{e_1}) = \sum_{i=1}^{4} \langle e_1, v_i \rangle D_i = D_1 - D_3$$

$$0 \sim div(\chi^{e_2}) = \sum_{i=1}^{4} \langle e_2, v_i \rangle D_i = D_2 + aD_3 - D_4.$$

Thus $D_3 \sim D_1$, $D_4 \sim D_2 + aD_3$, and

$$\mathrm{Pic}(\mathscr{F}_r) \simeq \{aD_3 + bD_4 \mid a, b \in \mathbb{Z}\}.$$

The maximal cones of Σ are σ_1, σ_2, σ_3 and σ_4 as in Figure 2.1. Let $D = aD_3 + bD_4$. We compute the m_{σ_i} to be

$$m_1 = (-a, 0), \quad m_2 = (-a, b), \quad m_3 = (rb, b), \quad m_4 = (0, 0).$$

Then by [2, Lemma 6.1.13], D is very ample if and only if $a, b > 0$. The nef cone of $\mathscr{F}_r$ is given by

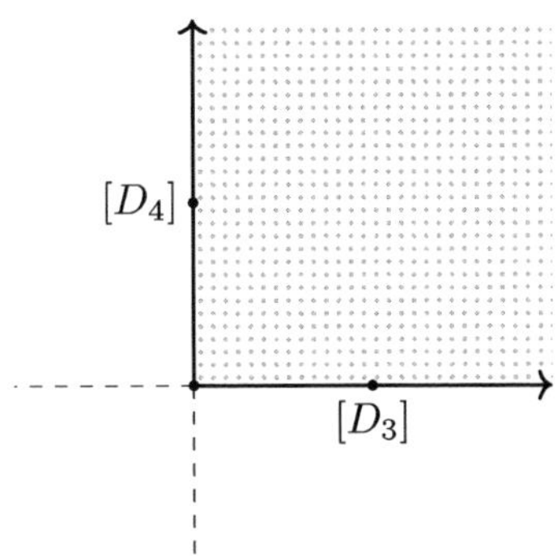

Fig. 2. The nef cone of $\mathscr{F}_r$

By [2, Lemma 10.4.1], we have $D_1^2 = D_3^2 = 0$, $D_2^2 = -a$, $D_4^2 = a$, $D_1 \cdot D_2 = D_2 \cdot D_3 = D_3 \cdot D_4 = D_4 \cdot D_1 = 1$, $D_1 \cdot D_3 = D_2 \cdot D_4 = 0$.

Finally, we will make use of the Hodge's Index Theorem:

Lemma 2.4 ([5, Theorem V.1.9]). *Let D be an ample divisor on a smooth projective surface S. If E is a divisor such that $D \cdot E = 0$, then $E^2 \leq 0$. The equality occurs if and only if E is numerically equivalent to* 0.

Corollary 2.2 ([5, Exercise V.1.9]). *Let D be an ample divisor on a smooth projective surface S and E an arbitrary divisor. Then*

$$(D \cdot E)^2 \geq D^2 E^2.$$

Proof. Since D is ample, $D^2 > 0$. Let $H = (D^2)E - (D \cdot E)D$. We have

$$D \cdot H = (D^2)E \cdot D - (D \cdot E)D^2 = 0.$$

Then by Lemma 2.4, we must have $H^2 \leq 0$. In other words,

$$\begin{aligned} 0 \geq & \left((D^2)E - (D \cdot E)D\right) \cdot \left((D^2)E - (D \cdot E)D\right) \\ = & D^4E^2 - 2(D \cdot E)^2(D^2) + D^2(D \cdot E)^2 \\ = & D^2 \left(D^2E^2 - (D \cdot E)^2\right). \end{aligned}$$

Since $D^2 > 0$, it follows that $(D \cdot E)^2 \geq D^2E^2$. □

3. Toric Surfaces and Lattice Polygons

In this section, we review and prove some lemmas on lattice polygons that we will use to the proof of Proposition 1.1.

Lemma 3.1 ([1, Lemma 1]). *Every lattice polygon with at least 5 edges has at least an interior lattice point.*

Lemma 3.2. *Let $v_1, \ldots, v_5$ be lattice points such that no three points are collinear. Then there exists a lattice point in* $\operatorname{conv}(v_1, \ldots, v_5) \backslash \{v_1, \ldots, v_5\}$.

Proof. Let the coordinates of v_i be (x_i, y_i) for $i = 1, \ldots, 5$. By the pigeonhole principle, there must be $i \neq j$ such that $x_i \equiv x_j \pmod 2$ and $y_i \equiv y_j \pmod 2$. Then the midpoint m of v_iv_j is a lattice point. Since no three points in $\{v_1, \ldots, v_5\}$ are collinear, it follows that $m \in \operatorname{conv}(v_1, \ldots, v_5) \backslash \{v_1, \ldots, v_5\}$. □

As a consequence, we obtain:

Lemma 3.3. *Let P be a lattice polygon that has at least* 5 *vertices and assume that one of its edges has lattice length* 4. *Then* $\operatorname{Vol}(P) \geq 9$.

Proof. It suffices to prove the lemma when P is a lattice pentagon. Let $P = \operatorname{conv}(v_1, \ldots, v_5)$, where $v_1, \ldots, v_5$ are ordered clockwise in M. Without loss of generality suppose that the lattice length of the edge joining v_1 and v_5 is 4; i.e., there are 3 other lattice points y_1, y_2, y_3 in between v_1 and v_5.

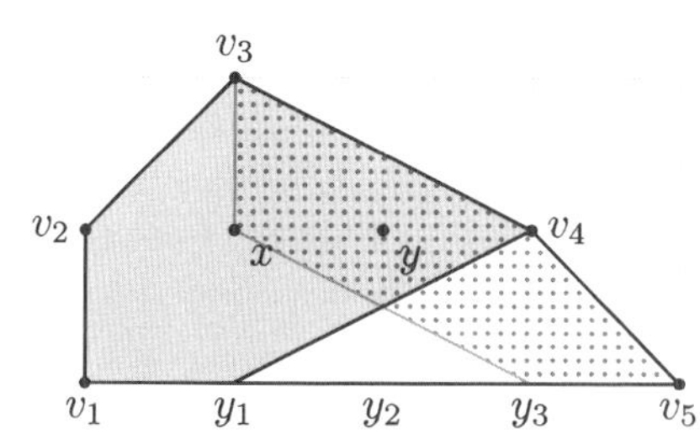

Fig. 3. A lattice pentagon that has an edge whose lattice length is 4

Consider the polytope $Q = \text{conv}(v_1, v_2, v_3, v_4, y_1)$. Then by Lemma 3.1, there must be a lattice point x in the interior of Q. Then x lies in at most one of the segments v_1v_3, y_1v_3, y_2v_3, y_3v_3, v_5v_3. If x lies in v_1v_3 or if x does not lie in any mentioned segments, consider the set of 5 points $\{x, v_3, v_4, v_5, y_1\}$. By Lemma 3.2, there must be another lattice point y in P that is not the same as the points listed before. If $y \in \partial P$, then $|\partial P \cap M| \geq 9$ and $|P^0 \cap M| \geq 1$. By Pick's theorem [14],

$$\text{Vol}(P) = |\partial P \cap M| + 2|P^0 \cap M| - 2 \geq 9.$$

If $y \in P^0$, then $|\partial P \cap M| \geq 8$ and $|P^0 \cap M| \geq 2$. Again, by Pick's theorem,

$$\text{Vol}(P) = |\partial P \cap M| + 2|P^0 \cap M| - 2 \geq 10.$$

If x lies in v_3y_1 or v_3y_2 then we get such a point y from $\text{conv}(x, v_3, v_4, v_5, y_3)$. If x lies in v_3y_3 or v_3v_5 then we get y from $\text{conv}(v_1, v_2, v_3, x, y_2)$. The same argument follows and we proved the lemma. □

We will also need the following lemmas for the proof of Proposition 1.1.

Lemma 3.4. *Let L be an ample line bundle over a smooth projective toric surface X. Let Σ be the fan of X. Suppose that Σ has $n \geq 5$ rays $\rho_1, \ldots, \rho_n$. Then for any integer $1 \leq i \leq n$,*

$$L^2 \geq L \cdot D_{\rho_i} + 4.$$

Proof. Let P be the polytope associated to L. By Pick's theorem ([14]) and since L is ample so that $L \cdot D_{\rho_i} \geq 1$ for all i,

$$\text{vol}(P) = \frac{L^2}{2} = \frac{|\partial P \cap M|}{2} + |P^0 \cap M| - 1, \tag{1}$$

where ∂P and P^0 are the sets of all boundary points and interior points of P, respectively. By Lemma 2.1,

$$|\partial P \cap M| = \sum_{j=1}^{n} L \cdot D_{\rho_j}. \tag{2}$$

Hence, combining (1) and (2) gives

$$L^2 = \sum_{j=1}^{n} L \cdot D_{\rho_j} + 2|P^0 \cap M| - 2 \geq L \cdot D_{\rho_i} + (n-1) + 2|P^0 \cap M| - 2.$$

Since $n \geq 5$, by Lemma 3.1, $|P^0 \cap M| \geq 1$. Therefore,

$$L^2 \geq L \cdot D_{\rho_i} + 4.$$

□

4. A Reider-type Result for Toric Surfaces

We will devote this section to prove Proposition 1.1. First of all, it is true for $X \cong \mathbb{P}^1 \times \mathbb{P}^1$.

Lemma 4.1. *Proposition 1.1 holds for $X \cong \mathbb{P}^1 \times \mathbb{P}^1$.*

Proof. Let Σ be the fan of $X = \mathbb{P}^1 \times \mathbb{P}^1$ as follows.

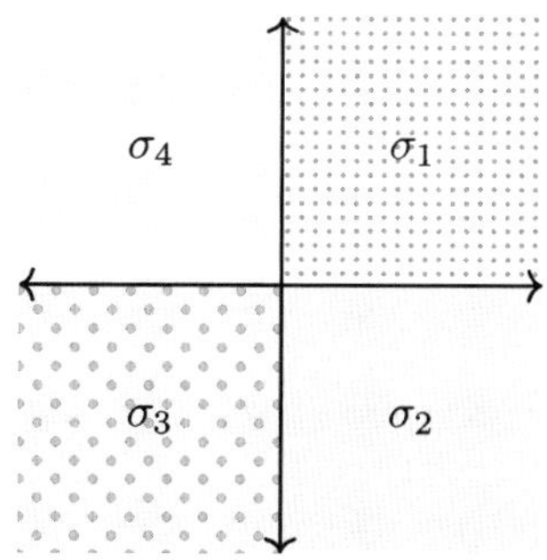

Fig. 4. The fan of $\mathbb{P}^1 \times \mathbb{P}^1$

By [2, Lemma 10.4.1], $D_\rho^2 = 0$ for all $\rho \in \Sigma(1)$. Thus, we need to show that there exists ρ such that $L \cdot D_\rho = 1$ in the first part and $L \cdot D_\rho \leq 2$ in the second part.

For any ample bundle L on X, if $L + K_X$ is not basepoint free, then there exists $\rho \in \Sigma(1)$ such that $(L + K_X) \cdot D_\rho < 0$. Then by Lemma 2.3,

$$(L + K_X) \cdot D_\rho = L \cdot D_\rho - D_\rho^2 - 2 < 0.$$

This implies $0 < L \cdot D_\rho < D_\rho^2 + 2 = 2$, so that $L \cdot D_\rho = 1$.

Now suppose that $L + K_X$ is not ample and $(L + K_X) \cdot D_\rho \leq 0$. Then by Lemma 2.3,

$$(L + K_X) \cdot D_\rho = L \cdot D_\rho - D_\rho^2 - 2 \leq 0.$$

This implies $1 \leq L \cdot D_\rho \leq D_\rho^2 + 2 = 2$. Hence, either $L \cdot D_\rho = 1$ and $D_\rho^2 = 0$ or $L \cdot D_\rho = 2$ and $D_\rho^2 = 0$. The conclusion follows. □

Secondly, we show that Proposition 1.1 holds for Hirzebruch surfaces.

Lemma 4.2. *Proposition 1.1 holds for $X \cong \mathscr{F}_a$, $a \geq 1$.*

Proof. Consider the Hirzebruch surface $X = \mathscr{F}_r = \mathbb{P}(\mathcal{O}_{\mathbb{P}^1} \oplus \mathcal{O}_{\mathbb{P}^1}(r))$, $r \geq 1$ as in Example 2.1. We have

$$\operatorname{Pic}(\mathscr{F}_r) \simeq \{aD_3 + bD_4 \mid a, b \in \mathbb{Z}\}.$$

The canonical divisor of X is given by

$$K_X = -(D_1 + D_2 + D_3 + D_4) \sim -(2-a)D_3 - 2D_4.$$

Recall that $D_1^2 = D_3^2 = 0$, $D_2^2 = -a$, $D_4^2 = a$, $D_1 \cdot D_2 = D_2 \cdot D_3 = D_3 \cdot D_4 = D_4 \cdot D_1 = 1$, and $D_1 \cdot D_3 = D_2 \cdot D_4 = 0$ (cf. [2, Lemma 10.4.1]).

Let L be an ample line bundle over $\mathscr{F}_r$. Then $L^2 > 0$. We have two cases as follows.

- If $r = 1$ then $K_X = -D_3 - 2D_4$. For L to be ample while $L + K_X$ is not nef, L has to be of the form $L \sim cD_3 + D_4$, $c > 0$. In this case, take $D = D_3$, then

  $$L \cdot D = 1 \text{ and } D^2 = 0.$$

  For L to be ample while $L + K_X$ is not ample, L has to be of the form $L \sim D_3 + cD_4$, or $L \sim cD_3 + D_4$, or $L \sim cD_3 + 2D_4$, where $c \geq 1$.

  (i) If $L \sim D_3 + cD_4$, take $D = D_2$, then

  $$L \cdot D = 1 \text{ and } D^2 = -1.$$

  (ii) If $L \sim cD_3 + D_4$, take $D = D_3$, then

  $$L \cdot D = 1 \text{ and } D^2 = 0.$$

  (iii) If $L \sim cD_3 + 2D_4$, take $D = D_3$, then

  $$L \cdot D = 2 \text{ and } D^2 = 0.$$

- $r \geq 2$: For L to be ample but $K_X + L$ is not nef, L has the form

  $$L \sim D_4 + cD_3 \quad (c \geq 0).$$

  Take $D = D_3$, then $L \cdot D = 1$ and $D^2 = 0$.
  For L to be ample but $K_X + L$ is not, L has the form $L \sim cD_3 + D_4$ or $L \sim cD_3 + 2D_4$, where $c \geq 1$.

  (i) If $L \sim cD_3 + D_4$, take $D = D_3$, then

  $$L \cdot D = 1 \text{ and } D^2 = 0.$$

  (ii) If $L \sim cD_3 + 2D_4$, take $D = D_3$, then

  $$L \cdot D = 2 \text{ and } D^2 = 0.$$

□

Finally, we will give the proof for the final case of Proposition 1.1, when X is an arbitrary blowup of $\mathbb{P}^1 \times \mathbb{P}^1$ or the Hirzebruch surface.

Proof of Proposition 1.1. The sufficiency trivially holds by Corollary 2.1. We now prove the necessity.

By the classification of smooth projective toric surfaces, the proofs for the cases of $\mathbb{P}^1 \times \mathbb{P}^1$ (Lemma 4.1) and $\mathscr{F}_a$ (Lemma 4.2), it suffices to prove the proposition in the case that the fan Σ of X has at least 5 rays.

We first prove part 1. Suppose that $K_X + L$ is not basepoint free. Then there exists $\rho \in \Sigma(1)$ such that $(K_X + L) \cdot D_\rho < 0$. Take $D = D_\rho$. By Lemma 2.3,

$$(L + K_X) \cdot D = L \cdot D - D^2 - 2 < 0.$$

This implies $L \cdot D < D^2 + 2$, so since L is ample,

$$0 \leq L \cdot D - 1 \leq D^2. \tag{3}$$

- If $D^2 \leq -1$, then $L \cdot D \leq 0$, which is a contradiction to the hypothesis that L is ample.
- If $D^2 = 0$, either $D \cdot L = 0$ or $D \cdot L = 1$. But $D \cdot L > 0$ since L is ample. Thus $D \cdot L = 1$. The proposition holds for this case.

It remains to show that D^2 cannot be positive. Since the fan of X contains at least 5 rays, by Lemma 3.4,

$$L^2 \geq L \cdot D + 4. \tag{4}$$

In addition, it follows from Corollary 2.2 that

$$(L \cdot D)^2 \geq L^2 \cdot D^2. \tag{5}$$

Combining (5) with (3) and (4) yields

$$(L \cdot D)^2 \geq (L \cdot D - 1)(L \cdot D + 4) = (L \cdot D)^2 + 3L \cdot D - 4.$$

This implies $L \cdot D \leq 1$. The only possibility is $L \cdot D = 1$. Then by (5), $D^2 = L^2 = 1$, which is impossible since $L^2 \geq L \cdot D + 4 = 5$. Therefore, it cannot be the case that $D^2 > 0$.

We now prove the second part of the proposition. Suppose that $K_X + L$ is not ample, so there exists $\rho \in \Sigma(1)$ such that $(K_X + L) \cdot D_\rho \leq 0$. Let $D = D_\rho$. By Corollary 2.1,

$$(L + K_X) \cdot D = L \cdot D - D^2 - 2 \leq 0.$$

This implies $L \cdot D \leq D^2 + 2$; hence,

$$1 \leq L \cdot D \leq D^2 + 2. \tag{6}$$

- If $D^2 = -1$, then $1 \leq L \cdot D \leq 1$, so $L \cdot D = 1$.
- If $D^2 = 0$, either $D \cdot L = 1$ or $D \cdot L = 2$.

Now we consider the case that $D^2 \geq 1$. Since the fan of X contains at least 5 rays, by Lemma 3.4,

$$L^2 \geq L \cdot D + 4. \tag{7}$$

By Corollary 2.2,

$$(L \cdot D)^2 \geq L^2 \cdot D^2 \tag{8}$$

Since $D^2 \geq 1$, then by (7), $L^2 \geq 5$. Thus by (8), $(L\cdot D)^2 \geq L^2\cdot D^2 \geq 5$, so $L\cdot D > 2$. It follows that $L\cdot D \geq 3$. Hence, $L\cdot D - 2 \geq 1$. This inequality combining with (6) and (7) yields

$$(L\cdot D)^2 \geq (L\cdot D-2)(L\cdot D+4) = (L\cdot D)^2 + 2L\cdot D - 8.$$

This implies $L\cdot D \leq 4$. The only possibilities are $L\cdot D = 3$ or $L\cdot D = 4$.

- If $D^2 = 1$ then $L\cdot D \leq 3$ by (6). Since $L\cdot D$ can only be either 3 or 4, $L\cdot D = 3$ in this case. Furthermore, suppose that $L^2 \geq 10$. If $L\cdot D = 3$ and $D^2 = 1$ then $9 = (L\cdot D)^2 < 10 \leq L^2\cdot D^2$, a contradiction to (8).
- Now assume that $D^2 \geq 2$. If $L\cdot D = 3$, then $L^2 \geq 7$ by (7), and $L^2\cdot D^2 \geq 7\cdot 2 = 14 > 9 = (L\cdot D)^2$, a contradiction to (8). Now assume that $L\cdot D = 4$. Then the polygon P_L associated to L has at least 5 vertices and one of its edges has lattice length 4 by Lemma 2.1. Hence, $L^2 \geq 9$ by Lemma 3.3. It follows that $16 = (L\cdot D)^2 < 18 \leq L^2\cdot D^2$, a contradiction to (8).

The proposition follows. □

5. Some Applications

The following corollary gives an affirmative answer for a stronger form of Fujita's conjecture (Conjecture 1.1) in case of smooth complete toric surfaces. Note that for n-dimensional toric varieties, the Fujita's conjecture is in fact a corollary of [4, Corollary 0.2] and [13, Theorem 1].

Corollary 5.1 ([4, 13]). *Let X be a smooth complete surface not isomorphic to $\mathbb{P}^2$. Let L be an ample line bundle on X such that $L\cdot C \geq 2$ for all toric invariant curve $C \subset X$. Then $\mathcal{O}_X(K_X+L)$ is globally generated. If $L^2 \geq 10$ and $L\cdot C \geq 3$ for all toric invariant curve $C \subset X$, then $\mathcal{O}_X(K_X+L)$ is very ample.*

Proof. Suppose that $\mathcal{O}_X(K_X+L)$ is not globally generated. By Proposition 1.1, there exists a toric invariant curve C such that $L\cdot C = 0$ or $L\cdot C = 1$, a contradiction. □

As a corollary, we have a stronger form of [8, Corollary 2.7] for smooth toric surfaces as follows.

Corollary 5.2. *If A is an ample line bundle on a smooth complete toric surface X not isomorphic to $\mathbb{P}^2$, then $|K_X+2A|$ is nef, and $|K_X+4A|$ is very ample.*

Proof. Take $L = 2A$, then for any toric invariant curve $C \subset X$, $L\cdot C = 2A\cdot C \geq 2$. By Proposition 1.1, $|K_X+2A|$ is nef. Similarly, take $L' = 4A$, then $(L')^2 = 16A^2 > 10$, and $L\cdot C = 4A\cdot C \geq 4$. By Proposition 1.1, $|K_X+4A|$ is very ample. □

Remark 5.1. It would be interesting to see if we can apply the classification in Proposition 1.1 to the study of Iskovskikh-Shokurov conjecture [6] for conic bundles over smooth toric surfaces.

References

[1] J.R. Arkinstall, *Minimal requirements for Minkowski's theorem in the plane i*, Bulletin of the Australian Mathematical Society **22** (1980), no. 2, 259–274.

[2] David A. Cox, John B. Little, and Henry K. Schenck, *Toric varieties (graduate studies in mathematics)*, American Mathematical Society, 2011.

[3] Takao Fujita, *On polarized manifolds whose adjoint bundles are not semipositive*, Algebraic geometry (Tadao Oda, ed.), Adv. Stud. Pure Math., vol. 10, Sendai, North-Holland, Amsterdam, 1985, pp. 167–178.

[4] Osamu Fujino, *Notes on toric varieties from mori theoretic viewpoint*, Tohoku Math. J. (2) **55** (2003), no. 4, 551–564.

[5] Robin Hartshorne, *Algebraic geometry*, Springer-Verlag, New York, 1977, Graduate Texts in Mathematics, No. 52. MR MR0463157 (57 #3116)

[6] V A Iskovskikh and V V Shokurov, *Birational models and flips*, Russian Mathematical Surveys **60** (2005), no. 1, 27–94.

[7] Robert Laterveer, *Linear systems on toric varieties*, Tohoku Math. J. (2) **48** (1996), no. 3, 451–458.

[8] R. Lazarsfeld, *Lectures on linear series*, eprint arXiv:alg-geom/9408011, August 1994.

[9] Anvar R. Mavlyutov, *Semiample hypersurfaces in toric varieties*, Duke Math. J. **101** (2000), no. 1, 85–116.

[10] B. G. Moishezon, *A projectivity criterion of complete algebraic abstract varieties*, Izv. Akad. Nauk SSSR Ser. Mat. **28** (1964), no. 1, 179–224.

[11] Mircea Mustaţă, *Vanishing theorems on toric varieties*, Tohoku Math. J. (2) **54** (2002), no. 3, 451–470.

[12] Yoshikazu Nakai, *A criterion of an ample sheaf on a projective scheme*, American Journal of Mathematics **85** (1963), no. 1, 14–26.

[13] Sam Payne, *Fujita's very ampleness conjecture for singular toric varieties*, Tohoku Mathematical Journal **58** (2006), no. 3, 447–459.

[14] G. Pick, *Geometrisches zur Zahlenlehre.*, Sonderabdr. Naturw.-medizin. Verein f. Böhmen "Lotos" Nr. 8, 9 S. 8° (1899), 1899.

[15] Igor Reider, *Vector bundles of rank 2 and linear systems on algebraic surfaces*, Annals of Mathematics **127** (1988), no. 2, 309–316.

Face enumeration on flag complexes and flag spheres

Hailun Zheng

Department of Mathematics
University of Michigan
Ann Arbor, MI 48109-1043, USA
E-mail: hailunz@umich.edu

We give a survey on the recent results and problems on the face enumeration of flag complexes and flag simplicial spheres, with an emphasis on the characterization of face vectors of flag complexes, several lower-bound type of conjectures including the Charney-Davis conjecture and Gal's conjecture, and the upper bound conjecture for flag spheres and pseudomanifolds.

Keywords: Flag complexes, face enumeration, γ-vector, cd-index, the flag upper bound conjecture.

1. Introduction

In this paper we will survey some fascinating properties of flag complexes and results on the face enumeration of flag simplicial spheres. A simplicial complex is called *flag* if all of the minimal non-faces have cardinality two, or equivalently, it is the clique complex of its graph.

We begin with a summary of major face enumeration results for general simplicial complexes and simplicial spheres. The Kruskal-Katona theorem [24, 28] fully characterizes the f-vector of simplicial complexes. In 1971, McMullen [31] conjectured a list of conditions to characterize the f-vector of simplicial polytopes. Around 1980, the work of Billera and Lee [7, 8] showed by construction the sufficiency of McMullen's conditions, and Stanley [40] proved the necessity, thus establishing the classical g-theorem. It is conjectured that the same characterization even holds for the f-vector of simplicial spheres. Other major results on face enumeration include the Upper Bound Theorem [31] (UBT, for short) and the Lower Bound Theorem [5] (LBT, for short), which state that among all simplicial d-polytopes with n vertices, cyclic polytopes simultaneously maximize all the face numbers while stacked polytopes simultaneously minimize all the face numbers. Both of the theorems extend to the class of simplicial spheres [5, 39], and have various generalizations in even larger classes of simplicial complexes, see, for example, [17, 23, 36].

However, none of the above results give tight bounds on the face numbers of flag complexes or spheres. For example, the clique complex of the graph of a stacked d-polytope and that of a cyclic d-polytope with n vertices are d-dimensional and $(n-1)$-dimensional respectively, and hence they cannot be flag $(d-1)$-spheres. It is natural to ask if there are flag analogs of the Kruskal-Katona theorem, the g-theorem, the UBT and the LBT, etc.

We know very little about the answers to these problems in general. In fact, we don't even have a plausible conjecture for the flag Kruskal-Katona theorem. The difficulty comes from the fact that the f-vector of a flag complex cannnot be changed "continuously": adding one edge to the graph of a given flag complex may result in a huge change in the f-vector of the corresponding clique complex. As to the flag lower bounds on the face numbers, various conjectures have been proposed in the last decade. Here we just mention one remarkable contribution of Gal: in [20] he defined the γ-vector which can be considered as the flag analog of the g-vector for general simplicial spheres, and he further conjectured that the γ-numbers are nonnegative for flag simplicial spheres. This conjecture has been verified in several important classes of flag simplicial spheres, including all 3-dimensional flag spheres [13] and the barycentric subdivision of all simplicial spheres (even of regular CW spheres) [25].

A upper bound conjecture (UBC, for short) for flag spheres was first proposed for odd-dimensional ones [30, 34]. It states that among all flag $(2k-1)$-spheres with n vertices (or more generally flag manifolds), the join of cycles of length as equal as possible is the unique maximizer of all the face numbers. This conjecture has been verified asymptotically [1] and for the edge number [45]. The even-dimensional flag UBC [44] turns out to be more complicated. As we can see from the dimension 4 case (where Gal's result gives the upper bounds of the face numbers), the maxmizer is at least not unique. So far this conjecture is wide open.

The paper is structured as follows. In Section 2 we give basic definitions, list several important properties of flag complexes, and provide the preliminary results on the lower bounds on f-, h- and g-vectors. in Section 3 we discuss how the f-vectors of flag complexes behave. Section 4 and Section 5 are devoted to the flag LBT and UBT respectively: in Section 4 we will see how the geometric meaning of flagness and introducing the *cd*-index help proving several important cases of the flag lower bound conjecture. We close with surveying current results on the flag UBC in Section 5.

2. Preliminaries

2.1. *Definitions and properties of flag complexes*

A *simplicial complex* Δ on a vertex set $V = V(\Delta)$ is a collection of subsets $\sigma \subseteq V$, called faces, that is closed under inclusion. For $\sigma \in \Delta$, let $\dim \sigma := |\sigma| - 1$ and define the *dimension* of Δ, $\dim \Delta$, as the maximal dimension of its faces. A *facet* is a maximal face under inclusion. We say Δ is *pure* if all of its facets have the same dimension. If Δ is a simplicial complex and σ is a face of Δ, the *link* of σ in Δ is $\mathrm{lk}(\sigma, \Delta) := \{\tau - \sigma \in \Delta : \sigma \subseteq \tau \in \Delta\}$. When the context is clear, we will abbreviate the notation and write it as $\mathrm{lk}(\sigma)$. If W is a subset of $V(\Delta)$, the *induced* subcomplex of Δ on W is the subcomplex $\Delta[W] = \{\tau \in \Delta : \tau \subseteq W\}$.

A simplicial complex Δ is a *simplicial sphere* (*simplicial manifold*, resp.) if the geometric realization of Δ, denoted as $||\Delta||$, is homeomorphic to a sphere (manifold, resp). A simplicial sphere is *polytopal* if it can be realized as the boundary complex of a polytope. Let $\mathbf{k}$ be a field and let $\tilde{H}_*(\Gamma, \mathbf{k})$ denote the reduced singular homology of $||\Gamma||$ with coefficients in $\mathbf{k}$. For a pure $(d-1)$-dimensional simplicial complex Δ and a field $\mathbf{k}$, we say that Δ is a $\mathbf{k}$-*homology manifold* if $\tilde{H}_*(\mathrm{lk}(\sigma), \mathbf{k}) \cong \tilde{H}_*(\mathbb{S}^{d-1-|\sigma|}, \mathbf{k})$ for every nonempty face $\sigma \in \Delta$. A $\mathbf{k}$-*homology sphere* is a $\mathbf{k}$-homology manifold that has the $\mathbf{k}$-homology of a sphere. Every simplicial manifold (resp. simplicial sphere) is a homology manifold (resp. homology sphere). A $(d-1)$-dimensional simplicial complex Δ is called a $(d-1)$-*pseudomanifold* if it is pure and every $(d-2)$-face (called ridge) of Δ is contained in exactly two facets. A $(d-1)$- pseudomanifold Δ is called a *normal* $(d-1)$-*pseudomanifold* if it is connected, and the link of each face of dimension $\leq d-3$ is also connected. For a fixed d, we have the following hierarchy:

$$\begin{gathered}\text{simplicial } (d-1)\text{-spheres} \subseteq \text{homology } (d-1)\text{-spheres}\\ \subseteq \text{homology } (d-1)\text{-manifolds}\\ \subseteq \text{normal } (d-1)\text{-pseudomanifolds} \subseteq (d-1)\text{-pseudomanifolds.}\end{gathered}$$

When $d = 3$, the first two classes complexes above coincide; so do the third and fourth classes. However, starting from $d = 4$, all the inclusions above are strict.

A simplicial complex is called *flag* if all of its minimal non-faces have cardinality two, or equivalently, it is the clique complex of its graph. Flag complexes have many properties which general simplicial complexes do not usually have, as we shall see in the following lemma [34, Lemma 5.2].

Lemma 2.1. *Let Δ be a flag complex on vertex set V.*

- *If $W \subseteq V$, then $\Delta[W]$ is also flag.*
- *If σ is a face in Δ, then $\mathrm{lk}(\sigma) = \Delta[V(\mathrm{lk}(\sigma))]$. In particular, all face links in a flag complex are also flag.*
- *Any edge $\{v, v'\}$ in Δ satisfies the link condition $\mathrm{lk}(v) \cap \mathrm{lk}(v') = \mathrm{lk}(\{v, v'\})$. More generally, any face $\sigma = \sigma_1 \cup \sigma_2$ in Δ satisfies $\mathrm{lk}(\sigma) = \mathrm{lk}(\sigma_1) \cap \mathrm{lk}(\sigma_2)$.*

The next property is about the connectivity of the graph of a complex. The graph of a simplicial complex Δ is said to be n-*connected* if Δ has at least $n+1$ vertices and the complex obtained from Δ by deleting any $n-1$ or fewer vertices and their incident faces is connected. It is known that every polytopal $(d-1)$-sphere, or more generally, every $(d-1)$-pseudomanifold is d-connected, see [46, theoerem 3.14] and [6, Corollary 5]. For flag complexes, a stronger statement holds [4, Theorem 1.2].

Lemma 2.2. *For every flag simplicial pseudomanifold Δ of dimension $d-1$, the graph $G(\Delta)$ is $(2d-2)$-connected.*

We introduce several ways to construct new flag complexes from given ones. If Δ and Γ are two simplicial complexes on disjoint vertex sets, then the *join* of Δ and Γ, denoted as $\Delta * \Gamma$, is the simplicial complex on vertex set $V(\Delta) \cup V(\Gamma)$ whose faces are $\{\sigma \cup \tau : \sigma \in \Delta, \tau \in \Gamma\}$. Hence by the definition, if Δ and Γ are flag complexes of dimension $d_1 - 1$ and $d_2 - 1$ respectively, then $\Delta * \Gamma$ is a flag complex of dimension $d_1 + d_2 - 1$.

To simplify the notation, in the following we write the one-vertex set $\{v\}$ as v. For a simplicial complex Δ and a face σ in it, let the *stellar subdivision* of Δ at σ be

$$\mathcal{S}(\sigma, \Delta) = \{\tau \in \Delta : \tau \cap \sigma = \varnothing\} \cup (v * \partial\sigma * \mathrm{lk}(\sigma, \Delta)),$$

where v is a new vertex. The stellar subdivision of Δ is always PL homeomorphic to Δ. However, since the stellar subdivision at face σ creates a missing face σ, $\mathcal{S}(\sigma, \Delta)$ is flag only when σ is an edge. The inverse of edge subdivision is called the edge contraction:

$$\mathcal{S}^{-1}(\{u, v\}, \Delta) = \{\tau \in \Delta : u \in \tau\} \cup \{(\tau \cup v)\backslash u : u \in F \in \Delta\}.$$

It is well-known that the edge contraction of a flag complex Δ is flag if and only if the contracted edge does not belong to any induced 4-cycles in Δ. The following lemma [30, Theorem 1.2] is the flag analog of the classical Alexander theorem in PL topology.

Lemma 2.3. *Two flag simplicial complexes are PL homeomorphic if and only if they can be connected by a sequence of edge subdivisions and their inverses such that all the complexes in the sequence are flag.*

2.2. *The lower bounds on f-, h- and g-vectors*

For a $(d-1)$-dimensional complex Δ, we let $f_i = f_i(\Delta)$ be the number of i-dimensional faces of Δ for $-1 \le i \le d-1$. The vector $(f_{-1}, f_0, \dots, f_{d-1})$ is called the *f-vector* of Δ. The h-vector of a $(d-1)$-dimensional simplicial complex Δ, $h(\Delta) = (h_0(\Delta), h_1(\Delta), \dots, h_{d-1}(\Delta))$, is defined by the equality

$$\sum_{i=0}^{d} h_i(\Delta) t^i = \sum_{i=0}^{d} f_{i-1}(\Delta) t^i (1-t)^{d-i}.$$

The polynomial which appears in the left-hand side of the above identity is the h-polynomial of Δ and is denoted by $h_\Delta(x)$. Define the g-vector $g(\Delta) = (g_0(\Delta), g_1(\Delta), \dots, g_{\lfloor d/2 \rfloor}(\Delta))$, where $g_0(\Delta) = 1$ and $g_i = h_i(\Delta) - h_{i-1}(\Delta)$ for $i \neq 0$. It is not hard to see that among all flag simplicial $(d-1)$-spheres, the octahedral $(d-1)$-sphere simultaneously minimizes all f-numbers. This even holds in the class of flag pseudomanifolds [4, Proposition 2.2].

Proposition 2.1. *Let Δ be a $(d-1)$-dimensional flag pseudomanifold. Then $f_{i-1}(\Delta) \ge 2^i \binom{d}{i}$ for all $0 \le i \le d$. Equality holds if and only if Δ is the octahedral $(d-1)$-sphere.*

For any flag complex Δ and vertex $v \in \Delta$, we have $h_i(\Delta) = h_i(\Delta\backslash v) + h_{i-1}(\mathrm{lk}(v))$ whenever $\dim(\Delta) = \dim(\Delta\backslash v)$. this identity together with the monotonicity of the

h-vectors of simplicial spheres [41] yields the lower bounds on the h-numbers [4, Theorem 1.3].

Proposition 2.2. *Let Δ be a flag homology $(d-1)$-sphere. Then $h_i(\Delta) \geq \binom{d}{i}$.*

Remark 2.1. There is a generalization of the above lemma for flag Buchsbaum* complexes, see Proposition 5.5 and Section 6 in [3].

Our next goal is to obtain lower bounds on the g-numbers. For g_2, this result is likely known to the experts. However, it seems to be missing from the literature. In the following we prove the lower bound on g_2 using the rigidity theory. Let $\mathbf{k}$ be an infinite field and assume that $V(\Delta) = [n]$. The *Stanley-Reisner ring* of Δ is $\mathbf{k}[\Delta] = \mathbf{k}[x_1, x_2, \ldots, x_n]/I_\Delta$, where the *Stanley-Reisner ideal* is

$$I_\Delta = (x_{i_1} \ldots x_{i_k} : \{i_1, \ldots, i_k\} \notin \Delta).$$

A $(d-1)$-dimensional complex is $\mathbf{k}$-*rigid* if for generic linear forms $\theta_1, \ldots, \theta_{d+1}$ and $1 \leq i \leq d+1$, the multiplication map $\cdot\theta_i : \mathbf{k}[\Delta]/(\theta_1, \ldots, \theta_{i-1})_1 \to \mathbf{k}[\Delta]/(\theta_1, \ldots, \theta_{i-1})_2$ is injective. For this interpretation of rigidity, see [29]. We also refer to [23, 43] for basics in the rigidity theory of frameworks and Stanley-Reisner ring respectively.

Lemma 2.4. *Let $d \geq 4$ and Δ be a flag normal $(d-1)$-pseudomanifold. Then for any vertex v, $\Delta \backslash v$ is $\mathbf{k}$-rigid.*

Proof. First we claim that $\Delta \backslash v = \cup_{u \notin \mathrm{st}(v)} \mathrm{st}(u)$. Indeed, if there is a facet F of $\Delta \backslash v$ with $V(F) \subseteq V(\mathrm{lk}(v))$, then by the flagness of Δ, $F \cup v \in \Delta$, a contradiction. Hence every facet of $\Delta \backslash v$ must contain at least one vertex from $\Delta \backslash v - \mathrm{lk}(v)$, i.e., $\Delta \backslash v \subseteq \cup_{u \notin \mathrm{st}(v)} \mathrm{st}(u)$. The other inclusion is obvious.

Since $\mathrm{lk}(v)$ is a normal $(d-2)$-pseudomanifold, it is $\mathbf{k}$-rigid. By the cone lemma [37, Lemma 5.3], $\mathrm{st}(v)$ is also $\mathbf{k}$-rigid. Furthermore, the vertices of $\Delta \backslash v - \mathrm{lk}(v)$ can be ordered as $(u_1, u_2, \ldots)$ so that the induced graph on any initial segment is connected, and hence $\mathrm{st}(u_i) \cap (\cup_{j<i} \mathrm{st}(u_j))$ contains a facet of $\Delta \backslash v$. So by the gluing lemma [37, Lemma 5.4], $\Delta \backslash v$ is $\mathbf{k}$-rigid. □

Proposition 2.3. *Let $d \geq 3$ and Δ be a flag normal $(d-1)$-pseudomanifold. Then $g_2(\Delta) \geq \binom{d}{2} - d$. Furthermore when $d \geq 4$, $g_2(\Delta) = \binom{d}{2} - d$ if and only if Δ is the octahedral $(d-1)$-sphere.*

Proof. We prove by induction on the dimension. Let $\mathbf{k}$ be an infinite field of characteristic zero. For $d = 3$, $g_2(\Delta) \geq 0$ follows from the fact that all triangulated 2-manifolds are $\mathbf{k}$-rigid. Let $d \geq 4$. For any vertex v of Δ, $\dim \Delta \backslash v = \dim \Delta$. By Lemma 4.1 in [4], $h_i(\Delta) = h_i(\Delta \backslash v) + h_{i-1}(\mathrm{lk}_\Delta v)$ for $i = 1, 2$, and hence $g_2(\Delta) = g_2(\Delta \backslash v) + g_1(\mathrm{lk}(v))$. Since Δ is a flag normal pseudomanifold, there is a vertex u such that $\{u, v\} \notin \Delta$. Let $\omega \in \mathbf{k}[x_1, \ldots, x_{f_0}]_1$ and Θ a generic linear system of parameters for Δ. Note that both $\Delta \backslash v$ and $\mathrm{st}(u)$ are induced subcomplexes of Δ and furthermore $\Delta\backslash v \supseteq \mathrm{st}(u)$. So there is a surjection $p : \mathbf{k}[\Delta\backslash v] \to \mathbf{k}[\mathrm{st}(u)] \cong \mathbf{k}[\mathrm{lk}(u)]$. In the following let $\mathbf{k}(\Gamma) = \mathbf{k}[\Gamma]/\Theta$ for $\Gamma = \Delta\backslash v$ or $\mathrm{lk}(u)$. The rigidity theory gives us the following commutative diagram.

$$\begin{array}{ccccccc}
\mathbf{k}(\Delta\backslash v)_1 & \xhookrightarrow{\varphi_1=\cdot\omega} & \mathbf{k}(\Delta\backslash v)_2 & \longrightarrow & \mathrm{coker}\varphi_1 & \longrightarrow & 0 \\
\downarrow{\scriptstyle \bar{p}_1} & & \downarrow{\scriptstyle \bar{p}_2} & & \downarrow{\scriptstyle q} & & \\
\mathbf{k}(\mathrm{lk}(u))_1 & \xhookrightarrow{\varphi_2=\cdot\omega} & \mathbf{k}(\mathrm{lk}(u))_2 & \longrightarrow & \mathrm{coker}\varphi_2 & \longrightarrow & 0.
\end{array}$$

Furthemore, since $\Delta\backslash v$ is $\mathbf{k}$-rigid,

$$\dim_{\mathbf{k}} \mathrm{coker}\varphi_1 = \dim_{\mathbf{k}} \mathbf{k}(\Delta\backslash v)_2 - \dim_{\mathbf{k}} \mathbf{k}(\Delta\backslash v)_1 = h_2(\Delta\backslash v) - h_1(\Delta\backslash v) = g_2(\Delta\backslash v).$$

Similarly, $\dim_{\mathbf{k}} \mathrm{coker}\varphi_2 = g_2(\mathrm{lk}(u))$. The fact that the above diagram commutes implies that q is also surjective. Hence by the inductive hypothesis, $g_2(\Delta\backslash v) \geq g_2(\mathrm{lk}(u)) \geq \binom{d-1}{2} - (d-1)$. Notice that $g_1(\mathrm{lk}(v)) \geq d-2$ for every vertex v, so $g_2(\Delta) \geq \binom{d-1}{2} - (d-1) + (d-2) = \binom{d}{2} - d$. When $d \geq 4$, the equality holds if and only if the g_2 of every vertex link is exactly $d-2$, i.e., it is octahedral. Hence Δ is the octahedral sphere. $\square$

What about the lower bounds on the other g-numbers? As we will see in Section 4, the following conjecture would follow if the nonnegative of the γ-vectors holds. However, even this conjecture is open.

Conjecture 2.1. *Let Δ be a flag homology $(d-1)$-sphere. Then $g_i(\Delta) \geq \binom{d}{i} - \binom{d}{i-1}$. Equality holds if and only if Δ is the boundary complex of the d-cross-polytope.*

3. Face vectors of flag complexes

Before we dive into the face vectors of flag spheres or manifolds, let's warm up by discussing the face vectors of flag complexes without placing any conditions on the geometric realization. For positive integers n and i, n can be expressed uniquely in the form

$$n = \binom{n_i}{i} + \binom{n_{i-1}}{i-1} + \cdots + \binom{n_j}{j},$$

where $n_i > n_{i-1} > \cdots > n_j \geq j \geq 1$. Define

$$n^{(i)} = \binom{n_i}{i+1} + \binom{n_{i-1}}{i} + \cdots + \binom{n_j}{j+1},$$

where $\binom{k}{l} = 0$ if $l > k$. Also we say a simplicial complex $\Delta \subseteq 2^{[n]}$ is *compressed* if its set of k-faces forms an initial segment with respect to the reverse lexicographic order on the $(k+1)$-subsets of $[n]$ for each k. Now we are ready to state the Kruskal-Katona theorem [24, 28] which characterizes the f-vector of simplicial complexes.

Theorem 3.1. *A vector $f = (f_{-1}, f_0, f_1, \ldots, f_{d-1})$ is the f-vector of a $(d-1)$-dimensional simplicial complex if and only if $f_{-1} = 1$ and*

$$f_{i+1} \leq f_i^{(i+1)}, \quad 0 \leq i \leq d-2.$$

Moreover f can be realized as the f-vector of a compressed simplicial complex.

Is there a Krukal-Katona type of theorem for flag complexes? Alternatively, given a graph with a fixed number of k-cliques, how many $(k+1)$-cliques can it have? Unlike the simplicial complexes, where we can add one face at a time and change the f-vector gradually, adding a single edge to a flag complex may cause the face vector of the resulting flag complex to vary dramatically. So it is natural to ask how many i-faces a flag $(d-1)$-dimensional complex with n vertices can have. The celebrated Túran's theorem states that given any graph G with n vertices and no cliques of size $r+1$, the Túran graph has the maximum number of edges. There are many generalization of this theorem regarding the number of cliques in graphs, see, for example, [15]. Another interesting conjecture which gives restriction on the range of f-vectors was proposed by Kalai (unpublished, see [43]) and Eckhoff [14] independently. Frohmader established it as a theorem [19]. Recall that a $(d-1)$-dimensional simplicial complex is called balanced if its graph is d-colorable.

Theorem 3.2. *The f-vector of a flag complex is the f-vector of a certain balanced complex.*

As a corollary, the f-vector of any flag complex must satisfy the Frankl-Füredi-Kalai inequalities [18]. However, the above theorem does not give a full characterization of the face vectors of flag complex. In fact, there exist many vectors that are f-vectors of balanced complexes but not of flag complexes. How to describe those non-admissible f-vectors remains unknown. We also remark that the constructive proof given by Frohmader doesn't solve another conjecture of Kalai, which is a stronger version of the above statement.

Conjecture 3.1. *The f-vector of any flag Cohen-Macaulay complex is the f-vector of a balanced Cohen-Macaulay complex.*

We end this section with a conjecture [10] that implies Conjecture 3.1. Some partial results were obtained in [11, 12].

Conjecture 3.2. *The set of h-vectors of flag Cohen-Macaulay simplicial complexes equals the set of f-vectors of flag simplicial complexes.*

4. The flag lower bound conjecture

In this section we will discuss various lower-bound type conjectures. Among them, the Charney-Davis conjecture comes first in the history. The geometric intuition behind this conjecture is from a long-standing conjecture in differential geometry.

Conjecture 4.1. *If M^{2n} is a closed Riemannian manifold of nonpositive sectional curvature, then $(-1)^{2n}\chi(M^{2n}) \geq 0$.*

To find a discrete version of the above conjecture, note that a simplicial complex Δ can be turned into a geodesic space by letting all the edges have length $\pi/2$. Gromov [22] showed that in this setting a cubical complex is non-positively curved if and only if the links are flag. (Gromov's interpretation of flagness is a very useful tool in studying flag simplicial complexes. For example, it is shown in [2] that the Hirsch conjecture holds for all flag homology manifolds by using the fact that every vertex star in the flag complex is geodesically convex.) Together with the fact that the Euler characteristic can be interpretted as $\chi(\Delta) = 1 + \sum_i (-1/2)^{i+1} f_i(\Delta)$, Conjecture 4.1 is rephrased as the Charney-Davis conjecture in [9]:

Conjecture 4.2. *Let Δ be a flag simplicial $(2m-1)$-sphere (or more generally, a flag homology $(2m-1)$-sphere). Then $(-1)^m h_\Delta(-1) \geq 0$.*

In the case of $m = 2$, the conjecture says that $f_1(\Delta) - 5f_0(\Delta) + 16 \geq 0$, which is proved in [13] by heavy topological machinery. In higher dimensions the conjecture remains open in general. What about even-dimensional flag spheres? Define a polynomial $\tilde{h}_\Delta(t) := \frac{1}{1+t} h_\Delta(t)$. Gal and Januszkiewicz [21] noticed that the Charney-Davis conjecture is equivalent to the following conjecture.

Conjecture 4.3. *Let Δ be a flag simplicial $2m$-sphere (or more generally, a flag homology $2m$-sphere). Then $(-1)^m \tilde{h}_\Delta(-1) \geq 0$.*

In the next two subsections we will discuss the proof of the Charney-Davis conjecture in some special cases by introducing the γ-vector and the cd-index.

4.1. *The lower bound conjectures around the γ-vector*

We start by discussing the classic Generalized Lower Bound Theorem (GLBT, for short) without a requirement that a polytope or sphere is flag. It turns out that instead of the f-vector, GLBT is better stated in terms of the g-vector.

Theorem 4.1. *Let P be a simplicial d-polytope. Then*

(i) $g_i(P) \geq 0$ for $1 \leq i \leq d/2$.
(ii) If $g_r(P) = 0$ for some $1 \leq r \leq d/2$, then P is $(r-1)$-stacked, i.e., there is a triangulation K of P all of whose faces of dimension at most $d-r$ are faces of P.

In fact we even know what the g-numbers count: as discovered by Lee [29], each g_r is the dimension of the affine stress spaces associated with the polytope.

Since $(i-1)$-stacked d-polytopes are not flag, the lower bounds given by Theorem 4.1 are not tight for flag polytopes or spheres. One would expect that there is an analogous vector for flag spheres that plays the same role as the g-vector for general spheres. By the Dehn-Sommerville relations, the h-vector of any flag homology $(d-1)$-sphere Δ is symmetric. Hence the h-polynomial of Δ has an integer expansion in the basis $\{t^i(1+t)^{d-2i} : 0 \le i \le d/2\}$, which shows that the following vector is well-defined.

Definition 4.1. The γ-*vector* of a flag homology $(d-1)$-sphere Δ is the vector $\gamma(\Delta) = (\gamma_0, \dots, \gamma_{\lfloor \frac{d}{2} \rfloor})$, whose entries are given by

$$h_\Delta(t) = \sum_{0 \le i \le d/2} \gamma_i(\Delta) t^i (1+t)^{d-2i}.$$

The polynomial $\gamma_\Delta(t) = \sum_{0\le i\le d/2} \gamma_i(\Delta)t^i$ is called the γ-*polynomial* of Δ. Alternatively, the γ-polynomial can be defined as a polynomial of degree at most $\lfloor \frac{d}{2} \rfloor$ such that

$$h_\Delta(t) = (1+t)^d \gamma_\Delta\left(\frac{t}{(1+t)^2}\right).$$

By Lemma 2.1, $\gamma_1 = f_0 - 2d \ge 0$. Also $\gamma_2 = f_1 - (2d-3)f_0 + 2d(d-2)$ is known to be nonnegative for $d = 3$ by the Davis-Okun theorem [13]. Gal proposed the following conjecture [20, Conjecture 2.1.7].

Conjecture 4.4. *If Δ is a flag homology sphere then $\gamma(\Delta)$ is nonnegative.*

In particular, Charney-Davis conjecture states that $\gamma_{\lfloor d/2 \rfloor} \ge 0$ and is a special case of Gal's conjecture. Conjecture 4.4 is known to hold for several important classes of flag spheres, including the order complexes of Gorenstein* complexes and Coxeter complexes. In an attempt of finding a flag analog of stackness, Nevo and Lutz proposed the following conjecture [30, Conjecture 6.1], which is still open.

Conjecture 4.5. *Let $d \ge 4$ be an integer and Δ be a flag simplicial $(d-1)$-sphere. Then the following are equivalent:*

(i) $\gamma_2(\Delta) = 0$.
(ii) There is a sequence of edge contractions from Δ to the boundary of the d-cross-polytope such that all complexes in the sequence are flag spheres, and the link of each edge contracted is the octahedral $(d-3)$-sphere.

However, we don't have a conjecture of which flag spheres can attain $\gamma_i = 0$ for $i \ge 3$. On the other hand, the γ-vector has an explicit combinatorial description in some cases (e.g., the γ_i of type A, B and D Coxeter complexeres are interpretted as the number of permutations in S_n that satisfies certain conditions, see [34]). We propose the following problems:

Problem 4.1.

(i) Find a characterization of flag homology $(d-1)$-spheres with $\gamma_i = 0$ for $3 \le i \le d/2$,

(ii) Find a combinatorial interpretation of the γ-numbers.

We end this section with a strenghening of Conjecture 4.4 suggested by Nevo and Petersen [34].

Conjecture 4.6. *Let Δ be a flag homology sphere.*

(i) $\gamma(\Delta)$ satisfies the Kruskal-Katona inequalities.
(ii) $\gamma(\Delta)$ satisfies the Frankl-Füredi-Kalai inequalities.

The first part of the conjecture is true but not sharp for flag 3- (or 4)-spheres. The second conjecture is more interesting; it is confirmed for barycentric subdivisions of simplicial homology spheres in [35], using enumerative properties of Eulerian polynomials. Later it is also proved for barycentric subdivisions of the boundary complexes of polytopes in [33] via *cd*-index.

4.2. *Proof idea: the cd-index*

In this section, we will discuss Karu's break-through result on the nonnegativity of the γ-vector. Let P be a finite graded poset of rank $n+1$ with $\hat{0}$ and $\hat{1}$, and the rank function is $\rho : P \to \{0, 1, \ldots, n+1\}$. For any $S \subseteq [n]$, the *S-rank selected subposet* of P is

$$P_S = \{x \in P : \rho(x) \in S\} \cup \{\hat{0}, \hat{1}\}.$$

The function $\alpha : 2^{[n]} \to \mathbb{Z}$, where each $\alpha(S)$ counts the number of maximal chains in P_S, is called the *flag f-vector* of P. Also define the *flag h-vector* of P as the function $\beta : 2^{[n]} \to \mathbb{Z}$,

$$\beta(S) = \sum_{T \subset S} (-1)^{|S|-|T|} \alpha(T).$$

The *cd*-index is a polynomial in variables c, d which encodes the flag h-vector of P. We assign each $S \subseteq [n]$ a noncommutative polynomial $u_S = u_1 u_2 \ldots u_n$ in the variables a, b by

$$u_i = \begin{cases} a & i \notin S \\ b & i \in S \end{cases}.$$

Let $\psi_P(a, b) = \sum_{S \subseteq [n]} \beta(S) u_S$. In particular, when P is an Eulerian poset, $\psi_P(a, b)$ can further be written as a polynomial $\varphi_P(c, d)$ in $c = a + b$ and $d = ab + ba$. This is the *cd-index* of P.

It is conjectured that the *cd*-index of a regular CW sphere, or more generally, a Gorenstein* poset (see [42] for the definition), is nonnegative. Many earlier works around this conjecture were based on recursive formulas of the *cd*-index, see, for example, [42]. We remark that it was also observed in [16] that the *cd*-index is a coalgebra map from the vector space spanned by isomorphism classes of Eulerian posets to the algebra of polynomials in the non-commutative variables c, d, which provides another tool to derive formulas for the *cd*-index. Karu gave a full proof of the conjecture in [25] by algebraic geometry tools.

Theorem 4.2. *If Δ is a Gorenstein* poset, then $\varphi_\Delta(c,d) \geq 0$, i.e., the coefficients of $\varphi_\Delta(c,d)$ are nonnegative.*

We sketch Karu's proof in the special case when Δ is a complete n-dimensional shellable fan. To each *cd*-monomial M we associate a polynomial in $t_1,\dots,t_n$ by replacing c with $1+t_i$ and d with t_i+t_{i+1}; for example, $cd = (t_1+1)(t_2+t_3)$. Let $P_n(\Delta) = \sum_S h_S(\Delta)t^S$ be the Poincaré polynomial of Δ. Let $\sigma_1,\dots,\sigma_n$ be a shelling of Δ and define $\sigma_i^- = \sigma_i \cap (\cup_{j=1}^{i-1}\sigma_j)$. The $(n-1)$-skeleton of Δ can be constructed from $\partial\sigma_n$ by attaching $\sigma_{n-1}^-, \sigma_{n-2}^-, \dots, \sigma_2^-$ in order. By Poincaré duality, if $P_{n-2}(\partial\sigma_i^-) = g_i(c,d)$ for some *cd*-polynomial of degree $n-2$, then $P_{n-2}(\sigma_i^-, \partial\sigma_i^-) = f_i(c,d) + g_i(c,d)t_{n-1}$ for some *cd*-polynomial f_i of degree $n-1$. So we have that

$$P_n(\mathrm{Skel}_{n-1}(\Delta)) = \sum_{i=2}^{n} P_{n-2}(\sigma_i^-, \partial\sigma_i^-) = \sum_{i=2}^{n} f_i + g_i t_{n-1}.$$

From this one can show that $P_n(\Delta) = \sum_{i=2}^n f_i c + g_i d$, and then the proof is done by induction on the dimension. The proof of the general case is more subtle and requires the sheaf theory, see [25] for the details.

The *cd*-index is closely related to the γ-vector. Let $\mathcal{S}(\Delta)$ be the barycentric subdivision of a regular CW-sphere Δ. Plugging in $c=1$ in the *cd*-index of the face poset of Δ, we get that

$$\varphi_\Delta(1,d) = \delta_0 + \delta_1 d + \cdots + \delta_{\lfloor \frac{n}{2} \rfloor} d^{\lfloor \frac{n}{2} \rfloor},$$

where δ_i is the sum of coefficients of monomials in $\varphi_\Delta(c,d)$ for which d appears i times. It follows from Theorem 4.2 that $\gamma_i(\mathcal{S}(\Delta)) = 2^i\delta_i \geq 0$. In other words, Karu's result settles Conjecture 4.4 in the class of order complexes of Gorenstein* posets.

Is there a combinatorial description of the *cd*-index? The following conjecture was first proposed by Murai and Nevo [33] and then rephrased by Karu [26]. Define the *multidegree* of a degree n *cd*-monomial M as a zero-one vector in $\mathbb{Z}^n$ obtained by replacing each c with 0 and each d with 10.

Conjecture 4.7. *Let $\mathbf{k}$ be a field. Let P be a Gorenstein* poset of rank $n+1$. Then there exists a standard $\mathbb{Z}^n$-graded $\mathbf{k}$-algebra $A = \oplus_v A_v$, such that for any $v \in \mathbb{Z}^n$, the coefficient of the cd-monomial M in $\varphi_P(c,d)$ with multidegree v is $\dim A_v$.*

The conjecture is known to hold for posets of Gorenstein* simplicial complexes [26, Theorem 1.3] but is open in general.

5. The flag upper bound conjecture

Denote by $C_d(n)$ the cyclic d-polytope with n vertices. The UBC for polytopes was proposed by Motzkin [32] and proved by McMullen [31], and it says that if P is a d-polytope with n vertices, then $f_i(P) \leq f_i(C_d(n))$ for all $1 \leq i \leq d-1$. The UBC has several important generalizations: Klee [27] proved the UBC for all Eulerian complexes with sufficiently many vertices, Stanley [39] verified it for simplicial spheres, and Novik [36] further extended it to all odd-dimensional homology manifolds and Eulerian even-dimensional homology manifolds. We remark that while the upper bounds on the face numbers for all classes of complexes mentioned above are the same, there are many polytopes or spheres that simultaneously maximize all face numbers; other classes of neighborly polytopes and neighborly spheres exist apart from the cyclic polytope. In fact, the current best lower bound for the number of combinatorial types of polytopes is attained by counting distinct neighborly polytopes [38].

The flag UBC turns out to be very different from the general UBT. Denote by $J_m(n)$ the simplicial $(2m-1)$-sphere on n vertices obtained as the join of m copies of the circle, each one a cycle with either $\lfloor \frac{n}{m} \rfloor$ or $\lceil \frac{n}{m} \rceil$ vertices. The following conjecture is proposed in [34, Conjecture 6.3] and [30, Conjecture 6.3].

Conjecture 5.1. *Let Δ be a flag $(2m-1)$-sphere with n vertices. Then $f_i(\Delta) \leq f_i(J_m(n))$ for all $1 \leq i \leq 2m-1$. Furthermore, if $f_i(\Delta) = f_i(J_m(n))$ for some i, then $\Delta = J_m(n)$.*

The above conjecture has been verified in several important cases. Adamaszek and Hladký [1] proved an asymptotic version of the conjecture in the class of flag triangulations of $(2m-1)$-manifolds. Using the extremal graph theory, they showed that not only the f-numbers, but also h-numbers, g-numbers and γ-numbers are maximized by $J_m(n)$ as long as the number of vertices is large enough. We sketch the proof below. Denote by $T_m(n)$ the clique complex generated by the complete m-partite Turán graph with n vertices and let Δ be a flag triangulated $(2m-1)$-manifold with n vertices. Fix a small number $a > 0$ and an integer $1 \leq i < m$. If $f_i(\Delta) \leq (1-a) f_i(T_m(n))$, then since $f_i(J_m(n)) = f_i(T_m(n)) + O(n^i) = O(n^{i+1})$ for any $1 \leq i < m$, it follows that the n^{i+1} term is also the dominant term in $f_i(\Delta)$. So with n large enough one can show that $f_j(\Delta) \leq f_j(J_m(n))$ for all j. Otherwise, since $f_m(\Delta) = o(n^{m+1})$, the Removal Lemma from extremal graph theory implies that a K_{m+1}-free subgraph G' can be obtained from $G(\Delta)$ by removing only $o(n^2)$ many edges. Thus as long as there are sufficiently many vertices in Δ, the number of i-faces in the clique complex of G' is at least $(1-2a) f_i(T_m(n))$. Then by applying other extremal graph theory results, one shows that $G(\Delta)$ can be obtained from the m-partite Turán graph by adding or deleting at most $O(n^2)$ edges. Finally since Δ triangulates a manifold, it follows that $f_i(\Delta)$ must be maximized by $f_i(J_m(n))$.

Zheng ([44], [45]) gave a proof of the conjecture for $i = 1$ case in the class of all flag normal pseudomanifolds. Given an edge $\{u, v\}$ in a flag pseudomanifold Δ, the inclusion-exclusion principle together with Lemma 2.1 yields that

$$f_0(\mathrm{lk}(u)) + f_0(\mathrm{lk}(v)) = f_0(\mathrm{lk}(u) \cup \mathrm{lk}(v)) - f_0(\mathrm{lk}(\{u,v\})) \le n - f_0(\mathrm{lk}(\{u,v\})).$$

More generally, using the inclusion-exclusion principle, one can show that for any facet σ of Δ,

$$\sum_{v\in\sigma} f_0(\mathrm{lk}(v)) \le 2(m-1)n + 4m.$$

The upper bound on the edge number then follows from the double counting. Note that while the upper bound even holds for flag pseudomanifolds, the maximizer of f_1 is not unique in this class: for example, both $J_2(16)$ and the join of C_8 and $C_4 \cup C_4$ have the same f-vector. However, the maximizer is unique in the class of normal pseudomanifolds.

What about the flag even-dimensional upper bound conjecture? In [20] Gal showed that the roots of the h-polynomial of a flag homology 4-sphere Δ with n vertices are all real for $d \le 5$. As a corollary, the face numbers of Δ are bounded by those of the suspension of $J_2(n-2)$. This motivates the following conjecture. Let $J_m^*(n) := \mathcal{S}^0 * C_1 * \cdots * C_m$, where each C_i is a circle of length either $\lceil \frac{n-2}{m} \rceil$ or $\lfloor \frac{n-2}{m} \rfloor$, and $f_0(J_m^*(n)) = n \ge 4m+2$. Denote by $\mathcal{S}_n$ the set of flag 2-spheres on n vertices, and define

$$\mathcal{J}_m^*(n) := \{S * C_2 * \cdots * C_m \mid S \in \mathcal{S}_{|V(C_1)|+2}\}.$$

Conjecture 5.2. *Let Δ be a flag homology $2m$-sphere on n vertices. Then $f_i(\Delta) \le f_i(J_m^*(n))$ for all $i = 1, \ldots, 2m$. If equality holds for some $1 \le i \le 2m$, then $\Delta \in \mathcal{J}_m^*(n)$.*

The conjecture is open except for the inequality part in the $m = 2$ case [20].

Acknowledgements

The author would like to thank Isabella Novik for helpful comments and discussions.

References

[1] M. Adamaszek and J. Hladký: Upper bound theorem for odd-dimensional flag triangulations of manifolds, *Mathematika* 62(2016), 909-928.

[2] K. Adiprasito and B. Benedetti: The Hirsch conjecture holds for normal flag complexes, *Math. Oper. Research* 39(2014), 1340-1348.

[3] C. A. Athanasiadis and V. Welker: Buchsbaum* complexes, *Math. Z.* 272 (2012), 131-149.

[4] C. A. Athanasiadis: Some combinatorial properties of flag simplicial pseudomanifolds and spheres, *Ark. Mat.*, 1(2011), 17-29.

[5] D. Barnette: A proof of the lower bound conjecture for convex polytopes, *Pacific J. Math.* 46(1973), 349-354.

[6] D. Barnette, Graph theorems for manifolds, *Israel J. Math.* 16(1973), 62-72.

[7] L. Billera and C. W. Lee: Sufficiency of McMullen's conditions for f-vectors of simplicial polytopes, *Bull. Amer. Math. Soc.*, 2(1980), 181-185.

[8] L. Billera and C. W. Lee: A proof of the sufficiency of McMullen's conditions for f-vectors of simplicial convex polytopes, *J. Combin. Theory Ser. A* 31(1981), 237-255.

[9] R. Charney and M. Davis: The Euler characteristic of a non-positively curved, piecewise Euclidean manifold, *Pacific J. Math.* 171(1995), 117-137.

[10] A. Constantinescu and M. Varbaro: On the h-vectors of Cohen-Macaulay flag complexes, *Math. Scand.* 112(2013), 86-111.

[11] G. Caviglia, A. Constantinescu and M. Varbaro: On a conjecture by Kalai, *Israel J. Math.* 204(2014), 469-475.

[12] D. Cook And U. Nagel: Cohen-Macaulay graphs and face vectors of flag complexes, *SIAM J. Discrete Math.*, 26(2012), 89-101.

[13] M. Davis and B. Okun: Vanishing theorems and conjectures for the ℓ^2-homology of right-angled Coxeter groups, *Geom. Topol.* 5(2001), 7-74.

[14] J. Eckhoff: Intersection properties of boxes. I. An upper-bound theorem, *Israel J. Math.* 62(1988), 283-301.

[15] J. Eckhoff: A new Turán-type theorem for cliques in graphs, *Discrete Mathematics*, 282(2004), 113-122.

[16] R. Ehrenborg and M. Readdy: Coproducts and the *cd*-index, *J. Algebr. Comb.* 8(1998), 273-299.

[17] A. Fogelsanger: The generic rigidity of minimal cycles. ProQuest LLC, Ann Arbor, MI, 1988. Thesis (Ph.D.)-Cornell University.

[18] P. Frankl, Z. Füredi and G. Kalai: Shadows of colored complexes, *Math. Scand.* 63(1988), 169-178.

[19] A. Frohmader: Face vectors of flag complexes, *Israel J. Math.* 164(2008), 153-164.

[20] Ś. R. Gal: Real root conjecture fails for five and higher dimensional spheres, *Discrete Comput. Geom.* 34(2005), 269-284.

[21] Ś. R. Gal and T. Januszkiewicz: Even- vs. odd-dimensional Charney–Davis conjecture, *Discrete Comput. Geom.* 44(2010), 802-804.

[22] M. Gromov: Hyperbolic groups in "Essays in Group Theory", M.S.R.I Publ. 8, Springer-Verlag, New York and Berlin, 1987, 75-264.

[23] G. Kalai: Rigidity and the lower bound theorem I, *Invent. Math.* 88(1987), 125-151.

[24] G. Katona: A theorem of finite sets, *Theory of graphs (Proc. Colloq., Tihany, 1966)*, 187-207. Academic Press, New York, 1968.

[25] K. Karu: The *cd*-index of fans and posets, *Compos. Math.* 142(2006), 701-718.

[26] K. Karu: M-vector analogue for the *cd*-index, *J. Combin. Theory. Ser. A* 137(2016), 257-272.

[27] V. Klee: The number of vertices of a convex polytope, *Canadian J. Math.* 16(1964), 702-720.

[28] J. Kruskal: The number of simplices in a complex, *Mathematical optimization techniques*, 251-278. Univ. of California Press, Berkeley, Calif., 1963.

[29] C. W. Lee: P.L.-spheres, convex polytopes, and stress, *Discrete Comput. Geom.* 15(1996), 389-421.

[30] F. Lutz and E. Nevo: Stellar theory for flag complexes, *Math. Scand.* 118(2016), 70-82.

[31] P. McMullen: The maximum numbers of faces of a convex polytope, *Mathematika* 17(1970), 179-184.

[32] T. S. Motzkin: Comonotone curves and polyhedra, *Bull. Amer. Math. Soc.* 63(1957), 35.

[33] S. Murai and E. Nevo: On the *cd*-index and γ-vector of S^*-shellable CW-spheres, *Math. Z.* 271(2012), 1309-1319.

[34] E. Nevo and T. K. Petersen: On γ-vectors satisfying the Kruskal-Katona inequalities, *Discrete Comput. Geom.* 45(2010), 503-521.

[35] E. Nevo, T. K. Petersen and B. Tenner: The γ-vector of a barycentric subdivision, *J. Combin. Theory Ser. A* 118(2011), 1364-1380.

[36] I. Novik: Upper bound theorems for homology manifolds, *Israel J. Math.* 108(1998), 45-82.

[37] I. Novik and E. Swartz: Applications of Klee's Dehn-Sommerville relations, *Discrete Comput. Geom.*, 42(2009), 261-276.

[38] A. Padrol: Many Neighborly Polytopes and Oriented Matroids, *Discrete Comput. Geom.* 50(2013), 865-902.

[39] R. Stanley: The upper bound conjecture and Cohen-Macaulay rings, *Studies in Applied Mathematics* 54(1975), 135-142.

[40] R. Stanley: The number of faces of a simplicial convex polytope, *Adv. in Math.*, 35(1980), 236-238.

[41] R. Stanley: A monotonicity property of h-vectors and h^*-vectors, *European J. Combinatorics* 14(1993), 251-258.

[42] R. Stanley: Flag f-vectors and the cd-index, *Math. Z.* 216(1994), 483-499.

[43] R. Stanley: *Combinatorics and Commutative Algebra*, Birkhäuser, Boston, 2. edition, 1996.

[44] H. Zheng: The flag upper bound theorem for 3- and 5-manifolds, *Israel J. Math.* 222(2017), 401-419.

[45] H. Zheng: The upper bounds on the edge numbers of flag odd dimensional normal pseudomanifolds, arXiv:1805.09179.

[46] G. Ziegler: Lectures on Polytopes, Graduate Texts in Mathematics 152, Springer-Verlag, New York, 1995.

Open problems from the 2018 Summer Workshop on Lattice Polytopes at Osaka University

Gabriele Balletti

Department of Mathematics, Stockholm University
SE-106 91, Stockholm, Sweden
E-mail: balletti@math.su.se

Federico Castillo

Mathematics Department, Kansas University
Lawrence, KS 66049, USA
E-mail: fcastillo@ku.edu

Liam Solus

Institutionen för Matematik, KTH Royal Institute of Technology
SE-100 44 Stockholm, Sweden
E-mail: solus@kth.se

Bach Le Tran

School of Mathematics, University of Edinburgh
James Clerk Maxwell Building, Peter Guthrie Tait Road, Edinburgh EH9 3FD, UK
E-mail: b.tran@sms.ed.ac.uk

Akiyoshi Tsuchiya

Department of Pure and Applied Mathematics
Graduate School of Information Science and Technology, Osaka University
Suita, Osaka 565-0871, Japan
E-mail: a-tsuchiya@ist.osaka-u.ac.jp

In the third week of the 2018 Summer Workshop on Lattice Polytopes at Osaka University, the authors lead an open problem session. In this note, we describe the open problems discussed at this session.

Keywords: Integer decomposition property, IDP, spanning polytope, unimodular cover, unimodular triangulation, Cayley sum, γ-positivity, gamma-positivity, symmetric, real-rooted, unimodal, log-concave, smooth polytope, h^*-polynomial, local h^*-polynomial.

1. Introduction

Lattice polytopes have recently become a central object of study across a wide variety of mathematical disciplines including algebra, discrete and convex geometry, optimization, statistics, algebraic geometry, as well as algebraic, geometric, and topological combinatorics. In the last century, key questions in such areas of mathematical research have aimed to develop a deep understanding of fundamental properties of lattice polytopes: these include, but are not limited to, their volume, their combinatorial structure with respect to the ambient lattice, and the properties of algebraic varieties associated to them. Important questions of this nature take a variety of forms: In discrete and convex geometry, a major goal is to provide useful classification results that allow researchers to better experiment with open conjectures on lattice polytopes. From this perspective, as well as that of optimization, researchers are also interested in understanding relaxations of lattice polytopes and the relationship between their volumes. In algebraic and geometric combinatorics, the generating functions we use to enumerate lattice points within lattice polytopes are highly investigated. Here, many important open questions pertain to the form of the generating function and how these functions change as we deform the geometric and/or combinatorial structure of the polytope. In algebra and algebraic geometry, lattice polytopes are treated as dictionaries from which we can read off the properties of an associated toric variety. As such, key questions ask us to use the structure of lattice polytopes to prove algebraic statements about toric varieties.

In the third week of the 2018 Summer Workshop on Lattice Polytopes at Osaka University, the participants held an open problem session. During this problem session, participants could present open problems and new directions for the investigation of lattice polytopes. A variety of problems, arising across the research areas given above, were discussed during this session. The purpose of this note is to summarize the problems presented. Each author of this note contributed an open problem, and each section of this paper is prepared by one author, in which that author has summarized their proposed problem. We, collectively, hope that these problems motivate future research that deepens our understanding of lattice polytopes and their significance across mathematics.

2. Unimodular Covers of IDP Polytopes - Gabriele Balletti

In the literature it is possible to find a vast multitude of hierarchies ordering lattice polytopes with a chain of progressively more restricting properties. Having the Integer Decomposition Property (being IDP for short) plays usually a central role in such a hierarchy, due to its importance in algebra and optimization. We now focus on the following properties, which are well-known and define a fertile area of investigation for research on topics pertaining to lattice polytopes.

Definition 2.1. Let $P \subset \mathbb{R}^d$ be a d-dimensional lattice polytope with respect to the lattice $\mathbb{Z}^d$.

(SP) We say that P is *spanning* if its lattice points affinely span $\mathbb{Z}^d$; i.e., if

$$\mathbb{Z}((P - \mathbf{p}) \cap \mathbb{Z}^d) = \mathbb{Z}^d,$$

for any $\mathbf{p} \in P \cap \mathbb{Z}^d$.

(VA) We say that P is *very ample* if for each vertex $\mathbf{v}$ of P the lattice points in the tangent cone $\mathbb{R}_{\geq 0}(P - \mathbf{v})$ are sums of lattice points of $P - \mathbf{v}$; i.e., if

$$\mathbb{R}_{\geq 0}(P - \mathbf{v}) \cap \mathbb{Z}^d = \mathbb{Z}_{\geq 0}((P - \mathbf{v}) \cap \mathbb{Z}^d).$$

(IDP) We say that P is *IDP* (or has the *integer decomposition property*, or is *integrally closed*) if for all $k \in \mathbb{Z}_{\geq 1}$ and for all $\mathbf{p} \in kP \cap \mathbb{Z}^d$, there exist $\mathbf{p}_1, \ldots, \mathbf{p}_k \in P \cap \mathbb{Z}^d$ such that $\mathbf{p} = \mathbf{p}_1 + \cdots + \mathbf{p}_k$.

(UC) We say that P *has a unimodular cover* if there exist unimodular lattice simplices $S_1, \ldots, S_n$ such that $P = S_1 \cup \cdots \cup S_n$.

(UT) We say that P *has a unimodular triangulation* if P admits a triangulation in unimodular lattice simplices.

It is an instructive exercise to verify that these properties satisfy the implications

$$\text{(UT)} \Rightarrow \text{(UC)} \Rightarrow \text{(IDP)} \Rightarrow \text{(VA)} \Rightarrow \text{(SP)}. \tag{1}$$

Moreover, one can note that in dimension two, all of these properties are equivalent. On the other hand, it is well-known that the opposite implications are false when there is no restriction on the dimension of P (see [2, Section 9]). In a recent paper [2], all equivalence classes of lattice polytopes in small dimensions and up to a fixed normalized volume are explicitly enumerated. Using such a classification we can focus on dimension three and try to extract minimal counterexamples to the implications opposite to those of (1). In particular, we note the following:

(i) The 3-dimensional polytope

$$\operatorname{conv}(\mathbf{0}, \mathbf{e}_1, \mathbf{e}_2, \mathbf{e}_3, \mathbf{e}_1 + \mathbf{e}_2 + 3\mathbf{e}_3)$$

is spanning but not very ample, where $\mathbf{0}$ denote the origin in $\mathbb{R}^d$ and $\mathbf{e}_1, \ldots, \mathbf{e}_d$ denote the standard basis vectors in $\mathbb{R}^d$.

(ii) The 3-dimensional polytope

$$\operatorname{conv}(\mathbf{0}, \mathbf{e}_1, \mathbf{e}_2, \mathbf{e}_1 + \mathbf{e}_2 + 2\mathbf{e}_3, \mathbf{e}_1 + \mathbf{e}_2 + 3\mathbf{e}_3, \mathbf{e}_1 - \mathbf{e}_3, \mathbf{e}_2 - \mathbf{e}_3, \mathbf{e}_3)$$

is very ample but not IDP.

(iii) The 3-dimensional polytope

$$\operatorname{conv}(\mathbf{0}, \mathbf{e}_1, \mathbf{e}_2, \mathbf{e}_1 + \mathbf{e}_2, \mathbf{e}_3, \mathbf{e}_1 + 2\mathbf{e}_2 - \mathbf{e}_3, 3\mathbf{e}_1 + \mathbf{e}_2 - \mathbf{e}_3, -2\mathbf{e}_1 - \mathbf{e}_2 + \mathbf{e}_3)$$

has a unimodular cover but does not have a unimodular triangulation.

No counterexample to the implication (IDP) $\Rightarrow$ (UC) has been found in dimension three. However, not all the enumerated 3-polytopes could be checked since the algorithm implemented for checking the existence of a unimodular cover is computationally expensive. Bruns and Gubeladze [17] gave examples of IDP polytopes not having a unimodular cover exist in dimension at least five, but nothing seems to be known about dimensions three and four. This leads to the following natural question:

Question 2.1. Is there a 3-dimensional IDP polytope that does not have a unimodular cover?

Stronger versions of this question appear in recent works, showing how little is known about unimodular covers. In [7], motivated by their work on Oda's Conjecture, the authors ask if centrally symmetric parallelepipeds of the form $\operatorname{conv}(D, -D)$, where D is a unit square, admit a unimodular cover. It also makes sense to wonder how fast a unimodular cover (or the lack thereof) can be detected.

Question 2.2. Is there a fast way to certify the (non-)existence of a unimodular cover for a lattice polytope P?

3. Deformations via Volumes - Federico Castillo

Let P be a d-polytope in some real vector space W with dual space W^*. We define its *support function* $h_P(\mathbf{u}) := \max_{\mathbf{x}\in P}\langle \mathbf{u}, \mathbf{x}\rangle$. We have that

$$P = \{\mathbf{x} \in W : \langle \mathbf{u}, \mathbf{x}\rangle \leq h_P(\mathbf{u}) \text{ for all } \mathbf{u} \in W^*\},$$

so that P and h_P determine each other uniquely. For any unit direction $\mathbf{u}$ we denote by $P^{\mathbf{u}}$ the face in direction $\mathbf{u}$, that is $P^{\mathbf{u}} := \{\mathbf{x} \in P : \langle \mathbf{u}, \mathbf{x}\rangle = h_P(\mathbf{u})\}$.

We say that Q is a *deformation* or *weak minkowski summand* of P if there exists a polytope R and a scalar λ such that $Q + R = \lambda P$. This happens if and only if the normal fan of P refines the normal fan of Q [22, Chapter 14].

The following two theorems are due to Minkowski. For a modern elementary proof see [27]. The first expresses a surprising way to distinguish polytopes by looking at their facet volumes.

Theorem 3.1 (Minkowski). *Let $P, Q \subset V$ be two d-polytopes. If $\operatorname{Vol}_{d-1}(P^{\mathbf{u}}) = \operatorname{Vol}_{d-1}(Q^{\mathbf{u}})$ for all $\mathbf{u} \in V$, then P is a translation of Q.*

Note that it suffices to consider only unit vectors for $\mathbf{u}$ in Theorem 3.1. There are infinitely many directions to check, although all but finitely many yield a zero volume. These volumes have to satisfy some linear relations called the Minkowski relations.

Theorem 3.2. *For a d-polytope P we have the following relations on its facet volumes:*

$$\sum_{\mathbf{u}\in V} \mathbf{u} \cdot \operatorname{Vol}(P^{\mathbf{u}}) = 0. \tag{2}$$

Furthermore, given unit vectors $\mathbf{u}_1, \cdots, \mathbf{u}_k$ and any relation

$$\mathbf{u}_1\alpha_1 + \cdots + \mathbf{u}_k\alpha_k = 0, \tag{3}$$

with $\alpha_1, \cdots, \alpha_k > 0$ then there exists a polytope P with unit facet normals $\mathbf{u}_i$ and such that $\operatorname{Vol}_{d-1}(P^{\mathbf{u}_i}) = \alpha_i$.

The sum in equation (2) is, in theory, infinite. However, only finitely many terms are nonzero, so it is well defined. One intuition for Equation (2) is that if we were to fill the polytope with a gas, it will exert a pressure on each facet in direction $\mathbf{u}$

proportional to $\mathrm{Vol}_{d-1}(P^{\mathbf{u}})$ and all this forces should cancel each other since the polytope will, presumably, not accelerate. The second part of the theorem, i.e. the existence statement, is more subtle. It comes down to finding the optimum to a convex program (see [27] for details).

Note that in light of Theorem 3.1 the polytope constructed in the second part of Theorem 3.2 is unique up to translation.

The set of all possible support functions h such that $h = h_Q$ for some deformation Q of P can be described using linear inequalities as done in [18]. Notice that h_P is a functional of degree one, in the sense that $h_{\lambda P} = \lambda^1 h_P$, whereas $V_P(\mathbf{u}) := \mathrm{Vol}_{d-1}(P^{\mathbf{u}})$ is a functional of degree $d-1$, since $V_{\lambda P} = \lambda^{d-1} V_P$.

Thus, we can fix a polytope P and consider the following:

Question 3.1. Fixing a polyope P, describe as explicitly as possible the set of all functions $v : W^* \longrightarrow \mathbb{R}$ such that $v = V_Q$ for some deformation Q of P.

Given a real vector space W, Theorem 3.2 guarantees that as long as a function v satisfies the Minkowski relations there will be a polytope Q with $v = V_Q$ but we cannot guarantee a priori that it will be a deformation of P. All we can say is that they have the same facet normals which is a much weaker condition. Ideally one could work out Question 3.1 for a particular polytope, for example for the permutohedron.

These functionals being of degree $d-1$ we expect the defining conditions to be an semialgebraic set whose defining equations are the Minkowski relations (as equalities) and some degree $d-1$ polynomial inequalities.

4. Thinking γ-Positively About Lattice Polytopes - Liam Solus

For a positive integer n, let $[n] := \{1,\dots,n\}$ and $[n]_0 := \{0,1,\dots,n\}$. Given a polynomial $p(z) = p_0 + p_1 z + \cdots + p_n z^n$ with $p_0,\dots,p_d$ nonnegative integers, combinatorists are often interested in identifying inequalities that hold among these coefficients. The polynomial $p(z)$ is called *unimodal* if $p_0 \leq p_1 \leq \cdots \leq p_t \geq \cdots \geq p_{n-1} \geq p_n$ for some $t \in [n]_0$. It is *log-concave* if $p_i^2 \geq p_{i-1}p_{i+1}$ for all $i \in [n-1]$, and it is *symmetric* with respect to $m \in \mathbb{Z}_{\geq 0}$ if $p_i = p_{m-i}$ for all $i \in [m]_0$. We say that $p(z)$ is *real-rooted* if all of its zeros are real numbers, or if $p \equiv 0$. If $p(z)$ is real-rooted then it is also log-concave and unimodal (see Theorem 1.2.1 in [16]). In recent years, these distributional properties have been heavily investigated for combinatorial generating polynomials associated to lattice polytopes in the context of Ehrhart theory [9].

If the polynomial $p(z)$ is symmetric, we can also ask if it satisfies a more mysterious distributional property: The space of polynomials that are symmetric with respect to d has the basis

$$\left\{z^k(z+1)^{d-2k}\right\}_{k=0}^{\lfloor \frac{d}{2} \rfloor}.$$

If we express the polynomial $p(z)$ in this basis as

$$p(z) = \sum_{k=0}^{\lfloor \frac{d}{2} \rfloor} \gamma_k z^k (z+1)^{d-2k},$$

then we call the coefficient sequence $\gamma(p) := (\gamma_0, \ldots, \gamma_{d/2})$ the *γ-vector* of $p(z)$. We say that $p(z)$ is *γ-positive* (or *γ-nonnegative*) if $\gamma_k \geq 0$ for all $k = 0, \ldots, d/2$. It is a short exercise to check the following well-known facts about γ-positive polynomials.

Proposition 4.1. *Let $p(z)$ be a polynomial with nonnegative coefficients.*

(i) If $p(z)$ is γ-positive then $p(z)$ is symmetric and unimodal.
(ii) If $p(z)$ is symmetric and real-rooted then $p(z)$ is γ-positive.

The first fact suggests the following natural question:

Question 4.1. If $p(z)$ is a symmetric and unimodal polynomial then is it also γ-positive?

In general, the answer to Question 4.1 is of course "no." However, in many cases where the polynomial $p(z)$ is combinatorially interesting, the answer turns out to be "yes." By our second fact, one way to address Question 4.1 for such cases would be to prove that the polynomial $p(z)$ is actually real-rooted. This method is very popular for answering γ-positivity questions in algebraic, geometric, and topological combinatorics [1, 13]. However, using the roots is not always possible. For instance, Gal's Conjecture [21] asserts that the h-polynomial of any flag homology sphere is γ-positive. Such an h-polynomial is well-known to be symmetric and unimodal, yet there exist examples of flag homology spheres whose h-polynomials are not real-rooted [21]. Thus, to address questions such as Gal's conjecture it becomes necessary to have a better understanding of algebraic/combinatorial interpretations of the γ-vector of the polynomials under consideration. One convenient testing ground for developing a better theory of the algebraic and combinatorial properties of γ-vectors is to analyze these properties for combinatorial polynomials we already know to be γ-positive by their real-rootedness.

Question 4.2. Suppose $p(z)$ is real-rooted and symmetric with nonnegative coefficients. Can we find combinatorial/algebraic interpretations of the γ-vector $\gamma(p)$?

Recently, a variety of symmetric combinatorial polynomials associated to lattice polytopes in the field of Ehrhart theory were shown to be real-rooted (see [23], [35], and Sections 3 and 4 of [15]). In this note, we will focus on the polynomials studied in [23] and [35], but similar questions can be asked for those in [15].

Let $P \subset \mathbb{R}^n$ be a d-dimensional convex lattice polytope; i.e., a convex polytope whose affine span has dimension d and all of whose vertices lie in $\mathbb{Z}^n$. For a nonnegative integer $t \in \mathbb{Z}_{\geq 0}$, the tth *dilate of* P is $tP := \{t\mathbf{p} \in \mathbb{R}^n : \mathbf{p} \in P\}$, and the

Ehrhart series of P is

$$\mathrm{Ehr}_P(z) := \sum_{t\geq 0} |tP\cap \mathbb{Z}^n|z^t = \frac{h_0^* + h_1^* z + \cdots + h_d^* z^d}{(1-z)^{d+1}}.$$

The polynomial in the numerator in the right-hand rational form is called the *(Ehrhart) h^*-polynomial* of P, and it is known to have only nonnegative integer coefficients [36]. We will also call a d-dimensional lattice polytope P *reflexive* whenever $h^*(P;z)$ is symmetric with respect to d.

In the case of lattice simplices, the h^*-polynomial has a related local-invariant known as its local h^*-polynomial [25], or box polynomial [4, 9, 33]. Let

$$\Delta := \mathrm{conv}(\mathbf{v}_0,\ldots,\mathbf{v}_d) \subset \mathbb{R}^n$$

be a lattice d-simplex defined as the convex hull of the $d+1$ affinely independent points $\mathbf{v}_0,\ldots,\mathbf{v}_d \in \mathbb{Z}^n$. The *open parallelpiped* of Δ is

$$\Pi_\Delta^\circ := \left\{\sum_{i=0}^{d} \lambda_i(\mathbf{v}_i,1) \in \mathbb{R}^{n+1} \; : 0<\lambda_i<1,\, i\in[d]_0\right\},$$

and the *local h^*-polynomial*, or *box polynomial*, of Δ is

$$\ell^*(\Delta;z) := \sum_{(x_1,\ldots,x_{n+1})\in\Pi_\Delta^\circ\cap\mathbb{Z}^n} z^{x_{n+1}}.$$

Unimodality of local h^*-polynomials of lattice simplices can sometimes be used to recover unimodality of h^*-polynomials of other lattice polytopes via a theorem of [8] (see [33] for details). It can be seen from the definition of $\ell^*(\Delta;z)$ that local h^*-polynomials are always symmetric with respect to $d+1$. In [23] and [35] the authors show that some combinatorially well-studied lattice simplices have unimodal local h^*-polynomials by showing that they are real-rooted. In [23] the *s-lecture hall simplices* were considered, and in [35] the author studied the *factoradic* and *base-r simplices*. The combinatorics of s-lecture hall simplices have been intensely investigated in the last two decades, [5, 6, 14, 26, 30–32] as have that of a family of simplices containing the factoradic and base-r simplices [10–12, 28, 34]. For each such local h^*-polynomial, we should therefore consider Question 4.2. The following problem summarizes three problems stated across [23] and [35].

Problem 4.1. *Identify combinatorial and/or algebraic interpretations of $\gamma(p)$ when*

(i) $p(z)$ is the local h^*-polynomial of an s-lecture hall simplex.
(ii) $p(z)$ is the local h^*-polynomial of the factoradic n-simplex.
(iii) $p(z)$ is the local h^*-polynomial of a base-r n-simplex.

The local h^*-polynomials of the s-lecture hall simplices were shown to be real-rooted in [23], and those of the factoradic and base-r n-simplices were shown to be real-rooted in [34]. Furthermore, each local h^*-polynomial in Problem 4.1 admits a combinatorial formula in terms statistics taken over various families of sequences. Given the recent and intense level of interest in these simplices within the combinatorics community, an answer to any part of Problem 4.1 would be interesting both in terms of the motivation for Question 4.2 and modern trends in geometric combinatorial theory. To give a teaser, we discuss the combinatorics of the factoradic n-simplex and its local h^*-polynomial in the following example.

Example 4.1 (Combinatorics of the factoradic n-simplex). Let $\mathfrak{S}_n$ denote the symmetric group on $[n]$. For a permutation $\pi = \pi_1\pi_2\cdots\pi_n$ we say that $i \in [n-1]$ is a *descent* if $\pi_i > \pi_{i+1}$. We let $\mathrm{Des}(\pi)$ denote the collection of all descents in π, and we let $\mathrm{des}(\pi)$ denote the cardinality of this collection. We then define two combinatorial polynomials:

$$A_n(z) := \sum_{\pi\in\mathfrak{S}_n} z^{\mathrm{des}(\pi)} \qquad \text{and} \qquad B_n(z) := \sum_{\pi\in\mathfrak{S}_n} z^{\max \mathrm{Des}(\pi)}.$$

Letting $B_n(z) = \sum_{k=0}^{n-1} b_{n,k}z^k$, the *factoradic n-simplex* is the convex hull

$$\Delta_n^! := \mathrm{conv}\left(\mathbf{e}_1, \ldots, \mathbf{e}_n, -\sum_{k=0}^{n} b_{n+1,k}\mathbf{e}_k\right) \subset \mathbb{R}^n .$$

In [34], the author showed that $h^*(\Delta_n^!; z) = A_{n+1}(z)$, thereby offering a geometric interpolation between the two polynomials $A_n(z)$ and $B_n(z)$.

The polynomial $A_n(z)$ is the well-known n^{th} *Eulerian polynomial*, which is real-rooted and symmetric. Furthermore, in consideration of Question 4.2, we note that $\gamma(A_n)$ already has a known combinatorial interpretation [20]. In [35], the author gave a combinatorial interpretation of $\ell^*(\Delta_n^!; z)$ and used it to prove this polynomial is also real-rooted: By treating each permutation $\pi = \pi_1\pi_2\cdots\pi_n$ as a string of digits, we can order them lexicographically as

$$\mathcal{S}_n := \left(\pi^{(0)}, \pi^{(1)}, \ldots, \pi^{(n!-1)}\right).$$

For a pair of positive integers k and m, we let $\overline{k}_m$ denote the equivalence class of all integers congruent to k modulo m. In [35], the author proved that

$$\ell^*(\Delta_n^!; z) = \sum_{b\in[(n+1)!]\cap\left(\overline{1}_6\cup\overline{5}_6\right)} z^{\mathrm{des}(\pi^{(b)})}.$$

Moreover, this polynomial is real-rooted and therefore γ-positive. However, unlike $A_n(z)$, this intriguing polynomial currently has no combinatorial interpretation for its associated γ-vector $\gamma(\ell^*(\Delta_n^!; z))$.

We make one final remark to further encourage the investigation of γ-positivity for symmetric generating polynomials associated to lattice polytopes. In particular, we consider two consequences derived in [23] from the real-rootedness of the local h^*-polynomials of s-lecture hall simplices. Recently, a family of lattice polytopes known as the *s-lecture hall order polytopes* were introduced in [14]. In Conjecture 5.3

of [14] the authors conjectured that a certain family of s-lecture hall order polytopes have γ-positive h^*-polynomials. A proof of this conjecture would then imply (by Proposition 4.1) that all such h^*-polynomials are symmetric and unimodal. Using the real-rootedness of the local h^*-polynomials of the s-lecture hall simplices, the authors of [23] showed that any reflexive s-lecture hall order polytope has a unimodal h^*-polynomial. Thus, with Question 4.1 in mind, we can make a similar conjecture to Conjecture 5.3 in [14]:

Conjecture 4.1. *The h^*-polynomial of a reflexive s-lecture hall order polytope is γ-positive.*

Using techniques developed in [37] and later in [25], one can also define a (symmetric) local h^*-polynomial $\ell^*(P; z)$ for any lattice polytope P, even those that are not simplices. In [23], the authors also showed that if P is an s-lecture hall order polytope then its local h^*-polynomial is unimodal. In the context of Question 4.1, we could then wager the following conjecture.

Conjecture 4.2. *The local h^*-polynomial of any s-lecture hall order polytope is γ-positive.*

A first step would be to decide if all such polynomials are real-rooted.

5. The Minimum Normalized Volume of Polytopes - Bach Le Tran

The problem of determining the minimal and maximal volumes of polytopes having some interior lattice points is a classical question in the study of lattice polytopes. In particular, for $d \geq 4$, it is conjectured [3] that the normalized volume of an arbitrary d-dimensional canonical Fano polytope P is bounded by

$$\mathrm{Vol}(P) \leq 2(s_d - 1)^2,$$

where $\mathrm{Vol}(P)$ denotes the normalized volume of P and s_i denotes the ith term of the Sylvester sequence:

$$s_1 := 2, \quad s_{i+1} := s_1 \cdots s_i + 1 \text{ for } i \in \mathbb{Z}_{\geq 1}\,.$$

The conjecture is true if P is a reflexive polytope [3, Corollary 1.2]. For a detailed introduction to the problem and other related results, please refer to [3] as well.

It is natural to ask the same question about what happens if the polytope P has more than one interior lattice point. Even more specifically, for a fixed set of interior lattice points, what can we say about the normalized volume of P?

Our original questions arose when we studied the case in which P has three interior lattice points that form the unit triangle. We wanted to find the minimum and maximum of the normalized volume of the polygons and 3-polytopes containing the unit triangle:

Question 5.1. Let $\Delta_2 = \mathrm{conv}(\mathbf{0}, \mathbf{e}_1, \mathbf{e}_2) \subset \mathbb{R}^2$. Let $P \subset \mathbb{R}^3$ be a lattice polytope such that $P^\circ \cap \mathbb{Z}^3 \cong \Delta_2 \cap \mathbb{Z}^2$. What are the minimum and maximum values of $\mathrm{Vol}(P)$?

Answers to the above question would be particularly useful for the classification of 3-polytopes containing exactly three interior lattice points.

With Question 5.1 in mind, it is natural to ask the same question in more general settings. We are particularly interested in the case of the lower bound of the normalized volume.

Definition 5.1. Let $P \subset \mathbb{R}^d$ be a lattice polytope. We let $\beta(P)$ denote the minimum normalized volume of all the lattice polytopes containing P in its interior. In other words,

$$\beta(P) := \min \{\mathrm{Vol}(Q) \mid Q \text{ is a lattice polytope such that } P \subseteq Q\}.$$

We also say that a polytope Q is a minimal bounding polytope of P if $P \subseteq Q$ and $\mathrm{Vol}(Q) = \beta(P)$.

Our next natural question is that whether the minimal bounding polytope of a given polytope P is unique.

Question 5.2. Let $P \subset \mathbb{R}^d$ be a lattice polytope. Is it true that minimal bounding polytopes of P are unique up to unimodular equivalence?

We can also ask the following question:

Question 5.3. Suppose that Q is a minimal bounding polytope of P. Is it true that $Q \cap \mathbb{Z}^d = P \cap \mathbb{Z}^d$; i.e., Q has no interior lattice points other than the lattice points of P?

Instinctively, one would think that the answer for the above question is true. A simple idea is to construct a new polytope inside Q containing P in its interior if $P \cap \mathbb{Z}^d \subsetneq Q^\cap \mathbb{Z}^d$. However, this approach does not work, even in dimension two.

Example 5.1. Consider the unimodular triangle $P = \mathrm{conv}((1,1),(1,2),(2,1))$. We first note that the triangle $\mathrm{conv}((0,0),(1,3),(3,1)) \subsetneq P'$ is a minimal bounding polytope of P. In particular, we have that $\beta(P) = 7$. Now let $Q = \mathrm{conv}((0,0),(0,3),(3,2),(2,0))$. Then $P \subset Q$, $P \cap \mathbb{Z}^2 \neq Q \cap \mathbb{Z}^2$, and $\mathrm{Vol}(Q) = 13$. Thus, while Q is not a minimal bounding polytope of P, there is no proper lattice subpolytope of Q containing P in its interior.

One natural candidate for a minimal bounding polytope of P is the polytope P' obtained by shifting each facet of P outward 1 lattice unit. Unfortunately, as we see with the next example, P' will not be a minimal bounding polytope even in dimension two.

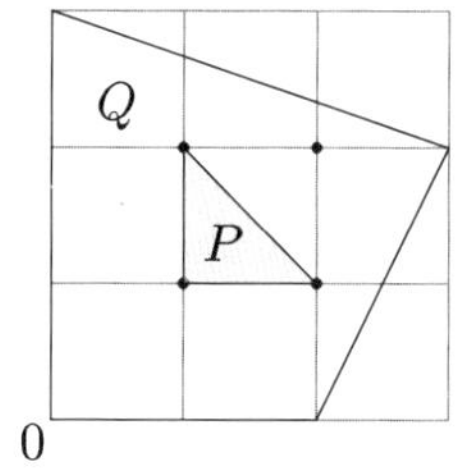

Fig. 1. The polytopes P and Q from Example 5.1.

Example 5.2. Consider the unimodular triangle $P = \text{conv}((1,1),(1,2),(2,1))$ again. Then $P' = \text{conv}((0,0),(4,0),(0,4))$ and $\text{Vol}(P') = 16$. Thus, referring back to Example 5.1, we see that P' is not a minimal bounding polytope of P.

However, one may ask the following question:

Question 5.4. Suppose that $Q \cap \mathbb{Z}^d \supseteq P \cap \mathbb{Z}^d$ and $\text{Vol}(Q) = \beta(P)$. What is the volume of P'? Is it true that $Q \subseteq P'$?

6. The Cayley Sum of Two Lattice Polygons - Akiyoshi Tsuchiya

For two lattice polytopes $P, Q \subset \mathbb{R}^d$, the polytope $P + Q := \{x + y : x \in P, y \in Q\}$ is called the *Minkowski sum* of P and Q. Similarly, given two subsets of lattice points $S, T \subset \mathbb{Z}^d$, we can define the set $S + T := \{x + y : x \in S, y \in T\}$. In [29], Oda posed the following fundamental question about Minkowski sums:

Question 6.1 ([29, Problem 1]). Let $P, Q \subset \mathbb{R}^d$ be lattice polytopes. When does the equation

$$(P \cap \mathbb{Z}^d) + (Q \cap \mathbb{Z}^d) = (P + Q) \cap \mathbb{Z}^d \tag{4}$$

hold?

In general, for any two lattice polytopes $P, Q \subset \mathbb{R}^d$, the equation (4) does not hold, not even in the special case $P = Q$. A lattice polytope P is called *smooth* if it

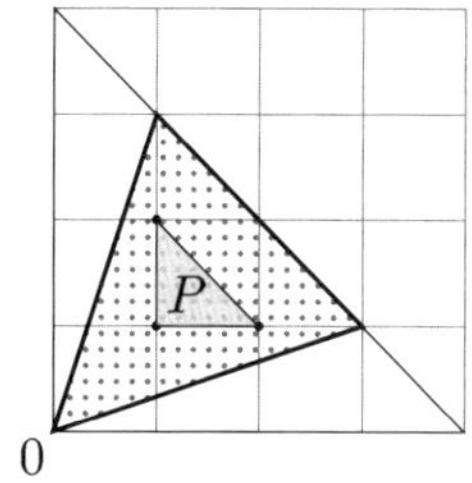

Fig. 2. The polytopes P and P' from Example 5.2, and a minimal bounding polytope of P.

is simple and at every vertex the primitive facet normals generate the dual lattice. From the perspective of algebraic geometry, Oda conjectured the following:

Conjecture 6.1 ([29]). *Let $P, Q \subset \mathbb{R}^d$ be lattice polytopes. If P is smooth and the normal fan of Q coarsens that of P, then equation (4) holds.*

The case $Q = nP$ for $n \in \mathbb{Z}_{>0}$ of Conjecture 6.1 is known as *Oda's Conjecture:*

Conjecture 6.2 (Oda's Conjecture). *Every smooth polytope is IDP.*

It is well known that every lattice polygon is IDP. Hence, Conjecture 6.2 is true for all lattice polygons. On the other hand, Fakhruddin [19] proved Conjecture 6.1 for the case of lattice polygons. Moreover, Haase, Nill and Paffenholz [24] proved Conjecture 6.1 for the case of lattice polygons without the smoothness assumption on P.

Theorem 6.1 ([24, Theorem 1.1]). *Let $P, Q \subset \mathbb{R}^2$ be lattice polygons. If the normal fan of Q coarsens that of P, then equation (4) holds.*

Given two lattice polytopes $P, Q \subset \mathbb{R}^d$, the *Cayley sum* of P and Q, denoted by $P*Q$, is the convex hull of $(P\times\{1\})\cup(Q\times\{0\}) \subset \mathbb{R}^{d+1}$. To better understand (and potentially settle) Question 6.1 and Conjecture 6.1, the following theorem suggests that we could focus on Cayley sums which have the *integer decomposition property;* i.e., which are *IDP* (see Definition 2.1):

Theorem 6.2 ([38, Theorem 0.4]). *Let $P, Q \subset \mathbb{R}^d$ be lattice polytopes. If $P * Q$ is IDP, then P, Q and $P + Q$ are IDP and equation (4) holds.*

Theorems 6.1 and 6.2 naturally inspired us to consider the following question.

Question 6.2. Let $P, Q \subset \mathbb{R}^2$ be lattice polygons. If the normal fan of Q coarsens that of P, then is $P * Q$ IDP?

From the proof of [38, Theorem 0.4], we notice that the Cayley sum $P * Q$ is IDP if and only if for any nonnegative integers a and b, $(aP \cap \mathbb{Z}^d) + (bQ \cap \mathbb{Z}^d) = (aP + bQ) \cap \mathbb{Z}^d$. Therefore, by using Theorem 6.1 we can prove this question. However, our original motivation is to give a stronger result than Theorem 6.1. In particular, we would like to answer Question 6.1 in dimensions higher than two. Hence the following is our main problem.

Problem 6.1. *Let $P, Q \subset \mathbb{R}^2$ be lattice polygons such that the normal fan of Q coarsens that of P.*
(1) Prove Question 6.2 without using Theorem 6.1.
(2) Prove that $P * Q$ has a unimodular cover or a unimodular triangulation.

If a lattice polytope possesses a unimodular cover or a unimodular triangulation, then it is IDP (see Definition 2.1). Hence Problem 6.1 (2) is a stronger problem than Question 6.2.

Acknowledgements. Liam Solus was supported by an NSF Mathematical Sciences Postdoctoral Research Fellowship (DMS - 1606407). Bach Le Tran wants to thank Alexander Kasperzyk for suggesting the 3-point polytopes classification question. The authors thank the organizers, Dr. Takayuki Hibi and Dr. Akiyoshi Tsuchiya, for their support during the 2018 Summer Workshop on Lattice Polytopes at Osaka University.

References

[1] C. A. Athanasiadis, Gamma-positivity in combinatorics and geometry, arXiv:1711.05983.

[2] G. Balletti, Enumeration of lattice polytopes by their volume, in preparation.

[3] G. Balletti, A. M. Kasprzyk and B. Nill, On the maximum dual volume of a canonical Fano polytope, arXiv:1612.08918.

[4] M. Beck and S. Robins. Computing the continuous discretely, Springer Science+ Business Media, LLC, 2007.

[5] M. Beck, B. Braun, M. Köppe, C. D. Savage and Z. Zafeirakopoulos, s-lecture hall partitions, self-reciprocal polynomials, and Gorenstein cones, *The Ramanujan Journal*, **36** (2015) 123–147.

[6] M. Beck, B. Braun, M. Köppe, C. D. Savage and Z. Zafeirakopoulos, Generating functions and triangulations for lecture hall cones, *SIAM Journal on Discrete Mathematics* **30** (2016), 1470–1479.

[7] M. Beck, C. Haase, A. Higashitani, J. Hofscheier, K. Jochemko, L. Katthän and M. Michałek, Smooth centrally symmetric polytopes in dimension 3 are IDP, arXiv:1802.01046.

[8] U. Betke and P. McMullen, Lattice points in lattice polytopes, *Monatshefte für Mathematik* **99** (1985), 253–265.

[9] B. Braun, Unimodality problems in Ehrhart theory, Recent trends in combinatorics, Springer, Cham, 2016, 687–711.

[10] B. Braun and R. Davis, Ehrhart series, unimodality, and integrally closed reflexive polytopes, *Annals of Combinatorics* **20** (2016), 705–717.

[11] B. Braun, R. Davis, and L. Solus, Detecting the integer decomposition property and Ehrhart unimodality in reflexive simplices, *Advances in Applied Mathematics* **100** (2018), 122-142.

[12] B. Braun and F. Liu, h^*-polynomials with roots on the unit circle, arXiv:1807.00105.

[13] P. Brändén, Unimodality, log-concavity, real-rootedness and beyond, Handbook of Enumerative Combinatorics (2015), 437–483.

[14] P. Brändén and M. Leander, Lecture hall P-partitions, arXiv:1609.02790.

[15] P. Brändén and L. Solus, Symmetric decompositions and real-rootedness, arXiv:1808.04141.

[16] F. Brenti, Unimodal log-concave and Pólya frequency sequences in combinatorics, No. 413. American Mathematical Soc., 1989.

[17] W. Bruns and J. Gubeladze, Normality and covering properties of affine semigroups, *J. Reine Angew. Math.* **510** (1999), 161–178.

[18] F. Castillo and F.Liu, Deformation Cones of nested Braid fans, arXiv:1710.01899.

[19] N. Fakhruddin, Multiplication maps of linear systems on smooth projective toric surfaces, arXiv:math.AG/0208178.

[20] D. Foata and V. Strehl, Euler numbers and variations of permutations, *Colloquio Internazionale sulle Teorie Combinatorie* **1** (1973), 119–131.

[21] R. S. Gal, Real root conjecture fails for five-and higher-dimensional spheres, *Discrete & Computational Geometry* **34** (2005), 269–284.

[22] B. Grünbaum, Convex Polytopes, Springer, Graduate texts in mathematics.

[23] N. Gustafsson and L. Solus, Derangements, Ehrhart theory, and local h-polynomials, arXiv:1807.05246.

[24] C. Haase, B. Nill and A. Paffenholz, Lattice points in Minkowski sums, *Electron. J. Comb.* **15** (2008), ♯N11.

[25] E. Katz and A. Stapledon, Local h-polynomials, invariants of subdivisions, and mixed Ehrhart theory, *Advances in Mathematics* **286** (2016), 181–239.

[26] F. Kohl and M. Olsen, Level algebras and s-lecture hall polytopes, arXiv:1710.10892.

[27] D. Klain, The Minkowski problem for polytopes. *Adv. Math.* **185** (2004), 270–288.

[28] F. Liu and L. Solus, On the relationship between Ehrhart unimodality and Ehrhart positivity, arXiv:1804.08258.

[29] T. Oda, Problems on Minkowski sums of convex lattice polytopes, arXiv:0812.1418.

[30] C. D. Savage, The mathematics of lecture hall partitions, *Journal of Combinatorial Theory, Series A* **144** (2016), 443–475.

[31] C. D. Savage and M. J. Schuster, Ehrhart series of lecture hall polytopes and Eulerian polynomials for inversion sequences, *Journal of Combinatorial Theory, Series A* **119** (2012), 850–870.

[32] C. D. Savage, M. Visontai, The s-Eulerian polynomials have only real roots, *Trans. Amer. Math. Soc.* **367** (2015), 1441–1466.

[33] J. Schepers and L. Van Langenhoven, Unimodality questions for integrally closed lattice polytopes, *Annals of Combinatorics* **17** (2013), 571–589.

[34] L. Solus, Simplices for numeral systems, *Trans. Amer. Math. Soc.* DOI: https://doi.org/10.1090/tran/7424 (2017).

[35] L. Solus, Local h^*-polynomials of some weighted projective spaces, To appear in the Proceedings of the 2018 Summer Workshop on Lattice Polytopes, Osaka University (2018).

[36] R. P. Stanley, Decompositions of rational convex polytopes, *Annals of discrete mathematics* **6** (1980), 333–342.

[37] R. P. Stanley, Subdivisions and local h-vectors, *Journal of the American Mathematical Society* **5** (1992), 805–851.

[38] A. Tsuchiya, Cayley sums and Minkowski sums of 2-convex-normal lattice polytopes, arXiv:1804.10538.

Author index